Fluid Mechanics

Joseph H. Spurk · Nuri Aksel

Fluid Mechanics

Third Edition

Joseph H. Spurk
Bad König, Hessen, Germany

Nuri Aksel
Lehrstuhl für Technische Mechanik
und Strömungsmechanik
Universität Bayreuth
Bayreuth, Germany

ISBN 978-3-030-30261-0   ISBN 978-3-030-30259-7   (eBook)
https://doi.org/10.1007/978-3-030-30259-7

1st & 2nd editions: © Springer-Verlag Berlin Heidelberg 1997, 2008
3rd edition: © Springer Nature Switzerland AG 2020

This Springer imprint is published by the registered company Springer Nature Switzerland AG
The registered company address is: Gewerbestrasse 11, 6330 Cham, Switzerland

Preface to the Third Edition

In this third edition, we completely revised the book. Whenever necessary, additional comments were made to deepen the subject. Also, connecting remarks between the chapters are made to homogenize the book. As far as possible, we eliminated the obvious typos. The index is expanded according to the text. N. A. thanks Dr. Markus Dauth for his valuable comments in the proofreading and for his help in the typesetting of the manuscript.

Bad König, Germany
Bayreuth, Germany
Summer 2019

Joseph H. Spurk
Nuri Aksel

Preface to the Second English Edition

The first English edition was the translation of the fourth German edition. In the meantime, the textbook has undergone several additions, mostly stimulated by consulting activities of the first author. Since the textbook continues to receive favorable reception in German-speaking countries and has been translated in other languages as well, the publisher suggested a second English edition. The additions were translated for the most part by Prof. L. Crane from Trinity College in Dublin, who has accompanied this textbook from the very beginning. Since the retirement of the first author, Prof. N. Aksel from the University of Bayreuth, Germany, the second author, was actively engaged in the sixth and the seventh editions. The additions were written by the first author who accepts the responsibility for any mistakes or omissions in this book.

Preface to the First English Edition

This textbook is the translation of the fourth edition of *Strömungslehre, Einführung in die Theorie der Strömungen*. The German edition has met with a favorable reception in German-speaking countries, showing that there was a demand for a book that emphasizes the fundamentals. In the English literature, there are books of the same nature, some excellent, and these have indeed influenced me to write this book. However, they cover different ground and are not aimed primarily at mechanical engineering students, which this book is. I have kept the original concept throughout all editions, and there is little to say that has not been said in the preface to the first German edition. There is now a companion volume *Solved Problems in Fluid Mechanics*, which alleviates the drawback of the first German edition, namely, the absence of problem exercises.

The book has been translated by Katherine Mayes during her stay in Darmstadt, and I had the opportunity to work with her daily. It is for this reason that I am solely responsible for this edition, too. My thanks also go to Prof. L. Crane from Trinity College in Dublin for his assistance with this book. Many people have helped, all of whom I cannot name, but I would like to express my sincere thanks to Ralf Münzing, whose dependable and unselfish attitude has been a constant encouragement during this work.

Darmstadt   J. H. Spurk
January 1997

ix

Contents

Chapter 1
The Concept of the Continuum and Kinematics

1.1 Properties of Fluids, Continuum Hypothesis

Fluid mechanics is concerned with the behavior of materials which deform without limit under the influence of shearing forces. Even a very small shearing force will deform a fluid body, but the velocity of the deformation will be correspondingly small. This property serves as the definition of a fluid: the shearing forces necessary to deform a fluid body go to zero as the velocity of deformation tends to zero. On the contrary, the behavior of a solid body is such that the deformation itself, not the velocity of deformation, goes to zero when the forces necessary to deform it tend to zero. To illustrate this contrasting behavior, consider a material between two parallel plates and adhering to them acted on by a shearing force F (Fig. 1.1).

If the extent of the material in the direction normal to the plane of Fig. 1.1 and in the x-direction is much larger than that in the y-direction, experience shows that for many solids (*Hooke's solids*), the force per unit area $\tau = F/A$ is proportional to the displacement a and inversely proportional to the distance between the plates h. At least one dimensional quantity typical for the material must enter this relation, and here this is the *shear modulus G*. The relationship

$$\tau = G\gamma(\gamma \ll 1) \qquad (1.1)$$

between the shearing angle $\gamma = a/h$ and τ satisfies the definition of a solid: the force per unit area τ tends to zero only when the deformation γ itself goes to zero. Often the relation for a solid body is of a more general form, e.g., $\tau = f(\gamma)$, with $f(0) = 0$.

If the material is a fluid, the displacement of the plate increases continually with time under a constant shearing force. This means there is no relationship between the displacement, or deformation, and the force. Experience shows here that with many fluids the force is proportional to the rate of change of the displacement, that is, to the velocity of the deformation. Again the force is inversely proportional to the distance between the plates. (We assume that the plate is being dragged at

© Springer Nature Switzerland AG 2020
J. H. Spurk and N. Aksel, *Fluid Mechanics*,
https://doi.org/10.1007/978-3-030-30259-7_1

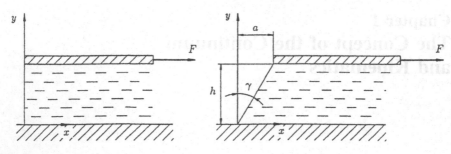

Fig. 1.1 Shearing between two parallel plates

constant speed, so that the inertia of the material does not come into play.) The dimensional quantity required is the *shear viscosity* η, and the relationship with $U = da/dt$ now reads

$$\tau = \eta \frac{U}{h} = \eta \dot{\gamma},\tag{1.2}$$

or, if the *shear rate* $\dot{\gamma}$ is set equal to du/dy,

$$\tau(y) = \eta \frac{du}{dy}.\tag{1.3}$$

$\tau(y)$ is the shear stress on a surface element parallel to the plates at point y. In so-called *simple shearing flow* (*rectilinear shearing flow*) only the x-component of the velocity is nonzero, and is a linear function of y.

The above relationship was known to Newton, and it is sometimes incorrectly used as the definition of a *Newtonian fluid*: there are also *non-Newtonian fluids* which show a linear relationship between the shear stress τ and the shear rate $\dot{\gamma}$ in this simple state of stress. In general, the relationship for a fluid reads $\tau = f(\dot{\gamma})$, with $f(0) = 0$.

While there are many substances for which this classification criterion suffices, there are some which show dual character. These include the glasslike materials which do not have a crystal structure and are structurally liquids. Under prolonged loads these substances begin to flow, that is to deform without limit. Under short-term loads, they exhibit the behavior of a solid body. Asphalt is an oftquoted example: you can walk on asphalt without leaving footprints (short-term load), but if you remain standing on it for a long time, you will finally sink in. Under very short-term loads, e.g., a blow with a hammer, asphalt splinters, revealing its structural relationship to glass. Other materials behave like solids even in the long-term, provided they are kept below a certain shear stress, and then above this stress they will behave like liquids. A typical example of these substances (*Bingham materials*) is paint: it is this behavior which enables a coat of paint to stick to surfaces parallel to the force of gravity.

The above definition of a fluid comprises both liquids and gases, since neither show any resistance to change of shape when the velocity of this change tends to zero. Now liquids develop a free surface through condensation, and in general do not fill up the whole space they have available to them, say a vessel, whereas gases completely fill the space available. Nevertheless, the behavior of liquids and gases is dynamically the same as long as their volume does not change during the course of the flow.

The essential difference between them lies in the greater compressibility of gases. When heated over the *critical temperature* T_c, liquid loses its ability to condense and it is then in the same thermodynamical state as a gas compressed to the same density. In this state even gas can no longer be "easily" compressed. The feature we have to take note of for the dynamic behavior, therefore, is not the state of the fluid (gaseous or liquid) but the resistance it shows to change in volume. Insight into the expected volume or temperature changes for a given change in pressure can be obtained from a graphical representation of the equation of state for a pure substance $F(p, T, v) = 0$ in the wellknown form of a p-v-diagram with T as the parameter (Fig. 1.2).

This graph shows that during dynamic processes where large changes of pressure and temperature occur, the change of volume has to be taken into account. The branch of fluid mechanics which evolved from the necessity to take the volume changes into account is called *gas dynamics*. It describes the dynamics of flows with large pressure changes as a result of large changes in velocity. There are also other branches of fluid mechanics where the change in volume may not be ignored, among these *meteorology*; there the density changes as a result of the pressure change in the atmosphere due to the force of gravity.

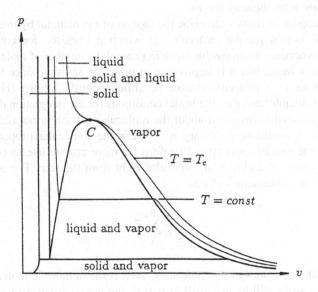

Fig. 1.2 *p-v*-diagram

The behavior of solids, liquids and gases described up to now can be explained by the molecular structure, by the thermal motion of the molecules, and by the interactions between the molecules. Microscopically the main difference between gases on the one hand, and liquids and solids on the other is the mean distance between the molecules.

With gases, the spacing at *standard temperature and pressure* (273.2 K; 1.013 bar) is about ten effective molecular diameters. Apart from occasional collisions, the molecules move along a straight path. Only during the collision of, as a rule, two molecules, does an interaction take place. The molecules first attract each other weakly, and then as the interval between them becomes noticeably smaller than the effective diameter, they repel strongly. The *mean free path* is in general larger than the mean distance, and can occasionally be considerably larger.

With liquids and solids the mean distance is about one effective molecular diameter. In this case there is always an interaction between the molecules. The large resistance which liquids and solids show to volume changes is explained by the repulsive force between molecules when the spacing becomes noticeably smaller than their effective diameter. Even gases have a resistance to change in volume, although at standard temperature and pressure it is much smaller and is proportional to the kinetic energy of the molecules. When the gas is compressed so far that the spacing is comparable to that in a liquid, the resistance to volume change becomes large, for the same reason as referred to above.

Real solids show a crystal structure: the molecules are arranged in a lattice and vibrate about their equilibrium position. Above the melting point, this lattice disintegrates and the material becomes liquid. Now the molecules are still more or less ordered, and continue to carry out their oscillatory motions although they often exchange places. The high mobility of the molecules explains why it is easy to deform liquids with shearing forces.

It would appear obvious to describe the motion of the material by integrating the equations of motion for the molecules of which it consists: for computational reasons this procedure is impossible since in general the number of molecules in the material is very large. But it is impossible in principle anyway, since the position and momentum of a molecule cannot be simultaneously known (Heisenberg's Uncertainty Principle) and thus the initial conditions for the integration do not exist. In addition, detailed information about the molecular motion is not readily usable and therefore it would be necessary to average the molecular properties of the motion in some suitable way. It is therefore far more appropriate to consider the average properties of a cluster of molecules right from the start. For example the macroscopic, or continuum, velocity

$$\vec{u} = \frac{1}{n} \sum_{1}^{n} \vec{c}_i, \tag{1.4}$$

where $\vec{c}_i$ are the velocities of the molecules and n is the number of molecules in the cluster. This cluster will be the smallest part of the material that we will consider, and we call it a *fluid particle*. To justify this name, the volume which this cluster of molecules occupies must be small compared to the volume occupied by the whole part of the fluid under consideration. On the other hand, the number of molecules in the cluster must be large enough so that the averaging makes sense, i.e., so that it becomes independent of the number of molecules. Considering that the number of molecules in one cubic centimeter of gas at standard temperature and pressure is 2.7×10^{19} (*Loschmidt's number*), it is obvious that this condition is satisfied in most cases.

Now we can introduce the most important property of a *continuum*, its *mass density* ρ. This is defined as the ratio of the sum of the molecular masses in the cluster to the occupied volume, with the understanding that the volume, or its linear measure, must be large enough for the density of the fluid particle to be independent of its volume. In other words, the mass of a fluid particle is a smooth function of the volume.

On the other hand the linear measure of the volume must be small compared to the macroscopic length of interest. It is appropriate to assume that the volume of the fluid particle is infinitely small compared to the whole volume occupied by the fluid. This assumption forms the basis of the *continuum hypothesis*. Under this hypothesis we consider the fluid particle to be a *material point* and the density (or other properties) of the fluid to be continuous functions of place and time. Occasionally we will have to relax this assumption on certain curves or surfaces, since discontinuities in the density or temperature, say, may occur in the context of some idealizations. The part of the fluid under observation consists then of infinitely many material points, and we expect that the motion of this continuum will be described by partial differential equations. However the assumptions which have led us from the material to the idealized model of the continuum are not always fulfilled. One example is the flow past a space craft at very high altitudes, where the air density is very low. The number of molecules required to do any useful averaging then takes up such a large volume that it is comparable to the volume of the craft itself.

Continuum theory is also inadequate to describe the structure of a shock (see Chap. 9), a frequent occurrence in compressible flow. Shocks have thicknesses of the same order of magnitude as the mean free path, so that the linear measures of the volumes required for averaging are comparable to the thickness of the shock.

We have not yet considered the role the thermal motion of molecules plays in the continuum model. This thermal motion is reflected in the macroscopic properties of the material and is the single source of viscosity in gases. Even if the macroscopic velocity given by (1.4) is zero, the molecular velocities $\vec{c}_i$ are clearly not necessarily zero. The consequence of this is that the molecules migrate out of the fluid particle and are replaced by molecules drifting in. This exchange process gives rise to the macroscopic fluid properties called *transport properties*. Obviously, molecules with other molecular properties (e.g. mass) are brought into the fluid particle. Take as an example a gas which consists of two types of molecule, say O_2 and N_2. Let the number of O_2 molecules per unit volume in the fluid particle be larger than that of the surroundings. The number of O_2 molecules which migrate out is proportional to the number density inside the fluid particle, while the number which drift in is proportional to that of the surroundings. The net effect is that more O_2 molecules drift in than drift out and so the O_2 number density adjusts itself to the surroundings. From the standpoint of continuum theory the process described above represents the *diffusion*.

If the continuum velocity $\vec{u}$ in the fluid particle as given by (1.4) is larger than that of the surroundings, the molecules which drift out bring their molecular velocities which give rise to $\vec{u}$ with them. Their replacements have molecular velocities with a smaller part of the continuum velocity $\vec{u}$. This results in momentum exchange through the surface of the fluid particle which manifests itself as a force on this surface. In the simple shearing flow (Fig. 1.1) the force per unit area on a surface element parallel to the plates is given by (1.3). The sign of this shear stress is such as to even out the velocity. However nonuniformity of the velocity is maintained by the force on the upper plate, and thus the momentum transport is also maintained. From the point of view of continuum theory, this momentum transport is the source of the internal friction, i.e., the *viscosity*. The molecular transport of momentum accounts for internal friction only in the case of gases. In liquids, where the molecules are packed as closely together as the repulsive forces will allow, each molecule is in the range of attraction of several others. The exchange of sites among molecules, responsible for the deformability, is impeded by the force of attraction from neighboring molecules. The contribution from these intermolecular forces to the force on surface elements of fluid particles having different macroscopic velocities is greater than the contribution from the molecular momentum transfer. Therefore the viscosity of liquids decreases with increasing temperature, since change of place among molecules is favored by more vigorous molecular motion. Yet the viscosity of gases, where the momentum transfer is basically its only source, increases with temperature, since increasing the temperature increases the thermal velocity of the molecules, and thus the momentum exchange is favored.

The above exchange model for diffusion and viscosity can also explain the third transport process: *conduction*. In gases, the molecules which drift out of the fluid particle bring with them their kinetic energy, and exchange it with the surrounding molecules through collisions. The molecules which migrate into the particle exchange their kinetic energy through collisions with the molecules in the fluid particle, thus equalizing the average kinetic energy (i.e. the temperature) in the fluid.

Thus, as well as the already mentioned differential equations for describing the motion of the continuum, the relationships which describe the exchange of mass (diffusion), of momentum (viscosity) and of kinetic energy (conduction) must be known. In the most general sense, these relationships establish the connection between concentration and diffusion flux, between forces and motion, and between temperature and heat flux. However these relations only reflect the primary reasons for "cause" and "effect". We know from the kinetic theory of gases, that an effect can have several causes. Thus, for example, the diffusion flux (effect) depends on the inhomogeneity of the concentration, the temperature and the pressure field (causes), as well as on other external forces. The above relationships must therefore occasionally permit the dependency of the effect on several causes. Relationships describing the connections between the causes and effects in a body are called *constitutive relations*. They reflect macroscopically the behavior of matter that is determined microscopically through the molecular properties. Continuum theory is however of a phenomenological nature: in order to look at the macroscopic behavior of the material, mathematical and therefore idealized models are developed. Yet this is necessary, since the real properties of matter can never be described exactly. But even if this possibility did exist, it would be wasteful to include all the material properties not relevant in a given technical problem. Thus the continuum theory works not with real materials, but with models which describe the behavior for the given application sufficiently accurately. The model of an ideal gas, for example, is evidently useful for many applications, although ideal gas is never encountered in reality.

In principle, models could be constructed solely from experiments and experiences, without consideration for the molecular structure. Yet consideration of the microscopic structure gives us insight into the formulation and limitations of the constitutive equations.

1.2 Kinematics

1.2.1 Material and Spatial Descriptions

Kinematics is the study of the motion of a fluid, without considering the forces which cause this motion, that is without considering the equations of motion. It is natural to try to carry over the kinematics of a mass-point directly to the kinematics of a fluid particle. Its motion is given by the time dependent position vector $\vec{x}(t)$ relative to a chosen origin.

In general we are interested in the motion of a finitely large part of the fluid (or the whole fluid) and this is made up of infinitely many fluid particles. Thus the single particles must remain identifiable. The shape of the particle is no use as an identification, since, because of its ability to deform without limit, it continually changes during the course of the motion. Naturally the linear measure must remain small in spite of the deformation during the motion, something that we guarantee by idealizing the fluid particle as a material point.

For identification, we associate with each material point a characteristic vector $\vec{\xi}$. The position vector $\vec{x}$ at a certain time t_0 could be chosen, giving $\vec{x}(t_0) = \vec{\xi}$. The motion of the whole fluid can then be described by

$$\vec{x} = \vec{x}\left(\vec{\xi}, t\right) \quad \text{or} \quad x_i = x_i\left(\xi_j, t\right) \tag{1.5}$$

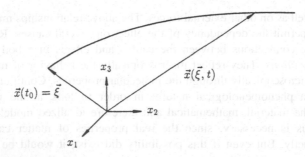

Fig. 1.3 Material description

(We use the same symbol for the vector function on the right side as we use for its value on the left.) For a fixed $\vec{\xi}$, (1.5) gives the path in space of the material point labeled by $\vec{\xi}$ (Fig. 1.3). For a different $\vec{\xi}$, (1.5) is the equation of the *pathline* of a different particle.

While $\vec{\xi}$ is only the particle's label we shall often speak simply of the "$\vec{\xi}$th" particle. The velocity

$$\vec{u} = \mathrm{d}\vec{x}/\mathrm{d}t$$

and the acceleration

$$\vec{a} = \mathrm{d}^2\vec{x}/\mathrm{d}t^2$$

of a point in the material $\vec{\xi}$ can also be written in the form

$$\vec{u}\left(\vec{\xi},t\right) = \left[\frac{\partial \vec{x}}{\partial t}\right]_{\vec{\xi}} \quad \text{or} \quad u_i(\xi_j,t) = \left[\frac{\partial x_i}{\partial t}\right]_{\xi_j}, \tag{1.6}$$

$$\vec{a}\left(\vec{\xi},t\right) = \left[\frac{\partial \vec{u}}{\partial t}\right]_{\vec{\xi}} \quad \text{or} \quad a_i(\xi_j,t) = \left[\frac{\partial u_i}{\partial t}\right]_{\xi_j}, \tag{1.7}$$

where "differentiation at fixed $\vec{\xi}$" indicates that the derivative should be taken for the "$\vec{\xi}$th" point in the material. Confusion relating to differentiation with respect to t does not arise since $\vec{\xi}$ does not change with time. Mathematically, (1.5) describes a mapping from the reference configuration to the actual configuration.

For reasons of tradition we call the use of the independent variables $\vec{\xi}$ and t the *material* or *Lagrangian description*, but the above interpretation of (1.5) suggests a more accurate name is *referential description*. $\vec{\xi}$ is called the *material coordinate*.

Although the choice of $\vec{\xi}$ and t as independent variables is obvious and is used in many branches of continuum mechanics; the material description is impractical in fluid mechanics (apart from a few exceptions). In most problems attention is focused on what happens at a specific place or in a specific region of space as time passes. The independent variables are then the place $\vec{x}$ and the time t. Solving Eq. (1.5) for $\vec{\xi}$ we get

$$\vec{\xi} = \vec{\xi}(\vec{x}, t) \tag{1.8}$$

This is the label of the material point which is at the place $\vec{x}$ at time t. Using (1.8) $\vec{\xi}$ can be eliminated from (1.6)

$$\vec{u}\left(\vec{\xi}, t\right) = \vec{u}\left[\vec{\xi}(\vec{x}, t), t\right] = \vec{u}(\vec{x}, t). \tag{1.9}$$

For a given $\vec{x}$, (1.9) expresses the velocity at the place $\vec{x}$ as a function of time. For a given t (1.9) gives the velocity field at time t. $\vec{x}$ is called the *field coordinate*, and the use of the independent variables $\vec{x}$ and t is called the *spatial* or *Eulerian description*.

With the help of (1.8) every quantity expressed in material coordinates can be expressed in field coordinates. Using (1.5) all quantities given in field coordinates can be converted into material coordinates. This conversion must be well defined, since there is only one material point $\vec{\xi}$ at place $\vec{x}$ at time t. The mapping (1.5) and the inverse mapping (1.8) must be uniquely reversible, and this is of course true if the *Jacobian* $J = \det(\partial x_i / \partial \xi_j)$ does not vanish.

If the velocity is given in field coordinates, the integration of the differential equations

$$\frac{d\vec{x}}{dt} = \vec{u}(\vec{x}, t) \quad \text{or} \quad \frac{dx_i}{dt} = u_i(x_j, t) \tag{1.10}$$

(with initial conditions $\vec{x}(t_0) = \vec{\xi}$) leads to the pathlines $\vec{x} = \vec{x}\left(\vec{\xi}, t\right)$.

If the velocity field and all other dependent quantities (e.g. the density or the temperature) are independent of time, the motion is called *steady*, otherwise it is called *unsteady*.

The Eulerian description is preferable because the simpler kinematics are better adapted to the problems of fluid mechanics. Consider a wind tunnel experiment to investigate the flow past a body. Here one deals almost always with steady flow. The paths of the fluid particles (where the particle has come from and where it is going to) are of secondary importance. In addition the experimental determination of the velocity as a function of the material coordinates (1.6) would be very difficult. But there are no difficulties in measuring the direction and magnitude of the velocity at any place, say, and by doing this the velocity field $\vec{u} = \vec{u}(\vec{x})$ or the pressure field $p = p(\vec{x})$ can be experimentally determined. In particular the pressure distribution on the body can be found.

1.2.2   Pathlines, Streamlines, Streaklines

The differential Eq. (1.10) shows that the path of a point in the material is always tangential to its velocity. In this interpretation the *pathline* is the tangent curve to the velocities of the same material point at different times. Time is the curve parameter, and the material coordinate $\vec{\xi}$ is the family parameter.

Just as the pathline is natural to the material description, so the *streamline* is natural to the Eulerian description. The velocity field assigns a velocity vector to every place $\vec{x}$ at time t and the streamlines are the curves whose tangent directions are the same as the directions of the velocity vectors. The streamlines provide a vivid description of the flow at time t.

If we interpret the streamlines as the tangent curves to the velocity vectors of different particles in the material at the same instant in time we see that there is no connection between pathlines and streamlines, apart from the fact that they may sometimes lie on the same curve.

By the definition of streamlines, the unit vector $\vec{u}/|\vec{u}|$ is equal to the *unit tangent vector* of the streamline $\vec{\tau} = d\vec{x}/|d\vec{x}| = d\vec{x}/ds$ where $d\vec{x}$ is a vector element of the streamline in the direction of the velocity. The differential equation of the streamline then reads

$$\frac{d\vec{x}}{ds} = \frac{\vec{u}(\vec{x}, t)}{|\vec{u}|}, \quad (t = \text{const}) \tag{1.11a}$$

or in index notation

$$\frac{dx_i}{ds} = \frac{u_i(x_j, t)}{\sqrt{u_k u_k}}, \quad (t = \text{const}). \tag{1.11b}$$

Integration of these equations with the "initial condition" that the streamline emanates from a point in space $\vec{x}_0$ ($\vec{x}(s = 0) = \vec{x}_0$) leads to the *parametric representation* of the streamline $\vec{x} = \vec{x}(s, \vec{x}_0)$. The curve parameter here is the arc length s measured from x_0, and the family parameter is $\vec{x}_0$.

The pathline of a material point $\vec{\xi}$ is tangent to the streamline at the place $\vec{x}$, where the material point is situated at time t. This is shown in Fig. 1.4. By definition the velocity vector is tangential to the streamline at time t and to its pathline. At another time the streamline will in general be a different curve.

In steady flow, where the velocity field is time-independent ($\vec{u} = \vec{u}(\vec{x})$), the streamlines are always the same curves as the pathlines. The differential equations for the pathlines are now given by $d\vec{x}/dt = \vec{u}(\vec{x})$, where time dependence is no longer explicit as in (1.10). The element of the arc length along the pathline is $d\sigma = |\vec{u}|dt$, and the differential equations for the pathlines are the same as for streamlines

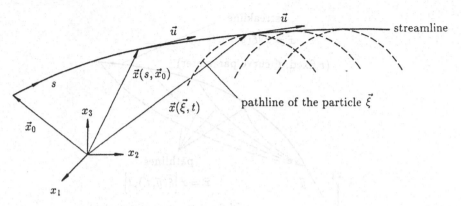

Fig. 1.4 Streamlines and pathlines

$$\frac{\mathrm{d}\vec{x}}{\mathrm{d}\sigma} = \frac{\vec{u}(\vec{x})}{|\vec{u}|}, \tag{1.12}$$

because how the curve parameter is named is irrelevant. Interpreting the integral curves of (1.12) as streamlines means they are still the tangent curves of the velocity vectors of different material particles at the same time t. Since the particles passing through the point in space $\vec{x}$ all have the same velocity there at all times, the tangent curves remain unchanged. Interpreting the integral curves of (1.12) as pathlines means that a material particle must move along the streamline as time passes, since it does not encounter velocity components normal to this curve.

What has been said for steady velocity fields holds equally well for unsteady fields where the direction of the velocity vector is time independent, that is for velocity fields of the form

$$\vec{u}(\vec{x}, t) = f(\vec{x}, t)\vec{u}_0(\vec{x}). \tag{1.13}$$

The *streakline* is also important, especially in experimental fluid mechanics. At a given time t a streakline joins all material points which have passed through (or will pass through) a given place $\vec{y}$ at any time t'. Filaments of color are often used to make flow visible. Colored fluid introduced into the stream at place $\vec{y}$ forms such a filament and a snapshot of this filament is a streakline. Other examples of streaklines are smoke trails from chimneys or moving jets of water.

Let the field $\vec{u} = \vec{u}(\vec{x}, t)$ be given, and calculate the pathlines from (1.10), solving it for $\vec{\xi}$. Setting $\vec{x} = \vec{y}$ and $t = t'$ in (1.8) identifies the material points $\vec{\xi}$ which were at place $\vec{y}$ at time t'.

The path coordinates of these particles are found by introducing the label $\vec{\xi}$ into the path equations, thus giving

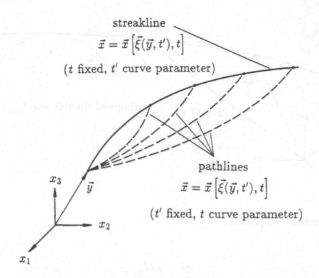

streakline

$$\vec{x} = \vec{x}\left[\vec{\xi}(\vec{y}, t'), t\right]$$

(t fixed, t' curve parameter)

pathlines

$$\vec{x} = \vec{x}\left[\vec{\xi}(\vec{y}, t'), t\right]$$

(t' fixed, t curve parameter)

Fig. 1.5 Streaklines and pathlines

$$\vec{x} = \vec{x}\left[\vec{\xi}(\vec{y}, t'), t\right]. \tag{1.14}$$

At a given time t, t' is the curve parameter of a curve in space which goes through the given point $\vec{y}$, and thus this curve in space is a streakline. In steady flows, streaklines, streamlines and pathlines all lie on the same curve (Fig. 1.5).

Surfaces can be associated with the lines introduced so far, formed by all the lines passing through some given curve C. If this curve C is closed, the lines form a tube (Fig. 1.6).

Streamtubes formed in this way are of particular technical importance. Since the velocity vector is by definition tangential to the wall of a streamtube, no fluid can pass through the wall. This means that pipes with solid walls are streamtubes.

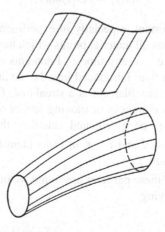

Fig. 1.6 Streamsheet and streamtube

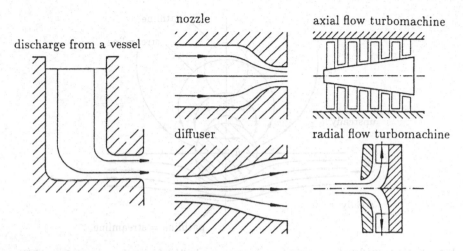

Fig. 1.7 Examples of streamtubes

Often the behavior of the whole flow can be described by the behavior of some "average" representative streamline. If the properties of the flow are approximately constant over the cross-section of the streamtube at the location where they are to be determined, we are led to a simple method of calculation: so-called *stream filament theory*. Since the streamtubes do not change with time when solid walls are present, the flow fields are, almost trivially, those where the direction of the velocity vector does not change. Consequently these flows may be calculated with relative ease.

Flows are often met in applications where the whole region of interest can be thought of as one streamtube. Examples are flows in tubes of changing cross-section, like in nozzles, in diffusers, and also in open channels. The space that the fluid occupies in turbomachines can often be taken as a streamtube, and even the flow between the blades of turbines and compressors can be treated approximately in this manner (Fig. 1.7).

The use of this "quasi-one-dimensional" view of the whole flow means that sometimes corrections for the higher dimensional character of the flow have to be introduced.

Steady flows have the advantage over unsteady flows that their streamlines are fixed in space, and the obvious convenience that the number of independent variables is reduced, which greatly simplifies the theoretical treatment. Therefore whenever possible we choose a reference system where the flow is steady. For example, consider a body moved through a fluid which is at rest at infinity. The flow in a reference frame fixed in space is unsteady, whereas it is steady in a reference frame moving with the body. Figure 1.8 demonstrates this fact in the example of a (frictionless) flow caused by moving a cylinder right to left. The upper half of the figure shows the unsteady flow relative to an observer at rest at time $t = t_0$ when the cylinder passes through the origin. The lower half shows the same flow relative to an observer who moves with the cylinder. In this system the flow is

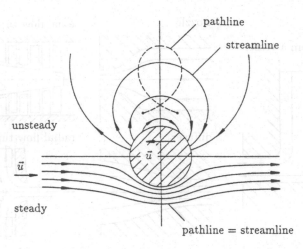

Fig. 1.8 Unsteady flow for a motionless observer; steady flow for an observer moving with the body

towards the cylinder from the left and it is steady. A good example of the first reference system is the everyday experience of standing on a street and feeling the unsteady flow when a vehicle passes. The second reference system is experienced by an observer inside the vehicle who feels a steady flow when he holds his hand out of the window.

1.2.3   Differentiation with Respect to Time

In the Eulerian description our attention is directed towards events at the place $\vec{x}$ at time t. However the rate of change of the velocity $\vec{u}$ at $\vec{x}$ is not generally the acceleration which the point in the material passing through $\vec{x}$ at time t experiences. This is obvious in the case of steady flows where the rate of change at a given place is zero. Yet a material point experiences a change in velocity (an acceleration) when it moves from $\vec{x}$ to $\vec{x} + \mathrm{d}\vec{x}$. Here $\mathrm{d}\vec{x}$ is the vector element of the pathline. The changes felt by a point of the material or by some larger part of the fluid and not the time changes at a given place or region of space are of fundamental importance in the dynamics. If the velocity (or some other quantity) is given in material coordinates, then the *material* or *substantial derivative* is provided by (1.6). But if the velocity is given in field coordinates, the place $\vec{x}$ in $\vec{u}(\vec{x}, t)$ is replaced by the path coordinates of the particle that occupies $\vec{x}$ at time t, and the derivative with respect to time at fixed $\vec{\xi}$ can be formed from

$$\frac{\mathrm{d}\vec{u}}{\mathrm{d}t} = \left\{ \frac{\partial \vec{u}\{\vec{x}(\vec{\xi}, t), t\}}{\partial t} \right\}_{\vec{\xi}}, \qquad (1.15a)$$

or

$$\frac{du_i}{dt} = \left\{ \frac{\partial u_i \{x_j(\xi_k, t), t\}}{\partial t} \right\}_{\xi_k}. \tag{1.15b}$$

The material derivative in field coordinates can also be found without direct reference to the material coordinates. Take the temperature field $T(\vec{x}, t)$ as an example: we take the total differential to be the expression

$$dT = \frac{\partial T}{\partial t} dt + \frac{\partial T}{\partial x_1} dx_1 + \frac{\partial T}{\partial x_2} dx_2 + \frac{\partial T}{\partial x_3} dx_3. \tag{1.16}$$

The first term on the right-hand side is the rate of change of the temperature at a fixed place: the *local change*. The other three terms give the change in temperature by advancing from $\vec{x}$ to $\vec{x} + d\vec{x}$. This is the *convective change*. The last three terms can be combined to give $d\vec{x} \cdot \nabla T$ or equivalently $dx_i \, \partial T / \partial x_i$. If $d\vec{x}$ is the vector element of the fluid particle's path at $\vec{x}$, then (1.10) holds and the rate of change of the temperature of the particle passing $\vec{x}$ (the material change of the temperature) is

$$\frac{dT}{dt} = \frac{\partial T}{\partial t} + \vec{u} \cdot \nabla T \tag{1.17a}$$

or

$$\frac{dT}{dt} = \frac{\partial T}{\partial t} + u_i \frac{\partial T}{\partial x_i} = \frac{\partial T}{\partial t} + u_1 \frac{\partial T}{\partial x_1} + u_2 \frac{\partial T}{\partial x_2} + u_3 \frac{\partial T}{\partial x_3}. \tag{1.17b}$$

This is quite a complicated expression for the material change in field coordinates, which leads to difficulties in the mathematical treatment. This is made clearer when we likewise write down the acceleration of the particle (the material change of its velocity)

$$\frac{d\vec{u}}{dt} = \frac{\partial \vec{u}}{\partial t} + (\vec{u} \cdot \nabla)\vec{u} = \frac{\partial \vec{u}}{\partial t} + (\vec{u} \cdot \mathrm{grad})\vec{u}, \tag{1.18a}$$

or

$$\frac{du_i}{dt} = \frac{\partial u_i}{\partial t} + u_j \frac{\partial u_i}{\partial x_j}. \tag{1.18b}$$

(Although the operator $d/dt = \partial/\partial t + (\vec{u} \cdot \nabla)$ is written in vector notation, it is here only explained in Cartesian coordinates. Now by appropriate definition of the Nabla operator, the operator d/dt is also valid for curvilinear coordinate systems, its application to vectors is difficult since the basis vectors can change. Later we will see a form for the material derivative of velocity which is more useful for orthogonal curvilinear coordinates since, apart from partial differentiation with respect to time, it is only composed of known quantities like the rotation of the velocity field and the gradient of the kinetic energy.)

It is easy to convince yourself that the material derivative (1.18) results from differentiating (1.15) with the chain rule and using (1.6).

The last three terms in the ith component of (1.18b) are nonlinear (quasilinear), since the products of the function $u_j(x,t)$ with its first derivatives $\partial u_i(\vec{x},t)/\partial x_j$ appear. Because of these terms, the equations of motion in field coordinates are nonlinear, making the mathematical treatment difficult. (The equations of motion in material coordinates are also nonlinear, but we will not go into details now.)

The view which has led us to (1.17) also gives rise to the *general time derivative*. Consider the rate of change of the temperature felt by a swimmer moving at velocity $\vec{w}$ relative to a fluid velocity of $\vec{u}$, i.e., at velocity $\vec{u}+\vec{w}$ relative to a fixed reference frame. The vector element $d\vec{x}$ of his path is $d\vec{x} = (\vec{u}+\vec{w})\,dt$ and the rate of change of the temperature felt by the swimmer is

$$\frac{dT}{dt} = \frac{\partial T}{\partial t} + (\vec{u}+\vec{w}) \cdot \nabla T, \tag{1.19}$$

where the operator $\partial/\partial t + (\vec{u}+\vec{w}) \cdot \nabla$, or $\partial/\partial t + (u_i+w_i)\partial/\partial x_i$ applied to other field quantities gives the rate of change of these quantities as experienced by the swimmer.

To distinguish between the general time derivative (1.19) and the material derivative we introduce the following symbol

$$\frac{D}{Dt} = \frac{\partial}{\partial t} + u_i \frac{\partial}{\partial x_i} = \frac{\partial}{\partial t} + (\vec{u} \cdot \nabla) \tag{1.20}$$

for the material derivative. (Mathematically, of course there is no difference between d/dt and D/Dt.)

Using the unit tangent vector to the pathline

$$\vec{t} = \frac{d\vec{x}}{|d\vec{x}|} = \frac{d\vec{x}}{d\sigma} \tag{1.21}$$

the convective part of the operator D/Dt can also be written

$$\vec{u} \cdot \nabla = |\vec{u}|\vec{t} \cdot \nabla = |\vec{u}|\frac{\partial}{\partial \sigma}, \tag{1.22}$$

so that the derivative $\partial/\partial\sigma$ is in the direction of $\vec{t}$ and that the expression

$$\frac{D}{Dt} = \frac{\partial}{\partial t} + |\vec{u}| \frac{\partial}{\partial \sigma} \tag{1.23}$$

holds. This form is used to state the acceleration vector in natural coordinates, that is in the coordinate system where the unit vectors of the accompanying triad of the pathline are used as basis vectors. σ is the coordinate in the direction of $\vec{t}, n$ is the coordinate in the direction of the principal normal vector $\vec{n}_\sigma = R d\vec{t}/d\sigma$, and b the coordinate in the direction of the binormal vector $\vec{b}_\sigma = \vec{t} \times \vec{n}_\sigma$. R is the radius of curvature of the pathline in the *osculating* plane spanned by the vectors $\vec{t}$ and $\vec{n}_\sigma$. Denoting the component of $\vec{u}$ in the $\vec{t}$-direction as $u, (u = |\vec{u}|)$, (1.23) then leads to the expression

$$\frac{D}{Dt}(u\vec{t}) = \left[\frac{\partial u}{\partial t} + u\frac{\partial u}{\partial \sigma}\right]\vec{t} + \frac{u^2}{R}\vec{n}_\sigma. \tag{1.24}$$

Resolving along the triad $(\vec{t}, \vec{n}_s, \vec{b}_s)$ of the streamline at time t, the convective acceleration is the same as in expression (1.24), since at the place $\vec{x}$ the streamline is tangent to the pathline of the particle found there. However the local change contains terms normal to the streamline, and although the components of the velocity u_b and u_n are zero here, their local changes do not vanish

$$\frac{\partial \vec{u}}{\partial t} = \frac{\partial u}{\partial t}\vec{t} + \frac{\partial u_n}{\partial t}\vec{n}_s + \frac{\partial u_b}{\partial t}\vec{b}_s. \tag{1.25}$$

Resolving the acceleration vector into the natural directions of the streamline then gives us

$$\frac{D\vec{u}}{Dt} = \left[\frac{\partial u}{\partial t} + u\frac{\partial u}{\partial s}\right]\vec{t} + \left[\frac{\partial u_n}{\partial t} + \frac{u^2}{R}\right]\vec{n}_s + \frac{\partial u_b}{\partial t}\vec{b}_s. \tag{1.26}$$

When the streamline is fixed in space, (1.26) reduces to (1.24).

1.2.4   State of Motion, Rate of Change of Line, Surface and Volume Elements

Knowing the velocity at the place $\vec{x}$ we can use the *Taylor expansion* to find the velocity at a neighboring place $\vec{x} + d\vec{x}$:

$$u_i(\vec{x} + \mathrm{d}\vec{x}, t) = u_i(\vec{x}, t) + \frac{\partial u_i}{\partial x_j} \mathrm{d}x_j. \tag{1.27a}$$

For each of the three velocity components u_i there are three derivatives in the Cartesian coordinate system, so that the velocity field in the neighborhood of $\vec{x}$ is fully defined by these nine spatial derivatives. Together they form a second order *tensor*, the *velocity gradient* $\partial u_i / \partial x_j$. The symbols $\nabla \vec{u}$ or grad $\vec{u}$ (defined by (A.40) in Appendix A) are used, and (1.27a) can also be written in the form

$$\vec{u}(\vec{x} + \mathrm{d}\vec{x}, t) = \vec{u}(\vec{x}, t) + \mathrm{d}\vec{x} \cdot \nabla \vec{u}. \tag{1.27b}$$

Using the identity

$$\frac{\partial u_i}{\partial x_j} = \frac{1}{2} \left\{ \frac{\partial u_i}{\partial x_j} + \frac{\partial u_j}{\partial x_i} \right\} + \frac{1}{2} \left\{ \frac{\partial u_i}{\partial x_j} - \frac{\partial u_j}{\partial x_i} \right\} \tag{1.28}$$

we expand the tensor $\partial u_i / \partial x_j$ into a *symmetric tensor*

$$e_{ij} = \frac{1}{2} \left\{ \frac{\partial u_i}{\partial x_j} + \frac{\partial u_j}{\partial x_i} \right\}, \tag{1.29a}$$

where this can be symbolically written, using (A.40), as

$$\mathbf{E} = e_{ij}\vec{e}_i\vec{e}_j = \frac{1}{2} \left[(\nabla \vec{u}) + (\nabla \vec{u})^{\mathrm{T}} \right], \tag{1.29b}$$

and an *antisymmetric tensor*

$$\Omega_{ij} = \frac{1}{2} \left\{ \frac{\partial u_i}{\partial x_j} - \frac{\partial u_j}{\partial x_i} \right\}, \tag{1.30a}$$

where this is symbolically (see A.40)

$$\mathbf{\Omega} = \Omega_{ji}\vec{e}_i\vec{e}_j = \frac{1}{2} \left[(\nabla \vec{u}) - (\nabla \vec{u})^{\mathrm{T}} \right]. \tag{1.30b}$$

Doing this we get from (1.27)

$$u_i(\vec{x} + d\vec{x}, t) = u_i(\vec{x}, t) + e_{ij}\mathrm{d}x_j + \Omega_{ij}\mathrm{d}x_j, \tag{1.31a}$$

or

$$\vec{u}(\vec{x} + d\vec{x}, t) = \vec{u}(\vec{x}, t) + d\vec{x} \cdot \mathbf{E} + d\vec{x} \cdot \mathbf{\Omega}. \tag{1.31b}$$

The first term in (1.31) arises from the translation of the fluid at place $\vec{x}$ with velocity u_i. The second represents the velocity with which the fluid in the neighborhood of $\vec{x}$ is deformed, while the third can be interpreted as an instantaneous local rigid body rotation. There is a very important meaning attached to the tensors e_{ij} and Ω_{ij}, which each describe entirely different contributions to the state of the motion. By definition the frictional stresses in the fluid make their appearance in the presence of deformation velocities, so that they cannot be dependent on the tensor Ω_{ij} which describes a local rigid body rotation. To interpret the tensors e_{ij} and Ω_{ij} we calculate the rate of change of a material line element dx_i. This is a vector element which always consists of a line distribution of the same material points. The material change is found, using

$$\frac{D}{Dt}(d\vec{x}) = d\left[\frac{D\vec{x}}{Dt}\right] = d\vec{u}, \tag{1.32}$$

as the velocity difference between the endpoints of the element. The vector component $d\vec{u}_E$ in the direction of the element is obviously the velocity with which the element is lengthened or shortened during the motion (Fig. 1.9). With the unit vector $d\vec{x}/ds$ in the direction of the element, the magnitude of this component is

$$d\vec{u} \cdot \frac{d\vec{x}}{ds} = du_i \frac{dx_i}{ds} = \left(e_{ij} + \Omega_{ij}\right)dx_j \frac{dx_i}{ds}, \tag{1.33}$$

and since $\Omega_{ij}dx_jdx_i$ is equal to zero (easily seen by expanding and interchanging the dummy indices), the extension of the element can only be caused by the symmetric tensor e_{ij}. e_{ij} is called the *rate of deformation tensor*. Other names are: *stretching*, *rate of strain*, or *velocity strain tensor*. We note that the stretching, for example, at

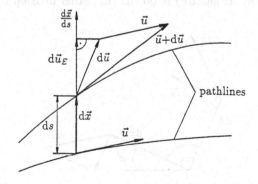

Fig. 1.9 The physical significance of the diagonal components of the deformation tensor

place $\vec{x}$ is the stretching that the particle experiences which occupies the place $\vec{x}$. For the rate of extension per instantaneous length ds we have from (1.33)

$$\frac{du_i}{ds}\frac{dx_i}{ds} = ds^{-1}\frac{D(dx_i)}{Dt}\frac{dx_i}{ds} = \frac{1}{2}ds^{-2}\frac{D(ds^2)}{Dt} \tag{1.34}$$

and using (1.33), we get

$$\frac{du_i}{ds}\frac{dx_i}{ds} = ds^{-1}\frac{D(ds)}{Dt} = e_{ij}\frac{dx_i}{ds}\frac{dx_j}{ds}. \tag{1.35}$$

Since $dx_i/ds = l_i$ is the ith component and $dx_j/ds = l_j$ is the jth component of the unit vector in the direction of the element, we finally arrive at the following expression for the rate of extension or the stretching of the material element

$$ds^{-1}\frac{D(ds)}{Dt} = e_{ij}l_i l_j. \tag{1.36}$$

(1.36) gives the physical interpretation of the diagonal elements of the tensor e_{ij}. Instead of the general orientation, let the material element $d\vec{x}$ be viewed when orientated parallel to the x_1-axis, so that the unit vector in the direction of the element has the components $(1,0,0)$ and, of the nine terms in (1.36), only one is nonzero. In this case, with $ds = dx_1$, (1.36) reads

$$dx_1^{-1}\frac{D(dx_1)}{Dt} = e_{11}. \tag{1.37}$$

The diagonal terms are now identified as the stretching of the material element parallel to the axes. In order to understand the significance of the remaining elements of the rate of deformation tensor, we imagine two perpendicular material line elements of the material $d\vec{x}$ and $d\vec{x}'$ (Fig. 1.10). The magnitude of the component $d\vec{u}_R$ perpendicular to $d\vec{x}$ (thus in the direction of the unit vector $\vec{l}' = d\vec{x}/ds'$ and in the plane spanned by $d\vec{x}$ and $d\vec{x}'$) is $d\vec{u} \cdot d\vec{x}'/ds'$. After division by ds we get the

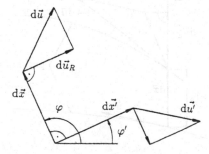

Fig. 1.10 The physical significance of the nondiagonal elements of the rate of deformation tensor

angular velocity with which the material line element rotates in the mathematically positive sense

$$\frac{D\varphi}{Dt} = -\frac{d\vec{u}}{ds} \cdot \frac{d\vec{x}'}{ds'} = -\frac{du_i}{ds} \frac{dx_i'}{ds'}. \tag{1.38}$$

Similarly we get the angular velocity with which $d\vec{x}'$ rotates

$$\frac{D\varphi'}{Dt} = -\frac{d\vec{u}'}{ds'} \cdot \left(-\frac{d\vec{x}}{ds}\right) = -\frac{du_i'}{ds'} \frac{dx_i}{ds}. \tag{1.39}$$

The difference between these gives the rate of change of the angle between the material elements $d\vec{x}$ and $d\vec{x}'$ (currently ninety degrees), and it gives a measure of the shear rate. Since

$$\frac{du_i}{ds} = \frac{\partial u_i}{\partial x_j} \frac{dx_j}{ds} \quad \text{and} \quad \frac{du_i'}{ds'} = \frac{\partial u_i}{\partial x_j} \frac{dx_j'}{ds'} \tag{1.40}$$

we get, for the difference between the angular velocities

$$\frac{D(\varphi - \varphi')}{Dt} = -\left\{\frac{\partial u_i}{\partial x_j} + \frac{\partial u_j}{\partial x_i}\right\} \frac{dx_i}{ds} \frac{dx_j'}{ds'} = -2e_{ij}l_i l_j'. \tag{1.41}$$

To do this, the dummy indices were relabeled twice. Choosing $d\vec{x}$ parallel to the x_2-axis, $d\vec{x}'$ parallel to the x_1-axis, so that $\vec{l} = (0,1,0)$ and $\vec{l}' = (1,0,0)$, and denoting the enclosed angle by α_{12}, (1.41) gives the element e_{12} as half of the velocity with which α_{12} changes in time

$$\frac{D\alpha_{12}}{Dt} = -2e_{12}. \tag{1.42}$$

The physical interpretation of all the other nondiagonal elements of e_{ij} is now obvious. The average of the angular velocities of the two material line elements gives the angular velocity with which the plane spanned by them rotates

$$\frac{1}{2}\frac{D}{Dt}(\varphi + \varphi') = -\frac{1}{2}\left\{\frac{\partial u_i}{\partial x_j} - \frac{\partial u_j}{\partial x_i}\right\} \frac{dx_j}{ds} \frac{dx_i'}{ds'} = \Omega_{ji}l_i'l_j. \tag{1.43}$$

Here again the dummy index has been relabeled twice and the property of the antisymmetric tensor $\Omega_{ij} = -\Omega_{ji}$ has been used. The Eq. (1.43) also yields the modulus of the component of the angular velocity $\vec{\omega}$ perpendicular to the plane spanned by $d\vec{x}$ and $d\vec{x}'$. The unit vector perpendicular to this plane

$$\frac{d\vec{x}'}{ds'} \times \frac{d\vec{x}}{ds} = \vec{l}' \times \vec{l} \tag{1.44}$$

can be written in index notation with the help of the *epsilon tensor* as $l_i' l_j \epsilon_{ijk}$, so that the right-hand side of (1.43) can be rewritten as follows

$$\Omega_{ji} l_i' l_j = \omega_k l_i' l_j \epsilon_{ijk}. \tag{1.45}$$

This equation assigns a vector to the antisymmetric tensor Ω_{ij}

$$\omega_k \epsilon_{ijk} = \Omega_{ji}. \tag{1.46}$$

Equation (1.46) expresses the well known fact that an antisymmetric tensor can be represented by an axial vector. Thus the contribution $\Omega_{ij} dx_j$ to the velocity field about the place $\vec{x}$ is the same as the ith component $\epsilon_{kji}\omega_k dx_j$ of the circumferential velocity $\vec{\omega} \times d\vec{x}$ produced at the vector radius $d\vec{x}$ by a rigid body at $\vec{x}$ rotating at angular velocity $\vec{\omega}$. For example, the tensor element Ω_{12} is then numerically equal to the component of the angular velocity perpendicular to the x_1-x_2-plane in the negative x_3-direction. Ω_{ij} is called the *spin tensor*. From (1.46) we can get the explicit representation of the vector component of $\vec{\omega}$, using the identity

$$\epsilon_{ijk} \epsilon_{ijn} = 2\delta_{kn} \tag{1.47}$$

(where δ_{kn} is the *Kronecker delta*) and multiplying by ε_{ijn} to get

$$\omega_k \epsilon_{ijk} \epsilon_{ijn} = 2\omega_n = \Omega_{ji} \epsilon_{ijn}. \tag{1.48}$$

Since e_{ij} is a symmetric tensor, then $\epsilon_{ijn} e_{ij} = 0$, and in general the following holds

$$\omega_n = \frac{1}{2} \frac{\partial u_j}{\partial x_i} \epsilon_{ijn}. \tag{1.49a}$$

The corresponding expression in vector notation

$$\vec{\omega} = \frac{1}{2} \nabla \times \vec{u} = \frac{1}{2} \operatorname{curl} \vec{u} \tag{1.49b}$$

introduces the *vorticity vector* curl $\vec{u}$, which is equal to twice the angular velocity $\vec{\omega}$. If this vorticity vector vanishes in the whole flow field in which we are interested, we speak of an irrotational flow field. The absence of vorticity in a field simplifies the mathematics greatly because we can now introduce a *velocity potential* Φ. The generally unknown functions u_i result then from the gradient of only one unknown scalar function Φ

$$u_i = \frac{\partial \Phi}{\partial x_i} \quad \text{or} \quad \vec{u} = \nabla \Phi. \tag{1.50}$$

This is the reason why irrotational flows are also called *potential flows*. The three component equations obtained from (1.50) are equivalent to the existence of a *total differential*

$$d\Phi = \frac{\partial \Phi}{\partial x_i} dx_i = u_i dx_i. \tag{1.51}$$

The necessary and sufficient conditions for its existence are that the following equations for the mixed derivatives should hold throughout the field

$$\frac{\partial u_1}{\partial x_2} = \frac{\partial u_2}{\partial x_1}, \quad \frac{\partial u_2}{\partial x_3} = \frac{\partial u_3}{\partial x_2}, \quad \frac{\partial u_3}{\partial x_1} = \frac{\partial u_1}{\partial x_3}. \tag{1.52}$$

Because of (1.50) these relationships are equivalent to the vanishing of the vorticity vector curl $\vec{u}$.

As with streamlines, in *rotational* flow vortex lines are introduced as tangent curves to the vorticity vector field, and similarly these can form vortex sheets and vortex tubes.

As is well known, symmetric matrices can be diagonalized. The same can be said for symmetric tensors, since tensors and matrices only differ in the ways that their measures transform, but otherwise they follow the same calculation rules. The reduction of a symmetric tensor e_{ij} to diagonal form is physically equivalent to finding a coordinate system where there is no shearing, only stretching. This is a so-called *principal axis system*. Since e_{ij} is a tensor field, the principal axis system is in general dependent on the place $\vec{x}$. If $\vec{l}$ (or l_i) is the unit vector relative to a given coordinate system in which e_{ij} is nondiagonal, the above problem amounts to determining this vector so that it is proportional to that part of the change in velocity given by e_{ij}, namely $e_{ij} dx_j$. We divide these changes by ds and since

$$\frac{du_i}{ds} = e_{ij} \frac{dx_j}{ds} = e_{ij} l_j \tag{1.53}$$

we are led to the *eigenvalue problem*

$$e_{ij} l_j = e \, l_i. \tag{1.54}$$

A solution of (1.54) only exists when the arbitrary constant of proportionality e takes on specific values, called the *eigenvalues* of the tensor e_{ij}. Using the Kronecker Delta symbol we can write the right-hand side of (1.54) as $e \, l_j \, \delta_{ij}$ and we are led to the homogeneous system of equations

$$\left(e_{ij} - e\,\delta_{ij}\right)l_j = 0. \tag{1.55}$$

This has nontrivial solutions for the unit vector we are searching for only when the determinant of the matrix of coefficients vanishes

$$\det\left(e_{ij} - e\,\delta_{ij}\right) = 0. \tag{1.56}$$

This is an equation of the third degree, and is called the *characteristic equation*. It can be written as

$$-e^3 + I_{1e}e^2 - I_{2e}e + I_{3e} = 0, \tag{1.57}$$

where I_{1e}, I_{2e}, I_{3e} are the first, second and third invariants of the rate of deformation tensor, given by the following formulae

$$I_{1e} = e_{ii}, \quad I_{2e} = \frac{1}{2}\left(e_{ii}\,e_{jj} - e_{ij}\,e_{ij}\right), \quad I_{3e} = \det\left(e_{ij}\right). \tag{1.58}$$

These quantities are invariants because they do not change their numerical values under change of coordinate system. They are called the *basic invariants* of the tensor e_{ij}. The roots of (1.57) do not change, and so neither do the eigenvalues of the tensor e_{ij}. The eigenvalues of a symmetric matrix are all real, and if they are all distinct, (1.54) gives three systems of equations, one for each of the components of the vector $\vec{l}$. With the condition that $\vec{l}$ is to be a unit vector, the solution of the homogeneous system of equations is unique. The three unit vectors of a real symmetric matrix are mutually orthogonal, and they form the principal axis system in which e_{ij} is diagonal. The statement of Eq. (1.31) in words is thus:

> The instantaneous velocity field about a place $\vec{x}$ is caused by the superposition of the translational velocity of the fluid there with stretching in the directions of the principal axes and a rigid rotation of these axes. (*fundamental theorem of kinematics*)

By expanding the first invariant I_{1e}, and using Eq. (1.37) and corresponding expressions, we arrive at the equation

$$e_{ii} = dx_1^{-1}\frac{D(dx_1)}{Dt} + dx_2^{-1}\frac{D(dx_2)}{Dt} + dx_3^{-1}\frac{D(dx_3)}{Dt}. \tag{1.59}$$

On the right is the rate of change of the material volume dV, divided by dV: it is the material change of this infinitesimal volume of the fluid particle. We can also write (1.59) in the form

$$e_{ii} = \nabla \cdot \vec{u} = dV^{-1}\frac{D(dV)}{Dt}. \tag{1.60}$$

Now, in flows where $D(dV)/Dt$ is zero, the volume of a fluid particle does not change, although its shape can. Such flows are called *volume preserving*, and the velocity fields of such flows are called *divergence free* or *source free*. The divergence $\nabla \cdot \vec{u}$ and the curl $\nabla \times \vec{u}$ are quantities of fundamental importance, since they can tell us a lot about the velocity field. If they are known in a simply connected space (where all closed curves may be shrunk to a single point), and if the normal component of $\vec{u}$ is given on the bounding surface, then, by a well known principle of vector analysis, the vector $\vec{u}(\vec{x})$ is uniquely defined at all $\vec{x}$. We also note the rate of change of a directional *material surface* element, $n_i dS$, which always consists of a surface distribution of the same fluid particles. With $dV = n_i dS dx_i$ we get from (1.60)

$$\frac{D}{Dt}(n_i dS dx_i) = n_i dS dx_i e_{jj}, \tag{1.61}$$

or

$$\frac{D}{Dt}(n_i dS) dx_i + du_i n_i dS = n_i dS dx_i e_{jj} \tag{1.62}$$

finally leading to

$$\frac{D}{Dt}(n_i dS) = \frac{\partial u_j}{\partial x_j} n_i dS - \frac{\partial u_j}{\partial x_i} n_j dS. \tag{1.63}$$

After multiplying by n_i and noting that $D(n_i n_i)/Dt = 0$ we obtain the specific rate of extension of the material surface element dS

$$\frac{1}{dS} \frac{D(dS)}{Dt} = \frac{\partial u_j}{\partial x_j} - e_{ij} n_i n_j. \tag{1.64}$$

Divided by the Euclidean norm of the rate of deformation tensor $(e_{lk} e_{lk})^{1/2}$, this can be used as a local measure for the "mixing"

$$\frac{D(\ln dS)}{Dt} / (e_{lk} e_{lk})^{1/2} = \left[\frac{\partial u_j}{\partial x_j} - e_{ij} n_i n_j\right] / (e_{lk} e_{lk})^{1/2}. \tag{1.65}$$

The higher material derivatives also play a role in the theory of the constitutive equations of non-Newtonian fluids. They lead to kinematic tensors which can be easily represented using our earlier results. From (1.35) we can read off the material derivative of the square of the line element ds as

$$\frac{D(ds^2)}{Dt} = 2 e_{ij} dx_i dx_j \tag{1.66}$$

and by further material differentiation this leads to the expression

$$\frac{D^2(ds^2)}{Dt^2} = \left\{\frac{D(2e_{ij})}{Dt} + 2e_{kj}\frac{\partial u_k}{\partial x_i} + 2e_{ik}\frac{\partial u_k}{\partial x_j}\right\}dx_i dx_j. \tag{1.67}$$

Denoting the tensor in the brackets as $A_{(2)ij}$ and $2e_{ij}$ as $A_{(1)ij}$, (symbolically $\mathbf{A}_{(2)}$ and $\mathbf{A}_{(1)}$), we find the operational rule for higher differentiation

$$\frac{D^n(ds^2)}{Dt^n} = A_{(n)ij}dx_i dx_j, \tag{1.68}$$

where

$$A_{(n)ij} = \frac{DA_{(n-1)ij}}{Dt} + A_{(n-1)kj}\frac{\partial u_k}{\partial x_i} + A_{(n-1)ik}\frac{\partial u_k}{\partial x_j} \tag{1.69}$$

gives the rule by which the tensor $\mathbf{A}_{(n)}$ can be found from the tensor $\mathbf{A}_{(n-1)}$ (*Oldroyd's derivative*). The importance of the tensors $\mathbf{A}_{(n)}$, also called the *Rivlin-Ericksen tensors*, lies in the fact that in very general non-Newtonian fluids, as long as the *deformation history* is smooth enough, the friction stress can only depend on these tensors. The occurrence of the above higher time derivatives can be disturbing, since in practice it is not known if the required derivatives actually exist. For kinematically simple flows, so called viscometric flows (the shearing flow in Fig. 1.1 is an example of these), the tensors $\mathbf{A}_{(n)}$ vanish in steady flows for $n > 2$. In many technically relevant cases, non-Newtonian flows can be directly treated as viscometric flows, or at least as related flows.

We will now calculate the kinematic quantities discussed up to now with an example of simple shearing flow (Fig. 1.11), whose velocity field is given by

$$u_1 = \dot{\gamma}x_2,$$
$$u_2 = 0, \tag{1.70}$$
$$u_3 = 0.$$

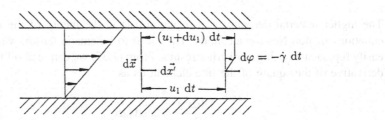

Fig. 1.11 Kinematics of simple shear flow

The material line element $d\vec{x}$ is rotated about $d\varphi = -(du_1/dx_2)dt$ in time dt, giving $D\varphi/Dt = -\dot{\gamma}$.

The material line element $d\vec{x}'$ remains parallel to the x_1-axis. The rate of change of the angle originally at ninety degrees is thus $-\dot{\gamma}$. The agreement with (1.41) can be seen immediately since $e_{12} = e_{21} = \dot{\gamma}/2$. Of the components of the tensor e_{ij}, these are the only ones which are nonzero. The average of the angular velocities of both material lines is $-\dot{\gamma}/2$, in agreement with (1.43). In order to work out the rotation of the element due to the shearing, we subtract the rigid body rotation $-\dot{\gamma}/2$ from the entire rotation calculated above ($-\dot{\gamma}\,dt$ and 0), and thus obtain $-(\dot{\gamma}/2)dt$ for the rotation of the element $d\vec{x}$ arising from shearing, and similarly $+(\dot{\gamma}/2)dt$ for the rotation of the element $d\vec{x}'$ due to shearing.

Now we can fully describe this flow: it consists of a translation of the point in common to both material lines along the distance $u_1\,dt$, a rigid body rotation of both line elements about an angle $-(\dot{\gamma}/2)dt$ and a shearing which rotates the element $d\vec{x}'$ about the angle $+(\dot{\gamma}/2)dt$ (so that its total rotation is zero) and the element $d\vec{x}$ about the angle $-(\dot{\gamma}/2)dt$ (so that its total rotation is $-\dot{\gamma}dt$). Since $A_{(1)ij} = 2e_{ij}$, the first Rivlin-Ericksen tensor has only two nonzero components: $A_{(1)12} = A_{(1)21} = \dot{\gamma}$. The matrix representation for $A_{(1)ij}$ thus reads

$$[\mathbf{A}_{(1)}] = \begin{bmatrix} 0 & \dot{\gamma} & 0 \\ \dot{\gamma} & 0 & 0 \\ 0 & 0 & 0 \end{bmatrix}. \tag{1.71}$$

Putting the components of $A_{(1)ij}$ in (1.71) we find there is only one nonvanishing component of the second Rivlin-Ericksen tensor $\left(A_{(2)22} = 2\dot{\gamma}^2\right)$, so that it can be expressed in matrix form as

$$[\mathbf{A}_{(2)}] = \begin{bmatrix} 0 & 0 & 0 \\ 0 & 2\dot{\gamma}^2 & 0 \\ 0 & 0 & 0 \end{bmatrix}. \tag{1.72}$$

All higher Rivlin-Ericksen tensors vanish.

An element $d\vec{x}$ whose unit tangent vector $d\vec{x}/ds$ has the components $(\cos\vartheta, \sin\vartheta, 0)$, thus making an angle ϑ with the x_1-axis ($l_3 = 0$), experiences, by (1.36), the stretching

$$\frac{1}{ds}\frac{D(ds)}{Dt} = e_{ij}l_il_j = e_{11}l_1l_1 + 2e_{12}l_1l_2 + e_{22}l_2l_2. \tag{1.73}$$

Since $e_{11} = e_{22} = 0$ the final expression for the stretching is

$$\frac{1}{ds}\frac{D(ds)}{Dt} = 2\frac{\dot{\gamma}}{2}\cos\vartheta\sin\vartheta = \frac{\dot{\gamma}}{2}\sin 2\vartheta. \tag{1.74}$$

The stretching reaches a maximum at $\vartheta = 45°$, $225°$ and a minimum at $\vartheta = 135°$, $315°$. These directions correspond with the positive and negative directions of the principal axes in the x_1-x_2-plane.

The eigenvalues of the tensor e_{ij} can be calculated using (1.57), where the basic invariants are given by $I_{1e} = 0$, $I_{2e} = -\dot{\gamma}^2/4$ and $I_{3e} = 0$. Since $I_{1e} = e_{ii} = \mathrm{div}\,\vec{u} = 0$ we see that this is a volume preserving flow. (The vanishing of the invariants I_{1e} and I_{3e} of the tensor e_{ij} is a necessary condition for *viscometric flows*, that is for flows which are locally simple shearing flows.) The characteristic Eq. (1.55) then reads $e(e^2 - \dot{\gamma}^2/4) = 0$ and it has roots $e^{(1)} = -e^{(3)} = \dot{\gamma}/2$, $e^{(2)} = 0$. The eigenvectors belonging to these roots, $\vec{n}^{(1)} = (1/\sqrt{2}, 1/\sqrt{2}, 0), \vec{n}^{(2)} = (0, 0, 1)$ and $\vec{n}^{(3)} = (1/\sqrt{2}, -1/\sqrt{2}, 0)$, give the principal rate of strain directions, up to the sign. (The otherwise arbitrary indexing of the eigenvalues is chosen so that $e^{(1)} > e^{(2)} > e^{(3)}$.) The second principal rate of strain direction is the direction of the x_3-axis, and the principal rate of strain $e^{(2)}$ is zero, since the velocity field is two-dimensional. The distortion and extension of a square shaped fluid particle is sketched in Fig. 1.12. In this special case the eigenvalues and eigenvectors are independent of place $\vec{x}$. The principal axis system is the same for all fluid particles, and as such Fig. 1.12 also holds for a larger square shaped part of the fluid.

We return now to the representation of the acceleration (1.18) as the sum of the local and convective accelerations. Transforming (1.20) into index notation and using the identity

$$\frac{Du_i}{Dt} = \frac{\partial u_i}{\partial t} + u_j \frac{\partial u_i}{\partial x_j} = \frac{\partial u_i}{\partial t} + u_j \left\{ \frac{\partial u_i}{\partial x_j} - \frac{\partial u_j}{\partial x_i} \right\} + u_j \frac{\partial u_j}{\partial x_i}, \tag{1.75}$$

and the definition (1.30), we are led to

$$\frac{Du_i}{Dt} = \frac{\partial u_i}{\partial t} + 2\Omega_{ij} u_j + \frac{\partial}{\partial x_i} \left\{ \frac{u_j u_j}{2} \right\}. \tag{1.76}$$

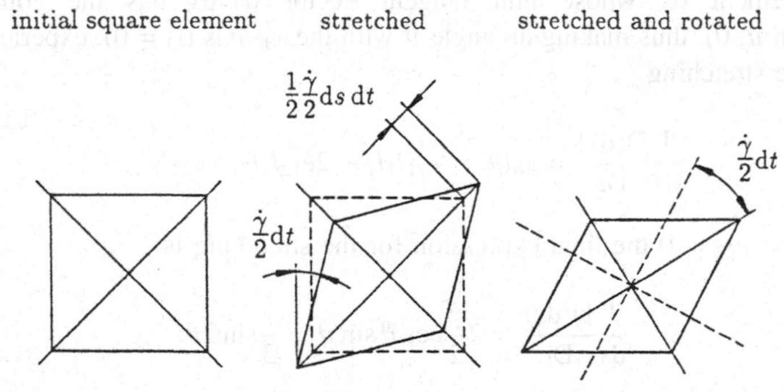

initial square element   stretched   stretched and rotated

Fig. 1.12 Deformation of a square of fluid in simple shearing flow

With (1.46), we finally obtain

$$\frac{Du_i}{Dt} = \frac{\partial u_i}{\partial t} - 2\epsilon_{ijk}\omega_k u_j + \frac{\partial}{\partial x_i}\left\{\frac{u_j u_j}{2}\right\}, \tag{1.77}$$

which written symbolically using (1.49b), is

$$\frac{D\vec{u}}{Dt} = \frac{\partial \vec{u}}{\partial t} - \vec{u} \times (\nabla \times \vec{u}) + \nabla\left[\frac{\vec{u} \cdot \vec{u}}{2}\right]. \tag{1.78}$$

This form shows explicitly the contribution of the rotation $\nabla \times \vec{u}$ to the acceleration field. In steady irrotational flow, the acceleration can be represented as the gradient of the kinetic energy (per unit mass).

We will often also use orthogonal curvilinear coordinate systems (e.g. cylindrical and spherical coordinates). In these cases the material derivative of the velocity in the form (1.78) is more useful than in (1.18), since the components of the acceleration in these coordinate systems are readily obtainable through the definition of the Nabla operator and by using the rules for calculation of the scalar and vector product. From (1.78) we can also get a dimensionless measure for the contribution of the rotation to the acceleration

$$W_D = \frac{|\vec{u} \times (\nabla \times \vec{u})|}{\left|\frac{\partial \vec{u}}{\partial t} + \nabla\left[\frac{\vec{u} \cdot \vec{u}}{2}\right]\right|}. \tag{1.79}$$

The ratio is called the *dynamic vortex number*. In general, it is zero for irrotational flows, while for nonaccelerating steady flows it takes the value 1. We can get a measure called the *kinematic vortex number* by dividing the Euclidean norm (the magnitude) of the rotation $|\nabla \times \vec{u}|$ by the Euclidean norm of the rate of deformation tensor

$$W_K = \frac{|\nabla \times \vec{u}|}{\sqrt{e_{ij}e_{ij}}}. \tag{1.80}$$

The kinematic vortex number is zero for irrotational flows and infinite for a rigid body rotation if we exclude the pure translation for which indeed both norms are zero.

Let us also compare the local acceleration with the convective acceleration using the relationship

$$S = \frac{\left|\frac{\partial \vec{u}}{\partial t}\right|}{\left|-\vec{u} \times (\nabla \times \vec{u}) + \nabla\left[\frac{\vec{u} \cdot \vec{u}}{2}\right]\right|}. \tag{1.81}$$

For steady flows we have $S = 0$, unless the convective acceleration is also equal to zero. $S = \infty$ is an important special case in unsteady flows, because the convective acceleration is then zero. This condition is the fundamental simplification used in acoustics and it is also used in the treatment of unsteady shearing flows.

1.2.5   Rate of Change of Material Integrals

From now on we shall always consider the same piece of fluid which is separated from the rest of the fluid by a closed surface. The enclosed part of the fluid is called a "body" and always consists of the same fluid particles (material points); its volume is therefore a material volume, and its surface is a material surface. During the motion, the shape of the material volume changes and successively takes up new regions in space. We will denote by $(V(t))$ the region which is occupied by our part of the fluid at time t. The mass m of the bounded piece of fluid is the sum of the mass elements dm over the set (M) of the material points of the body

$$m = \int_{(M)} dm. \tag{1.82}$$

Since in continuum theory, we consider the density to be a continuous function of position, we can also write the mass as the integral of the density over the region in space $(V(t))$ occupied by the body

$$m = \int_{(M)} dm = \iiint_{(V(t))} \rho(\vec{x}, t)\, dV. \tag{1.83}$$

Equivalently, the same holds for any continuous function ϕ, whether it is a scalar or a tensor function of any order

$$\int_{(M)} \varphi\, dm = \iiint_{(V(t))} \varphi\rho\, dV. \tag{1.84}$$

In the left integral we can think of φ as a function of the material coordinates $\vec{\xi}$ and t, and on the right we can think of it as a function of the field coordinates $\vec{x}$ and t. (Note that φ is not a property of the label $\vec{\xi}$, but a property of the material point labeled $\vec{\xi}$.) We are most interested in the rate of change of these material integrals and are led to a particularly simple derivation of the correct expression if we use the law of conservation of mass at this stage: the mass of the bounded part of the fluid must remain constant in time

$$\frac{Dm}{Dt} = 0. \tag{1.85}$$

This conservation law must also hold for the mass of the material point

$$\frac{D}{Dt}(dm) = 0, \tag{1.86}$$

since by (1.82) the mass is additive and the part of the fluid we are looking at must always consist of the same material points. Now taking the rate of change of the integral on the left side of (1.84) the region of integration is constant, and we have to differentiate the integral by the parameter t. Since φ and $D\varphi/Dt$ are continuous, the differentiation can be executed "under" the integral sign (*Leibniz's rule*), so that the equation now becomes

$$\frac{D}{Dt} \int_{(M)} \varphi \, dm = \int_{(M)} \frac{D\varphi}{Dt} dm. \tag{1.87}$$

The right-hand side can be expressed by an integration over the region in space $(V(t))$ and we get using (1.84)

$$\frac{D}{Dt} \int_{(M)} \varphi \, dm = \frac{D}{Dt} \iiint_{(V(t))} \varphi \rho dV = \iiint_{(V(t))} \frac{D\varphi}{Dt} \rho dV. \tag{1.88}$$

The result of the integration in the last integral does not change when, instead of a region varying in time $(V(t))$, we choose a fixed region (V), which coincides with the varying region at time t. We are really replacing the rate of change of the integral of φ over a deforming and moving body by the integral over a fixed region.

Although we got this result by the explicit use of the conservation of mass, the reduction of the material derivative of a volume integral to a fixed volume integral is purely kinematical. We recognize this when we apply the conservation of mass again and construct a formula equivalent to (1.88) where the density ρ does not appear. To this end we will consider the rate of change of a material integral over a fluid property related to volume, which we again call φ

$$\frac{D}{Dt} \iiint_{(V(t))} \varphi \, dV = \frac{D}{Dt} \int_{(M)} \varphi v \, dm = \int_{(M)} \frac{D}{Dt}(\varphi v) dm. \tag{1.89}$$

Here $v = 1/\rho$ is the *specific volume*. Carrying out the differentiation in the integrand, and replacing $Dv/Dt \, dm$ by $D(dV)/Dt$ (as follows from (1.86)) we get the equation

$$\frac{D}{Dt} \iiint\limits_{(V(t))} \varphi dV = \iiint\limits_{(V)} \frac{D\varphi}{Dt} dV + \iiint\limits_{(V)} \varphi \frac{D(dV)}{Dt}. \tag{1.90}$$

Without loss of generality we have replaced the time varying region on the right-hand side $(V(t))$ with a fixed region (V) which coincides with it at time t. This formula shows that the derivative of material integrals can be calculated by interchanging the order of integration and differentiation. From this general rule, Eq. (1.88) emerges immediately taking into account that, by (1.86), $D(\rho \, dV)/Dt = 0$ holds.

Another approach to (1.90), which also makes its pure kinematic nature clear is gained by using (1.5) and thereby introducing the new integration variables ξ_i instead of x_i. This corresponds to a mapping of the current domain of integration $(V(t))$ to the region (V_0) occupied by the fluid at the reference time t_0. Using the Jacobian J of the mapping (1.5) we have

$$dV = J \, dV_0,$$

and obtain

$$\frac{D(dV)}{Dt} = \frac{DJ}{Dt} dV_0 \tag{1.91a}$$

since V_0 is independent of time, from which follows, using (1.60), the material derivative of the Jacobian

$$\frac{DJ}{Dt} = e_{ii}J = \frac{\partial u_i}{\partial x_i}J, \tag{1.91b}$$

a formula known as *Euler's expansion formula*. From the last two equations we then have

$$\frac{D}{Dt} \iiint\limits_{(V(t))} \varphi dV = \iiint\limits_{(V_0)} \frac{D}{Dt}(\varphi J)dV_0 = \iiint\limits_{(V_0)} \left[\frac{D\varphi}{Dt}J + \varphi \frac{DJ}{Dt} \right] dV_0,$$

which under the inverse mapping leads directly to (1.90). Using (1.91b) and the inverse mapping the forms

$$\frac{D}{Dt} \iiint\limits_{(V(t))} \varphi \, dV = \iiint\limits_{(V)} \left[\frac{D\varphi}{Dt} + \varphi \frac{\partial u_i}{\partial x_i} \right] dV \tag{1.92}$$

and

$$\frac{D}{Dt} \iiint\limits_{(V(t))} \varphi \, dV = \iiint\limits_{(V)} \left[\frac{\partial \varphi}{\partial t} + \frac{\partial}{\partial x_i} (\varphi \, u_i) \right] dV \tag{1.93}$$

follow. If φ is a tensor field of any degree, which together with its partial derivatives is continuous in (V), then *Gauss' theorem* holds

$$\iiint\limits_{(V)} \frac{\partial \varphi}{\partial x_i} dV = \iint\limits_{(S)} \varphi \, n_i \, dS. \tag{1.94}$$

S is the directional surface bounding V, and the normal vector n_i is outwardly positive. Gauss' theorem relates a volume integral to the integral over a bounded, directional surface, provided that the integrand can be written as the "divergence" (in the most general sense) of the field φ. We will often make use of this important law. It is a generalization of the well known relationship

$$\int\limits_a^b \frac{df(x)}{dx} dx = f(b) - f(a). \tag{1.95}$$

The application of Gauss' law to the last integral in (1.93) furnishes a relationship known as *Reynolds' transport theorem*

$$\frac{D}{Dt} \iiint\limits_{(V(t))} \varphi \, dV = \iiint\limits_{(V)} \frac{\partial \varphi}{\partial t} dV + \iint\limits_{(S)} \varphi u_i n_i dS. \tag{1.96}$$

This relates the rate of change of the material volume integral to the rate of change of the quantity φ integrated over a fixed region (V), which coincides with the varying region $(V(t))$ at time t, and to the *flux* of the quantity φ through the bounding surfaces.

We note here that Leibniz's rule holds for a domain fixed in space: this means that differentiation can take place "under" the integral sign

$$\frac{\partial}{\partial t} \iiint\limits_{(V)} \varphi \, dV = \iiint\limits_{(V)} \frac{\partial \varphi}{\partial t} dV. \tag{1.97}$$

To calculate the expression for the rate of change of a directional material surface integral we change the order of integration and differentiation. If $(S(t))$ is a time varying surface region which is occupied by the material surface during the motion, in analogy to (1.90) we can write

$$\frac{D}{Dt} \iiint\limits_{(S(t))} \varphi \, n_i \, dS = \iint\limits_{(S)} \frac{D\varphi}{Dt} n_i \, dS + \iint\limits_{(S)} \varphi \frac{D}{Dt}(n_i \, dS). \tag{1.98}$$

For the integrals on the right-hand side, we can think of the region of integration $(S(t))$ as replaced by a fixed region (S) which coincides with the varying region at time t. After transforming the last integral with the help of (1.63) we get the formula

$$\frac{D}{Dt} \iiint\limits_{(S(t))} \varphi n_i \, dS = \iint\limits_{(S)} \frac{D\varphi}{Dt} n_i \, dS + \iint\limits_{(S)} \frac{\partial u_j}{\partial x_i} n_i \varphi dS - \iint\limits_{(S)} \frac{\partial u_j}{\partial x_i} n_j \varphi dS. \tag{1.99}$$

Let $(C(t))$ be a time varying one-dimensional region which is occupied by a material curve during the motion, and let φ be a (tensorial) field quantity. The rate of change of the material curve integral of φ can then be written as

$$\frac{D}{Dt} \int\limits_{(C(t))} \varphi \, dx_i = \int\limits_{(C)} \frac{D\varphi}{Dt} dx_i + \int\limits_{(C)} \varphi \, d \left[\frac{Dx_i}{Dt} \right] \tag{1.100}$$

from which we get using (1.10)

$$\frac{D}{Dt} \int\limits_{(C(t))} \varphi \, dx_i = \int\limits_{(C)} \frac{D\varphi}{Dt} dx_i + \int\limits_{(C)} \varphi \, du_i. \tag{1.101}$$

This formula has important applications when $\varphi = u_i$; in this case then

$$\varphi \, du_i = u_i \, du_i = d \left[\frac{u_i u_i}{2} \right] \tag{1.102}$$

is a total differential, and the last curve integral on the right-hand side of (1.101) is independent of the "path": it is only determined by the initial point I and the endpoint E. This obviously also holds for the first curve integral on the right-hand side, when the acceleration $D\varphi/Dt = Du_i/Dt$ can be written as the gradient of a scalar function

$$\frac{Du_i}{Dt} = \frac{\partial I}{\partial x_i}. \tag{1.103}$$

Then (and only then) is the first curve integral path independent

$$\int\limits_{(C)} \frac{D\varphi}{Dt} dx_i = \int\limits_{(C)} \frac{\partial I}{\partial x_i} dx_i = \int\limits_{(C)} dI = I_E - I_1. \tag{1.104}$$

The curve integral of u_i round a closed material curve (in the mathematically positive sense of direction)

$$\Gamma = \oint u_i \mathrm{d}x_i \qquad (1.105)$$

is called the *circulation*. Later we will discuss the conditions under which the acceleration may be written as the gradient of a scalar function, but now we will infer from (1.101) that then the rate of change of the circulation is zero. This follows directly from the fact that the initial and final points of a closed curve coincide and from our implicit assumption that I and u_i are continuous functions. The fact that the circulation is a conserved quantity, so that its rate of change is zero, often leads to an explanation for the strange and unexpected behavior of vortices and vortex motion.

Chapter 2
Fundamental Laws of Continuum Mechanics

2.1 Conservation of Mass, Equation of Continuity

Conservation of mass has already been postulated in the last chapter, and now we will make use of our earlier results and employ (1.83) and (1.93) to change the conservation law (1.85) to the form

$$\frac{D}{Dt} \iiint\limits_{(V(t))} \varrho \, dV = \iiint\limits_{(V)} \left[\frac{\partial \varrho}{\partial t} + \frac{\partial}{\partial x_i} (\varrho \, u_i) \right] dV = 0. \tag{2.1}$$

This equation holds for every volume that could be occupied by the fluid, that is, for arbitrary choice of the integration region (V). We could therefore shrink the integration region to a point, and we conclude that the continuous integrand must itself vanish at every $\vec{x}$. Thus we are led to the local or *differential form* of the law of conservation of mass

$$\frac{\partial \varrho}{\partial t} + \frac{\partial}{\partial x_i} (\varrho u_i) = 0. \tag{2.2}$$

This is the *continuity equation*. If we use the material derivative (1.20) we obtain

$$\frac{D\varrho}{Dt} + \varrho \frac{\partial u_i}{\partial x_i} = 0, \tag{2.3a}$$

or written symbolically

$$\frac{D\varrho}{Dt} + \varrho \nabla \cdot \vec{u} = 0. \tag{2.3b}$$

© Springer Nature Switzerland AG 2020
J. H. Spurk and N. Aksel, *Fluid Mechanics*,
https://doi.org/10.1007/978-3-030-30259-7_2

This also follows directly by using (1.86) together with (1.60). If

$$\frac{D\varrho}{Dt} = \frac{\partial \varrho}{\partial t} + u_i \frac{\partial \varrho}{\partial x_i} = 0 \tag{2.4}$$

holds, then the density of a single material particle does not vary during its motion. By (2.3a), (2.4) is equivalent to

$$\operatorname{div} \vec{u} = \nabla \cdot \vec{u} = \frac{\partial u_i}{\partial x_i} = 0, \tag{2.5}$$

i.e., the flow is volume preserving. This is also often called *incompressible flow*, by which is meant that the fluid, whether it is gas or liquid, can be viewed as incompressible. If (2.4) is satisfied, the continuity equation takes on the simpler form (2.5) where no derivative with respect to time appears, but which nevertheless holds for unsteady flows.

The conditions under which the assumption $D\varrho/Dt = 0$ is justified can only be properly discussed in the fourth chapter; it is enough to say here that in many technically important cases even gas flows may be regarded as incompressible.

As a rule the condition $D\varrho/Dt = 0$ is satisfied by liquids, but there are flows where even the change in volume in liquids is significant. This is the case in the unsteady flows which occur when valves on conduits are quickly opened or closed, or in supply pipes of hydraulic turbines when the gate settings are suddenly changed, but also in fuel injection systems when the injectors are opened or closed.

Incompressible flow does not mean that the density is the same for every particle. Consider the flow in the ocean which is incompressible ($D\varrho/Dt = 0$ holds), but where the density of particles differ from one to another as a result of different salt concentrations.

If the density is spatially constant, so $\nabla \varrho = 0$, we talk of a *homogeneous* density field. In incompressible flow, not only do the four terms in (2.4) sum to zero, but each term itself is now identically equal to zero.

Transforming the conservation of mass (1.85) with the help of Reynolds' transport theorem, we arrive at the *integral form of the continuity equation*

$$\frac{Dm}{Dt} = \frac{D}{Dt} \iiint\limits_{(V(t))} \varrho \, dV = \iiint\limits_{(V)} \frac{\partial \varrho}{\partial t} \, dV + \iint\limits_{(S)} \varrho \, u_i \, n_i \, dS = 0 \tag{2.6}$$

or

$$\iiint\limits_{(V)} \frac{\partial \varrho}{\partial t} dV = \frac{\partial}{\partial t} \iiint\limits_{(V)} \varrho dV = - \iint\limits_{(S)} \varrho u_i \, n_i \, dS. \tag{2.7}$$

In this equation we consider a fixed domain of integration, a so-called *control volume*, and we interpret this equation as follows: the rate of change of the mass in the control volume is equal to the difference between the mass entering and the mass leaving through the surface of the control volume per unit time. This very obvious interpretation often serves as a starting point for the elucidation of the mass conservation. In steady flow, $\partial\varrho/\partial t = 0$, and the integral form of the continuity equation reads

$$\iint\limits_{(S)} \varrho\, u_i\, n_i\, \mathrm{d}S = 0, \tag{2.8}$$

i.e., just as much mass enters as leaves the control volume per unit time.

2.2   Balance of Momentum

As the first law (axiom) of classical mechanics, accepted to be true without proof but embracing our experience, we state the momentum balance: in an inertial frame the rate of change of the momentum of a body is balanced by the force applied on this body

$$\frac{\mathrm{D}\vec{P}}{\mathrm{D}t} = \vec{F}. \tag{2.9}$$

What follows now only amounts to rearranging this axiom explicitly. The body is still a part of the fluid which always consists of the same material points. Analogous to (1.83), we calculate the momentum of the body as the integral over the region occupied by the body

$$\vec{P} = \iiint\limits_{(V(t))} \varrho\, \vec{u}\, \mathrm{d}V. \tag{2.10}$$

The forces affecting the body basically fall into two classes, *body forces*, and *surface* or *contact forces*. Body forces are forces with a long range of influence which act on all the material particles in the body and which, as a rule, have their source in fields of force. The most important example we come across is the earth's gravity field. The gravitational field strength $\vec{g}$ acts on every molecule in the fluid particle, and the sum of all the forces acting on the particle represents the actual gravitational force

$$\Delta\vec{F} = \vec{g}\sum_i m_i = \vec{g}\Delta m. \tag{2.11}$$

The force of gravity is therefore proportional to the mass of the fluid particle. As before, in the framework of the continuum hypothesis, we consider the body force as a continuous function of mass or volume and call

$$\vec{k} = \lim_{\Delta m \to 0} \frac{\Delta\vec{F}}{\Delta m} \tag{2.12}$$

the *mass* body force; in the special case of the earth's gravitational field $\vec{k} = \vec{g}$, we call it the gravitational force. The *volume body force* is the force referred to the volume, thus

$$\vec{f} = \lim_{\Delta V \to 0} \frac{\Delta\vec{F}}{\Delta V}, \tag{2.13}$$

(cf. Fig. 2.1), and in the special case of the gravitational force we get

$$\vec{f} = \lim_{\Delta V \to 0} \vec{g}\frac{\Delta m}{\Delta V} = \vec{g}\varrho. \tag{2.14}$$

Other technically important body forces appear because of electromagnetic fields, or are so-called *apparent forces* (like the centrifugal force), when the motion is referred to an *accelerating reference frame*.

The contact or surface forces are exerted from the surrounding fluid or more generally from other bodies on the surface of the fluid body under observation. If $\Delta\vec{F}$ is an element of the surface force, and ΔS is the surface element at $\vec{x}$ where the force is acting, we call the quantity

$$\vec{t} = \lim_{\Delta S \to 0} \frac{\Delta\vec{F}}{\Delta S} \tag{2.15}$$

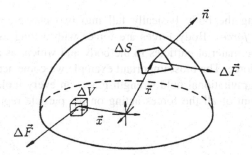

Fig. 2.1 Depiction of the volume and surface forces

the *stress* or *traction vector* at $\vec{x}$ (cf. Fig. 2.1). The stress vector is not only dependent on $\vec{x}$ and the time t, but also on the orientation of the surface element at $\vec{x}$, that is on the surface element's normal vector $\vec{n}$, and it is in general not directed parallel to the normal vector. More often we call the projection of $\vec{t}$ in the direction parallel to the normal the normal stress and the projection in the plane perpendicular to $\vec{n}$ the *tangential stress*.

We assume that the applied force is the sum of the two kinds of force and work out the whole force acting on the part of the fluid under observation by integration over the volume occupied by the fluid and over its surface to get

$$\vec{F} = \iiint\limits_{(V(t))} \varrho \vec{k} \, \mathrm{d}V + \iint\limits_{(S(t))} \vec{t} \, \mathrm{d}S \tag{2.16}$$

so that the balance of momentum takes the form

$$\frac{\mathrm{D}}{\mathrm{D}t} \iiint\limits_{(V(t))} \varrho \vec{u} \, \mathrm{d}V = \iiint\limits_{(V)} \varrho \vec{k} \, \mathrm{d}V + \iint\limits_{(S)} \vec{t} \, \mathrm{d}S. \tag{2.17}$$

As before, without loss of generality, we can replace the time varying domains of integration on the right with fixed domains. Then applying (1.88) to the left-hand side leads us to the form

$$\iiint\limits_{(V)} \frac{\mathrm{D}\vec{u}}{\mathrm{D}t} \varrho \, \mathrm{d}V = \iiint\limits_{(V)} \vec{k}\varrho \, \mathrm{d}V + \iint\limits_{(S)} \vec{t} \, \mathrm{d}S, \tag{2.18}$$

from which we reach an important conclusion: if we divide this equation by l^2, where l is a typical dimension of the domain of integration, say $l \sim V^{1/3}$, and take the limit $l \to 0$, the volume integrals vanish and we are left with

$$\lim_{l \to 0} \left[l^{-2} \iint\limits_{(S)} \vec{t} \, \mathrm{d}S \right] = 0. \tag{2.19}$$

Equation (2.19) means that the surface forces are locally balanced. Obviously (2.19) holds for nonvanishing $\vec{t}$, because $\vec{t}$ does not represent a field in the usual sense, but one which is dependent on $\vec{n}$ as well as $\vec{x}$. We will use this result to show the way that the stress vector is dependent on the normal vector $\vec{n}$ at the fixed place $\vec{x}$. Looking at the tetrahedron of Fig. 2.2, the normal vector to the inclined surface is $\vec{n}$, and the other surfaces are parallel to the coordinate planes; their normal vectors are then $-\vec{e}_1, -\vec{e}_2$ and $-\vec{e}_3$. If ΔS is the area of the inclined surface, then the other surface areas are $\Delta S \, n_1$, $\Delta S \, n_2$ and $\Delta S \, n_3$, respectively. For the stress vector belonging to the inclined surface we will write $\vec{t}^{(\vec{n})}$, and for the others $\vec{t}^{(-\vec{e}_1)}, \vec{t}^{(-\vec{e}_2)}$ and $\vec{t}^{(-\vec{e}_3)}$. Applying the local stress equilibrium (2.19) we arrive at

$$\lim_{l \to 0}\left[l^{-2} \iint_{(S)} \vec{t}\, dS \right] = \lim_{l \to 0}\left\{ \frac{\Delta S}{l^2}\left[\vec{t}^{(-\vec{e}_1)}n_1 + \vec{t}^{(-\vec{e}_2)}n_2 + \vec{t}^{(-\vec{e}_3)}n_3 + \vec{t}^{(\vec{n})} \right] \right\} = 0,$$

(2.20)

or

$$\vec{t}^{(\vec{n})} = -\vec{t}^{(-\vec{e}_1)}n_1 - \vec{t}^{(-\vec{e}_2)}n_2 - \vec{t}^{(-\vec{e}_3)}n_3,$$

(2.21)

since vanishes as l^2. In (2.21) all the stress vectors are to be taken at the same point, namely the origin of the coordinate system of Fig. 2.2. If we put $\vec{n} = \vec{e}_1$ we have $n_1 = 1, n_2 = n_3 = 0$, and (2.21) leads to

$$\vec{t}^{(\vec{e}_1)} = -\vec{t}^{(-\vec{e}_1)},$$

(2.22)

or more generally

$$\vec{t}^{(\vec{n})} = -\vec{t}^{(-\vec{n})}.$$

(2.23)

This means that the stress vectors on the opposite sides of the same surface elements have the same magnitudes and opposite signs. Then instead of (2.21) we write

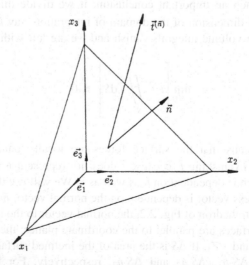

Fig. 2.2 The relationship between the normal vector and the stress vector

$$\vec{t}^{(\vec{n})} = \vec{t}^{(\vec{e}_1)} n_1 + \vec{t}^{(\vec{e}_2)} n_2 + \vec{t}^{(\vec{e}_3)} n_3. \tag{2.24}$$

Therefore, the stress vector is a linear function of the normal vector. The stress vector belonging to the surface with normal vector $\vec{e}_1$ can now be resolved into its components

$$\vec{t}^{(\vec{e}_1)} = \tau_{11}\vec{e}_1 + \tau_{12}\vec{e}_2 + \tau_{13}\vec{e}_3, \tag{2.25}$$

and we designate the first index as giving the direction of the normal vector, and the second as fixing the direction of the component. Similarly we can resolve the stress vectors of the other coordinate planes and of the inclined surfaces into their components, and insert them into Eq. (2.24). From the resulting equation

$$\vec{t}^{(\vec{n})} = \begin{matrix} \vec{t}_1\vec{e}_1 + \\ \vec{t}_2\vec{e}_2 + = \\ \vec{t}_3\vec{e}_3 \end{matrix} \begin{matrix} n_1(\tau_{11}\vec{e}_1 + \tau_{12}\vec{e}_2 + \tau_{13}\vec{e}_3) + \\ n_2(\tau_{21}\vec{e}_1 + \tau_{22}\vec{e}_2 + \tau_{23}\vec{e}_3) + \\ n_3(\tau_{31}\vec{e}_1 + \tau_{32}\vec{e}_2 + \tau_{33}\vec{e}_3) \end{matrix} \tag{2.26}$$

we can read off the component equation in the first direction

$$t_1 = \tau_{11}n_1 + \tau_{21}n_2 + \tau_{31}n_3, \tag{2.27}$$

where the superscript $\vec{n}$ has been and will continue to be dropped. The result for the ith direction is

$$t_i = \tau_{1i}n_1 + \tau_{2i}n_2 + \tau_{3i}n_3. \tag{2.28}$$

We can shorten (2.28) using *Einstein's summation convention*

$$t_i(\vec{x}, \vec{n}, t) = \tau_{ji}(\vec{x}, t)n_j \quad (i, j = 1, 2, 3); \tag{2.29a}$$

here we have indicated the dependency of $\vec{t}$ on $\vec{x}$, $\vec{n}$ and t explicitly. The nine quantities necessary to specify the stress vector on a surface element with some normal vector $\vec{n}$ at $\vec{x}$ form a second order tensor. The physical meaning of the general component τ_{ji} is given by (2.26): τ_{ji} is the magnitude of the ith component of the stress vector at the element of the coordinate plane with normal vector in the jth direction.

Although t_i is not a vector field in the usual sense, since it is linearly dependent on the vector $\vec{n}$ at $\vec{x}$, $\tau_{ji}(\vec{x}, t)$ is a field, or to be more precise, a tensor field. Expressed mathematically, (2.29a) is a linear homogeneous mapping of the normal vector $\vec{n}$ onto the vector $\vec{t}$. Symbolically we can write (2.29a) as

$$\vec{t} = \vec{n} \cdot \mathbf{T}, \tag{2.29b}$$

where the matrix representation of the *stress tensor* $\mathbf{T}$ is given below

$$[\mathbf{T}] = \begin{bmatrix} \tau_{11} & \tau_{12} & \tau_{13} \\ \tau_{21} & \tau_{22} & \tau_{23} \\ \tau_{31} & \tau_{32} & \tau_{33} \end{bmatrix}. \tag{2.30}$$

The main diagonal elements are the normal stresses and the nondiagonal elements are the shearing stresses. We will show later that the stress tensor is a symmetric tensor of the second order and it is therefore diagonalizable. At every $\vec{x}$ we can specify three mutually orthogonal surface elements on which only normal stresses act. These surface elements are parallel to the coordinate surfaces of the principal axis system. Just as we did in connection with the rate of deformation tensor, we find the normal vectors to these surface elements by looking for vectors which are parallel to the stress vectors, that is, those which satisfy the equation

$$t_i = \tau_{ji}n_j = \sigma\, n_i = \sigma\, n_j\delta_{ji}. \tag{2.31}$$

The characteristic equation of this homogeneous system of equations is

$$-\sigma^3 + I_{1\tau}\sigma^2 - I_{2\tau}\sigma + I_{3\tau} = 0, \tag{2.32}$$

where the invariants can be calculated in the same way as in (1.58). If this characteristic equation has three distinct roots (eigenvalues), there is only one principal axis system. In a fluid at rest, all the friction stresses vanish, by definition, and all three eigenvalues are equal: $\sigma^{(1)} = \sigma^{(2)} = \sigma^{(3)} = -p$. Now every orthogonal system of axes is a principal axis system and (2.31) holds for any $\vec{n}$. The state of stress is now *spherical*, i.e.,

$$\tau_{ji} = -p\delta_{ji} \tag{2.33}$$

and is called *hydrostatic*. The stress vector is, from (2.31)

$$t_i = \tau_{ji}n_j = -p\delta_{ji}n_j = -p\, n_i, \tag{2.34a}$$

or, written symbolically

$$\vec{t} = -p\,\vec{n}. \tag{2.34b}$$

The magnitude of this stress vector is the pressure p, which is a scalar quantity, independent of $\vec{n}$. Occasionally, although it is always possible, an arbitrary state of stress is decomposed as

$$\tau_{ij} = -p\delta_{ij} + P_{ij}, \tag{2.35}$$

and P_{ij} is called the friction stress tensor. It has the same principal axes as the tensor τ_{ij}. The mean normal stress $\bar{p}$ is defined by

$$\bar{p} = \frac{1}{3}\tau_{ii}, \tag{2.36}$$

which in general is not equal to the negative pressure. However if this is the case, then P_{ij} is a deviator.

If we put the expression (2.29a, 2.29b) for the stress vector into the momentum law (2.18), and transform the surface integral into a volume integral using Gauss' theorem, we get

$$\iiint\limits_{(V)} \left(\varrho \frac{Du_i}{Dt} - \varrho k_i - \frac{\partial \tau_{ji}}{\partial x_j} \right) dV = 0. \tag{2.37}$$

Because of the assumed continuity of the integrand and the arbitrary domain of integration (V), (2.37) is equivalent to the *differential form of the balance of momentum*

$$\varrho \frac{Du_i}{Dt} = \varrho k_i + \frac{\partial \tau_{ji}}{\partial x_j}, \tag{2.38a}$$

or written symbolically

$$\varrho \frac{D\vec{u}}{Dt} = \varrho \vec{k} + \nabla \cdot \mathbf{T}. \tag{2.38b}$$

This relationship is known as *Cauchy's first law of motion*. We can reach another form of it when we transform the left-hand side of (2.17) using Reynolds' transport theorem (1.93), and then conclude that

$$\frac{\partial}{\partial t}(\varrho u_i) + \frac{\partial}{\partial x_j}(\varrho u_i u_j) = \varrho k_i + \frac{\partial}{\partial x_j}(\tau_{ji}). \tag{2.39}$$

Cauchy's law of motion holds for every continuum, so it holds for every fluid, whatever its particular material properties are. It is the starting point for the calculation of fluid mechanical problems. Using the constitutive equation, that is, the relationship between the stress tensor and the motion (for example, the rate of deformation tensor), Cauchy's equation of motion is changed to a specific equation of motion for the material under observation.

If we are able to write the integrals as surface integrals, the *integral form of the balance of momentum* attains a considerable importance in technical applications. In order to do this, we first transform the balance of momentum (2.17) with Reynolds' transport theorem in the form of (1.96), and we obtain

$$\iiint\limits_{(V)} \frac{\partial(\varrho\,\vec{u})}{\partial t}\mathrm{d}V + \iint\limits_{(S)} \varrho\,\vec{u}(\vec{u}\cdot\vec{n})\mathrm{d}S = \iiint\limits_{(V)} \varrho\,\vec{k}\,\mathrm{d}V + \iint\limits_{(S)} \vec{t}\,\mathrm{d}S. \qquad (2.40)$$

The first integral on the left-hand side cannot be transformed into a surface integral. Therefore the balance of momentum in its integral form attains the afore mentioned importance only if this integral vanishes. This is the case in steady flows, or in unsteady flows whose time averaged values are steady, as happens in *turbulent* steady flows. (In steady turbulent flows, the time averaged momentum flux, which is different from the momentum flux formed with the average velocity, must be used in (2.40). We refer in this connection to Chap. 7.)

The first integral on the right-hand side can be written as a surface integral when the volume body force can be calculated as the gradient of a scalar function, that is, when the volume body force has a potential. We will write the *potential of the volume body force* as $\Omega\,(\vec{f} = \varrho\,\vec{k} = -\nabla\Omega)$, and the *potential of the mass body force* as $\psi\,(\vec{k} = -\nabla\psi)$. (To illustrate this, think of the most important potential: the gravitational potential $(\Omega = -\varrho\,g_i x_i, \psi = -g_i x_i)$. Analogous to our remarks about the velocity potential, $\nabla \times (\varrho\,\vec{k}) = 0$ is a necessary and sufficient condition for the existence of the potential of the volume body force. The most important case here is the one where ρ is constant and the mass body force $\vec{k}$ has a potential. Then the volume integral can be written as a surface integral

$$\iiint\limits_{(V)} \varrho\,\vec{k}\,\mathrm{d}V = -\iiint\limits_{(V)} \nabla\Omega\,\mathrm{d}V = -\iint\limits_{(S)} \Omega\,\vec{n}\,\mathrm{d}S, \qquad (2.41)$$

and the balance of momentum (2.40) now reads

$$\iint\limits_{(S)} \varrho\,\vec{u}(\vec{u}\cdot\vec{n})\,\mathrm{d}S = -\iint\limits_{(S)} \Omega\,\vec{n}\,\mathrm{d}S + \iint\limits_{(S)} \vec{t}\,\mathrm{d}S. \qquad (2.42)$$

We can get insight into the meaning of the balance of momentum when we consider that by knowing the momentum flux and the potential Ω we know the force on the surface of the control volume. Often we will only want to know the force which comes from the momentum flux. Then we take the balance of momentum in its most frequently used form

$$\iint\limits_{(S)} \varrho\, \vec{u}(\vec{u} \cdot \vec{n})\, \mathrm{d}S = \iint\limits_{(S)} \vec{t}\, \mathrm{d}S. \tag{2.43}$$

Conversely, the momentum flux is known from (2.43) if the force has been given. The often unknown, and sometimes incalculable, events right inside the control volume do not appear in the balance of momentum (2.43), and only the quantities on the surface are of importance. Since we can choose the control volume whatever way we want, we usually fix the surface so that the integrals are as easy as possible to evaluate. Often we can fix the surface so that the stress vector takes the same form as that in a fluid at rest, that is, $\vec{t} = -p\,\vec{n}$. Then it is possible to draw conclusions from (2.43) without having to refer to a particular constitutive law.

2.3   Balance of Angular Momentum

As the second general axiom of classical mechanics we shall discuss the angular momentum balance. This is independent of the balance of linear momentum. In an inertial frame, the rate of change of the angular momentum is equal to the moment of the external forces acting on the body

$$\frac{\mathrm{D}}{\mathrm{D}t}(\vec{L}) = \vec{M}. \tag{2.44}$$

We calculate the angular momentum $\vec{L}$ as the integral over the region occupied by the fluid body

$$\vec{L} = \iiint\limits_{(V(t))} \vec{x} \times (\varrho\, \vec{u})\, \mathrm{d}V. \tag{2.45}$$

The angular momentum in (2.45) is taken about the origin such that the position vector is $\vec{x}$, and so we must use the same reference point to calculate the moment of the applied forces

$$\vec{M} = \iiint\limits_{(V(t))} \vec{x} \times \left(\varrho\vec{k}\right)\, \mathrm{d}V + \iint\limits_{(S(t))} \vec{x} \times \vec{t}\, \mathrm{d}S, \tag{2.46}$$

recalling, however, that the choice of reference point is up to us. Therefore the law of angular momentum takes the form

$$\frac{D}{Dt} \iiint\limits_{(V(t))} \vec{x} \times (\varrho \vec{u}) \, dV = \iiint\limits_{(V(t))} \vec{x} \times \left(\varrho \vec{k}\right) dV + \iint\limits_{(S)} \vec{x} \times \vec{t} \, dS \qquad (2.47)$$

where, for the same reasons as before, we have replaced the time varying domain of integration on the right with a fixed domain. Now we wish to show that the differential form of the balance of angular momentum implies the symmetry of the stress tensor. We introduce the expression (2.29a, 2.29b) into the surface integral, which can then be written as a volume integral. In index notation this becomes

$$\iint\limits_{(S)} \epsilon_{ijk} x_j \, \tau_{lk} n_l \, dS = \iiint\limits_{(V)} \epsilon_{ijk} \frac{\partial}{\partial x_l} (x_j \tau_{lk}) dV, \qquad (2.48)$$

and after applying (1.88) to the left-hand side of (2.47) we get first

$$\iiint\limits_{(V)} \epsilon_{ijk} \left(\varrho \frac{D}{Dt} (x_j u_k) - \frac{\partial}{\partial x_l} (x_j \tau_{lk}) - x_j \varrho k_k \right) dV = 0, \qquad (2.49)$$

and after differentiation and combining terms

$$\iiint\limits_{(V)} \left[\epsilon_{ijk} x_j \left(\varrho \frac{Du_k}{Dt} - \frac{\partial \tau_{lk}}{\partial x_l} - \varrho \, k_k \right) + \varrho \, \epsilon_{ijk} u_j \, u_k - \epsilon_{ijk} \tau_{jk} \right] dV = 0. \qquad (2.50)$$

If the balance of momentum (2.38a, 2.38b) is satisfied, the expression in the middle brackets vanishes, thus eliminating position vector, $\vec{x}_j$ which then shows that the balance of angular momentum is indeed invariant with respect to reference point. The outer product $\epsilon_{ijk} u_j u_k$ vanishes also, since $\vec{u}$ is naturally parallel to itself, so the balance of angular momentum is reduced to

$$\iiint\limits_{(V)} \epsilon_{ijk} \tau_{jk} dV = 0. \qquad (2.51)$$

Since the tensor field τ_{jk} is continuous, (2.51) is equivalent to

$$\epsilon_{ijk} \tau_{jk} = 0, \qquad (2.52)$$

proving that τ_{jk} is a symmetric tensor

$$\tau_{jk} = \tau_{kj}. \tag{2.53}$$

Just as in the case of the integral form of the balance of momentum, so the integral form of the balance of angular momentum achieves special significance in technical applications. We are only interested in the moment which is due to the angular momentum flux through the control surface, and we restrict ourselves to steady flows, or unsteady flows which are steady in the manner discussed earlier. Using Reynolds' transport theorem (1.96), (2.47) yields the balance of angular momentum in a form where only surface integrals appear

$$\iint\limits_{(S)} \epsilon_{ijk} x_j u_k \varrho \, u_l n_l \, \mathrm{d}S = \iint\limits_{(S)} \epsilon_{ijk} x_j t_k \, \mathrm{d}S, \tag{2.54a}$$

or symbolically

$$\iint\limits_{(S)} \vec{x} \times \vec{u} \, \varrho \vec{u} \cdot \vec{n} \, \mathrm{d}S = \iint\limits_{(S)} \vec{x} \times \vec{t} \, \mathrm{d}S. \tag{2.54b}$$

There is a particular form of the balance of angular momentum (2.54a, 2.54b) called Euler's turbine equation (see Sect. 2.5) which forms the most important law in the theory of turbomachines.

2.4 Momentum and Angular Momentum in an Accelerating Frame

The balance of momentum and angular momentum that we have discussed so far are only valid in inertial reference frames. An *inertial reference frame* in classical mechanics could be a Cartesian coordinate system whose axes are fixed in space (relative, for example, to the fixed stars), and which uses the average solar day as a unit of time, the basis of all our chronology. All reference frames which move uniformly, i.e., not accelerating in this system, are equivalent and thus are inertial frames.

The above balances do not hold in frames which are accelerating relative to an inertial frame. But the *forces of inertia* which arise from nonuniform motion of the frame are often so small that reference frames can by regarded as being approximately inertial frames. On the other hand, we often have to use reference frames where such forces of inertia cannot be neglected.

To illustrate this we will look at a horizontal table which is rotating with angular velocity Ω. On the table and rotating with it is an observer, who is holding a string at the end of which is a stone, lying a distance R from the fulcrum of the table. The observer experiences a force (the centrifugal force) in the string. Since the stone is at rest in his frame, and therefore the acceleration in his reference frame is zero, the rate of change of momentum must also be zero, and thus, by the balance of momentum (2.9), the force in the string should vanish. The observer then correctly concludes that the balance of momentum does not hold in his reference frame. The rotating table must be treated as an noninertial reference frame. The source of the force in the string is obvious to an observer who is standing beside the rotating table. He sees that the stone is moving on a circular path and so it experiences an acceleration toward the center of the circle, and that according to the balance of momentum, there must be an external force acting on the stone. The acceleration is the *centripetal acceleration*, which is given here by $\Omega^2 R$. The force acting inwards is the *centripetal force* which is exactly the same size as the *centrifugal force* experienced by the rotating observer.

In this example the reference frame of the observer at rest, that is the earth, can be taken as an inertial reference frame. Yet in other cases deviations from what is expected from the balance of momentum appear. This is because the earth is rotating and therefore the balance of momentum strictly does not hold in a reference frame moving with the earth. With respect to a frame fixed relative to the earth we observe, for example, the deflection of a free falling body to the east, or the way that the plane of oscillation of *Foucault's pendulum* rotates. These examples, and many others, are not compatible with the validity of the balance of momentum in the reference frame chosen to be the earth. For most terrestrial events, however, a coordinate system whose origin is at the center of the earth, and whose axes are directed towards the fixed stars, is valid as an inertial reference frame. The easterly deflection mentioned above can then be explained by the fact that the body, in its initial position, has a somewhat higher circumferential speed because of the rotation of the earth than at the impact point nearer the center of the earth. To explain Foucault's pendulum, we notice that, in agreement with (2.9), the pendulum maintains its plane of oscillation relative to the inertial frame. The reference frame attached to the earth rotates about this plane, and an observer in the laboratory experiences a rotation of the plane of oscillation relative to his system with a period of twenty-four hours.

The description of the motion in the inertial reference frame is of little interest for the observer; it is far more important for him to be able to describe the motion in his own reference frame, since this is the only system where he can make measurements. In many applications the use of an accelerating reference frame is unavoidable, for example in meteorology we always want to know the motion of the wind relative to the earth, that is, in a rotating reference frame. It is often useful, and sometimes essential for the solution of technical problems, to use an accelerating frame.

If we want to calculate the motion of a spinning top, the earth is a good enough inertial reference frame. But in this system the tensor of the moments of inertia is time dependent, so it is better to choose a reference frame attached to the top, where, even though this is an accelerating reference frame, this tensor is constant in time. In problems in fluid mechanics it is a good idea to use an accelerating reference frame if the boundary of the flow region is at rest relative to this frame. Consider for example, the flow in the passages of a turbomachine. In a frame fixed to the rotor, and therefore rotating, not only are the blades forming the passages at rest, but the flow itself is more or less steady, making the analytical treatment of the problem much easier.

In what follows we shall formulate the balances for momentum and angular momentum so that they only contain quantities which can be determined in an accelerating system. We shall use the basic assumption that forces and moments are the same for all observers, whether they are in accelerating or inertial reference frames. The rate of change of the momentum or angular momentum, or the rate of change of the velocity is dependent on the reference frame, as is the change of any vector (with an exception, as we shall see).

First we shall turn towards the differential form of the balances of momentum and angular momentum in an accelerating system. Let us look at a system fixed in space (inertial reference frame) and a system accelerating with respect to it, which is carrying out a translation with velocity $\vec{v}(t)$ and a rotation with angular velocity $\vec{\Omega}(t)$ (Fig. 2.3). We shall denote the rate of change of the position vector $\vec{x}$ of a material particle in the *moving reference frame* with

$$\left[\frac{D\vec{x}}{Dt}\right]_A = \vec{w} \tag{2.55}$$

and we shall call $\vec{w}$ the *relative velocity*.

In the inertial reference frame the position vector of the particle under observation is $\vec{x} + \vec{r}$ and its rate of change is called the *absolute velocity*

$$\left[\frac{D}{Dt}(\vec{x} + \vec{r})\right]_I = \vec{c}. \tag{2.56}$$

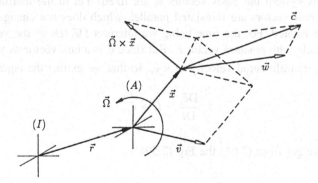

Fig. 2.3 Moving reference frame

Following the usual notation used in turbomachinery we shall denote the absolute velocity with $\vec{c}$. The absolute velocity results from the vector sum of the relative velocity $\vec{w}$, the velocity of the origin of the moving frame

$$\vec{v} = \left[\frac{D\vec{r}}{Dt}\right]_I \tag{2.57}$$

and the *circumferential velocity* $\vec{\Omega} \times \vec{x}$ arising from the rotation of the moving frame at the position $\vec{x}$, to give

$$\vec{c} = \vec{w} + \vec{\Omega} \times \vec{x} + \vec{v}. \tag{2.58}$$

From (2.55) to (2.58) we get the basic formula for the rate of change of the vector x in the two reference frames

$$\left[\frac{D\vec{x}}{Dt}\right]_I = \left[\frac{D\vec{x}}{Dt}\right]_A + \vec{\Omega} \times \vec{x}. \tag{2.59}$$

Obviously this formula does not only hold for the vector $\vec{x}$, but it holds in general. Consider the general vector $\vec{b}$ which has relative to the accelerating reference frame the Cartesian resolution

$$\vec{b} = b_1\vec{e}_1 + b_2\vec{e}_2 + b_3\vec{e}_3 = b_i\vec{e}_i. \tag{2.60}$$

Its observed change in the inertial reference frame is

$$\left[\frac{D\vec{b}}{Dt}\right]_I = \frac{Db_i}{Dt}\vec{e}_i + b_i\frac{D\vec{e}_i}{Dt}. \tag{2.61}$$

The first three terms represent the change of the vector $\vec{b}$ in the moving reference frame. In this system the basis vectors $\vec{e}_i$ are fixed. Yet in the inertial reference frame these unit vectors are translated parallel, which does not change them, but they are also rotated. For the time being we interpret $D\vec{e}_i/Dt$ as the velocity of a material particle with position vector $\vec{e}_i$. But since $\vec{e}_i$ is a unit vector its velocity can only be the circumferential velocity $\vec{\Omega} \times \vec{e}_i$, so that we extract the equation

$$\frac{D\vec{e}_i}{Dt} = \vec{\Omega} \times \vec{e}_i. \tag{2.62}$$

Using this we get from (2.61) the Eq. (2.59)

$$\left[\frac{D\vec{b}}{Dt}\right]_I = \left[\frac{D\vec{b}}{Dt}\right]_A + \vec{\Omega} \times \vec{b}. \tag{2.63}$$

If $\vec{b} = \vec{\Omega}$ the changes in the inertial reference frame and in the frame moving relative to it are equal

$$\left[\frac{D\vec{\Omega}}{Dt}\right]_I = \left[\frac{D\vec{\Omega}}{Dt}\right]_A = \frac{d\vec{\Omega}}{dt}. \tag{2.64}$$

This obviously holds only for the angular velocity $\vec{\Omega}$ or for vectors which are always parallel to $\vec{\Omega}$.

We will need the rate of change of the absolute velocity $[D\vec{c}/Dt]_I$ in Cauchy's Eq. (2.38a, 2.38b). As we have already noted, the right-hand side is frame invariant. If we use (2.58) we are led to the equation

$$\left[\frac{D\vec{c}}{Dt}\right]_I = \left[\frac{D\vec{w}}{Dt}\right]_I + \frac{\left[D\left(\vec{\Omega} \times \vec{x}\right)\right]}{Dt} + \left[\frac{D\vec{v}}{Dt}\right]_I, \tag{2.65}$$

to which we apply (2.63) and (2.64) to get

$$\left[\frac{D\vec{c}}{Dt}\right]_I = \left[\frac{D\vec{w}}{Dt}\right]_A + \vec{\Omega} \times \vec{w} + \vec{\Omega} \times \left(\left[\frac{D\vec{x}}{Dt}\right]_A + \vec{\Omega} \times \vec{x}\right) + \left[\frac{D\vec{\Omega}}{Dt}\right]_A \times \vec{x} + \left[\frac{D\vec{v}}{Dt}\right]_I. \tag{2.66}$$

If we write $(D\vec{v}/Dt)_I = \vec{a}$ for the translational acceleration of the frame and replace $(D\vec{x}/Dt)_B$ by $\vec{w}$ using (2.55), the acceleration in the inertial reference frame can be expressed in quantities of the accelerating frame

$$\left[\frac{D\vec{c}}{Dt}\right]_I = \left[\frac{D\vec{w}}{Dt}\right]_A + 2\vec{\Omega} \times \vec{w} + \vec{\Omega} \times \left(\vec{\Omega} \times \vec{x}\right) + \frac{d\vec{\Omega}}{dt} \times \vec{x} + \vec{a}. \tag{2.67}$$

Only the acceleration as seen from the inertial frame can enter Cauchy's equation, since it is only valid in this frame. But by using (2.67) this acceleration can be expressed in quantities seen from the accelerating system, so that we finally reach the equation

$$\varrho\left[\frac{D\vec{w}}{Dt}\right]_A = \varrho\vec{k} + \nabla\cdot\mathbf{T} - \left(\varrho\vec{a} + 2\varrho\,\vec{\Omega}\times\vec{w} + \varrho\,\vec{\Omega}\times\left(\vec{\Omega}\times\vec{x}\right) + \varrho\frac{d\vec{\Omega}}{dt}\times\vec{x}\right). \quad (2.68)$$

(Note here that (2.68) is a vector equation where $\vec{k}$ and $\nabla\cdot\mathbf{T}$ have meanings independent of frame, i.e., they are the same arrows in all frames. Where written as a matrix equation or in index notation the components must transform into the moving coordinate system, using the relationships in Appendix A.) Apart from the terms in the curved brackets, Eq. (2.68) has the same form as Cauchy's equation in the inertial reference frame. In the moving reference frame, these terms act as additional volume forces, which are added to the external forces. They are pure inertial forces which stem from the motion of the system relative to the inertial reference frame, and are therefore only "apparent" external forces hence their name *apparent* or *fictitious forces*.

The term $-\varrho\vec{a}$ is the apparent force due to the translational acceleration (per unit volume) and it vanishes when the origin of the relative system is at rest or is moving with constant velocity. The term $-2\varrho\,\vec{\Omega}\times\vec{w}$ is the *Coriolis force*, and it vanishes when the material point is at rest in the moving reference frame. The centrifugal force is represented by the term $-\varrho\,\vec{\Omega}\times(\vec{\Omega}\times\vec{w})$, and is also present when the material point is at rest in the moving reference frame. The fourth expression has no special name.

Equation (2.68) furnishes the differential form of the balance of momentum in a moving reference frame. If this law is satisfied, no rate of change of velocity appears in the differential form of the balance of angular momentum (cf. (2.50)), and this law remains valid in all reference frames, something that is expressed by the symmetry of the stress tensor in all reference frames. Thus the apparent forces appear only in the differential form of the balance of momentum and not in that of angular momentum.

The apparent forces that arise from the rotation of the earth can only influence events if the spatial extent of the motion under consideration is the order of the earth's radius, or if its duration is the order of hours. That means that their influence is barely noticed in rapid flow events of small extent, and can, in general, be ignored. However their influence is noticeable in the motion of the sea, and it is even larger in atmospheric flows. The earth rotates about 2π in one sidereal day (which with 861,64 s is somewhat shorter than a solar day of 86,400 s), so it moves with an angular velocity of $\Omega = 2\pi/86164 \approx 7.29\cdot10^{-5}\mathrm{s}^{-1}$. Since the angular velocity is constant, the last term of (2.68) vanishes. In addition, the effect of the rotation about the sun can be ignored, so that only the Coriolis and centrifugal forces act as apparent forces. The centrifugal force at the equator amounts to 0.3% of the earth's attraction. In measurements it is hardly possible to separate the two forces and it is actually the resultant of both forces that we call the gravity force $\vec{g}$. The vector $\vec{g}$ is normal to the geoid, and is not directed exactly at the center of the earth.

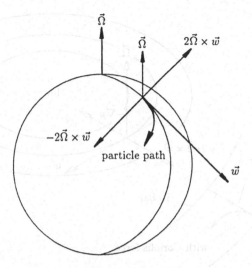

Fig. 2.4 The influence of the Coriolis force on the particle path

Now let us consider an air particle which moves in a north-south direction (Fig. 2.4). In the northern hemisphere the vector $\vec{\Omega}$ points out of the earth. The Coriolis force $-2\varrho\,\vec{\Omega} \times \vec{w}$ is perpendicular to $\vec{\Omega}$ and to $\vec{w}$, and forces the particle in the direction of its motion to the right. The same holds for a particle which is moving in a south-north direction: it is forced to the right seen in the direction of motion. As a rule, irrespective of the velocity direction, particles in the northern hemisphere are forced to the right and those in the southern hemisphere are forced to the left. Without allowing for the Coriolis force in Cauchy's equation, we would conclude that the air flows in the direction of the pressure gradient, and therefore normal to the isobars. If we ignore the friction, we get from (2.35)

$$\tau_{ij} = -p\,\delta_{ij}. \tag{2.69}$$

If in addition, we only consider motion parallel to the geoid, so that the force of gravity $\varrho\,\vec{g}$ has no component in the direction of motion, (2.68) reads in index notation

$$\varrho\frac{Dw_i}{Dt} = \frac{\partial(-p\,\delta_{ij})}{\partial x_j} = -\frac{\partial p}{\partial x_i}, \tag{2.70}$$

which means that the air is only accelerated in the direction of the pressure gradient, and so it flows radially into a low. Yet because of the Coriolis force, the air in the northern hemisphere is turned to the right, and it flows anticlockwise, almost tangential to the isobars, into the low (Fig. 2.5). Since the acceleration in the relative system is small compared to the Coriolis acceleration, the pressure gradient and the Coriolis force almost balance (*Buys-Ballot's rule*). A consequence of the Coriolis

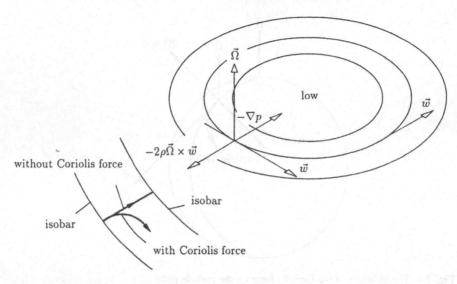

Fig. 2.5 Low in the northern hemisphere

force is the slightly higher water level in the right bank of rivers in the northern hemisphere and a tendency to deviate to the right. This phenomenon, called *Baer's law*, can also be observed in lakes into and out of which rivers flow. Definite erosion can even be seen on the right bank of some rivers. However, other influences, like the mutable resistance of the river bed, are morphologically more important.

Although on the earth the Coriolis force is very small, these examples show that is cannot always be ignored. Even with velocities of $u = 1000$ m/s, typical of artillery shells, the maximum Coriolis acceleration is only $2\Omega u \approx 2 \cdot (7.29 \cdot 10^{-5}) \cdot (1000\,\mathrm{ms}^{-2}) \approx 0.015\,g$. In spite of this its influence on the trajectory is quite noticeable.

In technical applications, the balances of momentum and of angular momentum in their integral form must often be used in reference frames attached to rotating machine parts. As already noted, the flow is then almost always steady. A starting point is the balance of momentum (2.17). The velocity appearing here is of course the absolute velocity $\vec{c}$

$$\frac{\mathrm{D}}{\mathrm{D}t}\left[\iiint\limits_{(V(t))} \varrho\, \vec{c}\, \mathrm{d}V\right]_I = \iiint\limits_{(V)} \varrho\, \vec{k}\, \mathrm{d}V + \iint\limits_{(S)} \vec{t}\, \mathrm{d}S. \tag{2.71}$$

We will apply the basic formula (2.63) to the rate of change of momentum in order to express this in quantities relative to the rotating reference frame. This leads to

$$\frac{D}{Dt}\left[\iiint\limits_{(V(t))} \varrho\,\vec{c}\,dV\right]_I + \vec{\Omega}\times\iiint\limits_{(V)} \varrho\,\vec{c}\,dV = \iiint\limits_{(V)} \varrho\,\vec{k}\,dV + \iint\limits_{(S)} \vec{t}\,dS, \qquad (2.72)$$

where, in the second integral on the left-hand side, we have replaced a time dependent domain with a fixed integration domain. Immediately we can apply Reynolds' transport theorem to the first term, since this theorem is purely kinematical, and therefore holds in all reference frames. Equation (2.72) now takes the form

$$\frac{\partial}{\partial t}\left[\iiint\limits_{(V(t))} \varrho\,\vec{c}\,dV\right]_A + \iint\limits_{(S)} \varrho\,\vec{c}(\vec{w}\cdot\vec{n})\,dS + \vec{\Omega}\times\iiint\limits_{(V)} \varrho\,\vec{c}\,dV = \iiint\limits_{(V)} \varrho\,\vec{k}\,dV + \iint\limits_{(S)} \vec{t}\,dS.$$
$$(2.73)$$

In this equation, both the absolute velocity $\vec{c}$ and the relative velocity $\vec{w}$ enter. The latter appears because the momentum in the relative system is transported through the surface of the control volume with the relative velocity $\vec{w}$. As mentioned, in applications the flow in the relative system is often steady, and the rotational velocity $\vec{\Omega}$ is constant, so that in many technically important cases the first term on the left-hand side drops out. If we restrict ourselves to the statement of the balance of momentum without volume body forces, we get from (2.73)

$$\iint\limits_{(S)} \varrho\,\vec{c}(\vec{w}\cdot\vec{n})\,dS + \iiint\limits_{(V(t))} \varrho\vec{\Omega}\times\vec{c}\,dV = \iint\limits_{(S)} \vec{t}\,dS, \qquad (2.74)$$

where we have brought the constant vector $\vec{\Omega}$ into the volume integral. For incompressible flow, the volume integral can be transformed into a surface integral. We shall not do this because in applications we are often only interested in the component of the momentum in the $\vec{\Omega}$ direction. If we take the inner product with the unit vector $\vec{e}_{\vec{\Omega}} = \vec{\Omega}/|\vec{\Omega}|$ the volume integral vanishes, since $\vec{\Omega}\times\vec{c}$ is always perpendicular to $\vec{e}_{\Omega}$. Therefore the component equation in the $\vec{\Omega}$ direction reads

$$\iint\limits_{(S)} \varrho\,\vec{e}_{\Omega}\cdot\vec{c}(\vec{w}\cdot\vec{n})\,dS = \iint\limits_{(S)} \vec{e}_{\Omega}\cdot\vec{t}\,dS. \qquad (2.75)$$

We note the appearance of both the relative and the absolute velocities. In applications this does not cause confusion and we refrain from replacing $\vec{c}$ using (2.58).

Now we shall apply the same considerations to the balance of angular momentum: using the formula (2.63) the rate of change in the inertial reference frame is expressed through the change in the relative system, and then Reynolds' transport theorem is applied to this. Let the flow in the relative system be steady. Neglecting the moment of the volume forces, the integral form of the balance of angular momentum then becomes

$$\iint\limits_{(S)} \varrho\,(\vec{x} \times \vec{c})(\vec{w} \cdot \vec{n})\,\mathrm{d}S + \vec{\Omega} \times \iiint\limits_{(V)} \varrho\,(\vec{x} \times \vec{c})\,\mathrm{d}V = \iint\limits_{(S)} \vec{x} \times \vec{t}\,\mathrm{d}S. \qquad (2.76)$$

The middle term contains a volume integral, but it is zero if the angular momentum vector $\vec{L}$ has the same direction as $\vec{\Omega}$. Turbomachines are designed so that this is the case. Only in very extreme operating conditions, near shutoff, is it possible that the flow is no longer rotationally symmetric to the axis of rotation. Then the angular momentum $\vec{L}$ is no longer in the direction of $\vec{\Omega}$. This corresponds to a dynamic imbalance of the rotor. If we consider only the component equation of angular momentum in the direction of $\vec{\Omega}$ (from which the torque on the rotor can be calculated) we always get an equation where the volume integral no longer appears

$$\vec{e}_\Omega \cdot \iint\limits_{(S)} \varrho\,(\vec{x} \times \vec{c})(\vec{w} \cdot \vec{n})\,\mathrm{d}S = \vec{e}_\Omega \cdot \iint\limits_{(S)} \vec{x} \times \vec{t}\,\mathrm{d}S. \qquad (2.77)$$

Here too both the absolute velocity $\vec{c}$ and the relative velocity $\vec{w}$ appear.

2.5 Applications to Turbomachines

Typical applications of the balances of momentum and of angular momentum can be found in the theory of turbomachines. The essential element present in all turbomachines is a rotor equipped with blades surrounding it, either in the axial or radial direction.

When the fluid exerts a force on the moving blades, the fluid does work. In this case we can also speak of *turbo force machines* (turbines, wind wheels, etc.). If the moving blades exert a force on the fluid, and thus do work on it, increasing its energy, we speak of *turbo work machines* (fans, compressors, pumps, propellers).

Often the rotor has an outer casing, called stator, which itself is lined with blades. Since these blades are fixed, no work is done on them. Their task is to direct the flow either towards or away from the *moving blades* attached to the rotor. These blades are called *guide blades* or *guide vanes*. A row of fixed blades together with a row of moving blades is called a *stage*. A turbomachine can be constructed with one or more of these stages. If the cylindrical surface of Fig. 2.6 at radius r through the stage is cut and straightened, the contours of the blade sections originally on the cylindrical surface form two *straight cascades*. The set up shown consists of a turbine stage where the fixed cascade is placed before the moving cascade seen in the direction of the flow.

Obviously the cascades are used to turn the flow. If the turning is such that the magnitude of the velocity is not changed, the cascade is a pure *turning* or *constant pressure cascade*, since then no change of pressure occurs through the cascade (only in the case of frictionless flow). In general the magnitude of the velocity changes with the turning and therefore also the pressure. If the magnitude of the velocity is increased we have an *acceleration cascade*, typically found in turbines, and if it is decreased we have a *deceleration cascade*, typically found in compressors. We shall consider the cascade to be a strictly periodic ordering of blades, that is, an infinitely long row of blades with exactly the same *spacing s* between blades along the cascade. Because of this the flow is also strictly periodic.

In the following the object is to calculate the force acting on the cascade or on a single blade for a given flow deflection and pressure drop through the cascade. We shall assume that the flow is a plane two-dimensional flow, that is, that the same flow is found in all sections parallel to the plane of Fig. 2.6. In reality the flow

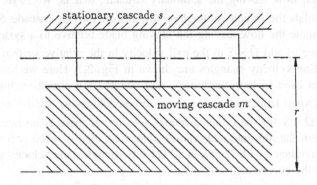

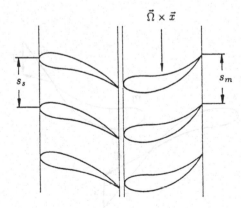

Fig. 2.6 Axial turbine stage

passages between blades become wider in the radial direction, so that the assumption of plane cascade flow represents the limit $r \to \infty$ with constant blade height. For the moving blades this also means that for a given constant circumferential velocity of $|\vec{\Omega} \times \vec{x}| = \Omega r$, the angular velocity tends to zero as r tends to infinity. Then the centrifugal acceleration $|\vec{\Omega} \times (\vec{\Omega} \times \vec{x})| = \Omega^2 r$ and the Coriolis acceleration $|2\vec{\Omega} \times \vec{w}|$ both tend to zero with Ω.

The assumption of a plane two-dimensional flow therefore means that the moving cascade is an inertial reference frame. This is also evident from the fact that in this approximation every point of the moving cascade moves with the same constant velocity. The balance of momentum in an inertial reference frame can therefore be applied both to the stationary and the moving cascade.

In dealing with the moving cascade, we observe that the approach flow to the moving cascade is not equal to the flow leaving the stationary cascade. If the moving cascade in Fig. 2.6 has a circumferential velocity of $\vec{\Omega} \times \vec{x}$ downwards, an observer in the reference frame of the moving cascade experiences an air-stream of the same magnitude blowing upwards $-\vec{\Omega} \times \vec{x}$. This velocity is to be added to the velocity of the flow leaving the stationary cascade, that is, we have to subtract $\vec{\Omega} \times \vec{x}$ to calculate the velocity of the flow towards the moving cascade. Similarly in order to calculate the flow leaving the moving blade relative to a system fixed in space, we have to add $\vec{\Omega} \times \vec{x}$ to the exit velocity in the relative system.

The resulting velocity triangles are shown in Fig. 2.7. Here we have used the notation often used in turbomachinery, and denoted the circumferential velocity $\vec{\Omega} \times \vec{x}$ by $\vec{u}$. (Apart from this section about turbomachines we shall continue to use the notation $\vec{\Omega} \times \vec{x}$ for the circumferential velocity. If there is no need to differentiate between the absolute and relative velocities, then u is the general velocity vector.) In accordance with (2.58), in all velocity triangles, the velocity vectors $\vec{c}, \vec{w}$ and $\vec{u}$ satisfy the equation

$$\vec{c} = \vec{w} + \vec{u}. \tag{2.78}$$

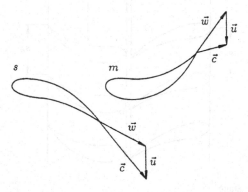

Fig. 2.7 Velocity triangle

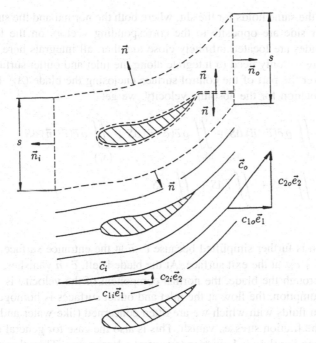

Fig. 2.8 Control volume for applying the momentum balance

This allows the construction of the velocity triangle without having to consciously change reference frames.

Now we shall consider a single cascade at rest Fig. 2.8. The equations which follow also hold for a rotating cascade in an axial turbomachine, since by the earlier arguments, every straight cascade of blades represents an inertial system. (The absolute velocity $\vec{c}$ is then only to be replaced by the relative velocity $\vec{w}$ measured in the moving reference frame.) At a large distance from the cascade the inlet velocity $\vec{c}_i$ and the outlet velocity $\vec{c}_o$ are constant in space, that is, homogeneous. Homogeneous conditions, especially behind the cascade, are strictly only true at infinite distances away from the cascade, although for practical purposes the flow evens out only a short distance away. To apply the momentum balance in the form (2.43) we use the control volume shown in Fig. 2.8. Inlet and outlet surface areas (per unit length of the cascade) A_i and A_o correspond to the spacing s. As the upper and lower boundaries we choose streamlines. The blade profile is excluded from the control volume by using a narrow, but otherwise arbitrary, slit. Instead of using streamlines as the upper and lower boundaries to the control volume we could have used any other lines such that the upper boundary is mapped onto the lower by a translation through the spacing s. Since the flow is periodic, we can be sure that at corresponding points on the upper and lower boundaries exactly the same flow conditions prevail. Since the normal vectors at these corresponding points are directed opposite to each other, and the same holds for the stress vectors (cf. Eq. (2.23)), all integrals along the upper and lower boundaries cancel each other out. Exactly the same holds for the slit, where both the normal and the stress vectors on the upper side are opposite to the corresponding vectors on the lower bank. Since both sides are located infinitely close together, all integrals here also cancel out. Therefore we only need to integrate along the inlet and outlet surfaces (A_i, A_o) as well as over the part of the control surface enclosing the blade (A_f). Putting into (2.43) our notation for the absolute velocity, we get

$$
\iint\limits_{(A_i)} \varrho \, \vec{c}(\vec{c} \cdot \vec{n}) \, \mathrm{d}S + \iint\limits_{(A_o)} \varrho \, \vec{c}(\vec{c} \cdot \vec{n}) \, \mathrm{d}S + \iint\limits_{(A_f)} \varrho \, \vec{c}(\vec{c} \cdot \vec{n}) \, \mathrm{d}S
$$

$$
= \iint\limits_{(A_i)} \vec{t} \, \mathrm{d}S + \iint\limits_{(A_o)} \vec{t} \, \mathrm{d}S + \iint\limits_{(A_f)} \vec{t} \, \mathrm{d}S.
\tag{2.79}
$$

This equation is further simplified because $\vec{c} \cdot \vec{n}$ at the entrance surface is given by $-c_{1i}$ and by $+c_{1o}$ at the exit surface. At the blade itself, $\vec{c} \cdot \vec{n}$ vanishes. Since there is no flow through the blade, the normal component of the velocity is in any case zero. By assumption, the flow at the inlet and outlet surfaces is homogeneous, and in Newtonian fluids with which we are often concerned (like water and gases) this means that the friction stresses vanish. This is also the case for general constitutive relations, when the flow is homogeneous over a larger area. Then the stress vector can be written as $\vec{t} = -p\vec{n}$. Finally, the last integral represents the force which the blade exerts on the flow (or the negative of the force applied to the blade by the flow). If we solve for the force (per unit height of the cascade), noting that the flow properties are constant over A_i and A_o, we get first

$$
\vec{F} = -\vec{c}_i \varrho_i c_{1i} s + \vec{c}_o \varrho_o c_{1o} s + p_i \vec{n}_i s + p_o \vec{n}_o s.
\tag{2.80}
$$

Resolving to get the components in the $\vec{e}_1$ and $\vec{e}_2$ directions, with $\vec{n}_i = -\vec{e}_1, \vec{n}_o = \vec{e}_1$ we extract the equations

$$
\vec{F} \cdot \vec{e}_1 = F_1 = -\varrho_i c_{1i}^2 s + \varrho_o c_{1o}^2 s - p_i s + p_o s,
\tag{2.81}
$$

$$
\vec{F} \cdot \vec{e}_2 = F_2 = -\varrho_i c_{1i} c_{2i} s + \varrho_o c_{1o} c_{2o} s.
\tag{2.82}
$$

The continuity equation for steady flow in integral form (2.8) leads to

$$
\iint\limits_{(A_i)} \varrho \, \vec{c} \cdot \vec{n} \, \mathrm{d}S + \iint\limits_{(A_o)} \varrho \, \vec{c} \cdot \vec{n} \, \mathrm{d}S = 0,
\tag{2.83}
$$

or, using the concept of *mass flux*, to

$$\dot{m} = \iint\limits_{(A_o)} \varrho \vec{c} \cdot \vec{n}\, dS = - \iint\limits_{(A_i)} \varrho \vec{c} \cdot \vec{n}\, dS. \qquad (2.84)$$

The notation $\dot{m}$ used in the literature is not very well chosen: it has nothing to do with the rate of change of the mass, which is of course zero, but with the flux of the mass through a surface, according to the definition in (2.84). An expression for the mass flux per unit height of the cascade follows from this definition

$$\dot{m} = \varrho_i c_{1i} s = \varrho_o c_{1o} s. \qquad (2.85)$$

In incompressible flow, and with the assumed homogeneity of the approach flow, the density is always constant $(\varrho_i = \varrho_o = \varrho)$, and from (2.85), with $\dot{V} = \dot{m}/\varrho$,

$$\dot{V} = c_{1i} s = c_{1o} s. \qquad (2.86)$$

$\dot{V}$ is the *volume flux* (per unit height of the cascade), and this is often used instead of the mass flux in incompressible flow. Finally, we get the expression below for the force components

$$F_1 = \dot{m}(c_{1o} - c_{1i}) + s(p_o - p_i), \qquad (2.87)$$

$$F_2 = \dot{m}(c_{2o} - c_{2i}), \qquad (2.88)$$

where, for our assumed incompressible flow, the first term on the right-hand side of (2.87) drops out.

If the integration path along the blade is omitted in Fig. 2.8, the control surface is again a closed line, which surrounds the blade profile, so that we can form the curve integral

$$\Gamma = \oint \vec{c} \cdot d\vec{x}, \qquad (2.89)$$

which has mathematically positive sense. We have already met this integral in (1.105). Even when this curve is fixed in space, and so is not a material curve, we call this curve integral the *circulation*, and again use the symbol Γ for it. To evaluate this integral, we note that at corresponding points on the upper and lower boundaries in Fig. 2.8, $\vec{c}$ has the same value, while the line element of the curve $d\vec{x}$ has opposite signs at corresponding points. Thus the contribution from the upper and lower boundaries to the curve integral cancels out. The straight sections yield the values $-c_{2i}s$ and $c_{2o}s$, so we get

$$\Gamma = (c_{2o} - c_{2i})s, \tag{2.90}$$

and therefore the following holds

$$F_2 = \varrho_i c_{1i} \Gamma = \varrho_o c_{1o} \Gamma. \tag{2.91}$$

Clearly one wishes to design cascades so that losses are as small as possible. Since losses originate through the friction stresses (ignoring the losses from heat conduction), one tries to build cascades so that they are as close as possible to being theoretically frictionless. Assuming frictionless flow, and to go only a small step further, potential flow, the component F_1 of the force can also be expressed by the circulation. We then arrive at the result that the whole force is proportional to the circulation. We shall not use this assumption here, because here we stress the general validity of the expressions for the momentum balance (2.87 and 2.91). Yet we point to the important fact that if the cascade spacing is given, the action of losses are restricted to the component F_1 of the force.

   As a second example, consider the calculation of the torque about the radial cascade of a single stage radial machine, using the balance of angular momentum in its integral form. Both force and work turbomachines have a similar design to that shown in Fig. 2.9. The flow in radial force turbomachines (Francis turbines, exhaust driven turbines) is predominantly radial and as a rule inward, i.e., towards the axis of rotation, whereas in *work machines* (pumps, compressors) it is always outward. Therefore, in work machines, the stationary cascade is placed behind the moving cascade in the direction of the flow. The sketched radial cascade is the cascade of a work machine. The cascade is fixed, and the reference system is an inertial reference frame, so that the balance of angular momentum can be used in the form (2.54a, 2.54b). The control volume is chosen as shown in Fig. 2.9: it starts at the outlet surface A_o, goes along the side of a narrow slit to a vane, and around the other side and back along the side of the slit to the outlet surface, and then on to the next vane. The outlet surface is connected to the inlet surface via the lateral surfaces of the guide vane ring, and so the control volume is closed. The wetted surfaces (vane and sides of the ring) are denoted as A_w. Because of the reasons given when we applied the balance of momentum earlier, integrating around the sides of the slit gives no contribution, and, replacing $\vec{u}$ with $\vec{c}$ we extract from (2.54a, 2.54b)

$$\iint\limits_{(A_i, A_o, A_w)} \varrho(\vec{x} \times \vec{c})(\vec{c} \cdot \vec{n}) dS = \iint\limits_{(A_i, A_o, A_w)} \vec{x} \times \vec{t} \, dS. \tag{2.92}$$

On the left there is no contribution to the integral from A_w, since there is no flow through the wetted surfaces. At the inlet and outlet surfaces, the velocity is homogeneous, so that the stress vector is given by $\vec{t} = -p\vec{n}$. However, this is not exactly true for radial cascades, because, among other things, the flow area increases with increasing r. The integration over the inlet and outlet surfaces on the

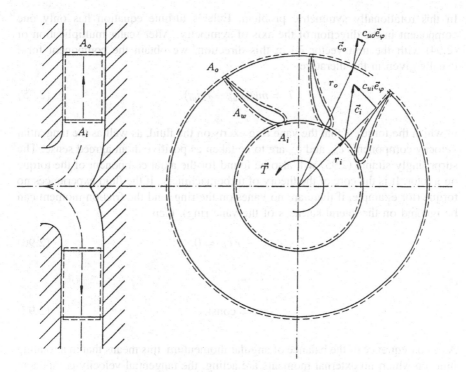

Fig. 2.9 Radial machine with control volume in the guiding cascade flow

right-hand side do not provide any contribution to the moment, since on these surfaces $\vec{n}$ is always parallel to $\vec{x}$. This can be directly understood: the stress vector $-p\vec{n}$ on these surfaces is directed towards the center of the cascade, so that there is no torque about the center. The term remaining on the right-hand side is the torque $\vec{T}$, which the wetted surface exerts on the flow. Therefore $-\vec{T}$ is the torque which the fluid exerts on the cascade. Thus we are led to

$$\iint\limits_{(A_i, A_o)} \varrho(\vec{x} \times \vec{c})(\vec{c} \cdot \vec{n})\mathrm{d}S = \vec{T}, \tag{2.93}$$

and we note that the vector $\vec{x} \times \vec{c}$ is constant over the inlet and outlet surfaces, and thus can be brought in front of the integral. Using the continuity equation in the form (2.84), we obtain the torque in the form of the famous *Euler's turbine equation*

$$\vec{T} = \dot{m}(\vec{x}_o \times \vec{c}_o - \vec{x}_i \times \vec{c}_i). \tag{2.94}$$

In this rotationally symmetric problem, Euler's turbine equation has only one component in the direction of the axis of symmetry. After scalar multiplication of (2.94) with the unit vector $\vec{e}_\Omega$ in this direction, we obtain the component form usually given in the literature

$$T = \dot{m}(r_o c_{uo} - r_i c_{ui}),\tag{2.95}$$

in which the torque T that the vane ring exerts on the fluid, as well as the tangential velocity components c_{uo} and c_{ui} are to be taken as positive in an agreed sense. The surprisingly simple (2.95) will also be found for the axial component of the torque on a rotor. It is the core of the theory of turbomachinery. If the fluid experiences no torque (for example, if there are no vanes on the ring, and the friction moment can be ignored on the lateral surfaces of the vane ring), then

$$r_o c_{uo} - r_i c_{ui} = 0,\tag{2.96}$$

or

$$rc_u = \text{const.}\tag{2.97}$$

As a consequence of the balance of angular momentum, this means that in a rotating fluid on which no external moments are acting, the tangential velocity component falls off as $1/r$.

In order to calculate the torque on the rotor, we use the balance of angular momentum relative to a rotating reference frame. In this system the flow is steady. We assume that at the inlet and outlet surfaces, and only there, the friction stresses can be ignored, for the reasons explained earlier. From (2.77) we obtain the component of the torque in the direction of the axis of rotation, as

$$\iint\limits_{(A_i,A_o)} \varrho\, \vec{e}_\Omega \cdot (\vec{x} \times \vec{c})(\vec{w} \cdot \vec{n}) dS + \iint\limits_{(A_i,A_o)} p\, \vec{e}_\Omega \cdot (\vec{x} \times \vec{n})\, dS = T.\tag{2.98}$$

T is the torque exerted on the fluid by the rotor; $-T$ is the torque exerted on the rotor by the fluid. The inlet and outlet surfaces are surfaces of rotation (Fig. 2.10), so that the vector $\vec{x} \times \vec{n}$ is perpendicular to $\vec{e}_\Omega$, and the pressure integrals, clearly, do not contribute to the torque. To continue we resolve the position vector and the velocity vector into components along the radial, circumferential and axis of rotation directions, thus

$$\vec{x} = r\, \vec{e}_r + x_\Omega\, \vec{e}_\Omega,\tag{2.99}$$

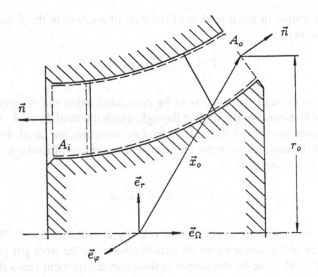

Fig. 2.10 Half axial rotor

$$\vec{c} = c_r\,\vec{e}_r + c_u\,\vec{e}_\varphi + c_\Omega\,\vec{e}_\Omega, \tag{2.100}$$

and so the cross product $\vec{x} \times \vec{c}$ becomes

$$\vec{x} \times \vec{c} = -x_\Omega c_u \vec{e}_r - (rc_\Omega - x_\Omega c_r)\vec{e}_\varphi + rc_u\vec{e}_\Omega, \tag{2.101}$$

from which the following expression for the component in the direction of the axis of rotation results

$$\vec{e}_\Omega \cdot (\vec{x} \times \vec{c}) = rc_u, \tag{2.102}$$

since the unit vectors $\vec{e}_r, \vec{e}_\varphi$ and $\vec{e}_\Omega$ are orthogonal. Therefore (2.98) simplifies to

$$\iint\limits_{(A_i, A_o)} \varrho\, rc_u(\vec{w} \cdot \vec{n})\, \mathrm{d}S = T. \tag{2.103}$$

If rc_u at A_i and A_o are constant, or if their variations are so small that they can be ignored, then, using the continuity equation in the reference frame fixed to the rotor

$$\dot{m} = \iint\limits_{(A_o)} \varrho\, \vec{w} \cdot \vec{n}\, \mathrm{d}S = -\iint\limits_{(A_i)} \varrho\, \vec{w} \cdot \vec{n}\, \mathrm{d}S \tag{2.104}$$

we write the torque in the direction of the axis of rotation in the form of Euler's turbine equation

$$T = \dot{m}(r_o c_{uo} - r_i c_{ui}).$$ (2.105)

The mass flux through the rotor is to be calculated using the component of the relative velocity normal to the surface through which the fluid flows $\vec{w} \cdot \vec{n}$. Often the normal components of the relative and absolute velocities are equal. For example, this occurs if the surfaces as in the above case are surfaces of rotation. The second term on the right-hand side of

$$\vec{c} \cdot \vec{n} = \vec{w} \cdot \vec{n} + \vec{u} \cdot \vec{n}$$ (2.106)

is then zero, because the circumferential velocity is orthogonal to $\vec{n}$. We interpret the component of the torque along the axis of rotation as the work per unit angle of rotation. The work done by the torque is therefore the moment times the angle of rotation, and the power P is this moment times the angular velocity. If we take into account the vectorial character of these quantities, we write the power as

$$P = \vec{T} \cdot \vec{\Omega} = \Omega \dot{m}(r_o c_{uo} - r_i c_{ui}).$$ (2.107)

If the vectors of the torque and of the angular velocity form an acute angle, the power of the rotor is delivered on the fluid and we have a work machine. Finally we calculate the force in the axial direction which is transferred to the fluid from the rotor, or to the rotor from the fluid. This force is usually supported by special thrust bearings. It is desirable to keep this axial force as small as possible. For this reason the sides of the rotor are often fully or partially acted on by the fluid. By properly choosing these wetted areas the axial force can be influenced as desired.

The control volume is then so shaped that these surfaces become components of the control surface. We shall take the control volume down along the rotor sides to some desired radius, and, forming a slit, back up to either the inlet or outlet surface (Fig. 2.11). Then, in an already familiar way the control volume is formed so that the wetted surfaces (blades and casings) are parts of the control surface. Starting from the momentum balance in the accelerating reference frame, we need to integrate the left-hand side of this equation only over the inlet and outlet surfaces, since there is no flow through the wetted surfaces including the wetted side surfaces and the surfaces A_s opposite to these. Assuming that the friction stresses can be ignored on A_i, A_o and A_s, we reach

$$\iint\limits_{(A_i, A_o)} \varrho \, \vec{e}_\Omega \cdot \vec{c}(\vec{w} \cdot \vec{n}) \, dS = - \iint\limits_{(A_i, A_o, A_s)} p \, \vec{e}_\Omega \cdot \vec{n} \, dS + F_a,$$ (2.108)

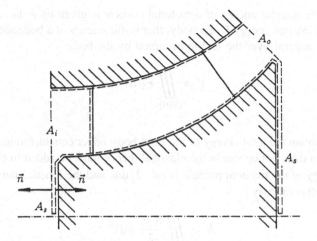

Fig. 2.11 Control volume for calculating the axial thrust

where F_a is the axial force exerted on the fluid by the rotor. Further simplifications are possible when the integrand is constant over the given surfaces, and often because in practical cases the momentum flux through the surfaces A_i and A_o is much smaller than the pressure forces.

2.6 Balance of Energy

The fact that mechanical energy can be changed into heat and heat can be changed into mechanical energy shows that the balance laws of mechanics we have discussed up to now are not enough for a complete description of the motion of a fluid. As well as the two laws we have already treated, therefore a third basic empirical law, the balance of energy, appears:

> The rate of change of the total energy of a body is equal to the power of the external forces plus the rate at which heat is transferred to the body.

This law can be "deduced" from the well known *first law of thermodynamics* together with a mechanical energy equation which follows from Cauchy's Eq. (2.38a, 2.38b). However here we prefer to postulate the balance of the total energy, and to infer from it the more restrictive statement of the first law of thermodynamics.

We shall assume the fundamentals of classical thermodynamics as known. Thermodynamics is concerned with processes where the material is at rest and where all quantities appearing are independent of position (homogeneous), and therefore are only dependent on time. An important step to the thermodynamics of *irreversible processes* as they appear in the motion of fluids, consists of simply applying the classical laws to a material particle. If e is the internal energy per unit mass, then the internal energy of a material particle is given by $e\,dm$, and we can calculate the internal energy E of a body, that is, the energy of a bounded part of the fluid, as the integral over the region occupied by the body

$$E = \iiint\limits_{(V(t))} e\varrho\,dV. \tag{2.109}$$

In order to obtain the total energy of the fluid body under consideration, the kinetic energy which does not appear in the classical theory must be added to (2.109). The kinetic energy of the material particle is $(u^2/2)\,dm$, and the kinetic energy K of the body is correspondingly

$$K = \iiint\limits_{(V(t))} \frac{u_i u_i}{2}\,\varrho\,dV. \tag{2.110}$$

The applied forces which appear are the surface and body forces which were discussed in the context of the balance of momentum. The power of the surface force $\vec{t}\,dS$ is $\vec{u}\cdot\vec{t}\,dS$, while that of the body force $\varrho\vec{k}\,dV$ is $\vec{u}\cdot\vec{k}\varrho\,dV$. The power of the applied forces is then

$$P = \iiint\limits_{(V(t))} \varrho\,u_i k_i\,dV + \iint\limits_{(S(t))} u_i t_i\,dS. \tag{2.111}$$

In analogy to the volume flow $\vec{u}\cdot\vec{n}\,dS$ through an element of the surface, we introduce the heat flux through an element of the surface with $-\vec{q}\cdot\vec{n}\,dS$ and denote $\vec{q}$ as the *heat flux vector*. The minus sign is chosen so that inflowing energy ($\vec{q}$ and $\vec{n}$ forming an obtuse angle) is counted as positive. From now we shall limit ourselves to the transfer of heat by conduction, although $\vec{q}$ can also contain other kinds of heat transfer, for example, heat transfer by radiation, via *Poynting's vector*.

The relationship between the heat flux vector $\vec{q}$ and the temperature field (or other quantities) depends on the material under consideration. Therefore it is a constitutive relation, which we leave open to be specified later. Using the amount of heat flowing into the body per unit time

$$\dot{Q} = - \iint\limits_{(S(t))} q_i n_i\,dS, \tag{2.112}$$

we can write for the energy balance

$$\frac{D}{Dt}(K+E) = P + \dot{Q},\tag{2.113}$$

or, more explicitly

$$\frac{D}{Dt}\iiint\limits_{(V(y))}\left[\frac{u_iu_i}{2}+e\right]\varrho\,dV = \iiint\limits_{(V)} u_i\,k_i\,\varrho\,dV + \iint\limits_{(S)} u_i\,t_i\,dS - \iint\limits_{(S)} q_i\,n_i\,dS.\tag{2.114}$$

On the right-hand side, we have already replaced the time varying domains with the fixed domains V and S. By applying (1.88), this is also possible on the left. If we express the stress vector in the first surface integral using the stress tensor, both surface integral can be transformed into volume integrals using Gauss' theorem. Thus Eq. (2.114) becomes

$$\iiint\limits_{(V)}\left\{\varrho\frac{D}{Dt}\left[\frac{u_iu_i}{2}+e\right] - \varrho\,k_i\,u_i - \frac{\partial}{\partial x_j}(\tau_{ji}u_i) + \frac{\partial q_i}{\partial x_i}\right\}dV = 0.\tag{2.115}$$

Since the integrand is assumed to be continuous, and the domain of integration is arbitrary, the integrand must vanish, and, after differentiating, we obtain the differential form of the energy balance

$$\varrho\,u_i\frac{Du_i}{Dt} + \varrho\frac{De}{Dt} = \varrho\,k_i\,u_i + u_i\frac{\partial\tau_{ji}}{\partial x_j} + \tau_{ji}\frac{\partial u_i}{\partial x_j} - \frac{\partial q_i}{\partial x_i}.\tag{2.116}$$

Using the expansion of the stress tensor (2.35), the definition of enthalpy

$$h = e + \frac{p}{\varrho}\tag{2.117}$$

and the continuity Eq. (2.3a, 2.3b), the energy equation can be recast in the often used form

$$\varrho\frac{D}{Dt}\left[\frac{u_iu_i}{2}+h\right] = \frac{\partial p}{\partial t} + \varrho\,k_i\,u_i + \frac{\partial}{\partial x_j}(P_{ji}\,u_i) - \frac{\partial q_i}{\partial x_i}.\tag{2.118}$$

If Eq. (2.38a, 2.38b) is satisfied, the terms in (2.116) which are multiplied with u_i, drop out, and we are led to the following equation for the rate of change of the internal energy of a material particle

$$\frac{De}{Dt} = \frac{\tau_{ji}}{\varrho}\frac{\partial u_i}{\partial x_j} - \frac{1}{\varrho}\frac{\partial q_i}{\partial x_i},$$
(2.119)

which is the continuum mechanical analogue to the first law of classical thermo-dynamics. In the first law,

$$de = \delta w + \delta q,$$
(2.120)

de is the change in the internal energy in the time dt, δw is the work done in this time, and δq is the gain of heat in this time (each per unit mass). In applying the classical law to a material particle, we replace the operator "d" by "D/Dt", and therefore we must replace δw on the right-hand side with the work done per unit time, which we shall denote by $\delta\dot{w}$. Similarly we replace δq with $\delta\dot{q}$, so that the first law of thermodynamics must be written in the form

$$\frac{De}{Dt} = \delta\dot{w} + \delta\dot{q}.$$
(2.121)

Just like (2.120), this equation holds without restrictions both for reversible and irreversible processes. In particular, for *reversible processes* we have the classical thermodynamic relations

$$\delta w = -p\,dv$$
(2.122)

and

$$\delta q = T\,ds,$$
(2.123)

or

$$\delta\dot{w} = -p\frac{Dv}{Dt},$$
(2.124)

$$\delta\dot{q} = T\frac{Ds}{Dt}.$$
(2.125)

Here, $v = 1/\varrho$ is the specific volume, and s is the specific entropy. By comparing (2.121) with (2.119), we can extract two formulas which are valid without restriction to calculate the work done

$$\delta\dot{w} = \frac{\tau_{ji}}{\varrho}\frac{\partial u_i}{\partial x_j}$$
(2.126)

and the heat added

$$\delta \dot{q} = -\frac{1}{\varrho}\frac{\partial q_i}{\partial x_i}, \tag{2.127}$$

where each are per unit time and mass. The work per unit time and mass can be split up into the reversible work as in (2.124), and the irreversible work. The latter contribution is irreversibly changed into heat by the action of the friction stresses. Replacing the stress tensor by its decomposition according to (2.35), we extract the following

$$\delta \dot{w} = -\frac{p}{\varrho}\frac{\partial u_i}{\partial x_i} + \frac{1}{\varrho}P_{ij}\,e_{ij}, \tag{2.128}$$

where the last term results from $P_{ji}\,\partial u_i/\partial x_j$, because the friction stress tensor P_{ij} is, like τ_{ij}, a symmetric tensor. This term represents the deformation work converted irreversibly into heat. In general, the deformation work per unit time and volume $P_{ij}e_{ij}$ is written as a *dissipation function* Φ, where

$$\Phi = P_{ij}e_{ij}. \tag{2.129}$$

The dissipation function depends on the relationship between the friction stresses and the motion, that is, on the constitutive relation, and therefore we defer calculating this explicitly until we know the constitutive relation. However this term is zero for frictionless flow, or for fluids at rest. We identify the first term, using the continuity Eq. (2.3a, 2.3b), as the reversible contribution to the work, known from (2.124)

$$-\frac{p}{\varrho}\frac{\partial u_i}{\partial x_i} = \frac{p}{\varrho^2}\frac{D\varrho}{Dt} = -p\frac{Dv}{Dt}, \tag{2.130}$$

so that we finally attain the expression for the work per unit time and mass

$$\delta \dot{w} = -p\frac{Dv}{Dt} + \frac{\Phi}{\varrho}. \tag{2.131}$$

2.7 Balance of Entropy

We begin with the equation

$$T\,ds = de + p\,dv, \tag{2.132}$$

which is known as *Gibbs' relation*. It is given here for the special case of single component material, in which there is no phase change and where no chemical reactions take place. It is to this that we wish to restrict our discussion. Apart from this, this equation holds without restriction for both reversible and irreversible processes. Its validity for reversible processes can be found from the first law of thermodynamics, in connection with (2.122) and (2.123). Its acceptance for irreversible processes is the fundamental assumption for the thermodynamics of these processes. We shall not justify this assumption further, except to say that its consequences agree with our experience. Gibbs' relation can also be obtained from kinetic theory, where the results of the kinetic theory of gases remain restricted to small deviations from thermodynamic equilibrium, and to a monatomic dilute gas. Therefore these results can neither be used as a "proof" of Gibbs' relation, nor do they have the general validity in which we shall apply this relation. Gibbs' relation for a material particle leads to the equation

$$T \frac{Ds}{Dt} = \frac{De}{Dt} + p \frac{Dv}{Dt},$$
(2.133)

in which we replace the material change of the internal energy using the energy equation (2.121), and (2.127) and (2.131), so that the following equation emerges

$$\varrho \frac{Ds}{Dt} = \frac{\Phi}{T} - \frac{1}{T} \frac{\partial q_i}{\partial x_i}.$$
(2.134)

Transforming the last term on the right-hand side using the identity

$$\frac{\partial}{\partial x_i} \left[\frac{q_i}{T} \right] = \frac{1}{T} \frac{\partial q_i}{\partial x_i} - \frac{q_i}{T^2} \frac{\partial T}{\partial x_i}$$
(2.135)

furnishes the balance of entropy

$$\varrho \frac{Ds}{Dt} = \frac{\Phi}{T} - \frac{q_i}{T^2} \frac{\partial T}{\partial x_i} - \frac{\partial}{\partial x_i} \left[\frac{q_i}{T} \right].$$
(2.136)

In this equation the rate of change of the entropy of a material particle is split up into two contributions: A rate of entropy production with the value

$$\varrho \frac{D}{Dt} s_{(irr)} = \frac{\Phi}{T} - \frac{q_i}{T^2} \frac{\partial T}{\partial x_i},$$
(2.137)

which is always greater or equal to zero, and a divergence of an *entropy flux vector* q_i/T, which can be greater than, equal to, or less than zero

$$\varrho \frac{D}{Dt} s_{(rev)} = -\frac{\partial}{\partial x_i} \left[\frac{q_i}{T} \right].$$
(2.138)

The first part arises via the irreversible actions of friction and heat conduction in the fluid particle. Sufficient for the inequality

$$\frac{D}{Dt} s_{(irr)} \geq 0 \tag{2.139}$$

are the conditions

$$\Phi \geq 0 \tag{2.140}$$

and

$$q_i \frac{\partial T}{\partial x_i} \leq 0. \tag{2.141}$$

The first inequality expresses the experience that during deformation mechanical energy is dissipated into heat by the action of friction, but that heat cannot be changed into mechanical energy by the action of friction during deformation. The second inequality states that the heat flux vector must form an obtuse angle with the temperature gradient, and reflects the fact that heat flows in the direction of falling temperature. Equation (2.138) represents the change in entropy which the particle experiences from its neighborhood, since the divergence of the entropy flux is the difference between the inflowing and outflowing entropy flux. This difference can clearly be positive, negative or zero.

Elimination of Φ/T between Eqs. (2.137) and (2.134) leads to a form of (2.139) known as *Clausius-Duhem's inequality*

$$\rho \frac{Ds_{(irr)}}{Dt} = \rho \frac{Ds}{Dt} + \frac{1}{T} \frac{\partial q_i}{\partial x_i} - \frac{q_i}{T^2} \frac{\partial T}{\partial x_i} \geq 0.$$

We can obtain the change of entropy of a bounded part of the fluid by integrating (2.136) over the domain occupied by the fluid. We shall apply Reynolds' transport theorem to the left-hand side, and transform the integral on the right-hand side using Gauss' theorem. Doing this, we extract the following equation for the balance of entropy of the fluid body

$$\frac{D}{Dt} \iiint_{(V(t))} s \varrho \, dV = \frac{DS}{Dt} = \iiint_{(V)} \left[\frac{\Phi}{T} - \frac{q_i}{T^2} \frac{\partial T}{\partial x_i} \right] dV - \iint_{(S)} \frac{q_i n_i}{T} \, dS. \tag{2.142}$$

As stated, the volume integral on the right-hand side is never negative, and therefore we can read off the *second law of thermodynamics*

$$\frac{DS}{Dt} \geq - \iint\limits_{(S)} \frac{q_i n_i}{T} \, dS. \tag{2.143}$$

The equality sign only holds when the process in the body is reversible. However all real processes in nature are irreversible, so the inequality sign must hold for these. If heat is neither added to nor taken away from the body, the surface integral on the right-hand side vanishes. The process taking place in the body is then *adiabatic*, and Eq. (2.142) expresses the following fact:

The entropy cannot decrease in an adiabatic process.

The second law of thermodynamics is of course, like the first law, a law of experience. In our discussion, the second law arises as a consequence of the assumptions in (2.140) and (2.141), which are based on experience. Had we postulated the second law we would have had to conclude that the integrand of the volume integral on the right-hand side of (2.142) is never negative. Equations (2.140) and (2.141) are sufficient for this.

2.8   Thermodynamic Equations of State

The principles we have discussed so far in Chap. 2 form the basis of continuum mechanics. These principles represent a summary of our experience of the behavior common to all bodies. All solids and fluids, whether Newtonian or non-Newtonian fluids, are subject to these universal laws. The distinguishing properties of solids and fluids are determined by the materials from which they consist. These properties are abstracted by constitutive relations. They define ideal materials, and therefore are models of the material's real behavior. Besides these constitutive relations in a narrow sense, i.e., those which establish the relationship between stress state and motion, or between heat flux vector and temperature, there are also the thermodynamic equations of state. We shall introduce the constitutive relations in the next chapter, but here we discuss how the equations of state known from classical thermodynamics are carried over to the moving continuum, and their application to determining the thermodynamic state of a material particle.

It is a fact of experience of classical thermodynamics that a thermodynamic state is uniquely defined by a certain number of independent *variables of state*. For the single component material to which we shall restrict ourselves, two independent variables of state are required. These two independent variables, which otherwise are of arbitrary choice, fix the value of every other variable of state. An equation of state, which can also be given in the form of a diagram or graph, is a relationship by which two variables of state, as independent variables, determine a third as a dependent variable. For a small class of materials, in particular for gases, equations of state can be found on the basis of specific molecular models from statistical mechanics and quantum mechanics. Here, however, we do not wish to go into the origin of the equations of state, and we shall consider them as given.

We call an equation of state between p, ϱ and T a *thermal equation of state*, and so we have

$$p = p(\varrho, T). \tag{2.144}$$

The equation of state

$$p = \varrho RT \tag{2.145}$$

defines, for example, the *thermally perfect gas*. If the so-called *caloric variables of state*, such as internal energy e, enthalpy h or entropy s, appear as dependent variable, we denote equations like, for example

$$e = e(\varrho, T) \tag{2.146}$$

as *caloric equations of state*. For a thermally perfect gas, the caloric equation of state takes the simple form

$$e = e(T) \tag{2.147}$$

or

$$h = h(T). \tag{2.148}$$

The equation of state $e = c_v T$ (or $h = c_p T$) with constant specific heat c_v (or c_p) therefore also defines the *calorically perfect gas*.

In general, however, one equation of state does not necessarily determine the other. There exist "reciprocity relations" between the thermal and caloric equations of state. Yet these are relations between partial differentials, so that the determination of the other equation of state requires an integration, where unknown functions appear as "constants" of integration. An equation of state, from which the other can be found by the processes differentiation and elimination alone, is called a *canonical* or *fundamental equation of state*. If we compare the differential of the canonical equation of state $e = e(s, v)$

$$de = \left[\frac{\partial e}{\partial s}\right]_v ds + \left[\frac{\partial e}{\partial v}\right]_s dv \qquad (2.149)$$

with Gibbs' relation (2.132), we read off

$$T = \left[\frac{\partial e}{\partial s}\right]_v, \qquad (2.150)$$

and

$$p = -\left[\frac{\partial e}{\partial v}\right]_s. \qquad (2.151)$$

The right-hand sides of (2.150) and (2.151) are functions of s and v. If we consider both relationships to be solved for s, the equations $s = s(v, T)$ and $s = s(p, v)$ arise. Elimination of s gives a relation between T, p and v, and thus the thermal equation of state.

The *Mollier diagram* known from applications is the graphical representation of the canonical equation of state $h = h(s, p)$, where h is plotted as a function of s, with p as family parameter. Specific volume and temperature may then be ascertained by comparing the differential of the canonical equation of state $h = h(s, p)$

$$dh = \left[\frac{\partial h}{\partial s}\right]_p ds + \left[\frac{\partial h}{\partial p}\right]_s dp \qquad (2.152)$$

with Gibbs' relation in the form

$$T\, ds = dh - v\, dp, \qquad (2.153)$$

which yields

$$v = \left[\frac{\partial h}{\partial p}\right]_s \qquad (2.154)$$

and

$$T = \left[\frac{\partial h}{\partial s}\right]_p. \qquad (2.155)$$

Now to obtain v, for example, note the values of h and p along an isentrope $s = \text{const}$, and numerically or graphically determine the slope of $h = h(p)$. For thermally and calorically perfect gas the canonical form of the enthalpy is easily explicitly given

$$h = \text{const} \cdot c_p \exp(s/c_p)p^{(R/c_p)}. \tag{2.156}$$

The essential step which leads from the classical thermodynamics of reversible homogeneous processes to the thermodynamics of irreversible processes of continuum mechanics is the assumption that exactly the same equations of state as hold for the material at rest also hold for a moving material point of the continuum. This means, for example, that the internal energy e of a material particle can be calculated from the values of s and v, irrespective of where the particle is or what its motion is. This assumption is equivalent to the assumption that Gibbs' relation is valid for irreversible processes. For from the material derivative of the relation $e = e(s, v)$ we have

$$\frac{De}{Dt} = \left[\frac{\partial e}{\partial s}\right]_v \frac{Ds}{Dt} + \left[\frac{\partial e}{\partial v}\right]_s \frac{Dv}{Dt}. \tag{2.157}$$

If this relationship always holds, and if we regard (2.150) and (2.151) as definitions of temperature and pressure, Gibbs' relation (2.133) follows immediately. This means then too, that the internal energy is given at every place and time if s and v are known at this place and time. Although the thermodynamic state changes from place to place, it is not dependent on the gradients of the variables of state.

Chapter 3
Constitutive Relations for Fluids

As already explained in the previous chapter on the fundamental laws of continuum mechanics, bodies behave in such a way that the universal balances of mass, momentum, energy and entropy are satisfied. Yet only in very few cases, like, for example, the idealizations of a point mass or of a rigid body without heat conduction, are these laws enough to describe a body's behavior. In these special cases, the characteristics of "mass" and "mass distribution" belonging to each body are the only important features. In order to describe a deformable medium, the material from which it is made must be characterized, because clearly, the deformation or the rate of deformation under a given load is dependent on the material. Because the balance laws yield more unknowns than independent equations, we can already conclude that a specification of the material through relationships describing the way in which the stress and heat flux vectors depend on the other field quantities is generally required. Thus the balance laws yield more unknowns than independent equations. The summarizing list of the balance laws of mass (2.2)

$$\frac{\partial \varrho}{\partial t} + \frac{\partial}{\partial x_i}(\varrho u_i) = 0,$$

of momentum (2.38)

$$\varrho \frac{Du_i}{Dt} = \varrho k_i + \frac{\partial \tau_{ji}}{\partial x_j},$$

of angular momentum (2.53)

$$\tau_{ij} = \tau_{ji}$$

© Springer Nature Switzerland AG 2020
J. H. Spurk and N. Aksel, *Fluid Mechanics*,
https://doi.org/10.1007/978-3-030-30259-7_3

and of energy (2.119)

$$\varrho \frac{De}{Dt} = \tau_{ij} \frac{\partial u_i}{\partial x_j} - \frac{\partial q_i}{\partial x_i}$$

yield 17 unknown functions $(\varrho, u_i, \tau_{ij}, q_i, e)$ in only eight independently available equations. Instead of the energy balance, we could also use the entropy balance (2.134) here, which would introduce the unknown function s instead of e, but by doing this the number of equations and unknown functions would not change. Of course we could solve this system of equations by specifying nine of the unknown functions arbitrarily, but the solution found is then not a solution to a particular technical problem.

It may happen that the "mechanical" balance laws for mass, momentum and angular momentum are decoupled from the energy equation. Then six constitutive relations are enough to complete the reduced system for ϱ, u_i and τ_{ij}. If the internal energy field is not required, it can be assumed arbitrarily without changing, for example, the velocity field. In these cases the internal energy is not counted as an unknown function, and the energy equation is superfluous.

Even if no proof for the uniqueness of the solution is available, we still expect that the solution of a physical problem is unique if the number of unknown functions is the same as the number of equations and the properly posed initial and boundary values are present. Further, we take as self evident that all equations are given by the problem itself, and that therefore only constitutive relations as they arise from the specification of the flowing material, appear along with the universal balance laws.

In principle, constitutive relations could be gotten from the molecular theory of gases and liquids. For structurally simple molecules, and in particular for gases, this theory provides constitutive relations which agree very well with experimental results. This has not been successful to the same extent for Newtonian liquids; even less so for non-Newtonian fluids. Yet the results found from the molecular theory do not contradict the phenomenological model of continuum theory. In fact, they show that this model provides a suitable framework for describing the material behavior of even non-Newtonian fluids. Indeed continuum theory has become for the most part a theory of the constitutive relations. It develops mathematical models from specific experimental observations which idealize the behavior of the actual material but which in more general circumstances do describe it as accurately as possible.

Let us adopt the viewpoint of an engineer who forecasts the flow of a given fluid from the balance laws on the basis of the constitutive relations. As with the thermodynamic constitutive relations (equations of state), we shall not go any further into the derivations, but will only note that certain axioms are of fundamental importance for the formulation of the constitutive relations. Some of these axioms have arisen during more recent developments of continuum mechanics, and are not satisfied by older constitutive relations, which were proposed to explain particular features of the behavior of materials. Constitutive relations which these axioms satisfy must, among other things

(a) be consistent with the balance laws and with the second law of thermodynamics (but they are not consequences of these laws),
(b) be valid in all coordinate systems (thus they must be formulated as tensor equations),
(c) be deterministic (the history of the motion and of the temperature of the body up to time t determines, for example, the stresses on the material particle at time t),
(d) hold locally (thus, for example, the stress at a material particle depends only on the motion of material particles in its immediate neighborhood),
(e) reflect the symmetry properties of the material, and
(f) be valid in all reference frames, i.e., be objective or frame independent.

The final condition is here of particular importance, since, as we know from Sect. 2.4, the equations of motion (momentum balance) are not frame independent in this sense. In accelerating reference frames, the apparent forces are introduced, and only the axiom of objectivity ensures that this remains the only difference for the transition from an inertial system to a relative system. However, it is clear that an observer in an accelerating reference frame detects the same material properties as an observer in an inertial system. To illustrate this, for a given deflection of a massless spring, an observer in a rotating reference frame would detect exactly the same force as in an inertial frame.

In so-called *simple fluids*, the stress on a material point at time t is determined by the history of the deformation involving only gradients of the first order or more exactly, by the relative deformation tensor (relative Cauchy-Green-tensor) as every fluid is isotropic. Essentially all non-Newtonian fluids belong to this group.

The most simple constitutive relation for the stress tensor of a viscous fluid is a linear relationship between the components of the stress tensor τ_{ij} and those of the rate of deformation tensor e_{ij}. Almost trivially, this constitutive relation satisfies all the above axioms. The material theory shows that the most general linear relationship of this kind must be of the form

$$\tau_{ij} = -p\delta_{ij} + \lambda^* e_{kk}\delta_{ij} + 2\eta\, e_{ij}, \qquad (3.1a)$$

or, using the *unit tensor* **I**

$$\mathbf{T} = (-p + \lambda^* \nabla \cdot \vec{u})\mathbf{I} + 2\eta\, \mathbf{E} \qquad (3.1b)$$

(*Cauchy-Poisson law*), so that noting the decomposition (2.35), the tensor of the friction stresses is given by

$$P_{ij} = \lambda^* e_{kk}\delta_{ij} + 2\eta\, e_{ij}, \qquad (3.2a)$$

or

$$\mathbf{P} = \lambda^* \nabla \cdot \vec{u}\, \mathbf{I} + 2\eta\, \mathbf{E}. \tag{3.2b}$$

We next note that the friction stresses at the position $\vec{x}$ are given by the rate of deformation tensor e_{ij} at $\vec{x}$, and are not explicitly dependent on $\vec{x}$ itself. Since the friction stress tensor P_{ij} at $\vec{x}$ determines the stress acting on the material particle at $\vec{x}$, we conclude that the stress on the particle only depends on the instantaneous value of the rate of deformation tensor and is not influenced by the history of the deformation. We remind ourselves that for a fluid at rest or for a fluid undergoing rigid body motion, $e_{ij} = 0$, and (3.1a) reduces to (2.33). The quantities λ^* and η are scalar functions of the thermodynamic state, typical to the material. Thus (3.1a, 3.1b) is the generalization of $\tau = \eta\, \dot{\gamma}$, which we have already met in connection with simple shearing flow and defines the Newtonian fluid.

The extraordinary importance of the linear relationship (3.1a, 3.1b) lies in the fact that it describes the actual material behavior of most technically important fluids very well. This includes practically all gases, in particular air and steam, gas mixtures and all liquids of low molecular weight, like water, and also all mineral oils.

As already noted, $e_{ij} = 0$ describes the stress state of a fluid at rest or in rigid body motion. The pressure p of compressible fluids is then determined by the thermal equation of state $p = p(\varrho, T)$. The same equation of state also holds for the moving material particle, thus the pressure is fixed for every position of the particle and for every instant by ϱ and T. In incompressible fluids, the pressure is not a function of the thermodynamic state, but is a fundamentally dependent variable. As is already clear from Cauchy's Eq. (2.38) in connection with (3.1a, 3.1b), and as we shall show explicitly later, only the gradient of the pressure appears in Cauchy's equation. In incompressible flow, an arbitrary constant may be added to the pressure without affecting the equations of motion. If the pressure is not fixed by a boundary condition it can only be determined up to this additive constant. Expressed otherwise, only pressure differences can be calculated from the theory of incompressible flow.

Using (2.36) and (3.1a), we extract the following equation for the sum of the mean normal stress and the pressure

$$\bar{p} + p = \frac{1}{3}\tau_{ii} + p = e_{ii}\left(\lambda^* + \frac{2}{3}\eta\right). \tag{3.3}$$

By (2.5), $e_{ii} = 0$ holds for incompressible flow, thus the mean normal stress is equal to the negative pressure. This only holds in compressible flow if the *bulk viscosity*

$$\eta_B = \lambda^* + \frac{2}{3}\eta \tag{3.4}$$

vanishes. Kinetic gas theory shows that the bulk viscosity arises because the kinetic energy of the molecules is transferred to the internal degrees of freedom. Therefore, the bulk viscosity of monatomic gases, which have no internal degrees of freedom, is zero. The bulk viscosity is proportional to the characteristic time in which the transfer of energy takes place. This effect can be important for the structure of shock waves, but is otherwise of lesser importance, and therefore, even for polyatomic gases, use is most often made of *Stokes' hypothesis*

$$\eta_B = 0. \tag{3.5}$$

The assumption of the constitutive relation now also allows the explicit calculation of the dissipation function Φ. Following (2.129) we obtain

$$\Phi = P_{ij}e_{ij} = \lambda^* e_{kk}e_{ii} + 2\eta\, e_{ij}\, e_{ij}, \tag{3.6a}$$

or, written symbolically

$$\Phi = \lambda^* (\text{sp}\mathbf{E})^2 + 2\eta\, \text{sp}\mathbf{E}^2, \tag{3.6b}$$

and we see, by expansion and relabelling the dummy indices, that the inequality (2.140) is satisfied, if the inequalities

$$\eta \geq 0, \eta_B \geq 0 \tag{3.7}$$

hold for the shear viscosity η and the bulk viscosity η_B.

As already noted, the viscosity depends on the thermodynamic state, so $\eta = \eta(p, T)$, where the dependency on pressure is small. The kinetic gas theory states that for dilute gases the only dependency is on the temperature: for the model of hard sphere molecules we have $\eta \sim \sqrt{T}$. In the phenomenological model, the dependency on p and T remains free and must be determined by experiment. The shear viscosity η often appears in the combination $\eta/\varrho = \nu$, which is known as the *kinematic* viscosity, and clearly depends strongly on the density or the pressure.

From kinetic gas theory, the viscosity η can be predicted quantitatively very well if a realistic molecular potential is used. The less developed kinetic theory for liquids can not yet furnish comparable viscosity data. In this case the temperature dependency of the viscosity is given by $\eta \sim \exp(\text{const}/T)$, that is, it decreases exponentially with temperature. This behavior has been experimentally confirmed qualitatively for most liquids, and so we see that liquids show a contrasting viscosity behavior to gases. The reason for this lies in the differing molecular structure, and has already been discussed in Sect. 1.1.

With the linear constitutive Eq. (3.1a, 3.1b) for the stress goes a linear constitutive relation for the heat flux vector. This linear relationship is known as *Fourier's law*, and for isotropic materials reads

$$q_i = -\lambda \frac{\partial T}{\partial x_i} \quad \text{or} \quad \vec{q} = -\lambda \nabla T. \tag{3.8}$$

Here λ is a positive function of the thermodynamic state, and is called the *thermal conductivity*. The minus sign here is in agreement with the inequality (2.141). Experiments show that this linear law describes the actual behavior of materials very well. The dependency of the thermal conductivity on p and T remains open in (3.8), and has to be determined experimentally. For gases the kinetic theory leads to the result $\lambda \sim \eta$, so that the thermal conductivity shows the same temperature dependence as the shear viscosity. (For liquids, one discovers theoretically that the thermal conductivity is proportional to the velocity of sound in the fluid.)

In the limiting case $\eta, \lambda^* = 0$, we extract from the Cauchy-Poisson law the constitutive relation for inviscid fluids

$$\tau_{ij} = -p\,\delta_{ij}. \tag{3.9}$$

Thus, as with a fluid at rest, the stress tensor is only determined by the pressure p. As far as the stress state is concerned, the limiting case $\eta, \lambda^* = 0$ leads to the same result as $e_{ij} = 0$. Also consistent with $\eta, \lambda^* = 0$ is the case $\lambda = 0$; ignoring the friction stresses implies that we should in general also ignore the heat conduction.

It would now appear that there is no technical importance attached to the condition $\eta, \lambda*, \lambda = 0$. Yet the opposite is actually the case. Many technically important, real flows are described very well using this assumption. This has already been stressed in connection with the flow through turbomachines. Indeed the flow past a flying object can often be predicted using the assumption of inviscid flow. The reason for this can be clearly seen when we note that fluids which occur in applications (mostly air or water) only have "small" viscosities. However, the viscosity is a dimensional quantity, and the expression "small viscosity" is vague, since the numerical value of the physical quantity "viscosity" may be arbitrarily changed by suitable choice of the units in the dimensional formula. The question of whether the viscosity is small or not can only be settled in connection with the specific problem, however this is already possible using simple dimensional arguments. For incompressible fluids, or by using Stokes' relation (3.5), only the shear viscosity appears in the constitutive relation (3.1a, 3.1b). If, in addition, the temperature field is homogeneous, no thermodynamic quantities enter the problem, and the incident flow is determined by the velocity U, the density ϱ and the shear viscosity η. We characterize the body past which the fluid flows by its typical length L, and we form the dimensionless quantity

$$Re = \frac{UL\varrho}{\eta} = \frac{UL}{\nu} \tag{3.10}$$

which is called the *Reynolds' number*. This is the most important dimensionless group of fluid mechanics, and is a suitable measure for the action of the viscosity. If

η tends to zero, then the Reynolds' number becomes infinite. The assumption of inviscid flow is thus only justified if the Reynolds' number is very large. If we have, for example, a cascade flow in a water turbine with the blade chord $L = 1$ m, inflow velocity $U = 10$ m/s, and kinematic viscosity of water of $v = 10^{-6}$ m^2/s, the Reynolds' number is already $Re = 10^7$, and so is indeed very large. It therefore can make sense to perform the calculation on the basis of an inviscid flow.

A further fact, which is important in connection with viscous flow, follows from simple dimensional analysis: let us consider, for example, the drag D on a body in a flow field. This drag may be made dimensionless using the data of the above problem, forming the *drag coefficient*

$$c_D = \frac{D}{\frac{\varrho}{2} U^2 L^2}. \tag{3.11}$$

The drag coefficient as a dimensionless number can only be dependent on other dimensionless variables, and the only one which can be formed using the above data is the Reynolds' number. Thus we are led inevitably to the relation

$$c_D = c_D(Re). \tag{3.12}$$

This relation has been confirmed in countless experiments. It represents perhaps the most convincing argument for the applicability of the constitutive relation (3.1a, 3.1b) to pure, low molecular fluids.

The constitutive relations for the linear viscous fluid (3.1a, 3.1b) and for the inviscid fluid (3.9) apply to most technical applications. In what follows, we shall deal almost exclusively with the flows of these fluids. Yet there is a series of technical applications where non-Newtonian fluids play a role, among these the manufacture of plastics, lubrication technology, food processing and paint production. Typical representatives of non-Newtonian fluids are liquids which are formed either partly or wholly of macromolecules (polymers), and two phase materials like, for example, high concentration suspensions of solid particles in a liquid carrier solution. For most of these fluids, the shear viscosity decreases with increasing shearrate, and we call them *shear-thinning* fluids. Here the shear viscosity can decrease by many orders of magnitude. This is a phenomenon which is very important in the plastics industry, since the aim is to process plastics at high shearrates in order to keep the dissipated energy small. If the shear viscosity increases with increasing shearrate, we speak of *shear-thickening* fluids. Note that this notation is not unique, and shear-thinning fluids are often called "pseudo-plastic", and shear-thickening fluids are called "dilatant".

In the simple shearing flow of incompressible fluids (Fig. 1.1), which conforms with the linear law (3.1a, 3.1b), the normal stresses (terms on the main diagonal of the matrix representation of the tensor **T**) are all equal. Expansion of the Eq. (3.1a, 3.1b) leads to

$$\tau_{11} = -p + 2\eta \partial u_1 / \partial x_1,$$
$$\tau_{22} = -p + 2\eta \partial u_2 / \partial x_2,$$
$$\tau_{33} = -p + 2\eta \partial u_3 / \partial x_3.$$

Since the velocity field is $u_1 = \dot{\gamma} x_2, u_2 = u_3 = 0$,

$$\tau_{11} = \tau_{22} = \tau_{33} = -p$$

follows. Obviously this also holds for more general flows with $u_1 = u_1(x_2)$, $u_2 = u_3 = 0$. Indeed, the normal stress differences vanish in all steady unidirectional flows which follow the linear law (3.1a, 3.1b).

In general this is not the case in non-Newtonian flows. They show *normal stress effects*, of which the best known is the *Weissenberg effect*. Contrary to what a Newtonian fluid does, some non-Newtonian fluids climb up a rotating rod which is inserted perpendicular to the free surface. This effect, which only takes place with a small enough rod radius, can be seen by stirring paint or cream. It is caused by the nonvanishing difference between the normal stresses. Another normal stress effect is the *extrudate swell*: as the liquid emerges from a capillary tube the diameter increases. This phenomenon is important in the extrusion of melted plastics, because, depending on the extrusion pressure, the diameter can be more than twice the diameter of the tube. (At smaller Reynolds' numbers, we see, even for Newtonian fluids, a small jet swell which has its origin in the rearrangement of the velocity profile at the exit.) The normal stress effects are an expression of a "fluid elasticity", which manifests itself in an elastic recovery when the load on the liquid is suddenly removed. These phenomena can be qualitatively explained by the structure of the polymeric fluid. Polymers are macromolecules consisting of long chains, whose single members have arisen from monomers and still show a similar structure. Silicon oil (polydimethylsiloxane), for example, consists of chain molecules of the form

$$\begin{array}{ccc} CH_3 & CH_3 & CH_3 \\ | & | & | \\ -Si-O-Si-O-Si-O- \\ | & | & | \\ CH_3 & CH_3 & CH_3 \end{array}$$

which arise, through polymerization, from monomers with the formula

$$\begin{array}{c} CH_3 \\ | \\ OH-Si-OH \\ | \\ CH_3 \end{array}.$$

These long chains can, in some cases, contain many thousand molecules, so that the molecular weight, that is, the weight of 6.0222×10^{23} molecules (Avogadro's number) is correspondingly large, and reaches values of up to 10^6 g/mol. Typical non-Newtonian effects are seen at molecular weights of over 10^3 g/mol. Polymeric fluids can have quite different physical properties from those of the corresponding monomeric fluid. This also comes from the fact that the chains themselves (which indeed are not all the same length) can easily become tangled. Because of the thermal motion, they continually undo and reform new tangles. Under shearing loads, the chains are straightened out, and this can serve as a rough model to explain the decreasing viscosity with increasing shearrate. The remaining viscosity when the shearrate vanishes is the so-called *null viscosity*, which is almost proportional to the molecular weight of the fluid. The aligned molecules try to retangle themselves and if this is hindered additional normal stresses arise. In the extrusion process the molecules in the tube are aligned. Following this highly ordered state, the molecules retangle themselves after exiting the tube, and thus cause an increase of the extrudate diameter. In accordance with the second law of thermodynamics, they try to reach a state of maximum disorder, that is, of maximum entropy. The elastic recovery mentioned above can be viewed similarly. We stress that this form of elasticity has a completely different character from the elasticity of a solid. By stretching a solid, the atoms are pulled away from each other. The work done by stretching is then retained by the solid as potential energy. On release, the solid immediately springs back into shape, if we ignore the inertia of the material. The elasticity of a polymer fluid is a consequence of the thermal motion (retangling) and therefore it needs a certain time, which is the reason why the extrudate swell does not necessarily begin directly after the material exits the tube.

As well as the phenomena we have already mentioned, non-Newtonian fluids exhibit a number of further, sometimes very surprising, effects, and therefore we do not expect that a single constitutive relation is enough to describe all these different phenomena. From a technical standpoint, shearthinning is particularly important, because many flows in applications are shear flows, or closely related flows. Thus the strong dependency of the viscosity on the shearrate can have a great influence. For example, this is the case in hydrodynamic lubrication flows and pipe flow of non-Newtonian fluids, as well as in the processing of plastics.

We have already described the constitutive relation $\tau = \tau(\dot{\gamma})$ for the simple shearing flow of non-Newtonian fluids, and we shall write this as

$$\tau = \eta(\dot{\gamma})\dot{\gamma}. \tag{3.13}$$

We obtain an extension of this relation for the general stress state if we allow a dependency of the shear viscosity on the rate of deformation tensor in (3.1a, 3.1b). Since η is a scalar, it can only be dependent on the invariants of this tensor. For incompressible flow, the first invariant (cf. (1.58)) $I_{1e} = e_{ii}$ is zero, the third invariant $I_{3e} = \det(e_{ij})$ vanishes for simple shearing flow, and for incompressible flow, the second invariant becomes $2\,I_{2e} = -e_{ij}e_{ij}$. Using these we introduce a generalized shearrate

$$\dot{\gamma} = \sqrt{-4I_{2e}},\tag{3.14}$$

so that, in agreement with (1.3), we have for simple shearing flow

$$\dot{\gamma} = du/dy.\tag{3.15}$$

The constitutive relation of the generalized Newtonian fluid then follows from the Cauchy-Poisson law (3.1a, 3.1b)

$$\tau_{ij} = -p\,\delta_{ij} + 2\eta(\dot{\gamma})e_{ij}.\tag{3.16}$$

In the literature, we find numerous empirical or semi-empirical models for the function $\eta(\dot{\gamma})$, of which we shall only mention the often used *power law*

$$\eta(\dot{\gamma}) = m|\dot{\gamma}|^{n-1},\tag{3.17}$$

(where the parameters m and n are determined experimentally), because in simple cases, this allows closed form solutions. Obviously m is a parameter whose dimension depends on the dimensionless parameter n. For $n > 1$ shear-thickening behavior is described, and for $n < 1$ we have shear-thinning behavior. For $\dot{\gamma} \to 0$, the yield function in the first case tends to zero, and in the second case it becomes infinite, so that then (3.17) is of no use if $\dot{\gamma} = 0$ is reached in the flow field. This difficulty can be overcome with a modification of the model (3.17) with three free parameters

$$\eta = \begin{cases} \eta_0 & \text{for } \dot{\gamma} \le \dot{\gamma}_0 \\ \eta_0|\dot{\gamma}/\dot{\gamma}_0|^{n-1} & \text{for } \dot{\gamma} > \dot{\gamma}_0 \end{cases}.\tag{3.18}$$

Here $\dot{\gamma}_0$ is the shearrate up to which Newtonian behavior with the null viscosity η_0 is found. The generalized Newtonian fluid shows no normal stress effects. These are only found in a more comprehensive model for steady shearing flow, which we shall not go into now, but which contains the generalized Newtonian fluid as a special case.

For unsteady flows, where fluid elasticity is particularly noticeable, linear viscoelastic models, whose origin goes back to Maxwell, are often used. The mechanical analogue to the linear viscoelastic model is the series arrangement of a spring and a damper (Fig. 3.1). We identify the deflection of the spring with the shearing γ_S, that of the damper with γ_D, and the force with τ_{21}, and so we obtain from the balance of forces

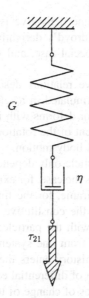

Fig. 3.1 Maxwell's model of a linear elastic fluid

$$\tau_{21} = G\gamma_S = \eta\dot{\gamma}_D. \qquad (3.19)$$

We write the complete deflection $\gamma_S + \gamma_D$ as γ, so that, from (3.19), the equation

$$\tau_{21} = \eta\dot{\gamma} - \frac{\eta}{G}\dot{\tau}_{21} \qquad (3.20)$$

arises. Using $\eta/G = \lambda_0$, $\dot{\gamma} = du/dy = 2e_{12}$ for the simple shearing flow, (3.20) can also be written in the form

$$\tau_{21} + \lambda_0\dot{\tau}_{21} = 2\eta\,e_{12}. \qquad (3.21)$$

The tensorial generalization of this equation is the constitutive relation of the linear viscoelastic fluid

$$P_{ij} + \lambda_0\frac{\partial P_{ij}}{\partial t} = 2\eta\,e_{ij}. \qquad (3.22)$$

We can call the characteristic time λ_0 the "memory span" of the fluid. As $\lambda_0 \to 0$, we obtain from (3.22) the constitutive relation valid for Newtonian fluids (3.2a, 3.2b), if we set there $e_{kk} = 0$ (incompressible flow).

In this sense the Newtonian fluid is a fluid without memory. But Eq. (3.22) is neither frame independent, nor describes the phenomena of shear-thinning or shear-thickening. However, the constitutive relation can be brought to a frame independent form if the partial time derivative is replaced by an objective time derivative, like that given by Oldroyd's derivative in (1.69), or by *Jaumann's derivative*, of which (2.63) is a special case, and in general it then describes also shear-thinning behavior.

This is so, because constitutive relations describe properties of the material point, and therefore should be formulated in a reference frame which moves and rotates with the material particle (or deforms with it). By doing this we ensure that the material behavior is independent of the rotation and translation of the particle, which indeed represent local rigid body motion.

If the stress on the material particle only depends on the instantaneous value of the rate of deformation tensor, as is the case, for example, in Cauchy-Poisson's law, then an observer-fixed reference frame, for the time being considered as fixed in space, is adequate. Because then the constitutive relation takes exactly the same form as in the frame co-rotating with the particle. We can immediately convince ourselves of this if we transform from one system into the other by the rules of Appendix A. If the deformation history enters the stress state, for example if the constitutive relations take the form of differential equations, then the frame fixed in space is not allowed, since the rates of change of tensors do not in general comply with the transformation rules in Appendix A: they are thus not frame independent or *objective tensors*. This is the name for tensors which comply with usual transformation rules even when the transformation matrix is time dependent. This is of course necessary for the constitutive relations to have the same form in all systems. Thus a constitutive relation of the form (3.22) only holds in systems which rotate and translate with the particle, where the translation is taken into account if the partial derivative in (3.22) is the material derivative. It would appear obvious at first to transform the equations of motion into the reference frame rotating with the material particle. There are many reasons why this is not practicable: apart from the fact that in general the angular velocities of different material particles are different, and the boundary conditions of a given problem would continually have to be transformed, it is also almost impossible to make measurements in the different rotating systems. As a rule, measurements and calculations are performed in a frame fixed in space, in which as a rule the boundary of the flow field is at rest. In fact it is this point which decides which reference frame we use. Therefore we attempt to express the constitutive relations which are only valid in the system rotating and translating with the material particle through quantities referred to the fixed frame. To do this, it is enough to interpret the partial time derivative in (3.22) as a material derivative of the components in the rotating system, and to represent this derivative in quantities and components of the frame fixed in space, since the other tensors are already in the fixed frame. We reach the required formula for the derivative if, starting with the transformation (A.29)

$$P_{ij} = a_{ik}a_{jl}P'_{kl}, \qquad (3.23)$$

where the P'_{kl} are the components in the rotating system, we form the material derivative

$$\frac{DP_{ij}}{Dt} = \left(\frac{Da_{ik}}{Dt}a_{jl} + a_{ik}\frac{Da_{jl}}{Dt}\right)P'_{kl} + a_{ik}a_{jl}\frac{DP'_{kl}}{Dt}. \tag{3.24}$$

It is the expression in parentheses which prevents the objectivity of the rate of change of the tensor. The time derivative of the orthogonal transformation matrix $a_{ij} = \vec{e}_i \cdot \vec{e}'_j(t)$ is found from (2.62), in which the angular velocity $\vec{\Omega}$ is now to be replaced by the angular velocity $\vec{\omega}$ of the particle, leading to

$$\frac{Da_{ij}}{Dt} = \vec{e}_i \cdot \left(\vec{\omega} \times \vec{e}'_j\right) = \vec{e}_i \cdot (\vec{\omega} \times \vec{e}_m)a_{mj}, \tag{3.25}$$

where the final expression follows from (A.23) and contains only terms in the system fixed in space. Writing the scalar triple product in index notation

$$\vec{e}_i \cdot (\vec{\omega} \times \vec{e}_m) = (\vec{e}_i)_k \varepsilon_{kln}\omega_l(\vec{e}_m)_n, \tag{3.26}$$

and noting that the kth component $\vec{e}_i \cdot \vec{e}_k = (\vec{e}_i)_k$ of the ith basis vector is the Kronecker-Delta, we extract from (1.46) the expression

$$\frac{Da_{ij}}{Dt} = \varepsilon_{ilm}\omega_l a_{mj} = -\Omega_{mi}a_{mj}, \tag{3.27}$$

which brings (3.24) to the form

$$a_{ik}a_{jl}\frac{DP'_{kl}}{Dt} = \frac{DP_{ij}}{Dt} + P_{mj}\Omega_{mi} + P_{im}\Omega_{mj}. \tag{3.28}$$

The right-hand side of (3.28) already is the required rate of change of the tensor P'_{kl} in the system rotating with the material particle, given in components of the system fixed in space. This derivative, Jaumann's derivative mentioned above, will be denoted with the symbol $\mathcal{D}/\mathcal{D}t$

$$\frac{\mathcal{D}P_{ij}}{\mathcal{D}t} = \frac{DP_{ij}}{Dt} + P_{mj}\Omega_{mi} + P_{im}\Omega_{mj}. \tag{3.29}$$

Jaumann's derivative of an objective tensor is another objective tensor, as can be read from (3.28) noting that the spin tensor vanishes in the co-rotating frame. Thus the reference frame which was denoted as fixed in space above can also be a relative system. The rate of change $(\mathcal{D}\mathbf{P}/\mathcal{D}t)_A$ in the relative system is the same as in the inertial system $(\mathcal{D}\mathbf{P}/\mathcal{D}t)_I$, while the components transform according to (A.28).

Constitutive relations in which only objective tensors appear are then valid in all reference frames, and satisfy the axiom of frame independence. They have the same form in relative and inertial systems. Closely related to Jaumann's derivative is Oldroyd's derivative (1.67), which, when applied to the friction stress tensor, leads to the expression

$$\frac{\delta P_{ij}}{\delta t} = \frac{DP_{ij}}{Dt} + P_{mj}\frac{\partial u_m}{\partial x_i} + P_{im}\frac{\partial u_m}{\partial x_j}. \tag{3.30}$$

This is also found when the objective symmetric tensor $P_{mj}\,e_{mi} + P_{im}\,e_{mj}$ is added to the right-hand side of (3.29). Then, besides the spin tensor, the rate of deformation tensor also appears. Indeed, Oldroyd's derivative represents the rate of change of a tensor in the "body fixed" frame, thus in a reference frame which translates, rotates and deforms with the particle, again decomposed into components of the frame fixed in space. Oldroyd's derivative of an objective tensor is also objective, and therefore the Rivlin-Ericksen tensors known from Sect. 1.2.4 are objective tensors. A relation between the stress tensor and the Rivlin-Ericksen tensors always expresses an objective constitutive relation.

The value of these objective derivatives (and indeed others) lies in the fact that they generalize material behavior measured in the frame fixed in space to arbitrarily large deformations. For sufficiently small deformation velocities, which in general also means small rotational velocities, (3.29) and (3.30) reduce back to the partial time derivatives, and therefore Eq. (3.22) serves to describe oscillatory fluid motions of small amplitude very well.

Both of the models discussed are examples from the many non-Newtonian fluid models, which are, as a rule, all of empirical nature. On the basis of a simple fluid, a number of these constitutive relations can be systematically ordered. We refer here to the more advanced literature, but shall mention two more models which have found numerous technological applications, because the general functional dependency of the friction stress tensor on the history of the relative deformation gradient has an explicit form in these cases. The *viscous stress tensor* is a tensor valued function of this history, with nine (or in the case of symmetry, six) components. The history is a function of the time t', which describes the course of the relative deformation gradient tensor. t' can lie between $-\infty$ and the current time t. The tensor of the friction stresses is therefore a tensor valued function, whose arguments are also tensor valued functions. We speak of a function of a function, or of a *functional*. The relative deformation gradient tensor $C_{ij}(\vec{x}, t, t')$ describes the deformation which the particle situated at $\vec{x}$ at time t has experienced at time t'. Consider the fluid motion $\vec{x} = \vec{x}(\vec{\xi}, t)$ and the position of the material point $\vec{\xi}$ at time $t' < t$, i.e., $\vec{x}' = \vec{x}\left(\vec{\xi}, t'\right)$. If we replace $\vec{\xi}$ here by $\vec{\xi} = \vec{\xi}(\vec{x}, t)$ to obtain

$$\vec{x}' = \vec{x}(\vec{x}, t, t')$$

we are actually using the current configuration as the reference configuration. For fixed current time t and the new parameter $t - t' \geq 0$, the relative motion is the *history* of the motion. The symmetric tensor

$$\frac{\partial x'_l}{\partial x_i} \frac{\partial x'_l}{\partial x_j}$$

formed with the relative deformation gradient

$$\frac{\partial x'_l}{\partial x_i}$$

is the above relative deformation gradient tensor, also called the relative right Cauchy-Green tensor (see also Eq. (3.45)).

We are considering the case in which the history $C_{ij}(\vec{x}, t, t')$ can be expanded into a Taylor series. The coefficients of the series are Rivlin-Ericksen tensors defined by (1.68), so that the following holds for the expansion

$$C_{ij}(\vec{x}, t, t') = \delta_{ij} + (t' - t)A_{(1)ij} + \frac{1}{2}(t' - t)^2 A_{(2)ij} + \dots \tag{3.31}$$

(To see the equivalence

$$A_{(n)ij} = \left[\frac{D^n C_{ij}}{Dt'^n}\right]_{t'=t}$$

we differentiate the square of the line element ds' with respect to t'

$$\frac{D^n ds'^2}{Dt'^n} = \frac{D^n}{Dt'^n}\left(\frac{\partial x'_l}{\partial x_i}\frac{\partial x'_l}{\partial x_j}\right)dx_i dx_j = \frac{D^n C_{ij}}{Dt'^n}\frac{\partial x_i}{\partial x'_k}\frac{\partial x_j}{\partial x'_m}dx'_k dx'_m.$$

On the other hand by (1.68)

$$\frac{D^n ds'^2}{Dt'^n} = A_{(n)ij}dx'_i dx'_j.$$

For $t' = t$ therefore

$$\left[\frac{D^n C_{ij}}{Dt'^n}\right]_{t'=t}\delta_{ik}\delta_{jm}dx'_k dx'_m = A_{(n)ij}dx'_i dx'_j,$$

hence the above equivalence.)

If we truncate the series at the nth term (either because the higher Rivlin-Ericksen tensors become very small, as according to (1.68) is the case if the change of the material line element occurs at a low enough rate, or if the kinematics is so restricted that the higher tensors vanish identically, as is the case in *steady unidirectional* or *viscometric flow* for $n > 2$), then the friction stress tensor is no longer a function of a function, but is a function of n Rivlin-Ericksen tensors. Then the constitutive relation reads

$$\tau_{ij} = -p\delta_{ij} + \varphi_{ij}\{A_{(1)kl}, \ldots, A_{(n)kl}\}, \tag{3.32a}$$

or, symbolically

$$\mathbf{T} = -p\,\mathbf{I} + \varphi\{\mathbf{A}_{(1)}, \ldots, \mathbf{A}_{(n)}\}, \tag{3.32b}$$

where φ is a tensor valued function of the n tensor variables $\mathbf{A}(1)$ to $\mathbf{A}_{(n)}$. For unidirectional flows in particular, the transition from the functional leads to the equation

$$\mathbf{T} = -p\,\mathbf{I} + \varphi\{\mathbf{A}_{(1)}, \mathbf{A}_{(2)}\}. \tag{3.33}$$

By unidirectional flows we understand flows in which in a certain (not necessarily Cartesian) coordinate system, only one velocity component is nonzero, and this varies only perpendicular to the direction of flow. Because of the particularly simple kinematics, this class of flows often leads to closed solutions, and will be treated further in Chap. 6.

If we denote the flow direction with the unit vector $\vec{e}_1$, the direction of velocity change with $\vec{e}_2$ and the direction orthogonal to these by $\vec{e}_3$, the first and second Rivlin-Ericksen tensors take on the form known from Sect. 1.2 of the simple shearing flow (1.71) and (1.72). Since the components of $\mathbf{A}_{(1)}$ and $\mathbf{A}_{(2)}$ are only functions of $\dot{\gamma}$, we extract from (3.33) the equation

$$\tau_{ij} = -p\delta_{ij} + \varphi_{ij}(\dot{\gamma}). \tag{3.34}$$

The stresses $\tau_{13} = \tau_{31}$ and $\tau_{23} = \tau_{32}$ are zero in all unidirectional flows, and the matrix representation of (3.34) reads

$$[\mathbf{T}] = \begin{bmatrix} \varphi_{11}(\dot{\gamma}) - p & \varphi_{12}(\dot{\gamma}) & 0 \\ \varphi_{12}(\dot{\gamma}) & \varphi_{22}(\dot{\gamma}) - p & 0 \\ 0 & 0 & \varphi_{33}(\dot{\gamma}) - p \end{bmatrix}. \tag{3.35}$$

In order to eliminate the undefined pressure in incompressible flow, we form the differences of the normal stresses

$$\tau_{11} - \tau_{22} = N_1(\dot{\gamma})$$
$$\tau_{22} - \tau_{33} = N_2(\dot{\gamma})' \qquad (3.36)$$

which, together with the shear stress

$$\tau_{12} = \tau(\dot{\gamma}) \qquad (3.37)$$

fully determine the behavior of the simple fluid in steady *unidirectional flows*. $N_1(\dot{\gamma})$ is called the *primary normal stress function*, $N_2(\dot{\gamma})$ the *secondary normal stress function* and $\tau(\dot{\gamma})$ the *shear stress function*. N_1 and N_2 are even functions of $\dot{\gamma}$, and τ is an odd function of $\dot{\gamma}$. Of course all of these functions depend on the material. However, two different fluids with the same normal and shearing stress functions can show completely different behavior in flows which are not unidirectional.

We consider now the case where the change of ds^2 in (1.68) is sufficiently slow. This occurs in slow and slowly varying motions, and we shall say that $\mathbf{A}_{(1)}$ is of the first order and $\mathbf{A}_{(2)}$ is of the second order

$$\mathbf{A}_{(n)} \sim O(\epsilon^n). \qquad (3.38)$$

If we restrict ourselves to terms of the first order in ϵ, (3.32a, 3.32b) can be written in the form

$$\mathbf{T} = -p\mathbf{I} + \eta \mathbf{A}_{(1)}, \qquad (3.39a)$$

or

$$\tau_{ij} = -p\,\delta_{ij} + \eta\, A_{(1)ij}. \qquad (3.39b)$$

Since $A_{(1)ij} = 2e_{ij}$, we recognize the Cauchy-Poisson law (3.1a, 3.1b) for incompressible Newtonian fluids, which we have reached here for the limiting case of very slow or slowly varying motions. However, "slow variations" implies a variation with a typical time scale large in comparison to the memory time of the fluid. As we already found in connection with (3.22), the Newtonian fluid has no memory, so that the time scale can be arbitrarily small in the sense of the approximation (3.39a, 3.39b).

If we consider terms up to the second order in ϵ, (3.32b) furnishes the definition of a second order fluid

$$\mathbf{T} = -p\,\mathbf{I} + \eta\,\mathbf{A}_{(1)} + \beta\,\mathbf{A}_{(1)}^2 + \gamma\,\mathbf{A}_{(2)}. \qquad (3.40)$$

The coefficients η, β and γ here are material dependent constants (where, from measurements, γ turns out to be negative and should not be confused with the shear angle). The validity of this constitutive relation is not kinematically restricted, and it can be used in general also for unsteady, three-dimensional flows. The restriction is the necessary "slowness" of the flow under consideration, where the meaning of "slow" is to be clarified in the given problem.

The second order fluid is the simplest model which shows two different normal stress functions in simple shearing flow, which increase with $\dot{\gamma}^2$ as they should. But the shear-thinning always seen in experiments on polymeric fluids is not described. In spite of this, this model is used in many applications, and it also predicts most non-Newtonian effects qualitatively, if not always quantitatively. Finally, this constitutive relation, which satisfies all the axioms stated at the beginning of this chapter, can be seen, separate from its derivation, as an admissible fluid model, whose agreement with actual material behavior is in any case to be checked experimentally (as is also done with the Cauchy-Poisson law).

The materials mentioned until now have been pure fluids, that is materials where the shearing forces vanish when the rate of deformation vanishes. As already said, we often have to deal with substances which have a dual character. Of these substances, we shall mention here the *Bingham material*, which can serve as a model for the material behavior of paint, or more generally, for high concentration suspensions of solid particles in Newtonian fluids. If the solid particles and the fluid are dielectrics, that is do not conduct electrically, then these dispersions can take on Bingham character under a strong electric field, even if they show only pure fluid behavior without electric field. These *electrorheological fluids*, whose material behavior can be changed very quickly and without much effort, can find applications, for example, in the damping of unwanted oscillations. Through appropriate measures the material can be made to self-adjust to changing requirements and may be formed into "intelligent" materials, which are found increasingly interesting. Even the behavior of grease used as a means of lubricating ball bearings, can be described with the Bingham model.

We can gain considerable insight into the behavior of Bingham materials behavior looking at the simple shearing flow: if the material flows, we have for the shear stress

$$\tau = \eta_1 \dot{\gamma} + \vartheta; \quad \tau \geq \vartheta. \tag{3.41}$$

Otherwise the material behaves like an elastic solid, i.e., the shear stress is

$$\tau = G\gamma; \quad \tau < \vartheta, \tag{3.42}$$

where ϑ is the yield stress and G is the shear modulus. In a general stress state, the yield stress becomes tensorial, and in place of ϑ, ϑ_{ij} appears, so that the criterion for flow is not immediately obvious. In what follows, we introduce the generalized *Bingham constitutive relation*, and first describe the elastic behavior. Our starting

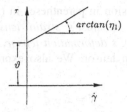

Fig. 3.2 Behavior of Bingham materials

point is Eqs. (1.5) and (1.8), where we now consider $\vec{\xi}$ as the particle position in a stress free state, and $\vec{x}$ as the position of the same particle in the deformed state. An undeformed material vector element has the following relation with the deformed element $d\vec{x}$ (Fig. 3.2)

$$dx_i = \frac{\partial x_i}{\partial \xi_j} d\xi_j, \tag{3.43}$$

which follows directly from (1.5) and where $\partial x_i / \partial \xi_j$ is the *deformation gradient*. Thus we write for the square of the element of length $|d\vec{x}|$

$$dx_i dx_i = \frac{\partial x_i}{\partial \xi_j} \frac{\partial x_i}{\partial \xi_k} d\xi_j d\xi_k \tag{3.44}$$

as well as for the difference

$$|d\vec{x}|^2 - \left|d\vec{\xi}\right|^2 = \left(\frac{\partial x_i}{\partial \xi_j} \frac{\partial x_i}{\partial \xi_k} - \delta_{jk}\right) d\xi_j d\xi_k \tag{3.45}$$

and we shall denote the half of the expression in parentheses as *Lagrangian strain tensor* E_{jk}. The obviously symmetric tensor $(\partial x_i / \partial \xi_j)(\partial x_i / \partial \xi_k)$ in (3.44) is called *Green's deformation tensor* or the *right Cauchy-Green tensor*. Using the intermediate step

$$|d\vec{x}|^2 - \left|d\vec{\xi}\right|^2 = \left(\frac{\partial x_i}{\partial \xi_j} \frac{\partial x_i}{\partial \xi_k} - \delta_{jk}\right) \frac{\partial \xi_j}{\partial x_l} dx_l \frac{\partial \xi_k}{\partial x_m} dx_m \tag{3.46}$$

equation (1.5) allows the representation of (3.45) in field coordinates

$$|d\vec{x}|^2 - \left|d\vec{\xi}\right|^2 = \left(\delta_{lm} - \frac{\partial \xi_k}{\partial x_l} \frac{\partial \xi_k}{\partial x_m}\right) dx_l dx_m. \tag{3.47}$$

We call the half of the expression in parentheses in (3.47) *Eulerian strain tensor* ϵ_{lm}; this is also known as *Almansi's strain tensor*. The symmetric tensor $(\partial \xi_k / \partial x_l)(\partial \xi_k / \partial x_m)$ is *Cauchy's deformation tensor*, and it is the Eulerian counterpart to Green's deformation tensor. We also express the deformation tensors using the displacement vector

$$\vec{y} = \vec{x} - \vec{\xi} \tag{3.48}$$

and extract, with Green's deformation tensor

$$\frac{\partial x_i}{\partial \xi_j}\frac{\partial x_i}{\partial \xi_k} = \frac{\partial y_i}{\partial \xi_j}\frac{\partial y_i}{\partial \xi_k} + \frac{\partial y_k}{\partial \xi_j} + \frac{\partial y_j}{\partial \xi_k} + \delta_{kj} \tag{3.49}$$

the following expression for the Lagrangian strain tensor

$$E_{jk} = \frac{1}{2}\left(\frac{\partial y_i}{\partial \xi_j}\frac{\partial y_i}{\partial \xi_k} + \frac{\partial y_j}{\partial \xi_k} + \frac{\partial y_k}{\partial \xi_j}\right), \tag{3.50}$$

which, for small enough deformations (ignoring the quadratic terms), simplifies to

$$E_{jk} = \frac{1}{2}\left(\frac{\partial y_j}{\partial \xi_k} + \frac{\partial y_k}{\partial \xi_j}\right). \tag{3.51}$$

From (3.48) follows

$$\frac{\partial y_j}{\partial x_k} = \delta_{kj} - \frac{\partial \xi_j}{\partial x_k} \tag{3.52a}$$

and for small deformations, i.e., $\partial y_j / \partial x_k \ll \partial \xi_j / \partial x_k$, we find

$$\frac{\partial \xi_j}{\partial x_k} \approx \delta_{kj}. \tag{3.52b}$$

Comparison of (3.46) and (3.47) furnishes

$$E_{jk}\frac{\partial \xi_j}{\partial x_l}\frac{\partial \xi_k}{\partial x_m} = \epsilon_{lm} \tag{3.53}$$

and we are led to

$$E_{lm} \approx \epsilon_{lm}. \tag{3.54}$$

In this case the difference between Lagrangian and Eulerian strain tensor vanishes. In what follows we shall restrict ourselves to small deformations, and we find from the substantial derivative of the deformation tensor $\epsilon_{lm} = 1/2(\partial y_l/\partial x_m + \partial y_m/\partial x_l)$ again the rate of deformation tensor (1.29a)

$$\frac{D\epsilon_{lm}}{Dt} = \frac{1}{2}\left(\frac{\partial u_l}{\partial x_m} + \frac{\partial u_m}{\partial x_l}\right) = e_{lm}. \tag{3.55}$$

In rheology it is usual to denote the negative mean normal stress as the pressure, and we shall follow this usage here, noting however that the mean normal stress in general includes isotropic terms which are dependent on the motion. (See Eq. (3.3) for the case of Newtonian fluid.) However, for incompressible materials, to which we shall restrict ourselves, the pressure is an unknown function which follows from the solution of the equations of motion only to within an additive constant, and the absolute value of the pressure is not important. Therefore, for the constitutive relation we shall write

$$\tau_{ij} = -p\,\delta_{ij} + \tau'_{ij}, \quad p = -\frac{1}{3}\tau_{kk}. \tag{3.56}$$

The tensor τ'_{ij} is, as above, a deviator, that is the trace of the tensor vanishes. If e'_{ij} and ϵ_{ij} are the deviators of the rate of deformation and the strain tensors, the following holds at the yield point

$$e'_{ij} = 0 \quad \text{and} \quad \tau'_{ij} = 2G\epsilon'_{ij} = \vartheta_{ij}. \tag{3.57}$$

We shall assume that yield occurs according to the *von Mises' hypothesis*, that is when the energy stored in the material as a result of the deviatoric stresses reaches a given value

$$\frac{1}{2}\epsilon'_{ij}\tau'_{ij} = \text{const.} \tag{3.58}$$

By (3.57), the potential energy at the yield point is then

$$\frac{1}{4G}\vartheta_{ij}\vartheta_{ij} = \text{const} = \frac{1}{2G}\vartheta^2, \tag{3.59}$$

so that we obtain the constitutive relation of the Bingham material in the form

$$\tau'_{ij} = 2\eta e'_{ij} \quad \text{if} \quad \frac{1}{2}\tau'_{ij}\tau'_{ij} \geq \vartheta^2, \tag{3.60}$$

and

$$\tau'_{ij} = 2G\epsilon_{ij} \quad \text{if} \quad \frac{1}{2}\tau'_{ij}\tau'_{ij} \leq \vartheta^2, \tag{3.61}$$

where

$$\eta = \eta_1 + \vartheta \Big/ \left(2e'_{ij}e'_{ij}\right)^{1/2}. \tag{3.62}$$

The incompressible Bingham material is determined by the three material constants G, ϑ and η_1. Wherever it flows it behaves as a fluid with variable viscosity η, which depends on the second invariant of the rate of deformation deviator I'_{2e}. Therefore here it behaves as a generalized Newtonian fluid. The yield criterion in (3.60) and (3.63) contains only the second invariant I'_{2e} of the stress deviator, so this is coordinate invariant. For simple shearing flow we have $\tau'_{ij}\tau'_{ij} = 2\tau^2_{xy}$, and Eqs. (3.60) and (3.61) reduce to Eqs. (3.41) and (3.42), since by (3.62) $e'_{xy} = \frac{1}{2}du/dy$. Often, instead of the elastic solid body behavior in the region where $\frac{1}{2}\tau'_{ij}\tau'_{ij} < \vartheta^2$, rigid body behavior is assumed. Then the constitutive relation takes on the form

$$\tau'_{ij} = 2\eta e'_{ij} \quad \text{if} \quad \frac{1}{2}\tau'_{ij}\tau'_{ij} \geq \vartheta^2 \tag{3.63}$$

and

$$\epsilon'_{ij} = 0 \quad \text{if} \quad \frac{1}{2}\tau'_{ij}\tau'_{ij} \leq \vartheta^2. \tag{3.64}$$

In numerical calculations, the Bingham constitutive relation is also approximated with a *two-viscosity model*, which is more easily dealt with numerically, and which also offers advantages in localizing the yield surfaces. In this model the rigid body character (3.64) is replaced by a Newtonian flow behavior with very high viscosity $\eta_0(\eta_0 \gg \eta_1)$. Then instead of (3.64) we have the law

$$\tau'_{ij} = 2\eta_0 e'_{ij} \quad \text{if} \quad \frac{1}{2}\tau'_{ij}\tau'_{ij} \leq \vartheta^2, \tag{3.65}$$

which, for $\eta_0 \to \infty$, i.e., $e'_{ij} \to 0$ becomes (3.64).

Chapter 4
Equations of Motion for Particular Fluids

We shall now specialize the universally valid equations, namely Cauchy's equation (2.38) and the energy equation (2.119) to the two most technically important cases: Newtonian fluids and inviscid fluids. The continuity equation (2.2) (mass balance) and the symmetry of the stress tensor (2.53) (angular momentum balance) remain unaffected by the choice of the constitutive relation.

4.1 Newtonian Fluids

4.1.1 The Navier-Stokes Equations

We start with a Newtonian fluid which is defined by the constitutive relation (3.1) and, by setting (3.1) and (1.29) into (2.38), we obtain the *Navier-Stokes equations*

$$\varrho \frac{Du_i}{Dt} = \varrho \, k_i + \frac{\partial}{\partial x_i} \left\{ -p + \lambda^* \frac{\partial u_k}{\partial x_k} \right\} + \frac{\partial}{\partial x_j} \left\{ \eta \left[\frac{\partial u_i}{\partial x_j} + \frac{\partial u_j}{\partial x_i} \right] \right\}, \qquad (4.1)$$

where we have used the exchange property of the Kronecker delta δ_{ij}.

With the linear law for the friction stresses (3.2) and the linear law for the heat flux vector (3.8), we specialize the energy equation to the case of Newtonian fluids

$$\varrho \frac{De}{Dt} - \frac{p}{\varrho} \frac{D\varrho}{Dt} = \Phi + \frac{\partial}{\partial x_i} \left[\lambda \frac{\partial T}{\partial x_i} \right], \qquad (4.2)$$

where the dissipation function Φ is given by (3.6). In the same way we deal with the forms (2.116) and (2.118) of the energy equation, which are often more appropriate. Another useful form of the energy equation arises by inserting the enthalpy $h = e + p/\varrho$ into (4.2). Because of

© Springer Nature Switzerland AG 2020
J. H. Spurk and N. Aksel, *Fluid Mechanics*,
https://doi.org/10.1007/978-3-030-30259-7_4

$$\varrho \frac{Dh}{Dt} = \varrho \frac{De}{Dt} - \frac{p}{\varrho} \frac{D\varrho}{Dt} + \frac{Dp}{Dt} \tag{4.3}$$

(4.2) can also be written as

$$\varrho \frac{Dh}{Dt} - \frac{Dp}{Dt} = \Phi + \frac{\partial}{\partial x_i} \left[\lambda \frac{\partial T}{\partial x_i} \right]. \tag{4.4}$$

As a consequence of Gibbs' relation (2.133), the entropy equation for Newtonian fluids can also appear in place of (4.2)

$$\varrho T \frac{Ds}{Dt} = \Phi + \frac{\partial}{\partial x_i} \left[\lambda \frac{\partial T}{\partial x_i} \right]. \tag{4.5}$$

If we choose the energy equation (4.2), together with the continuity equation and the Navier-Stokes equations we have five partial differential equations with seven unknown functions. But both the thermal equation of state $p = p(\varrho, T)$ and the caloric equation of state $e = e(\varrho, T)$ appear also. This set of equations forms the starting point for the calculation of frictional compressible flow.

By (4.1) the Navier-Stokes equations are given in Cartesian coordinates. However in many technical applications the geometry of the flow boundary suggests a curvilinear coordinate system (If we consider, for example, the unidirectional flow between rotating cylinders (Fig. 6.5), there is only one nonzero velocity component to consider in cylindrical coordinates, while in Cartesian coordinates there would be two components). It is then advisable to use the symbolic notation valid in all coordinate systems. In order to do this, we introduce the constitutive relation (3.1b) into Cauchy's equation (2.38b)

$$\varrho \frac{D\vec{u}}{Dt} = \varrho \vec{k} - \nabla p + \nabla(\lambda^* \nabla \cdot \vec{u}) + \nabla \cdot (2 \eta \mathbf{E}), \tag{4.6}$$

where now the use of the material derivative (1.78) is more expedient. In Eqs. (4.2)–(4.5), the operator $\partial / \partial x_i$, is to be replaced by the Nabla operator ∇, and the dissipation function is to be inserted in symbolic notation (3.6b). The most important curvilinear coordinate systems are orthogonal, and knowing the appropriate definition of the Nabla operator we can directly calculate the component equations of (4.6) in the chosen coordinate system. The method of calculation is explained in Appendix B, where the component forms of the Navier-Stokes equations (for incompressible flow) in the most often used coordinate systems can be found.

For isothermal fields, or by ignoring the temperature dependence of η and λ^*, the final term on the right-hand side of (4.1) can be put in a different form. In Cartesian index notation we have then

$$\frac{\partial}{\partial x_j}\left\{\eta\left[\frac{\partial u_i}{\partial x_j}+\frac{\partial u_j}{\partial x_i}\right]\right\}=\eta\left\{\frac{\partial^2 u_i}{\partial x_j\partial x_j}+\frac{\partial}{\partial x_i}\left[\frac{\partial u_k}{\partial x_k}\right]\right\},\tag{4.7}$$

where we have interchanged the order of differentiation in an intermediate step, so that from (4.1) the form cited by Navier and Stokes is obtained

$$\varrho\frac{Du_i}{Dt}=\varrho\,k_i-\frac{\partial p}{\partial x_i}+(\lambda^*+\eta)\frac{\partial}{\partial x_i}\left[\frac{\partial u_k}{\partial x_k}\right]+\eta\left[\frac{\partial^2 u_i}{\partial x_j\partial x_j}\right].\tag{4.8a}$$

In symbolic notation, this equation reads

$$\varrho\frac{D\vec{u}}{Dt}=\varrho\,\vec{k}-\nabla p+(\lambda^*+\eta)\nabla(\nabla\cdot\vec{u})+\eta\,\varDelta\vec{u}.\tag{4.8b}$$

In this $\varDelta=\nabla\cdot\nabla$ is the *Laplace operator*, whose explicit form in various coordinate systems may be found in Appendix B. In incompressible flow $(\partial u_k/\partial x_k=\nabla\cdot\vec{u}=0)$ (4.8a, 4.8b) is reduced to

$$\varrho\frac{Du_i}{Dt}=\varrho\,k_i-\frac{\partial p}{\partial x_i}+\eta\frac{\partial^2 u_i}{\partial x_k\partial x_k},\tag{4.9a}$$

or

$$\varrho\frac{D\vec{u}}{Dt}=\varrho\,\vec{k}-\nabla p+\eta\,\varDelta\vec{u}.\tag{4.9b}$$

Often the density distribution ϱ is homogeneous when the incompressible fluid is set in motion. Because $D\varrho/Dt=0$, this homogeneity remains for all time, so that the condition "incompressible flow" can be replaced by the condition "constant density". In what follows, we shall always assume this unless the opposite is explicitly stated (see also the discussion in Sect. 2.1). With (4.9a, 4.9b) and the continuity equation ($\partial u_i/\partial x_i=0$), we have four differential equations for the four unknown functions u_i and p, where p is now a dependent variable of the problem.

We interpret Eq. (4.9a, 4.9b) as follows: on the left is the product of the mass of the material particle (per unit volume) and its acceleration; on the right is the sum of the volume body force $\varrho\vec{k}$, the net pressure force per unit volume $-\nabla p$ (the difference between the pressure forces on the material particle, i.e., the divergence of the pressure stress tensor $-\nabla\cdot(p\mathbf{I})$, and the net viscous force per unit volume $\eta\,\varDelta\vec{u}$ (the difference between the viscous forces on the particle, i.e., the divergence of the viscous stress tensor in incompressible flow $2\eta\,\nabla\cdot\mathbf{E}$).

We next use the vector identity

$$\varDelta\vec{u}=\nabla(\nabla\cdot\vec{u})-\nabla\times(\nabla\times\vec{u}),\tag{4.10}$$

which is easily verified in index notation, and which reduces the application of the
Laplace operator to operations with ∇ even in curvilinear coordinates. Because
$\nabla \cdot \vec{u} = 0$, we then have

$$\eta \, \Delta \vec{u} = -2\eta \, \nabla \times \vec{\omega}. \qquad (4.11)$$

This equation makes it clear that in incompressible and irrotational flow
($\nabla \times \vec{u} = 2\vec{\omega} = 0$), the divergence of the viscous stress tensor vanishes. The viscous stresses themselves are not zero, it is only that they provide no contribution to
the acceleration of the particle. From the fact that the angular velocity appears on
the right-hand side, we may not conclude that the viscous stresses depend on $\vec{\omega}$
(which is of course impossible), but that $\Delta \vec{u}$ can be expressed by $-2\nabla \times \vec{\omega}$ in
incompressible flow.

4.1.2   Vorticity Equation

Since a viscous incompressible fluid behaves like an inviscid fluid in regions where
$\vec{\omega} = 0$, the question arises of what the differential equation for the distribution of $\vec{\omega}$
is. Of course this question does not arise if we consider the velocity field as given,
because then $\vec{\omega}$ can be calculated directly from the velocity field using Eq. (1.49).
To obtain the desired relation, we take the curl of the Eq. (4.9b). For reasons of
clarity, we shall use symbolic notation here. We assume further that $\vec{k}$ has a
potential ($\vec{k} = -\nabla \psi$), and use the identity (4.11) in Eq. (4.9b). In addition, we
make use of (1.78) to obtain the Navier-Stokes equations in the form

$$\frac{1}{2}\frac{\partial \vec{u}}{\partial t} - \vec{u} \times \vec{\omega} = -\frac{1}{2}\nabla \left[\psi + \frac{p}{\varrho} + \frac{\vec{u} \cdot \vec{u}}{2} \right] - \nu \nabla \times \vec{\omega}. \qquad (4.12)$$

The operation $\nabla \times$ applied to (4.12), along with the identity (easily verified in index
notation)

$$\nabla \times (\vec{u} \times \vec{\omega}) = \vec{\omega} \cdot \nabla \vec{u} - \vec{u} \cdot \nabla \vec{\omega} - \vec{\omega} \nabla \cdot \vec{u} + \vec{u} \nabla \cdot \vec{\omega} \qquad (4.13)$$

furnishes the new left-hand side $\partial \vec{\omega} / \partial t - \vec{\omega} \cdot \nabla \vec{u} + \vec{u} \cdot \nabla \vec{\omega}$, where we have already
noted that the flow is incompressible ($\nabla \cdot \vec{u} = 0$) and that the divergence of the curl
always vanishes

$$2\nabla \cdot \vec{\omega} = \nabla \cdot (\nabla \times \vec{u}) = 0. \qquad (4.14)$$

This can be shown in index notation or simply explained by the fact that the
symbolic vector ∇ is orthogonal to $\nabla \times \vec{u}$. On the right-hand side of (4.12), the
term in parantheses vanishes, since the symbolic vector ∇ is parallel to the gradient.
The remaining term on the right-hand side $-\nu \nabla \times (\nabla \times \vec{\omega})$ is recast using the
identity (4.10), and because $\nabla \cdot \vec{\omega} = 0$ from (4.14) we extract the new right-hand
side $\nu \, \Delta \vec{\omega}$. In this manner we arrive at the *vorticity equation*

$$\frac{\partial \vec{\omega}}{\partial t} + \vec{u} \cdot \nabla \vec{\omega} = \vec{\omega} \cdot \nabla \vec{u} + \nu \, \Delta \vec{\omega}. \tag{4.15}$$

Because $\partial/\partial t + \vec{u} \cdot \nabla = D/Dt$ we can shorten this to

$$\frac{D\vec{\omega}}{Dt} = \vec{\omega} \cdot \nabla \vec{u} + \nu \, \Delta \vec{\omega}. \tag{4.16}$$

This equation takes the place of the Navier-Stokes equation, and is often used as a starting point for, in particular, numerical calculations. Because $2\,\vec{\omega} = \text{curl}\,\vec{u}$, (4.16) represents a differential equation only in $\vec{u}$; the pressure term contained in (4.12) no longer appears. In two-dimensional flow $\vec{\omega} \cdot \nabla \vec{u}$ is zero, so that (4.16) can be written as

$$\frac{D\vec{\omega}}{Dt} = \nu \, \Delta \vec{\omega}. \tag{4.17}$$

For the time being we shall only treat the case of the *inviscid fluid* ($\nu = 0$), for which (4.16) takes the form

$$\frac{D\vec{\omega}}{Dt} = \vec{\omega} \cdot \nabla \vec{u} \tag{4.18a}$$

or in index notation

$$\frac{D\omega_i}{Dt} = \omega_k \frac{\partial u_i}{\partial x_k}. \tag{4.18b}$$

After expanding the material derivative, we can consider (4.18a, 4.18b) as a differential equation for the field $\vec{\omega}(\vec{x}, t)$, but also immediately as a differential equation for the angular velocity $\vec{\omega}\left(\vec{\xi}, t\right)$ of the material particle $\vec{\xi}$. If we view (4.18a, 4.18b) in this way it has a simple solution: instead of the unknown vector $\vec{\omega}\left(\vec{\xi}, t\right)$, we introduce with (1.5) $\left(x_i = x_i(\xi_j, t)\right)$ the unknown vector $\vec{c}\left(\vec{\xi}, t\right)$ with the mapping

$$\omega_i = c_j \frac{\partial x_i}{\partial \xi_j}. \tag{4.19}$$

The tensor $\partial x_i/\partial \xi_j$ is known from (3.43), where it provided the relation

$$dx_i = \frac{\partial x_i}{\partial \xi_j} d\xi_j \tag{4.20}$$

between the deformed element $\mathrm{d}\vec{x}$ and $\mathrm{d}\vec{\xi}$. This tensor is nonsingular since the Jacobian $J = \det(\partial x_i/\partial \xi_j)$ is not equal to zero, a fact which was used in Sect. 1.2 and in the discussion of the Bingham material in Chap. 3. The material derivative of (4.19) leads to the relation

$$\frac{\mathrm{D}\omega_i}{\mathrm{D}t} = \frac{\mathrm{D}c_j}{\mathrm{D}t}\frac{\partial x_i}{\partial \xi_j} + c_j \frac{\mathrm{D}}{\mathrm{D}t}\left[\frac{\partial x_i}{\partial \xi_j}\right], \tag{4.21}$$

whose final term we transform by interchanging the order of differentiation

$$c_j \frac{\mathrm{D}}{\mathrm{D}t}\left[\frac{\partial x_i}{\partial \xi_j}\right] = c_j \frac{\partial u_i}{\partial \xi_j}. \tag{4.22}$$

Here $\partial u_i/\partial \xi_j$ is the velocity gradient in the material description $\vec{u} = \vec{u}\left(\vec{\xi}, t\right)$. We take the velocity in material coordinates as given by (1.9), thus $\vec{u} = \vec{u}\left\{\vec{x}\left(\vec{\xi}, t\right), t\right\}$, so that after using the chain rule on (4.22), we obtain the equation

$$c_j \frac{\mathrm{D}}{\mathrm{D}t}\left[\frac{\partial x_i}{\partial \xi_j}\right] = c_j \frac{\partial u_i}{\partial x_k}\frac{\partial x_k}{\partial \xi_j}, \tag{4.23}$$

or, with (4.19) also

$$c_j \frac{\mathrm{D}}{\mathrm{D}t}\left[\frac{\partial x_i}{\partial \xi_j}\right] = \omega_k \frac{\partial u_i}{\partial x_k}. \tag{4.24}$$

Then by (4.21), instead of (4.18a, 4.18b) we can finally write

$$\frac{\mathrm{D}c_j}{\mathrm{D}t} = 0, \quad \text{or} \quad c_j = c_j\left(\vec{\xi}\right). \tag{4.25}$$

This means that for a material particle $\left(\vec{\xi} = \text{const}\right)$ the vector c_j does not change. We fix this still unknown vector from the initial condition for $\vec{\omega}$

$$\omega_i(t = 0) = \omega_{0i} = c_j \frac{\partial x_i}{\partial \xi_j}\bigg|_{t=0} = c_j \delta_{ij} = c_i \tag{4.26}$$

since $\vec{x}_i(t = 0) = \vec{\xi}_i$ and thus also obtain from (4.19) the desired solution

$$\omega_i = \omega_{0j}\frac{\partial x_i}{\partial \xi_j}, \tag{4.27}$$

which, compared to (4.20), shows us that the vector $\vec{\omega}$ obeys the same mapping as $d\vec{x}$. If we choose the vector $d\vec{\xi}$ to be tangential to $\vec{\omega}$, so that $d\vec{\xi}$ is simultaneously a vector element on the vorticity line, this comparison shows that the same material elements at the time t, denoted $d\vec{x}$, are still tangential to the vector of the angular velocity $\vec{\omega}$, and thus vorticity lines are material lines. Since the vector of the angular velocity $\vec{\omega}$ changes in exactly the same manner as the material line element $d\vec{x}$, the magnitude of the angular velocity must get larger when $|d\vec{x}|$ increases, i.e., when the material line element is stretched. Thus we deduce the following conclusion which is also important for the behavior of turbulent flows:

> The angular velocity of a vortex filament increases when it is stretched and decreases when it is compressed.

We shall go into this aspect of inviscid flow in more detail in connection with *Helmholtz's vortex theorems*, and shall infer from (4.27) the important fact that the angular velocity of a material particle remains zero for all times if it is zero at time $t = 0$. An inviscid flow thus remains (if $\vec{k}$ has a potential) irrotational for all times if it is irrotational at the reference time. We could also reach this conclusion from (4.18a, 4.18b) together with the initial condition, but (4.27) shows us clearly that the deformation gradient $\partial x_i / \partial \xi_j$ also must remain finite. A flow which develops discontinuities is in general no longer irrotational.

4.1.3 Effect of Reynolds' Number

In viscous flow, the term, $\nu \Delta \vec{\omega}$ in (4.16) represents the change in the angular velocity of a material particle which is due to its neighboring particles. Clearly, the particle is set into rotation by its neighbors via viscous torques, and it itself exerts torques on other neighboring particles, thus setting these into rotation. The particle only passes on the vector of angular velocity $\vec{\omega}$ on to the next one, just as temperature is passed on by heat conduction, or concentration by diffusion. Thus we speak of the "diffusion" of the angular velocity vector $\vec{\omega}$ or of the vorticity vector curl $\vec{u} = \nabla \times \vec{u} = 2\vec{\omega}$. From what we have said before, we conclude that angular velocity cannot be produced within the interior of an incompressible fluid, but gets there by diffusion from the boundaries of the fluid region. Flow regions where the diffusion of the vorticity vector is negligible can be treated according to the rules of inviscid and irrotational fluids.

As we know, equations which express physical relationships and which are *dimensionally homogeneous* (only these are of interest in engineering) must be reducible to relations between dimensionless quantities. Using the typical velocity U of the problem, the typical length L and the density ϱ, constant in incompressible flow, we introduce the dimensionless dependent variables

$$u_i^+ = \frac{u_i}{U} \qquad (4.28)$$

$$p^+ = \frac{p}{\varrho\, U^2} \qquad (4.29)$$

and the independent variables

$$x_i^+ = \frac{x_i}{L} \qquad (4.30)$$

$$t^+ = t\frac{U}{L} \qquad (4.31)$$

into the Navier-Stokes equations, and obtain (neglecting body forces)

$$\frac{\partial u_i^+}{\partial t^+} + u_j^+ \frac{\partial u_i^+}{\partial x_j^+} = -\frac{\partial p^+}{\partial x_i^+} + Re^{-1} \frac{\partial^2 u_i^+}{\partial x_j^+ \partial x_j^+}, \qquad (4.32)$$

where Re is the already known Reynolds' number

$$Re = \frac{UL}{\nu}.$$

Together with the dimensionless form of the continuity equation for incompressible flow

$$\frac{\partial u_i^+}{\partial x_i^+} = 0 \qquad (4.33)$$

and the dimensionless quantities which determine the shape of the flow boundary (for example, an airfoil), the problem is formulated in a mathematically proper way. The solutions found, the dimensionless velocity field u_i^+ and the dimensionless pressure field p^+ say, will then not be changed when the body exposed to the stream is enlarged in a geometrically similar manner, and the kinematic viscosity ν or the velocity U are simultaneously changed so that the Reynolds' number stays the same. As long as the Reynolds' number remains constant, nothing changes in the mathematical formulation. Thus the quantities calculated from the solution (for example the dimensionless drag c_D), do not change either. The coefficient of drag only changes if the Reynolds' number is changed in accordance with the law (3.12) obtained by dimensional considerations alone.

An important and largely unsolved problem of fluid mechanics is the dependency of the solution of the Navier-Stokes equations (4.32) and the continuity equation (4.33) on the Reynolds' number which only appears as a parameter. This difficulty is already evident in such simple flows as unidirectional flows to be discussed in Chap. 6. The *laminar* flows given there are only realized below a certain *critical Reynolds' number*. If this Reynolds' number is exceeded, for example by decreasing the viscosity, a completely different flow ensues. This flow is always unsteady, three-dimensional and rotational. If we measure the velocity at a fixed position, we observe that it varies irregularly about an average value: velocity and pressure are random quantities. We refer to such flows as *turbulent*. The calculation of turbulent flows has until now only been achieved using numerical integration of geometrically simple flows. The results of these numerical simulations allow important insights into the structure of turbulence. However for flows appearing in applications, the methods are computationally too difficult, and because of this we shall remain dependent on semi-empirical approximation methods for the conceivable future. These furnish only average flow quantities though these are the ones which are technically important.

We have introduced the Reynolds' number by way of dimensional consideration. But it can also be interpreted as the ratio of the typical inertial force to the typical viscous force. The typical inertial force is the (negative) product of the mass (per unit volume) and the acceleration, and so is the first term in the Navier-Stokes equation (4.1). The typical inertial term $\varrho\, u_1\, \partial u_1/\partial x_1$ is of the order of magnitude of $\varrho\, U^2/L$; the characteristic viscosity term $\eta\, \partial^2 u_1/\partial x_1^2$ has the order of magnitude of $\eta\, U/L^2$. The ratio of the two orders of magnitude is the Reynolds' number

$$(\varrho\, U^2/L)/(\eta\, U/L^2) = \varrho\, U L/\eta = U L/\nu = Re. \tag{4.34}$$

The Reynolds' number may also be interpreted as the ratio of the characteristic length L to the *viscous length* ν/U; this is an interpretation which is particularly useful if the inertia forces vanish identically, as is the case in steady unidirectional flow.

If the Reynolds' number tends to infinity or to zero, simplifications arise in the Navier-Stokes equations, and these are often the only way to make the solution of a problem possible. However these limiting cases are never reached in reality but lead to approximate solutions which are better the larger (or smaller) the Reynolds' number becomes (*asymptotic solutions*).

First we shall discuss the limiting case $Re \rightarrow 0$, which is realized

(a) if U is very small,
(b) if ϱ is very small (for example, flow of gases in evacuated tubes),
(c) if η is very large (thus generally in flows of very viscous fluids), or
(d) if the typical length is very small (flow past very small bodies, for example dust or fog particles. Such flows appear also in two phase flows if one phase is gaseous, and the other liquid or solid, but also if small solid particles are suspended in liquid. Flows through porous media, for example ground water flows, also fall into this category.)

From (4.34), $Re \rightarrow 0$ characterizes the dominance of the viscous forces over the inertial forces. The limit $Re \rightarrow 0$ in (4.32) shows this formally: the whole left-hand side of this equation can be ignored compared to the term $Re^{-1}\, \Delta\vec{u}$. The pressure gradient ∇p may not be neglected in general, because along with the velocity vector

$\vec{u}$, it is the other variable present in the differential equations (4.32) and (4.33). Only the solution for given boundary conditions resolves the relative role of the pressure, or more exactly the pressure difference because the pressure is determined by (4.32) and (4.33) only up to an additive constant. We also see directly from (4.29) that the pressure gradient tends to infinity as Re^{-1}, if the limit $Re \rightarrow 0$ is realized by $\varrho \rightarrow 0$.

Ignoring the inertia terms leads to an enormous simplification in the mathematical treatment, since these are the nonlinear terms in the equations. The equation arising from taking the limit in (4.32) is therefore linear, and reads in dimensional form

$$\frac{\partial p}{\partial x_i} = \eta \frac{\partial^2 u_i}{\partial x_j \partial x_j}. \tag{4.35}$$

For the second limiting case $Re \rightarrow \infty$, the viscous terms in (4.32) vanish. The resulting equation is known as *Euler's equation*, and it describes the inviscid flow. Later we shall discuss this equation in more detail (Sect. 4.2.1). If it were not for the experimental fact that a Newtonian fluid adheres to a wall, inviscid flow and flow at large Reynolds' numbers would be identical. If we assume at the outset that the flow be inviscid ($\nu = 0$), then in general the flow will be different from a viscous flow in the limit $\nu \rightarrow 0$. The reason for this singular behavior is that, mathematically, the highest derivative in Eq. (4.32) is lost for $\nu = 0$. We shall not go into the pure mathematical side of this problem here, but look at this condition through the following example. In simple shearing flow (or another steady unidirectional flow), the velocity field shown in Fig. 1.11 is entirely independent of the Reynolds' number (assuming we hold U constant, and the laminar flow does not change into turbulent flow). Theoretically this velocity distribution is maintained for $Re \rightarrow \infty$. Had we set $\nu = 0$, the shearing stress on the upper wall would be zero, and the flow could not be set into motion at all, i.e., the velocity of the fluid would be identically zero. Thus it remains to be clarified under which conditions a flow with large Reynolds' number corresponds to the flow calculated under the assumption of a completely inviscid fluid. The answer to this question depends on the given problem, and a generally valid answer cannot be given.

The influence of viscosity at large Reynolds' numbers is made clear by another simple example: a very thin plate coinciding with the positive x_1-axis is exposed to a steady uniform stream in the x_1-direction with velocity U. The material particles in the incident flow are taken as being irrotational, so that they remain so in inviscid flow (cf. (4.27)). Under the condition of zero viscosity, the plate does not impede the flow, although it does in viscous flow. The no-slip boundary condition leads to large velocity gradients near the wall and we expect the material particles to be set into rotation even if the viscosity is very small. From the discussion of the vorticity transport equation (4.16), we know that in viscous flow this can occur only through diffusion of the angular velocity $\vec{\omega}$ from the wall. The order of magnitude of the typical time τ for the diffusion of the angular velocity from the surface of the plate to a point at distance $\delta(x_1)$ can be estimated from (4.17)

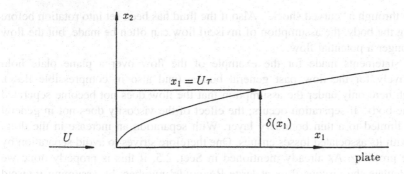

Fig. 4.1 Explanation of the boundary layer thickness

$$\frac{\omega}{\tau} \sim \nu \frac{\omega}{\delta^2(x_1)},$$

or, solving for τ

$$\tau \sim \frac{\delta^2(x_1)}{\nu}. \tag{4.36}$$

A particle not yet affected by the diffusion process that arrives exactly at the position $\delta(x_1)$ after this time, has covered the distance $U \tau = x_1$ (Fig. 4.1).

We extract the order of magnitude of the distance to which the diffusion can advance for a given x_1 from the equation

$$x_1 = U \tau \sim U \frac{\delta^2(x_1)}{\nu}, \tag{4.37}$$

or solving for $\delta(x_1)/x_1$

$$\delta(x_1)/x_1 \sim \sqrt{\nu/(U x_1)} = \sqrt{1/Re}. \tag{4.38}$$

Thus the effect of diffusion remains limited to a region whose extent grows as $\sqrt{x_1}$, but which becomes very narrow for large Reynolds' numbers. Apart from within this *boundary layer*, $2\vec{\omega} = \text{curl } \vec{u}$ is zero, and thus by (4.11), the viscous forces make no contribution to the acceleration, so that we might as well assume the flow to be inviscid potential flow. If we calculate the external flow under this assumption (leading us here to the trivial result $u(x_1, x_2) = U = \text{const}$), we incur a small error which vanishes with increasing Reynolds' number, because in reality the flow does not "feel" an infinitesimally thin plate but senses the boundary layer as a very slender obstacle by which it is somewhat deflected. In order to calculate the flow inside this boundary layer the viscosity certainly has to be taken into account.

It is of course possible that the outer flow may already be rotational for another reason, for example if the fluid particles in hypersonic flow are set into rotation by passing through a "curved shock". Also if the fluid has been set into rotation before reaching the body, the assumption of inviscid flow can often be made, but the flow is no longer a potential flow.

The statements made for the example of the flow over a plane plate hold qualitatively for the flow past general bodies (and also in compressible flow), although here only under the assumption that the flow does not become separated from the body. If separation occurs, the effect of the viscosity does not in general remain limited to a thin boundary layer. With separation an increase in the drag, along with its associated losses ensues. One therefore strives to avoid separation by suitable profiling. As already mentioned in Sect. 2.5, if this is properly done we may calculate the viscous flow at large Reynolds' numbers by assuming inviscid flow, in particular inviscid potential flow.

We are now in a position to give a more exact explanation of why with simple shearing flow (Fig. 1.11), even in the limiting case $Re \to \infty$, inviscid flow is not realized: at a distance x_2 from the lower plate the angular velocity of all the particles is the same, since the field only depends on x_2. The particle at the position (x_1, x_2) at the given instant in time thus carries as much angular velocity with it downstream as the particle which replaces it at this position has. The vorticity diffusing from the upper moving plate to the line x_2 is thus not carried (convected) downstream as in the case of the boundary layer flow, but permeates cross-stream to the lower wall, so that the flow in the whole gap is to be treated as a viscous flow, even for $Re \to \infty$.

Besides unidirectional flow, we could bring up many other examples which all would show that inviscid flow does not always correspond with viscous flow at large Reynolds' numbers. In every situation it is therefore necessary to check carefully whether a flow calculated under the assumption of zero viscosity is actually realized. On the other hand, the discussion here has shown that the assumption of inviscid flow often allows a realistic description of the flow field around a body.

4.2　Inviscid Fluids

4.2.1　Euler's Equations

As we have already seen in Sect. 4.1.3, Euler's equation emerges from the Navier-Stokes equation (4.8a, 4.8b) for $Re = \infty$. However Euler's equation is also a special case of Cauchy's equation (2.38) if we use the particular constitutive relation for inviscid fluids (3.9). *Euler's equation* then reads

$$\varrho \frac{Du_i}{Dt} = \varrho \, k_i + \frac{\partial}{\partial x_j} \left(-p \, \delta_{ij} \right) \tag{4.39}$$

or

$$\varrho \frac{Du_i}{Dt} = \varrho k_i - \frac{\partial p}{\partial x_i}, \tag{4.40a}$$

and it holds without restriction for all inviscid flows. In symbolic notation we write

$$\varrho \frac{D\vec{u}}{Dt} = \varrho \vec{k} - \nabla p. \tag{4.40b}$$

We derive Euler's equations in natural coordinates from (4.40b) by inserting the acceleration in the form (1.24). Relative to the basis vectors $\vec{t}$ in the direction of the pathline, $\vec{n}_\sigma$ in the principle normal direction and $\vec{b}_\sigma$ in the binormal direction, the vectors ∇p and $\vec{k}$ are

$$\nabla p = \frac{\partial p}{\partial \sigma} \vec{t} + \frac{\partial p}{\partial n} \vec{n}_\sigma + \frac{\partial p}{\partial b} \vec{b}_\sigma, \tag{4.41}$$

$$\vec{k} = k_\sigma \vec{t} + k_n \vec{n}_\sigma + k_b \vec{b}_\sigma, \tag{4.42}$$

and the component form of Euler's equation in natural coordinates, with $u = |\vec{u}|$, becomes

$$\frac{\partial u}{\partial t} + u \frac{\partial u}{\partial \sigma} = k_\sigma - \frac{1}{\varrho} \frac{\partial p}{\partial \sigma}, \tag{4.43}$$

$$\frac{u^2}{R} = k_n - \frac{1}{\varrho} \frac{\partial p}{\partial n} \tag{4.44}$$

$$0 = k_b - \frac{1}{\varrho} \frac{\partial p}{\partial b}. \tag{4.45}$$

As already noted, ignoring the viscosity is physically akin to ignoring the heat conduction, so that we write the constitutive relation for the heat flux vector in the form

$$q_i = 0. \tag{4.46}$$

By doing this we obtain from the energy equation (2.118) the energy equation of inviscid flow

$$\varrho \frac{D}{Dt} \left[\frac{1}{2} u_i u_i + h \right] = \frac{\partial p}{\partial t} + \varrho k_i u_i. \tag{4.47}$$

If, instead of the energy equation, the entropy equation (2.134) is used, this now reads

$$\frac{\mathrm{D}s}{\mathrm{D}t} = 0. \tag{4.48}$$

That is, the entropy of a material particle does not change in inviscid flow without heat conduction. (Here, as before, we have excluded other nonequilibrium processes which might arise through excitation of internal degrees of freedom of the fluid molecules or through chemical reactions.) The Eq. (4.48) characterizes an *isentropic flow*. If the entropy is homogeneous

$$\nabla s = 0, \tag{4.49}$$

we speak of *homentropic flow*. For the calorically perfect gas, (4.48) is replaced by

$$\frac{\mathrm{D}}{\mathrm{D}t}(p\,\varrho^{-\gamma}) = 0 \tag{4.50}$$

and (4.49) by

$$\nabla(p\,\varrho^{-\gamma}) = 0. \tag{4.51}$$

4.2.2 Bernoulli's Equation

Under mildly restricting assumptions it is possible to find so-called *first integrals* of Euler's equations, which then represent conservation laws. The most important first integral of Euler's equations is *Bernoulli's equation*. We assume that the mass body force has a potential $\left(\vec{k} = -\nabla\psi\right)$, i.e., $\psi = -g_i x_i$ for the gravitational force. We multiply Euler's equation (4.40a) by u_i, thus forming the inner product with $\vec{u}$, and obtain the relation

$$u_i \frac{\partial u_i}{\partial t} + u_i u_j \frac{\partial u_i}{\partial x_j} = -\frac{1}{\varrho} u_i \frac{\partial p}{\partial x_i} - u_i \frac{\partial \psi}{\partial x_i}. \tag{4.52}$$

After transforming the second term on the left-hand side and relabelling the dummy indices, this becomes

$$u_j \frac{\partial u_j}{\partial t} + u_j \frac{\partial}{\partial x_j}\left[\frac{u_i u_i}{2}\right] = -\frac{1}{\varrho} u_j \frac{\partial p}{\partial x_j} - u_j \frac{\partial \psi}{\partial x_j}. \tag{4.53}$$

We could, in principle, integrate this equation along an arbitrary smooth curve, but we arrive at a particularly simple and important result if we integrate along a streamline. With $u = |\vec{u}|$, from the differential equation for the streamline (1.11), we have

$$u_j = u \, dx_j/ds, \tag{4.54}$$

so that

$$u_j \frac{\partial}{\partial x_j} = u \frac{dx_j}{ds} \frac{\partial}{\partial x_j} = u \frac{d}{ds} \tag{4.55}$$

holds, and because $u_j \, \partial u_j/\partial t = u \, \partial u/\partial t$ we can write for (4.53)

$$\frac{\partial u}{\partial t} + \frac{d}{ds} \left[\frac{u^2}{2} \right] = -\frac{1}{\varrho} \frac{dp}{ds} - \frac{d\psi}{ds}. \tag{4.56}$$

Integration along the arc length of the streamline leads us to Bernoulli's equation in the form

$$\int \frac{\partial u}{\partial t} ds + \frac{u^2}{2} + \int \frac{dp}{\varrho} + \psi = C, \tag{4.57}$$

or integrating from the initial point A to the final point B we get the definite integral

$$\int_A^B \frac{\partial u}{\partial t} ds + \frac{1}{2} u_B^2 + \int_A^B \frac{1}{\varrho} \frac{dp}{ds} ds + \psi_B = \frac{1}{2} u_A^2 + \psi_A. \tag{4.58}$$

In order to evaluate the integrals, the integrands must in general appear as functions of the arc length s unless the integrand is a total differential. However, the first integral cannot be written as the integral of a total differential. Obviously, in incompressible flow of homogeneous density, dp/ϱ is a total differential. But this is also the case in *barotropic flow*, where the density is only a function of the pressure

$$\varrho = \varrho(p). \tag{4.59}$$

Then $dP = dp/\varrho(p)$ is a total differential, and the *pressure function*

$$P(p) = \int \frac{dp}{\varrho(p)} \tag{4.60}$$

can be calculated once and for all (if necessary, numerically). Clearly barotropic flows occur if the equation of state is given in the form $\varrho = \varrho(p, T)$ and the temperature field is homogeneous, or if we have the technically important case where the equation of state $\varrho = \varrho(p, s)$ is given and the flow is homentropic.

If gravity is the only mass body force appearing, Bernoulli's equation for incompressible flow of homogeneous density reads

$$\varrho \int \frac{\partial u}{\partial t}\,ds + \varrho \frac{u^2}{2} + p + \varrho\,g\,x_3 = C, \tag{4.61}$$

where we have assumed that the x_3-direction is antiparallel to the gravity vector $\vec{g}$. For steady, incompressible flow Bernoulli's equation reduces to

$$\varrho \frac{u^2}{2} + p + \varrho\,g\,x_3 = C. \tag{4.62}$$

Since for steady flows, streamlines and pathlines coincide, ϱ is constant along the streamline, even for inhomogeneous density fields $(\nabla \varrho \neq 0)$; this is because $D\varrho/Dt = 0$. Equation (4.62) therefore also holds for steady, incompressible flows when the density is inhomogeneous.

In compressible flows, the velocities are in general so large that the potential of the gravity force $\psi = g\,x_3$ only has to be taken into account if very large differences in altitude appear in the flow (meteorology). In technical applications, ψ in (4.57) can normally be neglected, and for barotropic flow this equation takes the form

$$\int \frac{\partial u}{\partial t}\,ds + \frac{u^2}{2} + P = C. \tag{4.63}$$

If, in addition, the flow is steady, (4.63) can be simplified further to

$$\frac{u^2}{2} + P = C. \tag{4.64}$$

In general, the constant of integration C differs from streamline to streamline. Therefore Bernoulli's equation only represents a relation between the flow quantities at position B on the streamline, and at position A on the same streamline. In order to apply Bernoulli's equation the streamline actually has to be known. Its calculation requires in general the knowledge of the velocity field, and this problem must be solved before Bernoulli's equation can be applied. Of course this restricts the application of Bernoulli's equation drastically. However this restriction vanishes in two technically very important cases:

The first case is the application of Bernoulli's equation to *stream filament theory* (see discussion in connection with Fig. 1.7). In this theory, the "representative" streamline is fixed by the shape of the streamtube which does not change in time. Therefore the streamline is known, and will be fixed in space even for unsteady flow (cf. (1.13)).

The second case is the application of Bernoulli's equation to *potential flow*. From the discussion in connection with the vorticity equation we have seen that in many practically important problems, inviscid flow is also irrotational. However in inviscid potential flows Bernoulli's constant has the same value on all streamlines: Bernoulli's equation (4.57) therefore holds between two arbitrary points A and B in the flow field. For the irrotational field we have

$$\text{curl } \vec{u} = 2\vec{\omega} = 0, \tag{4.65}$$

or because of (1.46)

$$\Omega_{ij} = \frac{1}{2}\left[\frac{\partial u_i}{\partial x_j} - \frac{\partial u_j}{\partial x_i}\right] = 0, \tag{4.66}$$

and so

$$\frac{\partial u_i}{\partial x_j} = \frac{\partial u_j}{\partial x_i}; \tag{4.67}$$

it follows that Euler's equation (4.40a) becomes

$$\frac{\partial u_i}{\partial t} + \frac{\partial}{\partial x_i}\left[\frac{u_j u_j}{2}\right] + \frac{1}{\varrho}\frac{\partial p}{\partial x_i} + \frac{\partial \psi}{\partial x_i} = 0. \tag{4.68}$$

After introducing the velocity potential Φ according to (1.50)

$$u_i = \frac{\partial \Phi}{\partial x_i},$$

Equation (4.68) yields

$$\frac{\partial^2 \Phi}{\partial x_i \partial t} + \frac{\partial}{\partial x_i}\left[\frac{1}{2}\frac{\partial \Phi}{\partial x_j}\frac{\partial \Phi}{\partial x_j}\right] + \frac{1}{\varrho}\frac{\partial p}{\partial x_i} + \frac{\partial \psi}{\partial x_i} = 0. \tag{4.69}$$

In barotropic flow, the whole left-hand side of this equation can be represented as the gradient of a scalar function

$$\frac{\partial}{\partial x_i}\left[\frac{\partial \Phi}{\partial t} + \frac{1}{2}\frac{\partial \Phi}{\partial x_j}\frac{\partial \Phi}{\partial x_j} + P + \psi\right] = \frac{\partial f}{\partial x_i}, \tag{4.70}$$

and the expression

$$df = \frac{\partial f}{\partial x_i}dx_i \tag{4.71}$$

is a total differential. Therefore the line integral

$$\int \frac{\partial}{\partial x_i}\left[\frac{\partial \Phi}{\partial t} + \frac{1}{2}\frac{\partial \Phi}{\partial x_j}\frac{\partial \Phi}{\partial x_j} + P + \psi\right]dx_i = \int df \tag{4.72}$$

is path independent, and we immediately obtain Bernoulli's equation for potential flow

$$\frac{\partial \Phi}{\partial t} + \frac{1}{2}\frac{\partial \Phi}{\partial x_i}\frac{\partial \Phi}{\partial x_i} + P + \psi = C(t). \tag{4.73}$$

Bernoulli's "constant" can, as pointed out, be a function of time. However this is unimportant since without loss of generality it can be incorporated into the potential

$$\Phi^* = \Phi - \int\limits_0^t C(t')\mathrm{d}t'. \tag{4.74}$$

Then $u_i = \partial \Phi^*/\partial x_i$, holds and from (4.73) we obtain

$$\frac{\partial \Phi^*}{\partial t} + \frac{1}{2}\frac{\partial \Phi^*}{\partial x_i}\frac{\partial \Phi^*}{\partial x_i} + P + \psi = 0. \tag{4.75}$$

Incidentally the Eq. (4.73) (or (4.75)) is also a first integral in viscous incompressible potential flow, since then, because of (4.12), the equation to be integrated corresponds with (4.68).

The progress achieved with Eq. (4.73) cannot be emphasized highly enough. In the theory of potential flow Bernoulli's equation takes the place of Euler's three nonlinear equations. Moreover in steady flow this even gives rise to a pure algebraic relationship between the velocity, the potential of the mass body force and the pressure function (in incompressible flow, the pressure). In order to apply Bernoulli's equation in *potential theory*, the streamlines do not need to be known. The simplifications thus found in the mathematical treatment and the practical significance of potential flows have made this an important area in fluid mechanics.

We have already seen that in technical applications, in particular in turbomachinery, reference frames rotating uniformly with $\vec{\Omega}$ are often introduced. We reach Euler's equation for these reference frames by inserting the constitutive relation for inviscid fluids (3.9) into Cauchy's equation (2.68), and expressing the relative acceleration using (1.78)

$$\left\{\frac{\partial \vec{w}}{\partial t} - \vec{w} \times (\nabla \times \vec{w}) + \nabla\left[\frac{\vec{w}\cdot\vec{w}}{2}\right]\right\} = -\left[\frac{\nabla p}{\varrho} - \vec{k} + 2\vec{\Omega}\times\vec{w} + \vec{\Omega}\times\left(\vec{\Omega}\times\vec{x}\right)\right]. \tag{4.76}$$

Instead of following the derivation of Bernoulli's equation as in (4.52), we immediately form the line integral along a streamline. If $\mathrm{d}\vec{x}$ is a vectorial line element along the streamline, $\{\vec{w}\times(\nabla\times\vec{w})\}\cdot\mathrm{d}\vec{x} = 0$ holds, and $\left\{2\vec{\Omega}\times\vec{w}\right\}\cdot\mathrm{d}\vec{x} = 0$, since $\vec{w}\times(\nabla\times\vec{w})$ and $\vec{\Omega}\times\vec{w}$ are orthogonal to $\vec{w}$ and thus orthogonal to $\mathrm{d}\vec{x}$. Therefore, the Coriolis force in particular has no component in the direction of the streamline. Using the relation

$$\vec{\Omega} \times \left(\vec{\Omega} \times \vec{x}\right) = -\nabla \left[\frac{1}{2}\left(\vec{\Omega} \times \vec{x}\right)^2\right], \tag{4.77}$$

(which may be proved using index notation), the centrifugal force can be written as the gradient of the scalar function $\frac{1}{2}\left(\vec{\Omega} \times \vec{x}\right)^2$ and thus has a potential. If we assume, as before, barotropy and a potential for the mass body force, the line integral of Euler's equation then reads

$$\int \frac{\partial \vec{\omega}}{\partial t} \cdot d\vec{x} + \int \left\{\nabla\left[\frac{\vec{w} \cdot \vec{w}}{2} - \frac{1}{2}\left(\vec{\Omega} \times \vec{x}\right)^2 + \psi\right] + \frac{\nabla p}{\varrho}\right\} \cdot d\vec{x} = 0. \tag{4.78}$$

With $|d\vec{x}| = ds$ and $|\vec{w}| = w$ we obtain Bernoulli's equation for a uniformly rotating reference frame

$$\int \frac{\partial w}{\partial t} ds + \frac{w^2}{2} + \psi + P - \frac{1}{2}\left(\vec{\Omega} \times \vec{x}\right)^2 = C. \tag{4.79}$$

A special form of this equation for incompressible flow arises if the mass body force is the gravitational force, the unit vector $\vec{e}_3$ is in the x_3-direction antiparallel to $\vec{g}$, and the reference frame rotates about the x_3-axis with $\Omega = \text{const}$ (Fig. 4.2). With $r^2 = x_1^2 + x_2^2$, the square of the cross product then reads

$$\left(\vec{\Omega} \times \vec{x}\right)^2 = (\Omega x_1 \vec{e}_2 - \Omega x_2 \vec{e}_1)^2 = \Omega^2 r^2, \tag{4.80}$$

and (4.79) reduces to

$$\int \frac{\partial w}{\partial t} ds + \frac{w^2}{2} + \frac{p}{\varrho} + g x_3 - \frac{1}{2}\Omega^2 r^2 = C. \tag{4.81}$$

Additionally, we note that a flow which is a potential flow in the inertial reference frame is no longer a potential flow in the rotating frame. The advantages connected with treating the flow using potential theory may outweigh those connected with

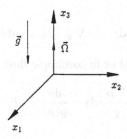

Fig. 4.2 Bernoulli's equation in a rotating reference frame

choosing a rotating reference frame, and it can sometimes be more useful to retain
the inertial frame.

4.2.3 Vortex Theorems

We shall now consider the circulation of a closed material line as it was introduced
by (1.105)

$$\Gamma = \oint_{(C(t))} \vec{u} \cdot d\vec{x}.$$

Its rate of change is calculated using (1.101) to give

$$\frac{D\Gamma}{Dt} = \frac{D}{Dt} \oint_{(C(t))} \vec{u} \cdot d\vec{x} = \oint_{(C)} \frac{D\vec{u}}{Dt} \cdot d\vec{x} + \oint_{(C)} \vec{u} \cdot d\vec{u}. \qquad (4.82)$$

The last closed integral vanishes, since $\vec{u} \cdot d\vec{u} = d(\vec{u} \cdot \vec{u}/2)$ is a total differential of a
single valued function, and the starting point of integration coincides with the end
point.

We now follow on with the discussion in connection with Eq. (1.102), and seek
the conditions for the time derivative of the circulation to vanish. It has already been
shown that in these circumstances the acceleration $D\vec{u}/Dt$ must have a potential I,
but this is not the central point of our current discussion.

Using Euler's equation (4.40a, 4.40b) we acquire the rate of change of the line
integral over the velocity vector in the form

$$\frac{D\Gamma}{Dt} = \oint_{(C)} \vec{k} \cdot d\vec{x} - \oint_{(C)} \frac{\nabla p}{\varrho} \cdot d\vec{x} \qquad (4.83)$$

and conclude from this that $D\Gamma/Dt$ vanishes if $\vec{k} \cdot d\vec{x}$ and $(\nabla p/\varrho) \cdot d\vec{x}$ can be
written as total differentials. If the mass body force $\vec{k}$ has a potential the first closed
integral is zero because

$$\vec{k} \cdot d\vec{x} = -\nabla \psi \cdot d\vec{x} = -d\psi. \qquad (4.84)$$

In a homogeneous density field or in barotropic flow, because of

$$\frac{\nabla p}{\varrho} \cdot d\vec{x} = \frac{dp}{\varrho(p)} = dP \qquad (4.85)$$

the second integral also vanishes. The last three equations form the content of *Thomson's vortex theorem* or *Kelvin's circulation theorem*

$$\frac{D\Gamma}{Dt} = 0. \tag{4.86}$$

In words:

> In an inviscid and barotropic fluid, the circulation of a closed material line remains constant for all times if the mass body force has a potential.

We use this theorem as a starting point for the explanation of the famous *Helmholtz's vortex theorems* which allow a vivid interpretation of vortex motions and in addition are of fundamental importance in aerodynamics.

Before we do this, we shall consider the origin of the circulation about an airfoil in two-dimensional inviscid potential flow, because Kelvin's circulation theorem seems to contradict the formation of this circulation.

In connection with Eq. (2.91) we have already referred to the fact that the force on an airfoil in two-dimensional potential flow is proportional to the circulation. We gain an insight into the relation between circulation and lift (force perpendicular to the undisturbed incident flow direction) by comparing a symmetric airfoil with an asymmetric airfoil (or a symmetric airfoil at an angle of attack) in plane two-dimensional flow. In the first case the flow is likewise symmetric, and for this reason we expect no force perpendicular to the incident flow direction. The contribution of the line integral about the upper half of the airfoil to the circulation has exactly the same size as the contribution about the lower half, but with opposite sign, that is, the total circulation about the symmetric airfoil is zero.

For the asymmetric airfoil shown in Fig. 4.3 the flow is likewise asymmetric, the contribution of the line integral about the upper half has an absolute value larger than that of the contribution about the lower half and therefore the circulation is nonzero. The velocity along a streamline which runs along the upper side of the airfoil is then larger on the whole than the velocity on the lower side. According to Bernoulli's equation (4.62), the pressure on the upper side is on the whole smaller than on the lower side (the term $\varrho\, g\, x_3$ is of no importance for the dynamic lift), so that in total a force upwards results.

If we first consider an airfoil in a fluid at rest, the circulation of a closed curve about the airfoil is clearly zero because the velocity is zero.

The circulation of this curve, which always consists of the same material particles, must remain zero by Kelvin's circulation theorem, even if the inviscid fluid is set into motion. Experience has shown us, however, that a lift acts on the airfoil. How can the airfoil acquire lift without Kelvin's law being contradicted? To answer this question, consider the airfoil in Fig. 4.4, a series of closed curves layed down in the fluid which is at rest.

The circulation is zero for all curves, and also for the surrounding line. We set the fluid into motion and, since all the curves are material lines, we obtain the configuration shown in Fig. 4.5. The airfoil "cuts through" the flow, and a dividing

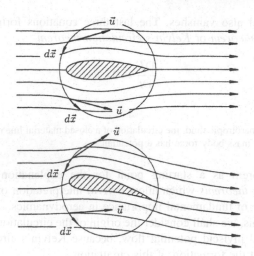

Fig. 4.3 Explanation of the circulation around an airfoil

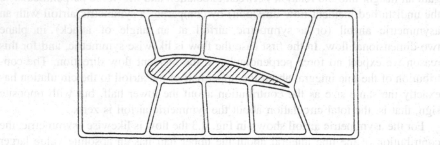

Fig. 4.4 Material curves for an airfoil at rest

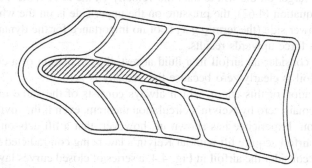

Fig. 4.5 Material curves after setting the airfoil into motion

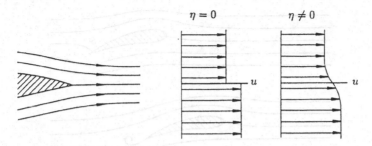

Fig. 4.6 Separation surfaces behind the airfoil

surface forms from the *trailing edge* as the fluid from the upper and lower sides flow together. For asymmetric airfoils the velocity above and below this dividing surface is different. A discontinuity appears, as sketched in Fig. 4.6.

The discontinuity surface is only possible in the limiting case of vanishing viscosity ($\eta = 0$). Even if there is only a small amount of viscosity, this discontinuity becomes evened out. In this region the rotation is nonzero. This does not contradict Kelvin's circulation theorem since the discontinuity surface or the wake are not part of the closed material curves. The discontinuity surface is in principle unstable: it rolls up into a vortex which keeps getting larger until the velocities at the trailing edge are equal; then the process of start-up is finished.

The formation of the discontinuity surface hinders the flow around the sharp edge which in real inviscid flow ($\eta = 0$) would produce infinitely large velocities.

In the first instant of start-up, the flow around the trailing edge is indeed at very high velocities, but it becomes separated from the upper side. Later we shall see that this is caused by the very large deceleration of the flow from the trailing edge (high velocity) to the stagnation point (zero velocity) which will be formed on the upper surface in the as yet circulation free flow. This flow separates from the upper surface even with very little viscosity ($\eta \to 0$) and forms the *wake*, which becomes the discontinuity surface in the limiting case $\eta = 0$. Apart from inside this wake, the flow is irrotational. Figure 4.7 shows the different phases of start-up.

A closed curve which surrounds the airfoil and vortex (Fig. 4.8) still has, by Kelvin's circulation theorem, a circulation of zero. A closed line which only surrounds the vortex has a certain circulation and must necessarily cross the discontinuity surface.

Therefore Kelvin's circulation theorem does not hold for this line. A curve which only surrounds the airfoil has the same circulation as the vortex, only with opposite sign, and therefore the airfoil experiences a lift. The vortex is called the *starting vortex*, and we associate the circulation about the airfoil with a vortex lying inside the airfoil, and call this vortex the *bound vortex*. (The seat of the circulation is actually the boundary layer and in the limit $\eta \to 0$ the thickness tends to zero while the vorticity in the layer tends to infinity.)

In addition we note that with every change in velocity the lift changes likewise, and consequentially a free vortex must form. (In a fluid with viscosity, circulation

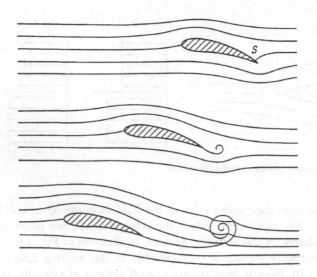

Fig. 4.7 Start-up

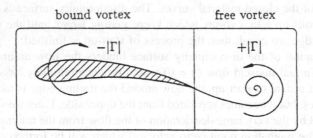

Fig. 4.8 The circulation of the starting vortex and the bound vortex are of equal magnitude

and vortices can arise in many ways, for example through boundary layer separation, without a sharp edge being necessary.)

Incidentally in the above discussion we have also used the obvious law that the circulation of a closed line is equal to the sum of the circulation of the meshed network bounded by the curve (Fig. 4.9)

$$\Gamma_{closed} = \sum \Gamma_i, \tag{4.87}$$

or else

$$\Gamma = \int d\Gamma. \tag{4.88}$$

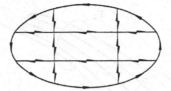

Fig. 4.9 Circulation of a meshed network

In order to discuss Helmholtz's vortex theorems, we need to make use of *Stokes' integral theorem*. Let S be a simply connected surface which is otherwise of arbitrary shape (i.e. any arbitrary closed curve on the surface can be shrunk to a single point), whose boundary is C, and let $\vec{u}$ be any arbitrary vector.

Stokes' theorem then reads:

The line integral $\int \vec{u} \cdot d\vec{x}$ about the closed curve C is equal to the surface integral $\int (\nabla \times \vec{u}) \cdot \vec{n}\, dS$ over any surface of arbitrary shape which has C as its boundary, therefore

$$\oint_{(C)} \vec{u} \cdot d\vec{x} = \iint_{(S)} (\operatorname{curl} \vec{u}) \cdot \vec{n}\, dS. \tag{4.89}$$

Stokes' theorem allows a line integral to be changed into a surface integral. The direction of integration is positive anticlockwise as seen from the positive side of the surface (Fig. 4.10).

Helmholtz's first vortex theorem reads:

The circulation of a vortex tube is constant along this tube.

In complete analogy to streamtubes, we shall form vortex tubes from vortex lines, which are tangential lines to the vorticity vector field curl $\vec{u}$ (or $\vec{\omega}$) (Fig. 4.11). The

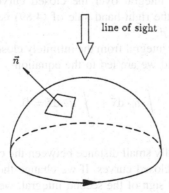

Fig. 4.10 Assigning the direction of integration in Stokes' integral theorem

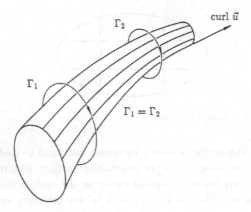

Fig. 4.11 Vortex tube

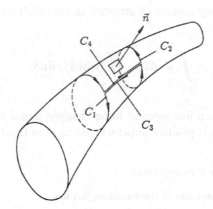

Fig. 4.12 Proof of Helmholtz's first vortex theorem

vortex lines which pass through a closed curve form a vortex tube. According to Stokes' theorem, the line integral over the closed curve in Fig. 4.12 vanishes, because the integrand on the right-hand side of (4.89) is zero, since curl $\vec{u}$ is by definition perpendicular to $\vec{n}$.

The contributions to the integral from the infinitely close pieces C_3 and C_4 of the curve cancel each other and we are led to the equation

$$\int\limits_{C_1} \vec{u} \cdot \mathrm{d}\vec{x} + \int\limits_{C_2} \vec{u} \cdot \mathrm{d}\vec{x} = 0. \tag{4.90}$$

Because of the infinitesimally small distance between the curves C_3 and C_4, we can consider C_1 and C_2 to be closed curves. If we change the direction of integration over C_2, thus changing the sign of the second integral, we obtain *Helmholtz's first vortex theorem*

$$\oint_{C_1} \vec{u} \cdot d\vec{x} = \oint_{C_2} \vec{u} \cdot d\vec{x}. \tag{4.91}$$

From this derivation the kinematic nature of this theorem is obvious. Another approach to this important theorem starts from Eq. (4.14) which shows that the divergence of the vorticity vector vanishes. We can therefore consider the vorticity vector field curl $\vec{u}$ as the velocity field of a new incompressible flow, i.e., the vortex tube becomes the streamtube of the new field. We apply the equation of continuity in its integral form (2.8) to a part of this streamtube, and at the same time replace $\vec{u}$ by curl $\vec{u}$. Since the flow is incompressible, quite generally

$$\iint_{(S)} (\text{curl}\,\vec{u}) \cdot \vec{n}\, dS = 0, \tag{4.92}$$

i.e., for every closed surface S, the flux of the vorticity vector is zero. We apply (4.92) to a part of the vortex tube whose closed surface consists of the surface of the tube and two arbitrarily oriented cross-sections A_1 and A_2, and find

$$\iint_{(A_1)} (\text{curl}\,\vec{u}) \cdot \vec{n}\, dS + \iint_{(A_2)} (\text{curl}\,\vec{u}) \cdot \vec{n}\, dS = 0, \tag{4.93}$$

since the integral over the tube surface vanishes. The integral $\iint (\text{curl}\,\vec{u}) \cdot \vec{n}\, dS$ is often called the *vortex strength*. It is clearly identical to the circulation, and in words the Eq. (4.93) reads:

The vortex strength of a vortex tube is constant.

Noting the sense of integration of the line integral, Stokes' theorem transforms equation (4.93) into Helmholtz's first theorem (4.91). We conclude from this representation that just like the streamtube, the vortex tube cannot come to an end within the fluid, since the amount of fluid which flows through the tube in unit time cannot simply vanish at the end of the tube. Either the tube must reach out to infinity, or end at the boundaries of the fluid, or else close around into itself and, in the case of a vortex tube, form a vortex ring.

Vortex filaments are of particular importance in aerodynamics. By a vortex filament we understand a very thin vortex tube. For a vortex filament the integrand of the surface integral in Stokes' theorem (4.89)

$$\oint_C \vec{u} \cdot d\vec{x} = \iint_{\Delta S} (\text{curl}\,\vec{u}) \cdot \vec{n}\, dS = \Gamma \tag{4.94}$$

can be taken in front of the integral and we get

$$(\operatorname{curl} \vec{u}) \cdot \vec{n} \, \Delta S = \Gamma \qquad (4.95)$$

or

$$2\vec{\omega} \cdot \vec{n} \, \Delta S = 2 \, \omega \, \Delta S = \text{const}, \qquad (4.96)$$

from which we conclude that the angular velocity increases with decreasing cross-section of the vortex filament.

We shall see later from Helmholtz's second vortex theorem that vortex tubes are material tubes. If we make use of this fact, then (4.96) leads to the same statement as (4.27): if the vortex filament is stretched, its cross-section becomes smaller and the angular velocity increases. The expression (4.27) was according to its derivation restricted to incompressible flow, while the conclusion we have drawn here (by using Helmholtz's second vortex theorem) holds in general for barotropic flow.

A frequently used idealized picture of a vortex filament is a vortex tube with infinitesimally small cross-section, whose angular velocity then, by (4.96), becomes infinitely large

$$\omega \, \Delta S = \text{const} \qquad (4.97)$$

for $\Delta S \to 0$ and $\omega \to \infty$.

Outside the vortex filament, the field is irrotational. Therefore if the position of a vortex filament and its strength Γ are known, the spatial distribution of curl $\vec{u}$ is fixed. In addition, if div $\vec{u}$ is given (e.g. div $\vec{u} = 0$ in incompressible flow), according to the already mentioned fundamental theorem of vector analysis, the velocity field $\vec{u}$ (which may extend to infinity) is uniquely determined if we further require that the normal component of the velocity vanishes asymptotically sufficiently fast at infinity and no internal boundaries exist. (On internal boundaries conditions have to be satisfied, and we will wait to Sect. 4.3 to introduce these.) (Fig. 4.13).

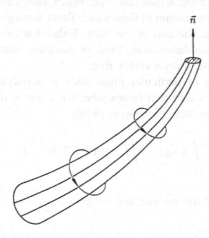

Fig. 4.13 Vortex filament

The assertion of the fundamental theorem of vector analysis is also purely kinematic in nature, and is therefore not restricted to inviscid fluids.

We split the vector $\vec{u}$ up into two parts

$$\vec{u} = \vec{u}_D + \vec{u}_R, \tag{4.98}$$

of which the first is an irrotational field, i.e.,

$$\text{curl } \vec{u}_D = \nabla \times \vec{u}_D = 0, \tag{4.99}$$

and the second is a solenoidal field, thus

$$\text{div } \vec{u}_R = \nabla \cdot \vec{u}_R = 0. \tag{4.100}$$

The combined field is therefore in general neither irrotational nor solenoidal. The field $\vec{u}_D$ is a potential flow, and thus by (1.50) we have $\vec{u}_D = \nabla \Phi$. We form the divergence of $\vec{u}$ and consider it to be a given function $q(\vec{x})$. Because of (4.100), we obtain

$$\text{div } \vec{u} = \nabla \cdot \vec{u}_D = q(\vec{x}) \tag{4.101}$$

or else

$$\nabla \cdot \nabla \Phi = \frac{\partial^2 \Phi}{\partial x_i \partial x_i} = q(\vec{x}). \tag{4.102}$$

(4.102) is an inhomogeneous *Laplace's equation* also called *Poisson's equation*. The theory of both these partial differential equations is the subject of *potential theory* which is as important in many branches of physics as in fluid mechanics. If we refer back to the results of this theory, the solution of (4.102) is given by

$$\Phi(\vec{x}) = -\frac{1}{4\pi} \iiint\limits_{(\infty)} \frac{q(\vec{x}')}{|\vec{x} - \vec{x}'|} \, dV', \tag{4.103}$$

where $\vec{x}$ is the place where the potential Φ is calculated, and $\vec{x}'$ is the abbreviation for the integration variables x_1', x_2' and x_3' $(dV' = dx_1' \, dx_2' \, dx_3')$. The domain (∞) implies that the integration is to be carried out over all space. We shall briefly sketch the manner of solution at the end of our consideration, but here we shall take the solution as given.

In order to calculate $\vec{u}_R$ we note that (4.100) is certainly satisfied if we represent $\vec{u}_R$ as the curl of a new, yet unknown, vector field $\vec{a}$

$$\vec{u}_R = \text{curl } \vec{a} = \nabla \times \vec{a}, \tag{4.104}$$

because, from Eq. (4.14), we have

$$\nabla \cdot (\nabla \times \vec{a}) = \nabla \cdot \vec{u}_R = 0. \tag{4.105}$$

We form the curl of $\vec{u}$ and, from (4.99), extract the equation

$$\nabla \times \vec{u} = \nabla \times (\nabla \times \vec{a}), \tag{4.106}$$

which by (4.10) is rewritten as

$$\nabla \times \vec{u} = \nabla(\nabla \cdot \vec{a}) - \Delta \vec{a}. \tag{4.107}$$

Up to now we have only required that the vector $\vec{a}$ satisfy (4.104). However this does not uniquely determine this vector, because we could always add the gradient of some other function f to $\vec{a}$ without changing (4.104) (since $\nabla \times \nabla f \equiv 0$). If, in addition, we require that the divergence of $\vec{a}$ vanishes ($\nabla \cdot \vec{a} = 0$), we obtain from (4.107) the simpler equation

$$\nabla \times \vec{u} = -\Delta \vec{a}. \tag{4.108}$$

In (4.108) we consider $\nabla \times \vec{u}$ as a given vector function $\vec{b}(\vec{x})$, which is determined by the choice of the vortex filament and its strength (circulation). Thus the Cartesian component form of the vector equation (4.108) leads to three Poisson's equations, namely

$$\Delta a_i = -b_i. \tag{4.109}$$

For each of the component equations, we can apply the solution (4.103) of Poisson's equation. We combine the results again vectorially, and write the solution from (4.108) in short form as

$$\vec{a} = +\frac{1}{4\pi} \iiint\limits_{(\infty)} \frac{\vec{b}(\vec{x}')}{|\vec{x} - \vec{x}'|} \, dV'. \tag{4.110}$$

By doing this, the calculation of the velocity field $\vec{u}(\vec{x})$ for a given distribution $q(\vec{x}) = \operatorname{div} \vec{u}$ and $\vec{b}(\vec{x}) = \operatorname{curl} \vec{u}$ is reduced to integration processes, which may have to be done numerically

$$\vec{u}(\vec{x}) = -\nabla \left\{ \frac{1}{4\pi} \iiint\limits_{(\infty)} \frac{\operatorname{div} \vec{u}(\vec{x}')}{|\vec{x} - \vec{x}'|} \, dV' \right\} + \nabla \times \left\{ \frac{1}{4\pi} \iiint\limits_{(\infty)} \frac{\operatorname{curl} \vec{u}(\vec{x}')}{|\vec{x} - \vec{x}'|} \, dV' \right\}. \tag{4.111}$$

For completeness, we shall sketch the path of solution for Eq. (4.103). Starting from Gauss' theorem (1.94)

$$\iiint\limits_{(V)} \frac{\partial \varphi}{\partial x_i} \, \mathrm{d}V = \iint\limits_{(S)} \varphi \, n_i \, \mathrm{d}S \tag{4.112}$$

we write, for the general function φ

$$\varphi = U \frac{\partial V}{\partial x_i} - V \frac{\partial U}{\partial x_i}, \tag{4.113}$$

where U and V are arbitrary functions which we only assume to be continuous to the degree which is necessary for the application of Gauss' theorem. Gauss' theorem then leads to the relation known as *Green's second formula*

$$\iint\limits_{(S)} \left[U \frac{\partial V}{\partial x_i} - V \frac{\partial U}{\partial x_i} \right] n_i \mathrm{d}S = \iiint\limits_{(V)} \left[U \frac{\partial^2 V}{\partial x_i \partial x_i} - V \frac{\partial^2 U}{\partial x_i \partial x_i} \right] \mathrm{d}V. \tag{4.114}$$

For U we now choose the potential function Φ, and for V

$$V = \frac{1}{|\vec{x} - \vec{x}'|} = \frac{1}{r}. \tag{4.115}$$

The function $1/r$ is a *fundamental solution* of Laplace's equation. It is so called because, as already shown by (4.103), with its help we can form general solutions through integration processes. The fundamental solution is also known as the *singular solution*, since it satisfies Laplace's equation everywhere except at a singularity, here for example at $r = 0$, where $1/r$ is discontinuous. Later we shall give the function $1/r$ an obvious meaning, and shall proceed to show by formal calculation that Laplace's equation is satisfied everywhere except at $\vec{x} = \vec{x}'$, ($r = 0$). Because $1/r$ is not continuous for $r = 0$, we have to exclude this point from the domain (V), as Gauss' theorem is only valid for continuous integrands.

As shown in Fig. 4.14, we surround the *singular point* with a small sphere (radius a) so that the surface domain of integration (S) consists of a very large sphere (radius $\to \infty$) and a very small sphere which surrounds the singularity. Now the integrand on the right-hand side of (4.114) is regular, and the first term vanishes everywhere in the domain of integration, since $V = 1/r$ satisfies Laplace's equation. In the second term, we replace $\Delta U = \Delta \Phi$ by $q(\vec{x})$ (because of (4.102)), so that the right-hand side now consists of the integral

$$-\iiint\limits_{(\infty)} \frac{q(\vec{x})}{|\vec{x} - \vec{x}'|} \mathrm{d}V.$$

On the left-hand side we shall first perform the integration over the large sphere and note that $(\partial V / \partial x_i) n_i$ is the derivative of V in the direction of the normal vector n_i of the sphere. Therefore we have

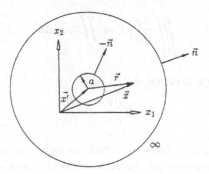

Fig. 4.14 Domain of integration

$$\left[\frac{\partial V}{\partial x_i}n_i\right]_{r\to\infty} = \left[\frac{\partial V}{\partial r}\right]_\infty = \left[\frac{\partial}{\partial r}\left(r^{-1}\right)\right]_\infty = \left[-r^{-2}\right]_\infty, \qquad (4.116)$$

and this vanishes as $1/r^2$. However the surface of integration increases as r^2, so that the dependency on r drops out. By assumption, $U = \Phi$ vanishes at infinity, and therefore there is no contribution from the first term on the left-hand side. The second term vanishes too because $(\partial\Phi/\partial x_i)n_i$ (the component of the vector $\vec{u}$ normal to the surface) die away fast enough for the second term to vanishes also. Therefore all that remains is the integral over the small sphere. However the normal vector of the small sphere points in the negative radial direction, giving us

$$\left[\frac{\partial V}{\partial x_i}n_i\right]_{r=a} = \left[-\frac{\partial V}{\partial r}\right]_a = +a^{-2}, \qquad (4.117)$$

and

$$\left[\frac{\partial\Phi}{\partial x_i}n_i\right]_{r=a} = \left[-\frac{\partial\Phi}{\partial r}\right]_a. \qquad (4.118)$$

We write $a^2\,d\Omega$ for the surface element, where $d\Omega$ is the surface element of the unit sphere. Then the left-hand side of (4.114) is

$$\iint_{(sphere)} \Phi a^{-2}a^2 d\Omega + \iint_{(sphere)} a^{-1}\frac{\partial\Phi}{\partial r}a^2 d\Omega. \qquad (4.119)$$

The second integral vanishes for $a \to 0$, the first yields $4\pi\,\Phi(\vec{x}')$, and then from (4.114), we extract

$$\Phi(\vec{x}') = -\frac{1}{4\pi}\iiint_{(V)} \frac{q(\vec{x})}{|\vec{x}-\vec{x}'|}\,dV. \qquad (4.120)$$

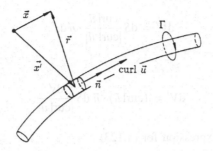

Fig. 4.15 On the Biot-Savart law

If we further replace $\vec{x}$ by $\vec{x}'$ which does not change the function

$$G(\vec{x}, \vec{x}') = -\frac{1}{4\pi} \frac{1}{|\vec{x} - \vec{x}'|}, \qquad (4.121)$$

we obtain the solution (4.103). We call $G(\vec{x}, \vec{x}')$ the *Green's function*, which appears here in the special form for infinite, unbounded space. In two-dimensional problems, the free space Green's function reads

$$G(\vec{x}, \vec{x}') = \frac{1}{2\pi} \ln|\vec{x} - \vec{x}'|. \qquad (4.122)$$

We now return to Eq. (4.111), and calculate the solenoidal term of the velocity $\vec{u}_R$. This is the only term in incompressible flow without internal boundaries. Since we are considering a field which is irrotational outside the vortex filament (Fig. 4.15), the velocity field outside the filament is given by

$$\vec{u}_R(\vec{x}) = \nabla \times \left[\frac{1}{4\pi} \iiint\limits_{\text{(filament)}} \frac{\text{curl } \vec{u}(\vec{x}')}{|\vec{x} - \vec{x}'|} dV' \right]. \qquad (4.123)$$

By assumption, the integration is only carried out over the volume of the vortex filament, whose volume element is

$$dV' = dS \, \vec{n} \cdot d\vec{x}', \qquad (4.124)$$

with $d\vec{x}' = \vec{n} \, ds'$ as the vectorial element of the vortex filament.

By simple manipulation and using

$$\vec{n} = \text{curl } \vec{u}/|\text{curl } \vec{u}|$$

we obtain

$$dV' = dS \frac{\text{curl } \vec{u}}{|\text{curl } \vec{u}|} \cdot \vec{n} \, ds', \tag{4.125}$$

therefore also

$$dV' = (\text{curl } \vec{u}) \cdot \vec{n} \, dS \frac{ds'}{|\text{curl } \vec{u}|} \tag{4.126}$$

and this leads to the expression for (4.123)

$$\vec{u}_R(\vec{x}) = \nabla \times \left[\frac{1}{4\pi} \iiint\limits_{\text{(filament)}} \frac{(\text{curl } \vec{u}) \cdot \vec{n} \, dS}{|\vec{x} - \vec{x}'|} d\vec{x}' \right]. \tag{4.127}$$

Here we have set

$$\frac{\text{curl } \vec{u} \, ds'}{|\text{curl } \vec{u}|} = \vec{n} \, ds' = d\vec{x}'. \tag{4.128}$$

First we integrate over the small cross-sectional surface ΔS and, for $\Delta S \to 0$, ignore the change of the vector $\vec{x}'$ over this surface thus taking $1/|\vec{x} - \vec{x}'|$ in front of the surface integral to obtain

$$\vec{u}_R(\vec{x}) = \nabla \times \left\{ \frac{1}{4\pi} \int \frac{1}{|\vec{x} - \vec{x}'|} \left[\iint (\text{curl } \vec{u}) \cdot \vec{n} \, dS \right] d\vec{x}' \right\}. \tag{4.129}$$

From Stokes' theorem, the surface integral is equal to the circulation Γ, and from Helmholtz's first vortex theorem this is constant along the vortex filament, and is therefore independent of $\vec{x}'$. From (4.129) we then find

$$\vec{u}_R(\vec{x}) = \frac{\Gamma}{4\pi} \nabla \times \int \frac{d\vec{x}'}{|\vec{x} - \vec{x}'|}. \tag{4.130}$$

The following calculation is more simply done in index notation, in which the right-hand side of (4.130) is written as

$$\frac{\Gamma}{4\pi} \epsilon_{ijk} \frac{\partial}{\partial x_j} \int \frac{1}{r} dx_k'.$$

We now see directly that the operator $\epsilon_{ijk} \partial / \partial x_j$ can be taken into the integral. The term $\partial(r^{-1})/\partial x_j$ (with $r_i = x_i - x_i'$ and $r = |\vec{r}|$) becomes

$$\frac{\partial(r^{-1})}{\partial x_j} = -\frac{1}{r^2} \frac{\partial r}{\partial x_j} = -\frac{1}{r^2} \left(x_j - x_j' \right) \frac{1}{r} = -r_j r^{-3}. \tag{4.131}$$

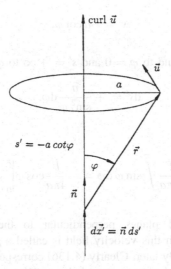

Fig. 4.16 Velocity induced by a straight vortex filament

If we replace (4.131) by the above expression in vector form, (4.130) finally leads to the famous *Biot-Savart law*

$$\vec{u}_R(\vec{x}) = \frac{\Gamma}{4\pi} \int\limits_{\text{(filament)}} \frac{\mathrm{d}\vec{x}' \times \vec{r}}{r^3}, \tag{4.132}$$

with $\vec{r} = \vec{x} - \vec{x}'$, which finds applications particularly in aerodynamics.

The Biot-Savart law is a pure kinematic law, which was originally discovered through experiments in electrodynamics. The vortex filament corresponds there to a conducting wire, the vortex strength to the current, and the velocity field to the magnetic field. The origin of this law also explains the terminology used in aerodynamics, that the vortex filament "induces" a velocity $\vec{u}$. As an illustration, we shall calculate the *induced velocity* from a straight infinitely long vortex filament, at a distance a from the vortex filament. The velocity $\vec{u}_R$ is always orthogonal to the plane spanned by $\mathrm{d}\vec{x}'$ and $\vec{r}$, and is therefore tangential to the circle with radius a in the plane orthogonal to the vortex filament. The magnitude of the induced velocity is found from (4.132), using the notation in Fig. 4.16 as being

$$|\vec{u}_R| = \frac{\Gamma}{4\pi} \int\limits_{-\infty}^{+\infty} \frac{\sin \varphi}{r^2} \mathrm{d}s'. \tag{4.133}$$

From Fig. 4.16 we deduce the relation

$$s' = -a \cot \varphi, \tag{4.134}$$

so that $s' = -\infty$ correspond to $\varphi = 0$ and $s' = +\infty$ to $\varphi = \pi$, and $\mathrm{d}s'$ becomes

$$\mathrm{d}s' = +\frac{a}{\sin^2 \varphi} \mathrm{d}\varphi. \tag{4.135}$$

With $r = a/\sin \varphi$ follows

$$|\vec{u}_R| = \frac{\Gamma}{4\pi a} \int_0^\pi \sin \varphi \, \mathrm{d}\varphi = -\frac{\Gamma}{4\pi a} \cos \varphi \Big|_0^\pi = \frac{\Gamma}{2\pi a}. \tag{4.136}$$

This result holds in all planes perpendicular to the vortex filament. The two-dimensional flow with this velocity field is called a *potential vortex*, and we shall discuss this more fully later. Clearly (4.136) corresponds with (2.97) which is a consequence of the angular momentum balance. We could have found the same result using the plausible assumption of constant velocity at radius a, and by calculating the circulation

$$\Gamma = \oint_a \vec{u}_R \cdot \mathrm{d}\vec{x} = \vec{u} \cdot \vec{e}_\varphi a \int_0^{2\pi} \mathrm{d}\varphi = |\vec{u}_R| a \, 2\pi. \tag{4.137}$$

We shall now calculate the contribution of a straight vortex filament of finite length to the induced velocity at the point P whose position is determined by the displacement a and the angles φ_1 and φ_2 (Fig. 4.17). After integrating from φ_1 to φ_2 we find from (4.136)

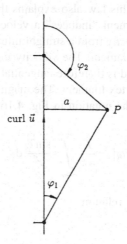

Fig. 4.17 Vortex filament of finite length

$$|\vec{u}_R| = \frac{\Gamma}{4\pi a}(\cos \varphi_1 - \cos \varphi_2). \tag{4.138}$$

For $\varphi_1 = 0$ and $\varphi_2 = \pi/2$ (semi-infinite vortex filament) the induced velocity in the orthogonal plane is given by

$$|\vec{u}_R| = \frac{\Gamma}{4\pi a}, \tag{4.139}$$

and it amounts to precisely half of the value for the infinitely long vortex filament, as we would expect for reasons of symmetry.

Such finite or semi-infinitely long pieces of a vortex filament cannot, by Helmholtz's first vortex theorem, exist alone, but must be parts of a vortex filament which is closed into itself, or which reaches to infinity on both sides. We saw in the discussion of Fig. 4.8 that the circulation about an airfoil in two-dimensional flow can be represented by using a bound vortex. We can imagine these bound vortices as straight, infinitely-long vortex filaments (potential vortices). As far as the lift is concerned we can think of the whole airfoil as being replaced by the straight vortex filament. The velocity field close to the airfoil is of course different from the field about a vortex filament in cross flow, but both fields become more similar the larger the distance from the airfoil.

In the same way, the starting vortex can be idealized as a straight vortex filament which is attached to the bound vortex at plus and minus infinity. The circulation of the vortex determines the lift, and the lift formula which gives the relation between circulation and lift per unit width in inviscid potential flow is the *Kutta-Joukowski theorem*

$$A = -\varrho \, \Gamma \, U_\infty, \tag{4.140}$$

where U_∞ is the so-called "undisturbed" approach velocity, i.e., the velocity which would appear if the body were removed. (By width or span of a wing we mean the extension normal to the plane drawn in Fig. 4.3, while the depth of the wing section is the chord of the wing section. The negative sign in all lift formulae arises since circulation is here defined as positive as in the mathematical sense.) The Kutta-Joukowski theorem can be derived from the momentum balance and Bernoulli's equation, in the same manner as was used to calculate the force on a blade in a cascade. Here we refrain from doing this since we wish to derive the Kutta-Joukowski formula by different means later (Fig. 4.18).

In this connection we expressly mention that the force on a single wing section in inviscid potential flow is perpendicular to the direction of the undisturbed stream and thus the airfoil experiences only lift and no drag. This result is of course contrary to our experience, and is due to ignoring the viscosity. The Kutta-Joukowski theorem in the form (4.140) with constant Γ only holds for wing sections in two-dimensional plane flow. All real wings are of finite span, but as long as the span is much larger than the chord of the wing section, the lift can be

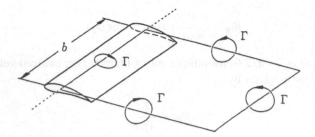

Fig. 4.18 Simplified vortex system of a finite airfoil

estimated using the assumption of constant circulation along the span. Approximately the lift of the whole wing with width b is given by

$$A = -\varrho\, \Gamma\, U_\infty\, b. \tag{4.141}$$

In reality however there is flow past the tips of the wing, because the pressure on the lower side of the wing is larger than that on the upper side, so that by Euler's equation the fluid flows under the influence of the pressure gradient from the lower to the upper side to even out the pressure difference. In this way the value of the circulation on the wing tips tends to zero, the circulation therefore varies over the span of the wing, and the lift is calculated by

$$A = -\varrho\, U_\infty \int\limits_{-b/2}^{+b/2} \Gamma(x)\mathrm{d}x, \tag{4.142}$$

if the origin is in the middle of the wing and x is measured along the span. Yet even when we assume that Γ is constant over the span of the wing, difficulties soon arise, because as far as the lift is concerned a wing cannot be replaced by a finite piece of a vortex filament.

According to Helmholtz's first vortex theorem, which being purely kinematic and therefore also holding for the bound vortex, isolated pieces of a vortex filament cannot exist. Neither can it be continued straight along into infinity, where the wing has not cut through the fluid and thus no discontinuity surface has been generated as is necessary for the formation of circulation. Therefore free vortices which are carried away by the flow must be attached at the wing tips. Together with the bound vortex and the starting vortex, these free vortices form a closed vortex ring which frame the fluid region cut by the wing. If a long time has passed since start-up, the starting vortex is at infinity, and the bound vortex and the tip vortices together form a *horseshoe vortex*, which, although it only represents a very rough model of a finite wing, can already provide a qualitative explanation for how a wing experiences a drag in inviscid flow, as already mentioned. The velocity w (*induced downwash*) induced in the middle of the wing by the two tip vortices amounts to double the

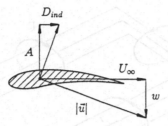

Fig. 4.19 Explanation of induced drag

velocity induced by a semi-infinite vortex filament at distance $b/2$. Therefore by (4.139) we have

$$w = 2\frac{\Gamma}{4\pi(b/2)} = \frac{1}{b}\frac{\Gamma}{\pi} \qquad (4.143)$$

and w is directed downwards. Thus the middle of the wing "experiences" not only the undisturbed velocity U_∞, but a velocity which arises from the superposition U_∞ and w (Fig. 4.19). In inviscid flow, the force vector is perpendicular to the actual approach direction of the stream, and therefore has a component parallel to the undisturbed flow, which manifests itself as the *induced drag* D_{ind}

$$D_{ind} = A\frac{w}{U_\infty}. \qquad (4.144)$$

But (4.144) only holds if the induced downwash from both vortices is constant over the span of the wing. However the downwash does change, because, at a distance x from the wing center, one vortex induces a downwash

$$\frac{\Gamma}{4\pi(b/2+x)},$$

the other

$$\frac{\Gamma}{4\pi(b/2-x)},$$

and together

$$w = \frac{\Gamma}{4\pi}\frac{b}{(b/2)^2-x^2},$$

from which we conclude that the downwash is smallest in the center of the wing (so we underestimate the drag with (4.144)) and tends to infinity at the wing tips. The unrealistic value there does not appear if the circulation distribution decreases

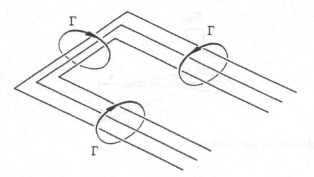

Fig. 4.20 Simplified vortex system of an airfoil

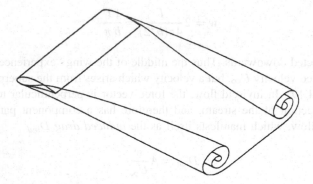

Fig. 4.21 The discontinuity surface rolls itself into vortices

towards the ends, as indeed it has to. For a semi-elliptical circulation distribution over the span of the wing, one finds a constant down-wash distribution, and (4.144) is applicable. Helmholtz's first vortex theorem further demands that, for an infinitesimal change in the circulation in the x-direction

$$d\Gamma = \frac{d\Gamma}{dx}dx,$$

a free vortex of the same infinitesimal strength must leave the trailing edge. In this way we are led to the improved vortex system of Fig. 4.20. The free vortices form a discontinuity surface in the velocity components parallel to the trailing edge, which rolls them into the vortices sketched in Fig. 4.21.

These vortices must be continually renewed as the wing moves forward, so that the kinetic energy in the vortices continually has to be newly delivered to them. The power needed to do this is the work done per unit time by the induced drag.

We can often see manifestations of Helmholtz's first vortex theorem in daily life. Recall the dimples seen on the free surface of coffee when the coffee spoon is suddenly moved forwards and then taken out (Fig. 4.22).

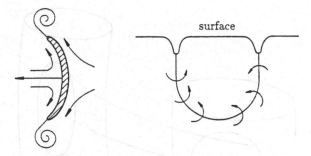

Fig. 4.22 Vortex on a coffee spoon

As the fluid flows together from the front and back, a surface of discontinuity forms along the rim of the spoon. The discontinuity surface rolls itself into a bow shaped vortex whose endpoints form the dimples on the free surface. Since the flow outside the vortex filament is a potential flow, Bernoulli's equation holds (4.62)

$$\frac{1}{2}\varrho\, u^2 + p + \varrho\, g\, x_3 = C.$$

This is valid not just along a streamline, but between any two points in the field. Everywhere on the free surface the pressure is equal to the ambient pressure p_0. At some distance from the vortex the velocity is zero and the free surface is not yet depressed and corresponds to $x_3 = 0$, say. Then Bernoulli's constant is equal to the ambient pressure ($C = p_0$), and we obtain

$$\frac{1}{2}\varrho\, u^2 + \varrho\, g\, x_3 = 0.$$

Near the endpoints of the vortex the velocity increases by the formula (4.139), and therefore x_3 must become negative, i.e., a depression of the free surface. The cross-sectional surface of the vortex filament is in reality not infinitesimally small, so that we cannot take the limit $a \to 0$ in (4.139), for which the velocity becomes infinite. However the induced velocity from the vortex filament is so large that it leads to the noticeable formation of dimples.

In this connection we note that an infinitesimally thin vortex filament cannot appear in actual flow because the velocity gradient of the potential vortex tends to infinity for $a \to 0$, so that the viscous stresses cannot be ignored any longer, even for very small viscosity. As we know from (4.11), viscous stresses make no contribution to particle acceleration in *incompressible potential flow*, but they do deformation work and thus provide a contribution to the dissipation. The energy dissipated in heat stems from the kinetic energy of the vortex. The idealization of a real vortex filament as a filament with an infinitesimally small cross-section is of course still useful.

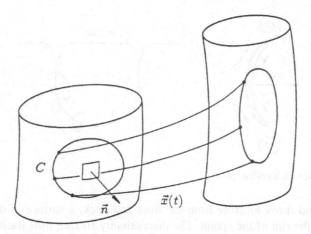

Fig. 4.23 Helmholtz's second vortex theorem

We shall now consider Helmholtz's second vortex theorem:

A vortex tube is always made up of the same fluid particles.

A vortex tube is therefore a material tube. This has already been proved for material coordinates by Eq. (4.27), but here we wish to represent it as a direct consequence of Kelvin's circulation theorem. We consider a vortex tube and an arbitrary closed curve on its surface at time t_0 (Fig. 4.23). By Stokes' integral theorem, the circulation of the closed curve is zero. The circulation of the curve, which is made up of the same material particles, still has the same value of zero at a later instant in time, by Kelvin's circulation theorem ($D\Gamma/Dt = 0$). By inverting the above reasoning it follows from Stokes' theorem that these material particles must be on the outer surface of the vortex tube.

If we consider smoke rings, the fact that vortex tubes are material tubes becomes obvious: the smoke clearly remains in the vortex ring and is transported with it, so that it is the smoke itself which carries the vorticity. This statement only holds under the restrictions of barotropy and zero viscosity. The slow disintegration seen in smoke rings is due to friction and diffusion. A vortex ring which consists of an infinitesimally thin vortex filament induces an infinitely large velocity on itself (similar to the horseshoe vortex already seen), so that the ring would move forward with infinitely large velocity. The induced velocity in the center of the ring remains finite (just as with the horseshoe vortex), and it is found from the Biot-Savart law (4.132) as

$$|\vec{u}| = \frac{\Gamma}{4\pi} \int_0^{2\pi} \frac{a^2 \mathrm{d}\varphi}{a^3} = \frac{\Gamma}{2a}.$$

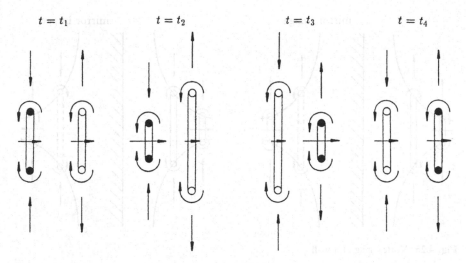

$t = t_1$ $t = t_2$ $t = t_3$ $t = t_4$

Fig. 4.24 Two vortex rings passing through one another

It is the assumption of an infinitesimally small cross-section that leads to the unrealistic infinitely large velocity on the vortex. If we assume a finite cross-section, then the velocity induced on itself, i.e., the velocity with which the ring moves forwards, remains finite. However the actual cross-section of the ring is not known, and probably depends on how the ring was formed.

In practice we notice that the ring moves forward with a velocity which is slower than the induced velocity in the center. It is well known that two rings moving in the same direction continually overtake each other whereby one slips through the one in front. This behavior, sketched in Fig. 4.24, is explained by the mutually induced velocities on the rings and the formula given above for the velocity in the center of the ring.

In the same manner it can be explained why a vortex ring moving towards a wall gets larger in diameter and at the same time reduces its velocity, while one moving away from the wall contracts and increases its velocity (Fig. 4.25).

The motion cannot be worked out without knowing the vortex cross-section, and the calculation for infinitesimally thin rings fails, because rings, like all curved vortex filaments, induce infinitely large velocities on themselves. For straight vortex filaments, i.e., for two-dimensional flow, a simple description of the "vortex dynamics" for infinitesimally thin filaments is possible, since here the self induced translation velocity vanishes. Because vortex filaments are material lines, it is sufficient to calculate the paths of the fluid particles which carry the rotation in the x-y-plane perpendicular to the filaments using (1.10); that is, to determine the paths of the vortex centers.

The magnitude of the velocity which a straight vortex filament at position $\vec{x}_{(i)}$ induces at position $\vec{x}$ is known from (4.136). As explained there, the induced velocity is perpendicular to the vector $\vec{a}_{(i)} = \vec{x} - \vec{x}_{(i)}$, and therefore has the direction $\vec{e}_z \times \vec{a}_{(i)} / |\vec{a}_{(i)}|$, so that the vectorial form of (4.136) reads

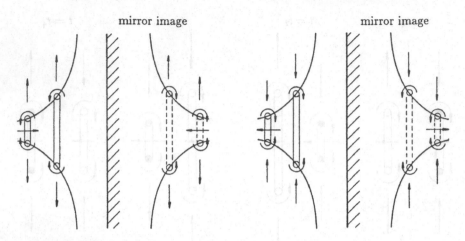

Fig. 4.25 Vortex ring at a wall

$$\vec{u}_R = \frac{\Gamma}{2\pi}\vec{e}_z \times \frac{\vec{x} - \vec{x}_{(i)}}{\left|\vec{x} - \vec{x}_{(i)}\right|^2}.$$

For $\vec{x} \to \vec{x}_{(i)}$ the velocity tends to infinity, but for reasons of symmetry the vortex cannot be moved by its own velocity field; the induced translational velocity is, as mentioned, zero. The induced velocity of n vortices with the circulation $\Gamma_{(i)}(i = 1 \ldots n)$ is

$$\vec{u}_R = \frac{1}{2\pi}\sum_i \Gamma_{(i)}\vec{e}_z \times \frac{\vec{x} - \vec{x}_{(i)}}{\left|\vec{x} - \vec{x}_{(i)}\right|^2}.$$

If there are no internal boundaries, or if the boundary conditions are satisfied by reflection, as in Fig. 4.25, the last equation describes the entire velocity field, and using (1.10), the "equation of motion" of the kth vortex reads

$$\frac{d\vec{x}_{(k)}}{dt} = \frac{1}{2\pi}\sum_{\substack{i \\ i \neq k}} \Gamma_{(i)}\vec{e}_z \times \frac{\vec{x}_{(k)} - \vec{x}_{(i)}}{\left|\vec{x}_{(k)} - \vec{x}_{(i)}\right|^2}. \qquad (4.145)$$

For the reasons given above, the vortex $i = k$ is excluded from the summation. With (4.145) the $2n$ equations for the path coordinates are given.

The dynamics of the vortex motion has invariants which are analogous to the invariants of a point mass system on which no external forces act. To start with, the conservation of the strengths of the vortices by Helmholtz's laws $\left(\sum \Gamma_{(k)} = \text{const}\right)$ corresponds to the conservation of the total mass of the point mass system. If we multiply the equation of motion (4.145) by $\Gamma_{(k)}$, sum over k and expand, we obtain

$$\sum_k \Gamma_{(k)} \frac{d\vec{x}_{(k)}}{dt} = \Gamma_{(1)} \frac{d\vec{x}_{(1)}}{dt} + \Gamma_{(2)} \frac{d\vec{x}_{(2)}}{dt} + \Gamma_{(3)} \frac{d\vec{x}_{(3)}}{dt} + \ldots =$$

$$\vec{e}_z \times \frac{1}{2\pi} \left\{ \Gamma_{(1)}\Gamma_{(2)} \frac{\vec{x}_{(1)} - \vec{x}_{(2)}}{|\vec{x}_{(1)} - \vec{x}_{(2)}|^2} + \Gamma_{(1)}\Gamma_{(3)} \frac{\vec{x}_{(1)} - \vec{x}_{(3)}}{|\vec{x}_{(1)} - \vec{x}_{(3)}|^2} + \ldots + \right.$$

$$+ \Gamma_{(2)}\Gamma_{(1)} \frac{\vec{x}_{(2)} - \vec{x}_{(1)}}{|\vec{x}_{(2)} - \vec{x}_{(1)}|^2} + \Gamma_{(2)}\Gamma_{(3)} \frac{\vec{x}_{(2)} - \vec{x}_{(3)}}{|\vec{x}_{(2)} - \vec{x}_{(3)}|^2} + \ldots +$$

$$\left. + \Gamma_{(3)}\Gamma_{(1)} \frac{\vec{x}_{(3)} - \vec{x}_{(1)}}{|\vec{x}_{(3)} - \vec{x}_{(1)}|^2} + \Gamma_{(3)}\Gamma_{(2)} \frac{\vec{x}_{(3)} - \vec{x}_{(2)}}{|\vec{x}_{(3)} - \vec{x}_{(2)}|^2} + \ldots \right\}.$$

We can see directly that the terms on the right-hand side cancel out in pairs, so that the equation

$$\sum_k \Gamma_{(k)} \frac{d\vec{x}_{(k)}}{dt} = 0$$

remains, which, when integrated, leads to

$$\sum_k \Gamma_{(k)}\vec{x}_{(k)} = \vec{x}_g \sum_k \Gamma_{(k)}. \tag{4.146}$$

For dimensional reasons, we have written the integration constants like a "center of gravity coordinate" $\vec{x}_g$. We interpret this result as

The center of gravity of the strengths of the vortices is conserved.

The corresponding law (conservation of momentum) for a system of mass points leads to the statement that the velocity of the center of gravity is a conserved quantity in the absence of external forces. For $\sum \Gamma_{(k)} = 0$ the center of gravity lies at infinity, so that, for example, two vortices with $\Gamma_{(1)} = -\Gamma_{(2)}$ must move along straight parallel paths (i.e. they turn about an infinitely distant point). If $\Gamma_{(1)} + \Gamma_{(2)} \neq 0$, the vortices turn about a center of gravity which is at a finite distance (Fig. 4.26).

Here the overtaking process of two straight vortex pairs is similar to the overtaking process of two vortex rings explained in Fig. 4.24. The paths of the vortex pairs are determined by numerical integration of (4.145) and are shown in Fig. 4.27.

The analogy of (4.146) is continued in the "balance of angular momentum of vortex systems" and can be carried over to a continuous vortex distribution. However we do not wish to go into this, but shall note the difference from the mechanics of mass points: (1.10) is the equation for the motion of a vortex under the influence of the remaining vortices of the system. The motion of a mass point under the influence of the rest of the system, that is, under the influence of the internal forces, is instead described by Newton's second law.

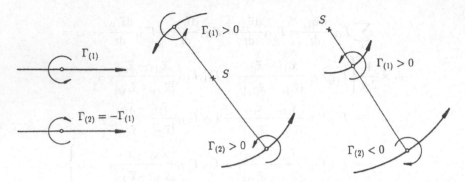

Fig. 4.26 Possible pathlines of a pair of straight vortices

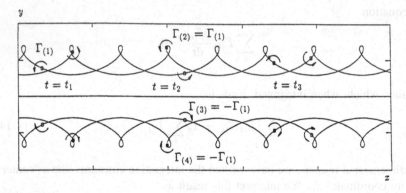

Fig. 4.27 Pathlines of two straight vortex pairs

Helmholtz's third vortex theorem reads:

The circulation of a vortex tube remains constant in time.

This follows immediately from Helmholtz's second law together with Kelvin's circulation theorem: a closed line generating the vortex tube (Fig. 4.11) is, by Helmholtz's second law, a material line whose circulation, by Kelvin's law, remains constant.

Helmholtz's second and third laws hold only for barotropic and inviscid fluids. The statements of these laws are also in Eq. (4.27), but there under the more restricting assumption of incompressible flow.

4.2.4 Integration of the Energy Equation

In steady and inviscid flow, when heat conduction can be ignored, an integral of the energy equation which is very useful may be found. We assume that k_i has a time independent potential, for example the mass body force of gravity. Then, since

$$\frac{D\psi}{Dt} = u_i \frac{\partial \psi}{\partial x_i} = -u_i k_i \qquad (4.147)$$

the work of the mass body force (per unit time) can also be written as the material derivative of the potential and, using $u = |\vec{u}|$, we obtain the energy equation (4.47) in the form

$$\varrho \frac{D}{Dt} \left[\frac{u^2}{2} + h + \psi \right] = 0. \qquad (4.148)$$

From this we conclude that the sum of the terms in brackets is a conserved quantity for a material particle, and therefore

$$\frac{u^2}{2} + h + \psi = C \qquad (4.149)$$

along a pathline. Because of our restriction to steady flows this also holds along a streamline. The constant of integration which appears differs in general from streamline to streamline. The value of this constant depends on how this flow arose, and is clearly the same for all streamlines if the energy is homogeneous at infinity. In most technically interesting flows this constant is equal for all streamlines, and these flows are thus called *homenergic*. In particular, homenergic flows do not have to be irrotational, and therefore they are kinematically not as restricted. On the other hand, as already mentioned, Bernoulli's constant is the same on every streamline only in irrotational fields (and also in fields where $\vec{\omega} \times \vec{u} = 0$, but these do not have the same technical importance as irrotational flows).

Equation (4.149) is mainly used in gas dynamics where the potential of the mass body force can often be ignored, and the energy equation assumes the form

$$\frac{u^2}{2} + h = h_t. \qquad (4.150)$$

This establishes an algebraic relation between velocity and enthalpy which always, independent of the specific problem, holds in steady and inviscid flow, and therefore in flows with chemical reactions where we have $Ds/Dt \neq 0$. If the enthalpy field is known the magnitude of the velocity in the field follows directly, and vice versa.

To find another form of the energy equation in which the dependency of the enthalpy does not expressly appear, the assumption of isentropic flow must be made explicitly. From Gibbs' relation (2.133) we find

$$\frac{De}{Dt} - \frac{p}{\varrho^2}\frac{D\varrho}{Dt} = 0 \tag{4.151}$$

or, using (4.3), also

$$\frac{Dh}{Dt} - \frac{1}{\varrho}\frac{Dp}{Dt} = 0. \tag{4.152}$$

Equations (4.152) and (4.148) then yield the energy equation in the form

$$\frac{D}{Dt}\left[\frac{u^2}{2} + \psi\right] + \frac{1}{\varrho}\frac{Dp}{Dt} = 0. \tag{4.153}$$

In steady flow, we can replace the operator D/Dt by $|\vec{u}|\partial/\partial\sigma$ or $|\vec{u}|\partial/\partial s$ (because of (1.23)). Integrating (4.153) along the pathline or streamline leads us again to Bernoulli's equation (4.57) in the form valid for steady flow

$$\frac{u^2}{2} + \psi + \int\frac{dp}{\varrho} = C. \tag{4.154}$$

In doing this we see that Bernoulli's equation is an energy equation. Indeed in the derivation of Bernoulli's equation (4.57) the inner product of the velocity $\vec{u}$ with the equation of motion was formed, thus making it a "mechanical energy equation". (The integral is to be taken along the streamline or pathline; if it is path independent (4.154) is called the "strong form" of Bernoulli's equation.) Incidentally, using the same assumptions, the "entropy equations" (4.151) and (4.152) are often used instead of the energy equation, although the kinetic energy does not appear explicitly in these formulae.

In order to clarify the relation between homenergic and irrotational flow mentioned above we shall need to use *Crocco's relation*, which only holds in steady flow. We can reach it by forming from the canonical equation of state $h = h(s, p)$ the gradient

$$\frac{\partial h}{\partial x_i} = \left[\frac{\partial h}{\partial s}\right]_p\frac{\partial s}{\partial x_i} + \left[\frac{\partial h}{\partial p}\right]_s\frac{\partial p}{\partial x_i}, \tag{4.155}$$

and using Eqs. (2.154) and (2.155) to get

$$-\frac{1}{\varrho}\frac{\partial p}{\partial x_i} = T\frac{\partial s}{\partial x_i} - \frac{\partial h}{\partial x_i}. \tag{4.156}$$

We introduce the formula into Euler's equation (4.40a), express the acceleration term there by (1.77), and extract the equation for steady flow known as Crocco's relation

$$-2\,\epsilon_{ijk}\,u_j\omega_k + \frac{\partial}{\partial x_i}\left[\frac{u_j\,u_j}{2} + h + \psi\right] = T\frac{\partial s}{\partial x_i}. \tag{4.157}$$

Here we have assumed that the mass body force has a potential. In homenergic flow the constant of integration C appearing in (4.149) has the same value on all streamlines, thus the gradient of C vanishes, and for this class of flows the following holds

$$\frac{\partial C}{\partial x_i} = \frac{\partial}{\partial x_i}\left[\frac{u^2}{2} + h + \psi\right] = 0. \tag{4.158}$$

Then for these flows it follows from Crocco's relation that irrotational flows must be homentropic. On the other hand we see that flows which are not homentropic but are homenergic must be rotational.

This case has already been discussed in 4.1.3 (curved shock) and is interesting because vorticity arises inside the flow field and not, as in incompressible flow, by diffusion from the boundaries inwards. By passing through a *curved shock* (Fig. 4.28), as in *hypersonic flow*, the entropy increases by a different amount on different streamlines. Therefore behind the shock surface the entropy is no longer homogeneous, and because of Crocco's relation, the flow can no longer be irrotational.

We also conclude from Crocco's relation that a two-dimensional homentropic (and homenergic) flow must necessarily be irrotational, because in two-dimensional flow $\vec{\omega}$ is always perpendicular to $\vec{u}$. Then the first term in (4.157) cannot vanish as it would if $\vec{\omega}$ and $\vec{u}$ were parallel vectors.

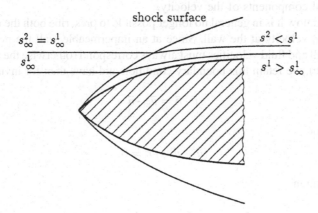

Fig. 4.28 Curved shock

4.3   Initial and Boundary Conditions

Up until now in Chap. 4 we have made general statements as they apply for every flow problem of Newtonian or inviscid fluids. Further progress in a given problem now demands that we make assertions about the shape of the flow boundary and about the conditions which the flow must satisfy at this boundary. Mathematically we shall deal here with the *boundary conditions*. In addition in unsteady flow problems the *initial conditions* are needed, i.e., the field quantities at the start of the time period of interest.

We shall first consider flow boundaries for the case of the impermeable wall (which we can generalize if necessary to permeable walls) and for the case of the free surface. Boundaries which are surfaces of discontinuity are also important. The best known example for this are shock surfaces, which we can only go into fully when the concept of a "shock" has itself been clarified.

We know from experience that Newtonian fluids adhere to walls. For an impermeable wall this means that both the tangential and the normal velocities of the fluid and of the wall must correspond at every point on the wall. The velocity vector $\vec{u}$ of the fluid at the wall must be equal to the vector of the wall velocity $\vec{u}_w$

$$\vec{u} = \vec{u}_w \quad \text{(at the wall)}. \tag{4.159}$$

The boundary condition when the wall is at rest ($\vec{u}_w = 0$) is

$$\vec{u} = 0 \tag{4.160}$$

at the wall, or alternatively

$$u_n = u_t = 0 \tag{4.161}$$

at the wall. Here the index n denotes the normal component and the index t denotes the tangential components of the velocity.

In inviscid flow it is in general no longer possible to prescribe both the normal and the tangential velocity at the wall. Since at an impermeable wall the normal component of wall and fluid velocities must always correspond (otherwise the wall would be permeable), we retain this boundary condition and have then for inviscid flow

$$\vec{u} \cdot \vec{n} = \vec{u}_w \cdot \vec{n} \tag{4.162a}$$

at the wall, or

$$(\vec{u} - \vec{u}_w) \cdot \vec{n} = 0, \tag{4.162b}$$

in index notation

$$(u_i - u_{i(w)})n_i = 0 \tag{4.162c}$$

at the wall. We call this condition the *kinematic boundary condition*, while (4.159) is called the *dynamic* or *physical boundary condition*. In inviscid flow we relax the dynamic boundary condition, since the derivatives in Euler's equation are of a lower order than in the Navier-Stokes equations. In Euler's equation the second order terms ($\eta \Delta \vec{u}$ in the incompressible case) are missing. It is known from the theory of ordinary differential equations that the order of the differential equation determines the number of boundary conditions which can be satisfied. In exactly the same way the order of a partial differential equation fixes how many functions can be satisfied on the boundary. Since only the boundary condition of the normal component of the velocity can be assigned in inviscid flow, in general different tangential components of the wall and fluid velocities arise: the dynamic boundary condition is therefore violated. Now we also understand why the viscous flow for $\nu \to 0$ does not turn into the solution with $\nu \equiv 0$: both flows satisfy different boundary conditions in which the viscosity ν does not appear explicitly and therefore are not affected by taking the limit $\nu \to 0$. In this connection we mention again that even in cases where the inviscid solution is a good approximation for the viscous flow at large Reynolds' numbers, this solution breaks down right next to the wall (in the boundary layer).

If the flow field around a finite sized body extends to infinity, the disturbances which originate from the body must die away at infinity. The degree to which the disturbances vanish depends on the given problem, and will be discussed only in connection with the specific problem (see Sect. 10.3).

The normal component of the wall velocity is required in the kinematic boundary condition. To find it consider the surface of the body given in implicit form by

$$F(\vec{x}, t) = 0, \tag{4.163}$$

where $\vec{x}$ is the position vector of a general point of the surface. The normal vector to the surface is (up to the sign)

$$\vec{n} = \frac{\nabla F}{|\nabla F|}, \tag{4.164}$$

so that we can write the kinematic boundary condition in the form

$$\vec{u} \cdot \nabla F = \vec{u}_w \cdot \nabla F \quad (\text{at } F(\vec{x}, t) = 0). \tag{4.165}$$

By definition a point on the surface with position vector $\vec{x}$ satisfies the Eq. (4.163) for all times. For an observer on the surface whose position vector is $\vec{x}$ (4.163) does not change, so it follows that

$$\frac{\mathrm{d}F}{\mathrm{d}t} = 0. \tag{4.166}$$

This time derivative is the general time derivative introduced with Eq. (1.19), since the observer on the surface moves with velocity $\vec{u}_w$ which is not equal to the velocity of a material particle at the same place. By (4.162a) only the normal components are equal. From

$$\frac{\mathrm{d}F}{\mathrm{d}t} = \frac{\partial F}{\partial t} + \vec{u}_w \cdot \nabla F = 0 \tag{4.167}$$

we first extract, by division with $|\nabla F|$, a convenient formula for the calculation of the normal velocity of a body

$$\vec{u}_w \cdot \frac{\nabla F}{|\nabla F|} = \vec{u}_w \cdot \vec{n} = -\frac{1}{|\nabla F|}\frac{\partial F}{\partial t}. \tag{4.168a}$$

In index notation this is

$$u_{i(w)}n_i = \frac{-\partial F/\partial t}{\left(\partial F/\partial x_j\, \partial F/\partial x_j\right)^{1/2}}. \tag{4.168b}$$

We are led to a particularly revealing form of the *kinematic boundary condition* if we insert (4.167) into (4.165)

$$\vec{u} \cdot \nabla F = -\frac{\partial F}{\partial t} \quad (\text{at } F(\vec{x}, t) = 0). \tag{4.169}$$

Using the definition of the material derivative (1.20) we then obtain

$$\frac{\partial F}{\partial t} + \vec{u} \cdot \nabla F = \frac{\mathrm{D}F}{\mathrm{D}t} = 0 \quad (\text{at } F(\vec{x}, t) = 0). \tag{4.170}$$

This final equation yields the following interpretation: the position vector $\vec{x}$ of a fluid particle on the surface of a body satisfies the Eq. (4.163) for the surface at all times, thus the material particle always remains on the surface.

This is *Lagrange's theorem*:

   The surface is always made up of the same fluid particles.

This at first surprising statement is the logical consequence of the condition that the normal components of the surface velocity and the fluid velocity at the surface be the same.

   The kinematic boundary condition also holds at the free surface and at interfaces between two fluids or more generally on material discontinuity surfaces.

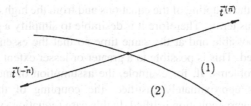

Fig. 4.29 Stress vector at an interface

Since the shape of the free surface is unknown beforehand, problems with free surfaces are mostly difficult to solve. Apart from the kinematic boundary condition, a *dynamic boundary condition* which expresses the continuity of the stress vector must be satisfied.

The stress vectors $\vec{t}_{(1)}$ and $\vec{t}_{(2)}$ at the same point of the interface with the normals $\vec{n}_{(1)} = \vec{n}$ in fluid (1) and $\vec{n}_{(2)} = -\vec{n}$ in fluid (2) must satisfy (2.23) (Fig. 4.29)

$$\vec{t}_{(1)}^{(\vec{n})} = -\vec{t}_{(2)}^{(-\vec{n})}. \tag{4.171}$$

Because of $\vec{n}_{(1)} = \vec{n} = -\vec{n}_{(2)}$ and using (2.29b) we also have

$$\vec{n} \cdot \mathbf{T}_{(1)} = \vec{n} \cdot \mathbf{T}_{(2)} \quad (\text{at } F(\vec{x}, t) = 0). \tag{4.172}$$

In inviscid fluid ($\mathbf{T} = -p\,\mathbf{I}$) we extract from (4.172) a condition for the pressure on the interface

$$p_{(1)} = p_{(2)} \quad (\text{at } F(\vec{x}, t) = 0). \tag{4.173}$$

Since we cannot fix a boundary condition for the tangential component of the velocity in inviscid flow, a jump in the tangential velocity generally arises at an interface and we speak of a "tangential discontinuity surface". The discontinuity surface behind an airfoil which we discussed earlier is of this kind.

4.4 Simplification of the Equations of Motion

Previously in this chapter we have stated the equations and boundary conditions with which the flow of a Newtonian fluid can, in principle, be calculated for general geometries of the flow boundary. Equations (4.1), (4.2) and (2.3) represent a system of coupled partial differential equations whose solution in general turns out to be a very difficult problem. The difficulties in the integration are based, firstly, on the fact that these equations, unlike most partial differential equations in physics, are non-linear. This means that solutions which have been found cannot be "superimposed" to form a new solution, as is possible with linear systems and as we have already seen in the example of Poisson's equation. Secondly the system is of a very high order, arising from the coupling of the equations and from the high derivatives which appear in the viscous terms. Therefore it is desirable to simplify a given problem so that a solution is possible and at the same time so that the essential aspect of the problem is preserved. This is possible to a greater or lesser extent in most technical fluid mechanics problems. If, for example, the assumption of incompressible and isothermal flow is approximately justified, the coupling of the Navier-Stokes equations and the energy equation is lifted. In this case (equation system (4.9a, 4.9b) and (2.5)) a class of exact solutions is known, and some of these are of fundamental importance in technical applications. Exact solutions arise either if the nonlinear terms identically vanish for kinematic reasons, as happens in unidirectional flow or if because of symmetry in the problem, the independent variables always appear in one combination which can then be written as a new independent variable, allowing the system of partial differential equations to become a system of ordinary differential equations (*similarity solution*). However the number of exact solutions is small, and we should not anticipate that future developments will increase the number of exact solutions significantly.

An essentially different situation appears when we consider numerical methods. Here we can expect that through the rapidly progressive developments very efficient methods of solution will appear, often making restrictive simplifications of the problem unnecessary. Indeed this development also justifies the detailed account of the general principles in the previous chapters.

We do not wish to go any further into numerical methods but shall note that the numerical solution of these equations also gives rise to substantial difficulties and certainly does not represent a "solved problem", even if all the complications involved with turbulent flows are excluded. Even if stable algorithms for numerical calculations do exist, for time and financial reasons all the simplifications which the problem allows should be exploited. Finally, the processes of simplification, abstraction and concentration on essential aspects of a problem are prerequisites for the understanding of every physical process.

In the following chapters flows will be considered which have all been idealized or specialized in certain ways, and we shall only consider the most important aspects of the flow in the given circumstances. The idealizations arise from the simplifying assumptions from the Eqs. (4.1), (4.2) and (2.3) for Newtonian fluids, or also from the more general Eqs. (2.38), (2.119), (2.3) and the corresponding constitutive relations in the case of non-Newtonian fluids.

The "theories" of fluid mechanics emerge from such simplifying assumptions. In this way, ignoring the viscosity and the heat conduction leads to the "theory of inviscid flows" which is described by Euler's equations (Sect. 4.2). Further simplifications divide this theory into incompressible and compressible inviscid flows. Finally, depending on the ratio of the typical flow velocity U to the *speed of sound* a, flows can be classified as subsonic, transonic and supersonic flows.

It is desirable to fit possible simplifications into some order, which both allows a classification of the given problem as well as giving an indication of the allowable and suitable simplifications for the problem. Such a scheme can follow from simplifications in

(a) the constitutive relation,
(b) the dynamics, or
(c) the kinematics.

Included in class (a) is ignoring the viscosity and the heat conduction as discussed already, as well as the assumptions of incompressible flow (which obeys the particular equations of state $D\varrho/Dt = 0$), barotropy and isentropy.

To (b) belong the simplifications which arise from the assumption of steady flow and the limiting cases of $Re \to \infty$ or $Re \to 0$. In addition the assumptions leading to subsonic, transonic, supersonic and hypersonic flows all fit in here.

In (c) we have, for example, irrotationality curl $\vec{u} = 0$. Additional kinematic simplifications arise from symmetry properties: in *rotational symmetry* the number of necessary spatial coordinates can be reduced using the cylindrical coordinate system to two $\left(r = \left(x_1^2 + x_2^2 \right)^{1/2}, x_3 \right)$, so that it can be treated as a two-dimensional problem. Steady *spherically symmetric* problems are one-dimensional, since in a spherical coordinate system we only have one independent coordinate $\left(r = \left(x_j x_j \right)^{1/2} \right)$.

Flows which are independent of one coordinate in a Cartesian coordinate system and whose velocity component in this direction vanishes are particularly important in applications. In the above sense they are *two-dimensional flows* but they are additionally *plane flows*. In a suitable coordinate system the same flow quantities are met in all planes, say $x_3 = $ const. Although two-dimensional flows never appear in nature, they are often good approximations to physical problems.

Belonging to (c) are also the simplifying assumptions of *stream filament theory*, which leads to a quasi-one-dimensional description, as well as the *theory of thin bodies*, in which the ratio of typical lengths (for example, the thickness ratio D/L of a body, or the inclination α of the streamlines) is very small. Of course combinations of these various criteria also appear: the *Mach number* $M = U/a > 1$ characterizes for example a supersonic flow, $D/L \ll 1$ a thin body, and $MD/L \ll 1$ a *linear supersonic flow*. The limiting value $\alpha \, Re \to 0$ denotes the simplification which leads to *hydrodynamic lubrication theory*.

Now this is not a comprehensive list of examples, nor is the classification into these three groups unique. For instance the case of incompressible flow with the equation of state $D\varrho/Dt = 0$ can be classified under (a), but because of the kinematic restrictions given by div $\vec{u} = 0$, also under (c). In the same manner the incompressibility can be grouped under (b), because the limiting case $U/a \to 0$ in steady flow corresponds with, as we shall see, the case of incompressible flow.

Many of the possible simplifications are immediately obvious, while others, for example the assumption of inviscid fluid, need careful justification. Apart from the assumption that the flow be inviscid the most incisive simplification is the assumption of incompressibility, because even for liquids, this assumption is not justified in certain circumstances; the examples in connection with Eq. (2.5) show this. We are lead to criteria for the admissibility of this simplification if we first form from the equation of state $p = p(\varrho, s)$ the expression

$$\frac{Dp}{Dt} = a^2 \frac{D\varrho}{Dt} + \left[\frac{\partial p}{\partial s}\right]_\varrho \frac{Ds}{Dt},$$

(4.174)

where it is known from thermodynamics that the state variable $(\partial p/\partial \varrho)_s$ is equal to the square of the speed of sound a

$$\left[\frac{\partial p}{\partial \varrho}\right]_s = a^2.$$

(4.175)

We bring (4.174) to a dimensionless form by multiplying with the typical convection time L/U and then dividing by ϱ to get

$$\frac{1}{\varrho} \frac{L}{U} \frac{D\varrho}{Dt} = \frac{L}{U} \frac{1}{\varrho a^2} \frac{Dp}{Dt} - \frac{L}{U} \frac{1}{\varrho a^2} \left[\frac{\partial p}{\partial s}\right]_\varrho \frac{Ds}{Dt}.$$

(4.176)

We see that the relative change in the density of a fluid particle can be ignored if the right-hand side vanishes. Unless by some chance both terms cancel out, in general each term on the right-hand side must vanish by itself. First we note that in the case of strong external heating, the internal irreversible production of entropy according to (2.137) is unimportant, and the change in entropy here is given by (2.138). This term alone is then so large that the relative change in density can not be ignored.

If the heating is by dissipation the irreversible production of entropy (2.137) becomes significant, and we estimate the final term in (4.176) using the assumption of the calorically perfect gas. By simple calculation, the relation

$$\left[\frac{\partial p}{\partial s}\right]_\varrho = \frac{R}{c_v} T \varrho$$

(4.177)

follows. For gases the dimensionless number

$$Pr = \frac{c_p \eta}{\lambda}$$

(4.178)

(the *Prandtl's number*) is approximately equal to one. For $Pr \approx 1$ the terms Φ/T and $T^{-2} q_i \partial T/\partial x_i$ in (2.137) are of the same order of magnitude, and we look at the term Φ/T. (In liquids which are not liquid metals $Pr \gg 1$, and the second term on the right-hand side of (2.137) is correspondingly small compared to the first one.) Using (4.177) we extract the equation

$$\frac{L}{U} \frac{1}{\varrho a^2} \left[\frac{\partial p}{\partial s}\right]_\varrho \frac{Ds}{Dt} = \frac{L}{U} \frac{R}{c_v} \frac{\Phi}{\varrho a^2}.$$

(4.179)

If L is the characteristic length of the problem then from $O(\Phi) = O(\eta\, U^2/L^2)$ we estimate

$$\frac{L}{U}\frac{R}{c_v}\frac{\Phi}{\varrho\, a^2} \sim \frac{L}{U}\frac{\nu\, U^2}{L^2\, a^2} = \frac{M^2}{Re}, \tag{4.180}$$

where M is the *Mach number* $M = U/a$ formed with the typical flow velocity and the speed of sound. In real flows M^2/Re is usually very small, and this term can be neglected. (If the typical length in the dissipation function Φ is the boundary layer thickness δ, the term in question in this equation is of the order M^2, as shall be shown later in Chap. 12.)

Since Dp/Dt is the change in pressure experienced by the material particle, the remaining term on the right-hand side can in general only vanish if a^2 becomes suitably large. To estimate this term qualitatively, we do not need to take the viscosity into account. We then assume irrotational flow and calculate $\varrho^{-1}\, Dp/Dt = DP/Dt$ from Bernoulli's equation in the form (4.75). First the term $D\psi/Dt$ arises, which we estimate for the most important case of the mass body force of gravity. The change in the quantity $\psi = -g_i\, x_i$ experienced by a material particle only originates from the convection of the particle, since the gravity field is time independent. Therefore the typical time of the change is the convection time L/U. Accounting for the factor L/U in Eq. (4.176) we are led to the following relation between orders of magnitude

$$\frac{L}{U}\frac{1}{a^2}\frac{D\psi}{Dt} \sim \frac{L}{U}\frac{U}{L}\frac{gL}{a^2} = \frac{gL}{a^2}. \tag{4.181}$$

Therefore a necessary condition for this contribution to vanish is

$$\frac{g\,L}{a^2} \ll 1. \tag{4.182}$$

This condition is satisfied if the typical length L in the problem is much smaller than a^2/g. For air under standard atmospheric conditions we have $a^2/g = 11,500$ m, and (4.182) is satisfied for all flows in technical applications, but not for problems which might arise in meteorology.

The next contribution to $\varrho^{-1}\, Dp/Dt$ from Bernoulli's equation is the term

$$\frac{1}{2}\frac{D}{Dt}\left[\frac{\partial \Phi^*}{\partial x_i}\right]^2 = \frac{1}{2}\frac{Du^2}{Dt}.$$

In steady flow the typical time of the change is again the convection time L/U, so that we estimate the contribution of this term to the first term on the right-hand side of (4.176) as having the order of magnitude

$$\frac{L}{U}\frac{1}{a^2}\frac{1}{2}\frac{D(u^2)}{Dt} \sim \frac{L}{U}\frac{1}{a^2}\frac{U}{L}U^2 = \frac{U^2}{a^2}. \tag{4.183}$$

From this the second necessary condition for ignoring compressibility follows

$$\frac{U^2}{a^2} = M^2 \ll 1. \tag{4.184}$$

In unsteady flow, besides the convection time L/U a further typical time generally appears as a measure of the rate of change, for example f^{-1} if f is the typical frequency of the motion. The restrictions arising from this are dealt with by the third contribution to $\varrho^{-1}\,Dp/Dt$ from Bernoulli's equation, that is $D(\partial\Phi^*/\partial t)/Dt$. From

$$\Phi^* = \int \nabla\Phi^* \cdot d\vec{x} = \int \vec{u} \cdot d\vec{x} \tag{4.185}$$

Φ^* has an order of magnitude $U\,L$, and if the typical time is given by the convection time L/U, using the estimation

$$\frac{L}{U}\frac{1}{a^2}\frac{D(\partial\Phi^*/\partial t)}{Dt} \sim \frac{L}{U}\frac{U^2}{L^2}\frac{UL}{a^2} \sim \frac{U^2}{a^2}, \tag{4.186}$$

the same restrictions arise as from (4.184). However if the typical time is f^{-1}, using

$$\frac{L}{U}\frac{1}{a^2}\frac{D(\partial\Phi^*/\partial t)}{Dt} \sim \frac{L}{U}f^2\frac{UL}{a^2} \sim \frac{L^2 f^2}{a^2} \tag{4.187}$$

a third necessary condition arises

$$\frac{L^2 f^2}{a^2} \ll 1. \tag{4.188}$$

In general all three necessary conditions must be satisfied if the assumption of incompressible flow is to be justified. Most important is the condition (4.184), which for steady flows, encountered in technical applications, is also sufficient. After this the Mach number of the flow must be small enough so that the compressibility effects can be ignored. We note that the condition (4.188) is not satisfied in acoustics. In *sound waves* the typical length L is equal to the wavelength λ and we have

$$\frac{\lambda f}{a} = 1. \tag{4.189}$$

Therefore acoustics belongs to the area of compressible flow.

Chapter 5
Hydrostatics

5.1 Hydrostatic Pressure Distribution

Hydrostatics is concerned with the behavior of fluids at rest. The *state of rest* is kinematically the most restricted state and problems in hydrostatics are among the simplest in fluid mechanics. We can obtain the laws of hydrostatics by setting

$$\vec{u} \equiv 0 \tag{5.1}$$

into the balance laws. From mass conservation it then follows directly that

$$\frac{\partial \varrho}{\partial t} = 0, \tag{5.2}$$

that is, the density must be constant in time, as is made particularly clear if we consider the integral form of mass conservation (2.7). Instead of using the balance laws we could go directly to the first integrals of Chap. 4. The velocity field in hydrostatics in trivially irrotational, so that Bernoulli's constant has the same value everywhere in the field, and directly from (4.79) we infer the fundamental general relation between pressure function and potential of the mass body force in a rotating reference frame in which the fluid is at rest

$$\psi + P - \frac{1}{2}\left(\vec{\Omega} \times \vec{x}\right)^2 = C. \tag{5.3}$$

This relation can easily be generalized for the case in which the origin of the reference frame moves with acceleration $\vec{a}$. To do this consider the potential $\vec{a} \cdot \vec{x}$ of the mass body force $-\vec{a}$ (an apparent force which has a potential because curl $\vec{a} = 0$) added to ψ. We note that (5.3) is only valid under the assumptions which also led to (4.79): the total mass body force has a potential, and the pressure p is a unique function of the density $p = p(\varrho)$ (barotropy). This means that lines of equal pressure are also lines of constant density, or expressed differently, that pressure and density gradients are parallel. As a consequence of the thermal equation of state (e.g. $p = \varrho RT$ for a thermally perfect gas), lines of equal pressure are then also lines of equal temperature. It is only under these conditions that *hydrostatic equilibrium* can exist. If these conditions are not satisfied then the fluid is necessarily set in motion.

We deduce this important statement from the corresponding differential form of (5.3), which results from Cauchy's equation (2.38) together with the spherically symmetric stress state (2.33), or straight from the Navier-Stokes equations (4.1) or Euler's equations (4.40) when we set $\vec{u} \equiv 0$

$$\nabla p = \varrho \vec{k}. \tag{5.4}$$

If we take the curl of (5.4) the left-hand side vanishes and we are led to the condition

$$\nabla \times \left(\varrho \vec{k} \right) = \nabla \varrho \times \vec{k} + \varrho \nabla \times \vec{k} = 0. \tag{5.5}$$

As noted in connection with (2.42), this is a necessary and sufficient condition for the existence of a potential Ω of the volume body force $(\vec{f} = \varrho \vec{k} = -\nabla \Omega)$. Clearly (5.5) is satisfied if the mass body force $\vec{k}$ has a potential $(\vec{k} = -\nabla \psi)$ and if $\nabla \varrho$ is parallel to $\vec{k}$ (or is zero). Because of (5.4) $\nabla \varrho$ is then parallel to ∇p and we have again reached the above statement.

An example of this is the natural convection from a radiator. The air close to the vertical surface of the radiator is warmed by heat conduction. Temperature and density gradients are then perpendicular to the radiator surface, and therefore perpendicular to the force of gravity. The hydrostatic equilibrium condition is then violated, and the air is set into motion. (The motion of the air improves heat transfer, and it is only because of this that rooms can be heated at all in this manner.)

In applying Eq. (5.3) to the pressure distribution in the atmosphere, we first note that the centrifugal force is already included in the gravity force (cf. Sect. 2.4). We choose a Cartesian coordinate system (thus ignoring the curvature of the earth) whose x_3-axis is directed away from the surface of the earth. We shall often denote the Cartesian coordinates x_i $(i = 1, 2, 3)$ as x, y and z, so that the *potential of the force of gravity* is $\psi = g\,z$. Equation (5.3) then reads

$$z_2 - z_1 = -\frac{1}{g} \int_{p_1}^{p_2} \frac{dp}{\varrho}. \tag{5.6}$$

Let us consider the case where the barotropy is a consequence of a homogeneous temperature distribution, so for thermally perfect gases we have

$$z_2 - z_1 = \frac{RT}{g} \int\limits_{p_2}^{p_1} \frac{dp}{p} = \frac{RT}{g} \ln \frac{p_1}{p_2}, \tag{5.7}$$

or

$$p_2 = p_1 \exp\left[-\frac{1}{RT} h g\right], \tag{5.8}$$

where the altitude difference $z_2 - z_1$ is denoted by h. Equation (5.8) is known as the *barometric altitude formula*. If the barotropy is a consequence of the homentropy (4.49), then since

$$\frac{p_1}{p} = \left[\frac{\varrho_1}{\varrho}\right]^\gamma \tag{5.9}$$

the formula corresponding to (5.7) reads

$$z_2 - z_1 = \frac{RT_1}{g} p_1^{-\left(\frac{\gamma-1}{\gamma}\right)} \int\limits_{p_2}^{p_1} p^{-1/\gamma} \, dp \tag{5.10}$$

or

$$z_2 - z_1 = \frac{\gamma}{\gamma - 1} \frac{RT_1}{g} \left\{ 1 - \left[\frac{p_2}{p_1}\right]^{\left(\frac{\gamma-1}{\gamma}\right)} \right\}, \tag{5.11}$$

where we have also made use of the thermal equation of state. With

$$\left[\frac{p_2}{p_1}\right]^{\left(\frac{\gamma-1}{\gamma}\right)} = \frac{T_2}{T_1} \tag{5.12}$$

we can also express (5.11) with the temperature difference made explicit

$$z_2 - z_1 = -\frac{\gamma}{\gamma - 1} \frac{R}{g} (T_2 - T_1). \tag{5.13}$$

Not all density distributions in the atmosphere which are statically possible are also stable. A necessary condition for *stability* is that the density decrease with increasing height. However this condition is not sufficient: the density must also decrease at least as strongly as in homentropic density stratification. This constitutes a neutral stratification: if, by some disturbance, a parcel of air is raised (friction and heat conduction being negligible), this air expands to the new pressure, its density decreases at constant entropy just so that the density and the temperature correspond to the new ambient pressure. If the density in the new position is lower, then the air parcel moves up further and the stratification is unstable. If, however, the density is higher the air parcel sinks down again and the stratification is stable. From (5.13) we calculate the temperature gradient of the neutral stratification to be

$$\frac{dT}{dz} = -\frac{\gamma - 1}{\gamma}\frac{g}{R} = -9.95 \cdot 10^{-3}\,\mathrm{K/m} \tag{5.14}$$

(for air with $R \approx 287\,\mathrm{J/(kg\,K)}, \gamma \approx 1.4$), that is, the temperature decreases about 1 K per 100 m. The stratification is unstable if the temperature decreases faster and is stable if it decreases slower. If the temperature increases with increasing height, as happens for example if a warm mass of air moves over colder ground air, we have *inversion*. This represents a particularly stable atmospheric stratification and has the consequence that polluted air remains close to the ground.

In what follows we restrict ourselves to homogeneous density fields and in particular to liquids. In the coordinate system of Fig. 4.2 Bernoulli's equation is applicable in the form (4.81), from which we conclude that for $w = 0$ the hydrostatic pressure distribution in a fluid of homogeneous density is

$$\frac{p}{\varrho} + gz - \frac{1}{2}\Omega^2 r^2 = C. \tag{5.15}$$

In the inertial system here ($\Omega = 0$) the pressure distribution therefore reads

$$p = p_0 - \varrho g z, \tag{5.16}$$

where p_0 is the pressure at height $z = 0$. We see that the pressure linearly increases with increasing depth ($z < 0$).

At points of equal height the pressure is the same. From this follows the law of communicating tubes: in communicating tubes (Fig. 5.1) the level of the fluid is the same everywhere because the pressure is equal to the ambient pressure p_0 everywhere on the surface of the fluid.

Pascal's paradox is a further consequence of (5.16). The bases of the vessels shown in Fig. 5.2 are at equal pressure. If the bases are of equal size, then so are the forces, independent of the total weight of the fluid in the vessels. Equation (5.16) also explains how the often used U-tube manometer works (Fig. 5.3). The pressure p_C in the container is found by first determining the intermediate pressure p_Z in the manometer fluid at depth Δh from p_0

$$p_Z = p_0 + \varrho_M g\,\Delta h. \tag{5.17}$$

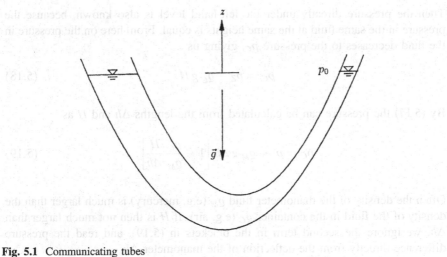

Fig. 5.1 Communicating tubes

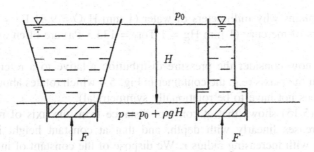

Fig. 5.2 Pascal's paradox

$$p = p_0 + \rho g H$$

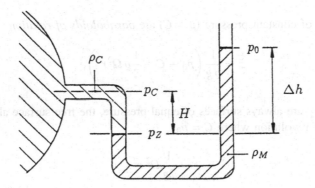

Fig. 5.3 U-tube manometer

Then the pressure directly under the left-hand level is also known, because the pressure in the same fluid at the same heights is equal. From here on the pressure in the fluid decreases to the pressure p_C, giving us

$$p_C = p_Z - \varrho_C \, g \, H. \tag{5.18}$$

By (5.17) the pressure can be calculated from the lengths Δh and H as

$$p_C = p_0 + \varrho_M \, g \, \Delta h \left[1 - \frac{\varrho_C}{\varrho_M} \frac{H}{\Delta h} \right]. \tag{5.19}$$

Often the density of the manometer fluid ϱ_M (e.g. mercury) is much larger than the density of the fluid in the container ϱ_C (e.g. air). If H is then not much larger than Δh, we ignore the second term in the brackets in (5.19), and read the pressure difference directly from the deflection of the manometer Δh

$$p_C - p_0 = \varrho_M \, g \, \Delta h \tag{5.20}$$

This also explains why millimeters of water (1 mm H_2O = 9.81 Pa = 9.81 N/m^2) or millimeters of mercury (1 mm Hg = 1 Torr = 133.3 Pa) are often used as units of pressure.

We shall now consider the pressure distribution relative to a reference frame rotating about the z-axis (e.g. the container in Fig. 5.4 which rotates about the z-axis but which does not have to be rotationally symmetrical).

Equation (5.15) shows that at constant distance from the axis of rotation, the pressure increases linearly with depth, and that at constant height it increases quadratically with increasing radius r. We dispose of the constant of integration in (5.15), by putting the pressure $p = p_0$ at $z = 0$, $r = 0$, and then write

$$p = p_0 - \varrho \, g \, z + \frac{1}{2} \varrho \, \Omega^2 r^2. \tag{5.21}$$

The surfaces of constant pressure ($p = C$) are *paraboloids of rotation*

$$z = \frac{1}{\varrho \, g} \left(p_0 - C + \frac{1}{2} \varrho \, \Omega^2 r^2 \right), \tag{5.22}$$

and since they are always surfaces of equal pressure, the free surface also forms a paraboloid of revolution where $C = p_0$

$$z = \frac{1}{2g} \Omega^2 \, r^2. \tag{5.23}$$

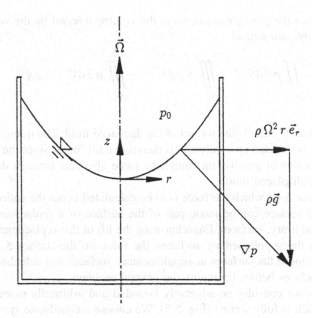

Fig. 5.4 Free surface on a rotating container

5.2 Hydrostatic Lift, Force on Walls

In liquids, in particular in water, the density is so high that the loads on container walls, dams, etc. from the hydrostatic pressure distribution become important. Using the pressure distribution (5.15), the force on a surface S can be calculated from

$$\vec{F} = - \iint\limits_{(S)} p\,\vec{n}\,\mathrm{d}S,\qquad(5.24)$$

if necessary numerically, by adding the vectors $-p\,\vec{n}\,\mathrm{d}S$ until the whole surface is exhausted. However using Gauss' theorem, the calculation of forces on surfaces (particularly on curved surfaces) can be reduced to finding the *buoyancy force*, which is given by *Archimedes' principle*:

A body in a fluid experiences an apparent reduction in weight (lift) equal to the weight of the displaced fluid.

This important law follows directly from Gauss' theorem and the Eq. (5.4): if the body is fully immersed then S is a closed surface and the total hydrostatic force is given by (5.24). Instead of calculating the surface integral directly, we transform it to a volume integral using Gauss' theorem. Now we consider the immersed body to be replaced by fluid which is of course in balance with its surroundings. Then, using

(5.4) we replace the pressure gradients in the volume integral by the volume body force of gravity, and extract

$$\vec{F} = -\iint\limits_{(S)} p\,\vec{n}\,\mathrm{d}S = -\iiint\limits_{(V)} \nabla p\,\mathrm{d}V = -\iiint\limits_{(V)} \varrho\,\vec{g}\,\mathrm{d}V = -\varrho\,\vec{g}V. \qquad (5.25)$$

The term on the far right is the weight of the displaced fluid. The minus sign shows that this force is directed upwardly and is therefore a lift force. Since the weight acts through the center of gravity, the buoyancy force also acts through the center of gravity of the displaced fluid.

If the surface S on which the force is to be calculated is not the entire surface of the body, this surface can be made part of the surface of a *replacement body* by using other, arbitrary, surfaces. From knowing the lift of this replacement body and the forces on the supplementary surfaces, the force on the surface S can be calculated. We choose flat surfaces as supplementary surfaces and calculate the forces on the flat surfaces before beginning the general problem.

To do this we consider an arbitrarily bounded and arbitrarily orientated plane surface A which is fully wetted (Fig. 5.5). We choose a coordinate system x', y', z' originating at the centroid of the surface, whose z'-axis is normal to the surface, whose y'-axis lying in the surface runs parallel to the free surface (and is therefore perpendicular to the mass body force), and whose x'-axis is chosen so that x', y' and z' form a right-handed coordinate system. In this primed coordinate system the potential of the mass body force reads

$$\psi = -\vec{g}\cdot\vec{x} = -(g'_x x' + g'_z z'), \qquad (5.26)$$

since $\vec{g}$ has no component in the y'-direction. As earlier, we obtain the hydrostatic pressure distribution from Bernoulli's equation where we set the velocity to zero. Beginning from (4.57), for an incompressible fluid, we obtain

$$p + \varrho\psi = C, \qquad (5.27)$$

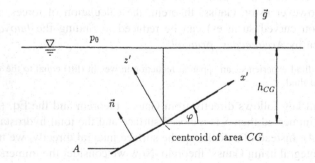

Fig. 5.5 The force on a plane surface

or

$$p - \varrho\left(g'_x x' + g'_z z'\right) = p_{CG},\tag{5.28}$$

where p_{CG} is the pressure at the centroid of the plane ($x' = y' = z' = 0$), which, by (5.16) is

$$p_{CG} = p_0 + \varrho g h_{CG}.\tag{5.29}$$

The component of $\vec{g}$ in the x'-direction is $g'_x = g \sin \varphi$ and the pressure on the plane A ($z' = 0$) is then

$$p = p_{CG} - \varrho g \sin \varphi x';\tag{5.30}$$

therefore the force is

$$\vec{F} = -\iint\limits_{(S)} p \vec{n} \, dS = -\vec{n} \iint\limits_{(A)} (p_{CG} - \varrho g \sin \varphi x') \, dA\tag{5.31}$$

or

$$\vec{F} = -\vec{n} \left[p_{CG} A - \varrho g \sin \varphi \iint\limits_{(A)} x' \, dA \right].\tag{5.32}$$

Since the origin of the coordinate system lies on the centroid of the plane surface ($x'_{CG} = y'_{CG} = 0$) and the centroid coordinates are, by definition, given by

$$A x'_{CG} = \iint\limits_{(A)} x' \, dA,\tag{5.33}$$

$$A y'_{CG} = \iint\limits_{(A)} y' \, dA,\tag{5.34}$$

the integral in (5.32) vanishes and for the force we extract

$$\vec{F} = -n p_{CG} A, \text{ i.e.}\tag{5.35}$$

The magnitude of force on a plane surface is the product of the pressure at the centroid of the surface and its area.

We shall also calculate the moment of the pressure distribution relative to an arbitrary point $P\left(\vec{x}_p = x_p'\vec{e}_x + y_p'\vec{e}_y\right)$ on the surface A

$$\vec{M}_p = -\iint\limits_{(A)} \left(\vec{x}' - \vec{x}_p\right) \times \vec{n}\, p\, dA. \tag{5.36}$$

Evaluating the cross product and with $\vec{n} = \vec{e}_z$ we obtain

$$\vec{M}_p = \iint\limits_{(A)} \left[\left(x' - x_p'\right)\vec{e}_y - \left(y' - y_p'\right)\vec{e}_x\right] p(x')\, dA. \tag{5.37}$$

Introducing the pressure distribution from (5.30), and noting the definitions of the centroid (5.33), (5.34) and $x_{CG}' = y_{CG}' = 0$ furnishes the equation

$$\vec{M}_p = \left[\varrho\, g\, \sin\varphi \iint\limits_{(A)} x'\, y'\, dA + y_p' p_{CG} A\right] \vec{e}_x +$$
$$\tag{5.38}$$
$$- \left[\varrho\, g\, \sin\varphi \iint\limits_{(A)} x'^2\, dA + x_p' p_{CG} A\right] \vec{e}_y.$$

The area moments of the second order appearing in (5.38) are

   (i) the area moment of inertia relative to the y'-axis

$$I_{y'} = \iint\limits_{(A)} x'^2\, dA; \tag{5.39}$$

   (ii) the mixed moment of inertia

$$I_{x'y'} = \iint\limits_{(A)} x'\, y'\, dA. \tag{5.40}$$

These correspond to the quantities known as polar moment of inertia and product of inertia from the theory of bending and torsion. Using these definitions we also write (5.38) as

$$\vec{M}_p = \left(\varrho\, g\, \sin\varphi\, I_{x'y'} + y_p' p_{CG} A\right)\vec{e}_x - \left(\varrho\, g\, \sin\varphi\, I_{y'} + x_p' p_{CG} A\right)\vec{e}_y. \tag{5.41}$$

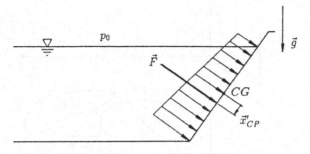

Fig. 5.6 Centroid and pressure point

The moment $\vec{M}_p$ vanishes relative to a particular point called the *pressure point CP* (Fig. 5.6), which is the point through which the force F acts. By setting the moment to zero we calculate the pressure point's coordinates as

$$x'_{CP} = -\frac{\varrho\, g\, \sin\varphi\, I_{y'}}{p_{CG}\, A} \tag{5.42a}$$

and

$$y'_{CP} = -\frac{\varrho\, g\, \sin\varphi\, I_{x'y'}}{p_{CG}\, A}\,. \tag{5.42b}$$

In order to now calculate the force on a general curved surface S, we complete S to a closed surface, by dropping perpendicular lines from every point on the boundary C of S to the fluid surface (Fig. 5.7). We now use the result (5.25); there S corresponds to the entire surface which here is made up of the general curved surface and the supplementary surfaces M and A_z. M is the surface generated by dropping the perpendicular, and A_z on the free surface closes the *replacement volume*. From (5.25) we have then

$$- \iint\limits_{(S+M+A_z)} p\,\vec{n}\,dS = -\iint\limits_{(S)} p\,\vec{n}\,dS - \iint\limits_{(A_z)} p\,\vec{n}\,dA - \iint\limits_{(M)} p\,\vec{n}\,dS = -\varrho\,\vec{g}\,V\,. \tag{5.43}$$

From (5.43) we obtain the component of the force on S in the positive z-direction as

$$F_z = -\iint\limits_{(S)} p\,\vec{n}\cdot\vec{e}_z\,dS = \iint\limits_{(A_z)} p\,\vec{n}\cdot\vec{e}_z\,dA + \iint\limits_{(M)} p\,\vec{n}\cdot\vec{e}_z\,dS - \varrho\,\vec{g}\cdot\vec{e}_z V\,. \tag{5.44}$$

On A_z, $\vec{n} = \vec{e}_z$ and $p = p_0$; on M, $\vec{n}\cdot\vec{e}_z = 0$, since $\vec{n}$ is perpendicular to $\vec{e}_z$. We also have $-\vec{g}\cdot\vec{e}_z = g$, and are led directly to the component of the force in the z-direction

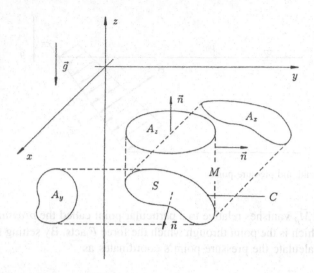

Fig. 5.7 The force on a curved surface

$$F_z = p_0\,A_z + \varrho\,g\,V\ .\tag{5.45}$$

For the component of the force in the x-direction we obtain

$$F_x = -\iint p\,\vec{n}\cdot\vec{e}_x\,dS = -\mathrm{sgn}(\vec{n}\cdot\vec{e}_x)\iint\limits_{(A_x)} p\,dA\ ,\tag{5.46}$$

where A_x is the projection of the surface S in the x-direction and the signum function determines the sign of the force. (If the sign of $\vec{e}_x\cdot\vec{n}$ changes on the surface, the surface is to be cut along the line $\vec{e}_x\cdot\vec{n}=0$ in two surfaces, which are treated separately.)

But the problem to calculate the force on a plane surface has already been done through Eqs. (5.35) and (5.42a, 5.42b). Analogously the component of the force in the y-direction follows

$$F_y = -\iint\limits_{(S)} p\,\vec{n}\cdot\vec{e}_y\,dS = -\mathrm{sgn}\left(\vec{n}\cdot\vec{e}_y\right)\iint\limits_{(A_y)} p\,dA\ .\tag{5.47}$$

The force components F_x and F_y do not appear in the second law of equilibrium (that the sum of the moments is zero) on the replacement body, since they are balanced by the corresponding force components on the surface M. The weight $\varrho\,g\,V$, the force p_0 A_z and F_z all lie in a vertical plane since they must balance separately.

The line of action of the buoyancy force (through the center of gravity of the displaced fluid) and the force $p_0 A_z$ (through the centroid of the surface A_z) determines this plane. Taking moments, for example about the center of gravity, we obtain the line of action of F_z. The lines of action of the two horizontal components F_x and F_y are to be calculated using the corresponding projections A_x and A_y from (5.42). These three lines of action do not in general meet at the same point.

5.3 Free Surfaces

Liquids form *free surfaces*, and these exhibit the phenomenon of surface or capillary tension. This surface tension can be important in technical problems under circumstances to be described presently.

From a microscopic standpoint this phenomenon is due to the fact that molecules on the free surface, or on an interface between two different fluids are in a different environment than those molecules within a fluid. The forces between the molecules are attractive forces at the average distances we are dealing with (cf. Sect. 1.1), (but can in certain circumstances be repulsive). A molecule within the fluid experiences the same attraction on all sides from its neighboring molecules. On the free surface, a molecule is pulled inwards in the same manner by its neighbors because the forces of attraction on the free side are missing, or at least are different. Therefore there are only as many molecules on the free surface as are absolutely necessary for its formation, and the free surface is always trying to contract.

Macroscopically, this manifests itself as if a tension were acting in the free surface, very much like the stress in a soap bubble. The capillary force on a line element is

$$\Delta \vec{F} = \vec{\sigma} \, \Delta l, \tag{5.48}$$

where $\vec{\sigma}$ is the stress vector of the surface tension, defined by

$$\vec{\sigma} = \lim_{\Delta l \to 0} \frac{\Delta \vec{F}}{\Delta l} = \frac{\mathrm{d}\vec{F}}{\mathrm{d}l}. \tag{5.49}$$

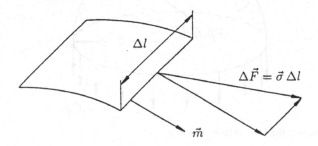

Fig. 5.8 Explanation of surface tension

In general, the stress vector lying in the surface has components both normal and tangential to the line element (Fig. 5.8). If the fluid particles which form the free surface are at rest, the tangential component vanishes and we have

$$\vec{\sigma} = C\vec{m}, \tag{5.50}$$

where $\vec{m}$ is the vector normal to the line element dl lying in the free surface. The magnitude of the surface tension vector, the *capillary constant C* is independent of $\vec{m}$, but dependent on the pairing liquid-gas, or in the case of an interface, liquid-liquid.

The best known manifestation of surface tension is the spherical shape of small drops. If we consider the surface of the drop to be a soap bubble under internal pressure p_i, then on one hand we have the force due to the pressure difference $p_i - p_0$ acting on one half of the surface and on the other hand the force due to the surface tension acting on the circumferential cut (Fig. 5.9). The force due to the surface tension is $2\pi r C \vec{m}$ and the equilibrium condition furnishes

$$2\pi r C\vec{m} - \iint\limits_{(S)} (p_0 - p_i)\vec{n}\,dS = 0. \tag{5.51}$$

If we form the component equation in the direction of $\vec{m}$ (for symmetry reasons this is the only nonzero component), with $\vec{m} \cdot \vec{n}\,dS = -dA$ we obtain

$$2\pi r C + (p_0 - p_i)\,\pi r^2 = 0 \tag{5.52}$$

or

$$\Delta p = p_i - p_0 = 2\,C/r. \tag{5.53}$$

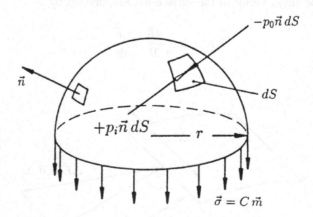

Fig. 5.9 Balance on the free surface of a drop

For very small drops the pressure drop over the surface can be quite considerable. For a general surface it is readily shown that the pressure drop is given by

$$\Delta p = C \left[\frac{1}{R_1} + \frac{1}{R_2} \right] , \tag{5.54}$$

where R_1 and R_2 are the *principal radii of curvature*, i.e., the extrema of the radii curvature at a point on the surface. The quantity $(1/R_1 + 1/R_2)$ is called the mean curvature and is a scalar, contrary to the curvature itself. For a plane surface, $(R_1 = R_2 \rightarrow \infty)$ the pressure drop vanishes. Therefore capillarity effects appear only if the surfaces are curved.

Curvature of the free surface often appears on boundaries if three different fluids meet, or if two fluids and a solid wall meet, as in Fig. 5.10, where the interface between fluids (1) and (2) touches a wall. We write

$$z = z(x, y) \tag{5.55}$$

for the explicit representation of the interface, and for the pressure drop across the surface we obtain

$$p_2 - p_1 = (\varrho_1 - \varrho_2) g\, z(x, y) . \tag{5.56}$$

Using (5.54) we also write this as

$$C \left[\frac{1}{R_1} + \frac{1}{R_2} \right] = (\varrho_1 - \varrho_2) g\, z(x, y) . \tag{5.57}$$

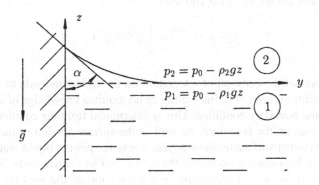

Fig. 5.10 Surface of a heavy fluid

We shall restrict ourselves to the plane case, that is $z = z(y)$, $R_1 \to \infty$, $R_2 = R$ and we further assume that fluid (2) is a gas, i.e., $\varrho = \varrho_1 \gg \varrho_2$. Then (5.57) simplifies to

$$C/R = \varrho\, g\, z(y)\,. \tag{5.58}$$

From this equation we extract a quantity a with the dimension of length

$$a = \sqrt{\frac{C}{\varrho\, g}}\,. \tag{5.59}$$

Therefore we can expect that the capillarity effects are particularly noticeable when the typical size of the flow region is of the order of this length. The quantity a, called *capillary length* or *Laplace's length*, has a value for water of about 0.3 cm. This explains why water flows straight out of a garden hose held high, while it cannot flow freely under the influence of gravity if the diameter of the hose is comparable to Laplace's length. The water then remains in the tube in the form of plugs. With the known expression

$$R^{-1} = \left(z'^2 + 1\right)^{-3/2} z''\,, \tag{5.60}$$

for the curvature R^{-1} of a curve $z(y)$, where the dash above z means the derivative with respect to y, we obtain from (5.58) an ordinary differential equation of the second order for the unknown shape $z(y)$ of the surface

$$\left(z'^2 + 1\right)^{-3/2} z'' - a^{-2} z = 0. \tag{5.61}$$

The particular integral of this equation requires two boundary conditions. Integrating once brings us to the equation

$$\left(z'^2 + 1\right)^{-1/2} + \frac{1}{2} a^{-2} z^2 = 1, \tag{5.62}$$

where we have set the constant of integration on the right-hand side to 1 using the boundary condition $z(\infty) = 0$. Integrating again requires knowledge of the angle of contact α as the boundary condition. This is determined from the equilibrium of the capillary stresses on the boundary. As well as the surface tension of the liquid-gas pair C_{12}, two further surface tensions appear due to the pairing liquid-wall (C_{13}) and gas-wall (C_{23}). Equilibrium normal to the wall is not of interest since the wall can take up arbitrary stresses. Equilibrium in the direction of the wall (cf. Fig. 5.11) leads to

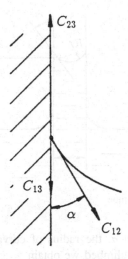

Fig. 5.11 Angle of contact

$$C_{23} = C_{13} + C_{12} \cos\alpha \qquad (5.63a)$$

or

$$\cos\alpha = \frac{C_{23} - C_{13}}{C_{12}}. \qquad (5.63b)$$

The fluid climbs or slides down the wall until the condition (5.63a) is satisfied. However if $C_{23} - C_{13}$ is larger than C_{12}, equilibrium cannot be satisfied, and the fluid coats the whole wall (e.g. petrol and silicon oil in containers). With the boundary condition $z'(y = 0) = -\cot\alpha$ the solution of (5.62) then reads in implicit form

$$y/a = \text{arccosh}\,(2a/z) - \text{arccosh}\,(2a/h) + \sqrt{4 - (h/a)^2} - \sqrt{4 - (z/a)^2}, \qquad (5.64)$$

where the square of the height climbed $h = z(y = 0)$ is to be taken from (5.62) as $h^2 = 2a^2(1 - \sin\alpha)$.

Another phenomenon often seen is the capillary rise in small tubes (Fig. 5.12). Obviously the pressure drop Δp over the surface must be equal to $\varrho\, g\, h$. If we take the shape of the surface to be spherical, because of $R_1 = R_2 = R$ we have from (5.54)

$$2\frac{C}{R} = \varrho\, g\, h. \qquad (5.65)$$

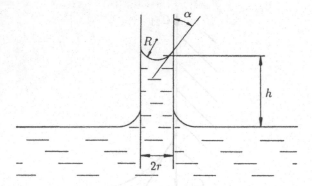

Fig. 5.12 Capillary rise in a small tube

For a known angle of contact α, the radius of curvature R can be replaced by $r/\cos\alpha$, so that for the height climbed we obtain

$$h = \frac{2C\cos\alpha}{r\varrho g}. \tag{5.66}$$

For very small r the height climbed can become very large and this explains why moisture rises so high in a porous wall. If we have $\alpha > \pi/2$, the capillary rise becomes negative, so that the fluid slides downwards. The best known example of this action is mercury.

Chapter 6
Laminar Unidirectional Flows

Quite important simplifications in the equations of motion arise for the class of unidirectional flows and these allow closed form solutions even for non-Newtonian fluids. As has already been discussed in Sect. 4.4, this solvability rests on the particularly simple kinematics of these flows.

Here we shall restrict ourselves to incompressible flows for which only pressure differences can be calculated unless there is a boundary condition on the pressure, e.g., the presence of a free surface. On a free surface the absolute value of the pressure enters the problem through the boundary condition (4.171) for the stress vector. Without free surfaces the influence of the mass body force can be removed from the problem if we limit ourselves to calculating pressure differences relative to the hydrostatic pressure distribution. We shall demonstrate this by way of the Navier-Stokes equations, and shall set the pressure as

$$p = p_{st} + p_{dyn}, \tag{6.1}$$

where the *hydrostatic pressure* p_{st} satisfies the hydrostatic relation (5.4). By (4.9b) we then have

$$\varrho \frac{D\vec{u}}{Dt} = \varrho \vec{k} - \nabla p_{st} - \nabla p_{dyn} + \eta \Delta \vec{u}, \tag{6.2}$$

which, because of (5.4), becomes

$$\varrho \frac{D\vec{u}}{Dt} = -\nabla p_{dyn} + \eta \Delta \vec{u}. \tag{6.3}$$

The mass body force no longer appears in this equation. p_{dyn} is the pressure difference $p - p_{st}$ and originates only from the motion of the fluid. From here on we shall write p in place of p_{dyn}, and shall understand that in all problems without free

© Springer Nature Switzerland AG 2020
J. H. Spurk and N. Aksel, *Fluid Mechanics*,
https://doi.org/10.1007/978-3-030-30259-7_6

surfaces, p means the pressure difference $p - p_{st}$. If the problem being dealt with does contain free surfaces, we shall, without further explanation, make use of the equations of motion in which the mass body force, if present, appears explicitly.

6.1 Steady Unidirectional Flow

6.1.1 Couette Flow

Simple shearing flow or *Couette flow* is a two-dimensional flow whose velocity field has already been commented on several times. The velocity components u, v, w in a Cartesian coordinate system with axes x, y, z read (cf. Fig. 6.1a)

$$u = \frac{U}{h}y, \quad v = 0, \quad w = 0. \tag{6.4}$$

Therefore the flow field is identical in all planes (z = const). The property common to all unidirectional flows, that the only nonvanishing velocity component (in this case u) only varies perpendicular to the flow direction, is a consequence of the continuity Eq. (2.5)

$$\nabla \cdot \vec{u} = \frac{\partial u}{\partial x} + \frac{\partial v}{\partial y} + \frac{\partial w}{\partial z} = 0. \tag{6.5}$$

From this, because $v = w = 0$, we obtain

$$\frac{\partial u}{\partial x} = 0 \quad \text{or} \quad u = f(y), \tag{6.6}$$

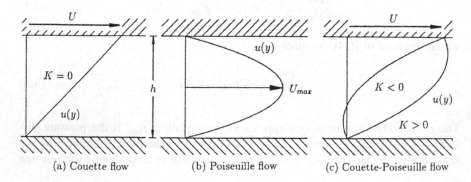

(a) Couette flow   (b) Poiseuille flow   (c) Couette-Poiseuille flow

Fig. 6.1 Plane unidirectional flow

of which (6.4) is a special case. The x-component of the Navier-Stokes equations reads

$$u\frac{\partial u}{\partial x}+v\frac{\partial u}{\partial y}+w\frac{\partial u}{\partial z}=-\frac{1}{\varrho}\frac{\partial p}{\partial x}+\nu\left[\frac{\partial^2 u}{\partial x^2}+\frac{\partial^2 u}{\partial y^2}+\frac{\partial^2 u}{\partial z^2}\right]. \tag{6.7}$$

Because of (6.4), all the convective (nonlinear) terms on the left-hand side vanish. This is the case in all unidirectional flows. Of course since we are dealing with a two-dimensional flow we could have set all derivatives with respect to z equal to zero, and indeed we shall want to do this in the future.

Since in this special case of Couette flow u is a linear function of y, all the terms in the brackets on the right-hand side of (6.7) vanish and we are led to the equation

$$\frac{\partial p}{\partial x}=0 \quad \text{or} \quad p=f(y). \tag{6.8}$$

The component of the Navier-Stokes equations in the y-direction

$$u\frac{\partial v}{\partial x}+v\frac{\partial v}{\partial y}=-\frac{1}{\varrho}\frac{\partial p}{\partial y}+\nu\left[\frac{\partial^2 v}{\partial x^2}+\frac{\partial^2 v}{\partial y^2}\right] \tag{6.9}$$

directly leads us to

$$\frac{\partial p}{\partial y}=0, \tag{6.10}$$

which, together with (6.8), furnishes the final result

$$p=\text{const.} \tag{6.11}$$

The field (6.4) satisfies the boundary condition (4.159), and therefore we have found the most simple nontrivial exact solution of the Navier-Stokes equations.

6.1.2 Couette-Poiseuille Flow

A generalization of simple shearing flow is suggested by (6.6): we consider the velocity field

$$u = f(y), \quad v = w = 0. \tag{6.12}$$

The x-component of the Navier-Stokes equations then reduces to

$$\frac{\partial p}{\partial x} = \eta \frac{\partial^2 u}{\partial y^2}, \tag{6.13}$$

and the y-component reads

$$0 = -\frac{1}{\varrho} \frac{\partial p}{\partial y}. \tag{6.14}$$

A consequence of the last equation is that p can only be a function of x. However since by assumption the right-hand side of (6.13) is not a function of x, neither is the left-hand side, i.e., $\partial p/\partial x$ is not a function of x. Therefore $\partial p/\partial x$ is a constant which we shall call $-K$. From (6.13) we then extract a differential equation of the second order for the desired function $u(y)$

$$\eta \frac{d^2 u}{dy^2} = -K. \tag{6.15}$$

Integrating (6.15) twice leads us to the general solution

$$u(y) = -\frac{K}{2\eta} y^2 + C_1 y + C_2. \tag{6.16}$$

We specialize the general solution to flow through a plane channel whose upper wall moves with velocity U in the positive x–direction. The function we are looking for, $u(y)$, must by (4.159), satisfy the two boundary conditions

$$u(0) = 0, \tag{6.17a}$$

and

$$u(h) = U, \tag{6.17b}$$

so that we determine the constants of integration as

$$C_1 = \frac{U}{h} + \frac{K}{2\eta} h, \quad C_2 = 0. \tag{6.18}$$

Thus the solution of the boundary value problem is

$$\frac{u(y)}{U} = \frac{y}{h} + \frac{K h^2}{2\eta U}\left[1 - \frac{y}{h}\right]\frac{y}{h}. \tag{6.19}$$

For $K = 0$ we get the simple shearing flow again; for $U = 0$ and $K \neq 0$ we obtain a parabolic velocity distribution (two-dimensional *Poiseuille flow*); the general case ($U \neq 0$, $K \neq 0$) yields the *Couette-Poiseuille flow* (Fig. 6.1).

As is directly obvious from (6.19), the general case is a superposition of Couette flow and Poiseuille flow. Since the unidirectional flows are described by linear differential equations, the superposition of other unidirectional flows is also possible.

The volume flux per unit depth is

$$\dot{V} = \int\limits_0^h u(y)\,\mathrm{d}y, \tag{6.20}$$

so that the *average velocity* defined by the equation

$$\overline{U} = \frac{\dot{V}}{h} \tag{6.21}$$

for the Couette-Poiseuille flow is

$$\overline{U} = \frac{U}{2} + \frac{K h^2}{12\eta}. \tag{6.22}$$

The maximum velocity for pure pressure driven flow is calculated from (6.19) as

$$U_{\max} = \frac{K h^2}{8\eta} = \frac{3}{2}\overline{U}.$$

Since these flows extend to infinity in the x-direction and are two-dimensional, they are never actually realized in applications, but they can often be used as good approximations. Thus we encounter simple shearing flow in the flow between two "infinitely" long cylinders as we take the limit $h/R \to 0$. Although the flow in Fig. 6.2 may be determined without taking the limit $h/R \to 0$ since it is also a unidirectional flow, the shearing flow is considerably easier to calculate. Incidentally this flow is approximately realized in journal bearings where the condition $h/R \to 0$ is well satisfied. The friction torque and the friction power per unit bearing depth can then be immediately estimated

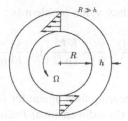

Fig. 6.2 Concentrically rotating journal

Fig. 6.3 Eccentrically rotating journal

$$T_{friction} \approx 2\pi R^2 \eta \frac{du}{dy} = 2\pi R^2 \eta \frac{U}{h} = 2\pi R^3 \eta \frac{\Omega}{h}, \qquad (6.23)$$

$$P_{friction} \approx 2\pi R^3 \eta \Omega^2 / h. \qquad (6.24)$$

However Fig. 6.2 is not the correct depiction of a bearing. Since the journal here rotates concentrically, for symmetry reasons it can support no load. Equation (6.8) states that the pressure in the x-direction (circumferential direction) is constant, and so no net force can act on the journal. Under a load the journal takes on an eccentric position in the bush (Fig. 6.3). The flow in the "lubricant film" is locally a Couette-Poiseuille flow, as we shall show in Chap. 8. The pressure distribution in this case gives rise to a net force which is in balance with the load on the bearing.

6.1.3   Flow Down an Inclined Plane

Closely related to Couette-Poiseuille flow is flow down an inclined plane, although in this case we deal with a free surface (Fig. 6.4). Here the volume body force plays the same role as the pressure gradient $\partial p / \partial x$ in Couette-Poiseuille flow, which as we shall see is here zero. The flow is not driven by the pressure gradient but by the volume body force of gravity, whose components are

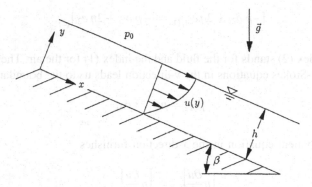

Fig. 6.4 Flow down an inclined plane

$$f_x = \varrho\, k_x = \varrho\, g \sin\beta, \tag{6.25a}$$

$$f_y = \varrho\, k_y = -\varrho\, g \cos\beta. \tag{6.25b}$$

Because of (6.6) and $v = 0$ the Navier-Stokes equations (4.9b) are simplified to

$$\frac{\partial p}{\partial x} - \varrho\, g \sin\beta = \eta\frac{\partial^2 u}{\partial y^2} \tag{6.26}$$

and

$$\frac{\partial p}{\partial y} = -\varrho\, g \cos\beta. \tag{6.27}$$

Therefore we obtain two differential equations for the unknown functions u and p. The no slip condition

$$u(0) = 0 \tag{6.28}$$

is to be satisfied at the wall ($y = 0$), while the condition (4.172) is to be satisfied at the *free surface*, which we write in index notation as

$$n_j\tau_{ji(1)} = n_j\tau_{ji(2)}. \tag{6.29}$$

From (3.1) with $n_j = (0, 1, 0)$ the boundary condition follows in the form

$$[-p\, \delta_{2i} + 2\eta\, e_{2i}]_{(1)} = [-p\, \delta_{2i} + 2\eta\, e_{2i}]_{(2)},\tag{6.30}$$

where the index (2) stands for the fluid and the index (1) for the air. The component of the Navier-Stokes equations in the y-direction leads us to the boundary condition

$$p_{(1)} = p_{(2)} = p_0,\tag{6.31}$$

and the component equation in the x-direction furnishes

$$\left[\eta\frac{\partial u}{\partial y}\right]_{(1)} = \left[\eta\frac{\partial u}{\partial y}\right]_{(2)}.\tag{6.32}$$

If we ignore the effect of the friction in the air, the left-hand side of (6.32) vanishes and this boundary condition reads

$$0 = \eta\frac{\partial u}{\partial y}\bigg|_{y=h}.\tag{6.33}$$

From integrating (6.27) we obtain

$$p = -\varrho\, g\, y\, \cos\beta + C(x),\tag{6.34}$$

and with the boundary condition (6.31) $p_{(2)} = p(y = h) = p_0$ also

$$p = p_0 + \varrho\, g\, \cos\beta(h - y).\tag{6.35}$$

Therefore p is not a function of x, and Eq. (6.26) simplifies to

$$-\varrho\, g\, \sin\beta = \eta\frac{\partial^2 u}{\partial y^2}.\tag{6.36}$$

This is the same differential equation as (6.13), if we replace $\partial p/\partial x$ by $-\varrho\, g\, \sin\beta$. Therefore we read the general solution off from (6.16) (with $K = \varrho\, g\, \sin\beta$)

$$u = -\frac{\varrho\, g\, \sin\beta}{2\eta}y^2 + C_1 y + C_2\tag{6.37}$$

and determine the constants from the boundary conditions (6.28) and (6.33) as

$$C_2 = 0, \quad C_1 = \frac{\varrho g \sin \beta}{\eta} h.$$

(6.38)

The solution of the boundary value problem is therefore

$$u(y) = \frac{\varrho g \sin \beta}{2\eta} h^2 \left[2 - \frac{y}{h} \right] \frac{y}{h}.$$

(6.39)

In the literature the *Nusselt film flow* became the common name for gravity-driven films down a flat incline.

6.1.4 Flow Between Rotating Concentric Cylinders

A cylindrical coordinate system r, φ, z with the velocity components u_r, u_φ, u_z is most suitable for this flow because the boundaries of the flow field are then given by the coordinate surfaces $r = R_i$ and $r = R_O$. In the axial direction the flow extends to infinity. Changes in flow quantities in the axial direction must therefore vanish or be periodic so that these quantities do not take on infinite values at infinity. We shall exclude the case of periodicity here, and shall set $\partial/\partial z = 0$ and $u_z = 0$. At all planes $z = $ const the flow is identical. Since the normal component of the velocity (i.e. u_r at $r = R_i$ and $r = R_O$) must vanish because of the kinematic boundary condition, we set $u_r = 0$ everywhere. Also the change in the circumferential direction must either vanish or be periodic: again we shall restrict ourselves to the first case. Because of $\partial/\partial z = \partial/\partial \varphi = 0$ and $u_r = u_z = 0$ we obtain from the Navier-Stokes equations in cylindrical coordinates (see Appendix B) the following for the r-component

$$\varrho \frac{u_\varphi^2}{r} = \frac{\partial p}{\partial r},$$

(6.40)

and for the φ-component

$$0 = \eta \left[\frac{\partial^2 u_\varphi}{\partial r^2} + \frac{1}{r} \frac{\partial u_\varphi}{\partial r} - \frac{u_\varphi}{r^2} \right],$$

(6.41)

while the z-component vanishes identically. The term u_φ^2/r in (6.40) arises from the material change of the component u_φ and corresponds to the centripetal acceleration. Clearly the pressure distribution $p(r)$ develops so that the centripetal force is balanced. Equation (6.40) is coupled with (6.41): if the velocity distribution is given by (6.41), then the pressure distribution corresponding to it follows from

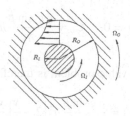

Fig. 6.5 Flow between rotating concentric cylinders

(6.40). Equation (6.41) is a linear ordinary differential equation with variable coefficients of the Eulerian type. It is solved by the substitution

$$u_\varphi = r^n.$$

From (6.41) we then have $n = \pm 1$, so that the general solution reads

$$u_\varphi = C_1 r + \frac{C_2}{r}. \tag{6.42}$$

The inner cylinder rotates with angular velocity Ω_I, the outer with Ω_O (Fig. 6.5). Then, from the no slip condition

$$u_\varphi(R_I) = \Omega_I R_I, \quad u_\varphi(R_O) = \Omega_O R_O \tag{6.43}$$

the constants are determined as

$$C_1 = \frac{\Omega_O R_O^2 - \Omega_I R_I^2}{R_O^2 - R_I^2}, \quad C_2 = \frac{(\Omega_I - \Omega_O) R_I^2 R_O^2}{R_O^2 - R_I^2}. \tag{6.44}$$

For the special case $C_1 = 0$, i.e.

$$\Omega_O/\Omega_I = (R_I/R_O)^2, \tag{6.45}$$

the velocity distribution from (6.42) is that of a potential vortex. Thus the angular velocities of the inner and outer cylinders must have a particular relation to one another in order that the flow in the gap be irrotational.

Another important special case for applications, namely the problem of the rotating cylinder with infinite gap height, arises if we allow R_O to go to infinity in (6.45); Ω_O then tends to zero. In these cases the potential vortex satisfies not only the Navier-Stokes equations (this is so for all incompressible potential flows), but also the no slip condition at the wall. Therefore we are dealing with an exact solution of the flow problem: boundary layers where the velocity distribution differs from the value given by potential theory do not arise. For $\Omega_I = 0$, $r = R_I + y$ and $y/R_I \to 0$ we obtain, from (6.42) and (6.44) the Couette flow (6.4).

6.1.5 Hagen-Poiseuille Flow

The flow through a straight circular pipe or *Hagen-Poiseuille flow* is the most important of all unidirectional flows and it is the rotationally symmetric counterpart to channel flow. Again cylindrical coordinates are suited to this problem where they describe the wall of the circular pipe by the coordinate surface $r = R$ (Fig. 6.6). At the wall $u_r = u_\varphi = 0$, and we set u_r and u_φ identically to zero in the whole flow field; moreover the flow is rotationally symmetric $(\partial/\partial_\varphi = 0)$. The continuity equation in cylindrical coordinates (see Appendix B) then gives

$$\frac{\partial u_z}{\partial z} = 0 \quad \text{or} \quad u_z = u_z(r). \tag{6.46}$$

The r-component of the Navier-Stokes equations leads us to

$$0 = \frac{\partial p}{\partial r} \quad \text{or} \quad p = p(z). \tag{6.47}$$

All terms of the Navier-Stokes equation in the φ-direction vanish identically, while the z-component equation becomes

$$0 = -\frac{\partial p}{\partial z} + \eta \left[\frac{\partial^2 u_z}{\partial r^2} + \frac{1}{r} \frac{\partial u_z}{\partial r} \right]. \tag{6.48}$$

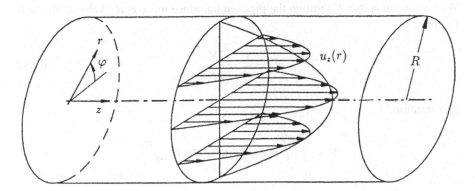

Fig. 6.6 Flow in a straight circular pipe

We see directly from (6.48) that $\partial p/\partial z$ does not depend on z and therefore the pressure p is a linear function of z. As before we set $\partial p/\partial z = -K$ and write (6.48) in the form

$$-\frac{K}{\eta} = \frac{1}{r}\frac{d}{dr}\left[r\frac{du_z}{dr}\right],\qquad(6.49)$$

which, integrated twice, gives

$$u_z(r) = -\frac{Kr^2}{4\eta} + C_1\ln r + C_2.\qquad(6.50)$$

Since $u_z(0)$ is finite, $C_1 = 0$ immediately follows. The no slip condition implies

$$u_z(R) = 0,\qquad(6.51)$$

thus

$$C_2 = \frac{KR^2}{4\eta}.\qquad(6.52)$$

Dropping the index z, the solution reads

$$u(r) = \frac{K}{4\eta}\left(R^2 - r^2\right).\qquad(6.53)$$

The maximum velocity is reached at $r = 0$, and therefore we write

$$u(r) = U_{max}\left\{1 - (r/R)^2\right\}.\qquad(6.54)$$

With the volume flux $\dot{V}$ through the pipe we introduce the average velocity through the pipe

$$\overline{U} = \frac{\dot{V}}{A} = \frac{\dot{V}}{\pi R^2},\qquad(6.55)$$

and because

$$\dot{V} = \int_0^{2\pi}\int_0^R u(r)r\,dr\,d\varphi = 2\pi\,U_{max}\frac{R^2}{4}\qquad(6.56)$$

we also find that

$$\overline{U} = \frac{1}{2} U_{\max},$$ (6.57)

i.e.,

$$\overline{U} = \frac{K R^2}{8\eta}.$$ (6.58)

Since the pressure gradient is constant, we may write

$$K = \frac{\Delta p}{l} = \frac{p_1 - p_2}{l}$$ (6.59)

and mean by Δp the *pressure drop* in the pipe over the length l. The pressure drop is positive if the pressure gradient $\partial p / \partial z$ is negative. It is appropriate to represent this pressure drop in a dimensionless form

$$\zeta = \frac{\Delta p}{\frac{\varrho}{2} \overline{U}^2}.$$ (6.60)

Using (6.58), the so-called *loss factor* ζ, can also be written in the form

$$\zeta = \frac{16 l \eta}{R^2 \varrho \overline{U}} = 64 \frac{l}{d} \frac{\eta}{\varrho d \overline{U}},$$ (6.61)

where $d = 2R$ and we have set the dimensional quantities into two dimensionless groups l/d and $\varrho d \overline{U}/\eta = Re$. In particular, in pipe flows the *friction factor* λ is often introduced

$$\lambda = \zeta \frac{d}{l},$$

so that the dimensionless form of the *resistance law* of a straight circular pipe arises

$$\zeta = \frac{l}{d} \frac{64}{Re} \quad \text{or} \quad \lambda = \frac{64}{Re}.$$ (6.62)

The *Hagen-Poiseuille equation* follows from (6.55), (6.58) and (6.59)

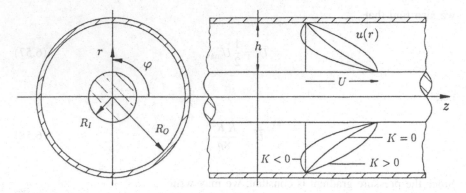

Fig. 6.7 Generalized Hagen-Poiseuille flow

$$\dot{V} = \frac{\pi R^4}{8\eta}\frac{\Delta p}{l}. \tag{6.63}$$

The proportionality of the volume flux to the fourth power of the radius has been experimentally confirmed to a high degree of accuracy which serves as a confirmation of the no slip condition (4.160). The Hagen-Poiseuille equation (6.63) is also the basis for measuring the shear viscosity η.

We are led to a generalized Hagen-Poiseuille flow if we subject the general solution (6.50) to the boundary conditions (Fig. 6.7)

$$u(R_O) = 0, \tag{6.64a}$$

and

$$u(R_I) = U. \tag{6.64b}$$

The resulting flow is clearly the Couette-Poiseuille flow in a ring gap, and is given by

$$u(r) = \frac{K}{4\eta}\left\{R_O^2 - r^2 - \left[R_O^2 - R_I^2 - \frac{4\eta\,U}{K}\right]\frac{\ln(r/R_O)}{\ln(R_I/R_O)}\right\}. \tag{6.65}$$

This can be superimposed with the velocity field (6.42) and then describes the case in which the cylinder is also rotating.

We could convince ourselves that with $R_O - R_I = h$ and $R_O - r = y$ and in the limit $h/R_O \to 0$, two-dimensional Couette-Poiseuille flow (6.19) results. For pure pressure driven flow ($U = 0$), by (6.55) we find the average velocity

$$\overline{U} = \frac{K}{8\eta}\left[R_O^2 + R_I^2 + (R_O^2 - R_I^2)\frac{1}{\ln(R_I/R_O)}\right], \tag{6.66}$$

which, for $R_I \to 0$ agrees with the known result (6.58).

For conduits which do not have a circular cross-section, we introduce the *equivalent* or *hydraulic diameter* d_h,

$$d_h = \frac{4A}{s}, \tag{6.67}$$

where A is the cross-sectional area and s is the wetted circumference of the cross-section. $d_h = d$ for the circular cross-section, and for the ring cross-section we have

$$d_h = \frac{4\pi(R_O^2 - R_I^2)}{2\pi(R_O + R_I)} = d_o - d_I. \tag{6.68}$$

We first write the loss factor ζ in the form

$$\zeta = \frac{\Delta p(d_O - d_I)^2}{\frac{\varrho}{2}\overline{U}^2 d_h^2}, \tag{6.69}$$

into which we replace one $\overline{U}$ by (6.66) (from (6.59)) and extract

$$\zeta = \frac{64}{\varrho\,\overline{U}d_h}\frac{\eta}{d_h}\frac{l}{d_h}\frac{\left[1 - \dfrac{d_I}{d_o}\right]^2 \ln\left[\dfrac{d_I}{d_o}\right]}{1 - \left[\dfrac{d_I}{d_o}\right]^2 + \ln\left[\dfrac{d_I}{d_o}\right]\left\{1 + \left[\dfrac{d_I}{d_o}\right]^2\right\}}. \tag{6.70}$$

Using the Reynolds' number $Re = \varrho\,\overline{U}d_h/\eta$ this becomes

$$\zeta = \frac{64}{Re}\frac{l}{d_h}f(d_I/d_o). \tag{6.71}$$

The dimensionless factor $f(d_I/d_o)$ is a measure of the deviation of the loss factor of a noncircular conduit from the friction factor of the circular pipe, if the hydraulic diameter is used as the reference length. For $d_I/d_o = 0$ we have $f(d_I/d_o) = 1$, and for $d_I/d_o = 1$, corresponding to channel flow, we extract $f(d_I/d_o) = 1.5$ after repeated application of l'Hôpital's rule. This result can be easily confirmed if, starting with (6.22) we construct the formula (6.71).

As can be seen, the pressure drop for the circular tube is very different from the pressure drop for the ring gap, even when the hydraulic diameter is used as the reference length. This is not the case for turbulent flows: the loss factor of the ring gap is practically identical to that of the circular pipe. This also holds for conduits with rectangular cross-sections and for most other technically interesting cross-sectional shapes, such as triangular crosssections, if the angles are not too small.

6.1.6   Flow Through Noncircular Conduits

In the treatment of laminar flows in infinitely long straight conduits with noncircular cross-sections, the same kinematic simplifications as in Hagen-Poiseuille flows arise. The only nonvanishing velocity component is the one in the axial direction. This component is independent of the coordinate in this direction, so that the nonlinear terms drop out in the equations of motion. Since a locally valid coordinate system where the stress tensor has the form (3.35) can be given for every point in the cross-section, we find ourselves dealing with a unidirectional flow. In a coordinate system whose z-axis runs parallel to the axis of the conduit, Poisson's equation

$$\Delta u = -\frac{K}{\eta},$$
(6.72)

follows from (6.3) for the only nonvanishing velocity component (which we shall denote by u) in steady flow. Since $K = -\partial p/\partial z = \text{const}$, the inhomogeneous term here is again a constant. This form of Poisson's equation appears in many technical problems, among these in the torsion of straight rods and in loaded membranes. Thus we can directly transfer results known from the theory of elasticity. Solutions of this equation in the form of polynomials describe, among others the torsion of rods with triangular cross-sections, and these correspond therefore to flows through pipes with triangular cross-sections. Using elementary integration methods, cross-sections whose boundaries are coordinate surfaces can be dealt with if Poisson's equation is separable in these coordinate systems.

As a typical example, we shall sketch the path of a solution for the technically important case of a conduit with a rectangular cross-section (Fig. 6.8). With $u_z(x, y) = u(x, y)$ we get from (6.72) the differential equation

$$\frac{\partial^2 u}{\partial x^2} + \frac{\partial^2 u}{\partial y^2} = -\frac{K}{\eta},$$
(6.73)

with the boundary conditions

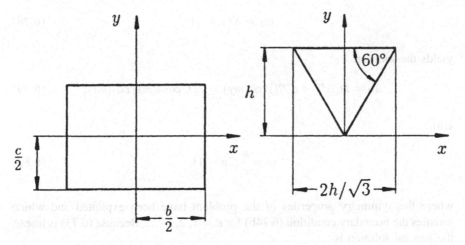

Fig. 6.8 Channels with rectangular and triangular cross-section

$$u\left(\pm\frac{b}{2}, y\right) = 0, \tag{6.74a}$$

and

$$u\left(x, \pm\frac{c}{2}\right) = 0. \tag{6.74b}$$

To solve the linear Eq. (6.73) we set

$$u = u_P + u_H, \tag{6.75}$$

where u_H satisfies the homogeneous equation and u_P is a particular solution. If we set, for example $u = u_P(y)$, the solution follows directly from (6.73)

$$u_P = -\frac{K}{2\eta}y^2 + C_1 y + C_2, \tag{6.76}$$

into which we introduce the boundary condition (6.74b), so that

$$u_P = \frac{K}{2\eta}\left[\frac{1}{4}c^2 - y^2\right] \tag{6.77}$$

arises. Using a *separation of variables* solution of the form

$$u_H = X(x)Y(y) \tag{6.78}$$

yields the solution

$$u_H = D_n(e^{mx} + e^{-mx})\cos(my) = 2D_n \cosh(mx)\cos(my), \tag{6.79}$$

with

$$m = \frac{\pi}{c}(2n - 1), \tag{6.80}$$

where the symmetry properties of the problem have been exploited and which satisfies the boundary condition (6.74b) for $n = 1, 2, 3, \ldots$. Because (6.73) is linear, the general solution is

$$u = \sum_{n=1}^{\infty} 2D_n \cosh(mx) \ \cos(my) + u_P(y). \tag{6.81}$$

The boundary conditions (6.74a) lead to the equation

$$\sum_{n=1}^{\infty} 2D_n \cosh(m\,b/2)\cos(my) + u_P(y) = 0. \tag{6.82}$$

In order to determine the coefficients D_n, u_P must also be represented as a Fourier series, whose coefficients are given by

$$a_n = \frac{2}{c} \int_{-c/2}^{c/2} \frac{K}{2\eta}\left[\frac{1}{4}c^2 - y^2\right]\cos(my)\,\mathrm{d}y. \tag{6.83}$$

Integrating leads to the Fourier expansion

$$u_P = -\frac{2K}{\eta c}\sum_{n=1}^{\infty}\left[\frac{c}{m^2}\cos(m\,c/2) - \frac{2}{m^3}\sin(m\,c/2)\right]\cos(m\,y). \tag{6.84}$$

Because

$$\frac{m\,c}{2} = (2n - 1)\frac{\pi}{2},$$ (6.85)

the first term in brackets in (6.84) vanishes, and the second reads

$$-2m^{-3}\sin(m\,c/2) = 2m^{-3}(-1)^n.$$ (6.86)

A comparison between (6.84) and (6.82) furnishes

$$D_n = \frac{2K}{\eta\ c\,m^3}\frac{(-1)^n}{\cosh(m\,b/2)},$$ (6.87)

and therefore the solution is

$$u = \frac{K}{2\eta}\left\{\frac{c^2}{4} - y^2 + \frac{8}{c}\sum_{n=1}^{\infty}\frac{(-1)^n}{m^3}\frac{\cosh(m\,x)}{\cosh(m\,b/2)}\cos(m\,y)\right\},$$ (6.88)

from which we find the average velocity, according to (6.55), as

$$\bar{U} = \frac{K\,c^2}{4\eta}\left\{\frac{1}{3} - \frac{c}{b}\frac{64}{\pi^5}\sum_{n=1}^{\infty}\frac{\tanh(m\,b/2)}{(2n-1)^5}\right\}.$$ (6.89)

The loss factor based on the hydraulic diameter

$$d_h = \frac{2\,b\,c}{b+c}$$ (6.90)

is

$$\zeta = \frac{64}{Re}\frac{l}{d_h}f(c/b),$$ (6.91)

with

$$f(c/b) = \left\{2\left[\frac{c}{b} + 1\right]^2\left[\frac{1}{3} - \frac{c}{b}\frac{64}{\pi^5}\sum_{n=1}^{\infty}\frac{\tanh(m\,b/2)}{(2n-1)^5}\right]\right\}^{-1}.$$ (6.92)

Two-dimensional channel flow corresponds to $c/b = 0$, and we have $f(c/b) = 3/2$. For $c/b = 1$ we obtain $f(c/b) = 0.89$.

For an equilateral triangle of height h (Fig. 6.8), the velocity distribution is

$$u = \frac{K}{\eta} \frac{1}{4h} (y - h)(3x^2 - y^2) \tag{6.93}$$

and the average velocity

$$\overline{U} = \frac{1}{60} \frac{K h^2}{\eta}. \tag{6.94}$$

Using the hydraulic diameter

$$d_h = \frac{2}{3} h \tag{6.95}$$

we obtain the loss factor

$$\zeta = \frac{64}{Re} \frac{l}{d_h} \frac{5}{6}. \tag{6.96}$$

The velocity distribution in an elliptic pipe whose cross-section is given by the equation of the ellipse

$$\left[\frac{x}{a}\right]^2 + \left[\frac{y}{b}\right]^2 = 1 \tag{6.97}$$

reads

$$u = \frac{K}{2\eta} \frac{a^2 b^2}{a^2 + b^2} \left[1 - \frac{x^2}{a^2} - \frac{y^2}{b^2}\right]. \tag{6.98}$$

From this equation we can see directly that the no slip condition is satisfied at the wall. The average velocity here is

$$\overline{U} = \frac{K}{4\eta} \frac{a^2 b^2}{a^2 + b^2}. \tag{6.99}$$

Since the perimeter of the ellipse cannot be represented in a closed form (elliptic integral) we do not introduce the hydraulic diameter. Instead it is recommended that the pressure drop be calculated directly from (6.99).

6.2   Unsteady Unidirectional Flows

6.2.1   Flow Due to a Wall Which Oscillates in Its Own Plane

The solutions so far can be extended to the unsteady case. First of all we shall consider harmonic time functions. From these we can build general time functions using a Fourier representation. The simple shear flow then corresponds to the flow between two plane infinitely extending plates (with separation distance h), one of which (the lower) is set into oscillation in its plane. The wall velocity is given by

$$u_w = U(t) = \hat{U} \cos(\omega t). \tag{6.100}$$

Using complex notation the wall velocity reads

$$u_w = U(t) = \hat{U}\, e^{i\omega t}, \tag{6.101}$$

where only the real part $\Re(e^{i\omega t})$ has physical meaning. Instead of (6.12) we now have

$$u = f(y,t), \quad v = 0 \tag{6.102}$$

and instead of (6.13)

$$\frac{\partial u}{\partial t} = -\frac{1}{\varrho}\frac{\partial p}{\partial x} + \nu \frac{\partial^2 u}{\partial y^2}. \tag{6.103}$$

We set $\partial p/\partial x = 0$, i.e., the flow is only kept in motion by the wall velocity through the no slip condition

$$u(0,t) = u_w = \hat{U}\, e^{i\omega t}. \tag{6.104a}$$

On the upper wall the no slip condition reads

$$u(h,t) = 0. \tag{6.104b}$$

We shall only be interested in the steady state oscillation after the initial transients have died away, so that the initial condition $u(y, 0)$ is superfluous. The boundary condition (6.104a) suggests that the solution is of the form

$$u(y,t) = \hat{U}\, e^{i\omega t} g(y), \tag{6.105}$$

where $g(y)$ satisfies the boundary conditions

$$g(0) = 1, \quad \text{and} \tag{6.106a}$$

$$g(h) = 0. \tag{6.106b}$$

Using the form (6.105), the partial differential Eq. (6.103) reduces to an ordinary differential equation with constant (complex) coefficients

$$g'' - \frac{i\omega}{\nu} g = 0, \tag{6.107}$$

where $g'' = d^2g/dy^2$. From the solution $g = e^{\lambda y}$ we obtain the characteristic polynomial

$$\lambda^2 - \frac{i\omega}{\nu} = 0, \tag{6.108}$$

with the roots

$$\lambda = \sqrt{i}\sqrt{\omega/\nu} = \pm(1+i)\sqrt{\frac{\omega}{2\nu}}. \tag{6.109}$$

The general solution can then be written in the form

$$g(y) = A \sinh\left\{(1+i)\sqrt{\omega/2\,\nu}\,y\right\} + B \cosh\left\{(1+i)\sqrt{\omega/2\,\nu}\,y\right\}, \tag{6.110}$$

from which, using the boundary condition (6.106a, 6.106b), we find the special solution

$$g(y) = \frac{\sinh\left\{(1+i)\sqrt{\omega/2\,\nu}\,(h-y)\right\}}{\sinh\left\{(1+i)\sqrt{\omega/2\,\nu h}\right\}}, \tag{6.111}$$

and finally by (6.105) the velocity distribution

$$u(y,t) = \hat{U}\Re\left\{e^{i\omega t} \frac{\sinh\left\{(1+i)\sqrt{\omega h^2/2\nu}\,(1-y/h)\right\}}{\sinh\left\{(1+i)\sqrt{\omega h^2/2\nu}\right\}}\right\}. \tag{6.112}$$

We discuss the two limiting cases

$$\omega h^2/\nu \ll 1,\tag{6.113}$$

$$\omega h^2/\nu \gg 1,\tag{6.114}$$

and note that h^2/ν is the typical time for the diffusion of the rotation across the channel height h. In the first case this time is much smaller than the typical oscillation time $1/\omega$, i.e., the diffusion process adjusts at every instant the velocity field to the steady shearing flow with the instantaneous wall velocity u_w (t). This is what is called *quasi-steady* flow.

Using the first term of the expansion of the hyperbolic sine function for small arguments we have

$$u = \hat{U}\Re\left\{e^{i\omega t}\frac{\sqrt{\omega h^2/2\nu}(1+i)(1-y/h)}{\sqrt{\omega h^2/2\nu}(1+i)}\right\},\tag{6.115}$$

and deduce that

$$u = \hat{U}\cos(\omega t)(1-y/h) = U(1-y/h).\tag{6.116}$$

Equation (6.116) corresponds to (6.4) where the upper plate represents the moving wall. We also obtain this limiting case if the kinematic viscosity ν tends to infinity. As is clear from (6.103), the unsteady term then vanishes. This limiting case $\nu \to \infty$ for fixed η also corresponds to taking the limit $\varrho \to 0$, thus ignoring the inertia terms, and therefore falls into group b) of the classification discussed in Sect. 4.4.

In the limit $\omega h^2/\nu \gg 1$ we use the asymptotic form of the hyperbolic sine function and write (6.112) in the form

$$u = \hat{U}\Re\left[e^{-\sqrt{\omega/2\nu}\,y}e^{i\left(\omega t-\sqrt{\omega/2\nu}\,y\right)}\right],\tag{6.117}$$

or

$$u = \hat{U}e^{-\sqrt{\omega/2\nu}\,y}\cos\left(\omega t - \sqrt{\omega/2\nu}\,y\right).\tag{6.118}$$

The separation h no longer appears in (6.118). Measured in units $\lambda = \sqrt{2\nu/\omega}$ the upper wall is at infinity. Relative to the variable y the solutions also have a wave form; we call these *shearing* or *transversal waves* of wavelength λ (Fig. 6.9). The wavelength λ also describes a penetration depth of the waves. The penetration depth increases with decreasing ω (low-frequencies) and higher viscosities. In the literature this phenomenon is called the *Skin-effect*.

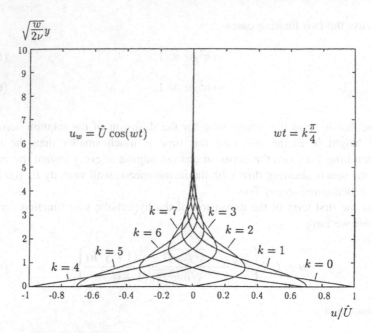

Fig. 6.9 Velocity distribution above the oscillating wall

6.2.2 Flow Due to a Wall Which Is Suddenly Set in Motion

Using (6.118), we could in principle form the solution for the wall which is suddenly accelerated to velocity U. However it is more instructive to take a different path which starts directly with the partial differential equation

$$\frac{\partial u}{\partial t} = \nu \frac{\partial^2 u}{\partial y^2}. \tag{6.119}$$

This differential equation also describes the unsteady one-dimensional heat conduction (where ν is then replaced by the coefficient of heat conduction a), and so the desired solution also appears in heat conduction problems. The no slip condition at the wall furnishes

$$u(0,t) = U \quad \text{for} \quad t > 0. \tag{6.120}$$

The second boundary condition is replaced by the condition

$$u(y,t) = 0 \quad \text{for} \quad y \to \infty. \tag{6.121}$$

In addition we have the initial condition

$$u(y, t) = 0 \text{ for } t \leq 0. \tag{6.122}$$

Equation (6.119) is a linear equation and since U enters the problem only linearly from the boundary condition (6.120), the field $u(y, t)$ must be proportional to U, so that the solution has to be of the form

$$u/U = f(y, t, \nu). \tag{6.123}$$

Since the function on the left-hand side is dimensionless, f must also be dimensionless, which is only possible if the argument of the function is dimensionless. However the only linearly independent dimensionless quantity is the combination $y^2/(\nu t)$. We set

$$\eta = \frac{1}{2} \frac{y}{\sqrt{\nu t}} \tag{6.124}$$

and are now dealing with a *similarity variable* η, because the solution cannot change if y and t are changed such that η remains constant. We note that $\sqrt{\nu t}$ represents a penetration length (a boundary layer thickness) which grows with time. We will discuss boundary layers later in Chap. 12. Instead of (6.123) we now write

$$u/U = f(\eta), \tag{6.125}$$

and from (6.119) we obtain the ordinary differential equation

$$-2\eta f' = f'' \tag{6.126}$$

with $f' = \mathrm{d}f/\mathrm{d}\eta$. Integrating twice gives the general solution

$$f = C_1 \int_0^\eta e^{-\eta^2} \mathrm{d}\eta + C_2. \tag{6.127}$$

For $y = 0$ we have $\eta = 0$, and the boundary condition (6.120) becomes

$$f(0) = 1, \tag{6.128}$$

and therefore it follows that $C_2 = 1$. If we subject (6.127) with $C_2 = 1$ to the "boundary condition" (6.121),

$$1/C_1 = -\int_0^\infty e^{-\eta^2} d\eta \qquad (6.129)$$

must hold. The improper integral has the value $\frac{1}{2}\sqrt{\pi}$, and therefore

$$C_1 = -2/\sqrt{\pi}; \qquad (6.130)$$

thus the solution reads

$$u/U = 1 - 2/\sqrt{\pi} \int_0^\eta e^{-\eta^2} d\eta \quad \text{for } t \geq 0. \qquad (6.131)$$

The integral

$$\text{erf}(\eta) = 2/\sqrt{\pi} \int_0^\eta e^{-\eta^2} d\eta \qquad (6.132)$$

is the *error function*. For $t = 0$ we have $\eta \to \infty$ and $u/U = 0$; thus the initial condition is satisfied.

6.3 Unidirectional Flows of Non-Newtonian Fluids

6.3.1 Steady Flow Through a Circular Pipe

In order to calculate the flow of non-Newtonian fluids we shall return to Cauchy's equations. As with the flow of Newtonian fluids, for kinematic reasons the only nonvanishing velocity component is that in the axial direction and this only depends on r. Therefore we are dealing with a unidirectional flow, and the stress tensor has the form (3.35) in cylindrical coordinates, where the index 1 corresponds to the z-direction, the index 2 to the r-direction and the index 3 to the φ-direction. Since the tensor valued function φ_{ij} in (3.35) corresponds to the friction stress tensor P_{ij} (which only depends on $\dot{\gamma} = du/dr$, that is, on r), we write the stress tensor in the following matrix form

$$[\mathbf{T}] = \begin{bmatrix} P_{zz} - p & P_{rz} & 0 \\ P_{zr} & P_{rr} - p & 0 \\ 0 & 0 & P_{\varphi\varphi} - p \end{bmatrix}. \qquad (6.133)$$

The material derivative Du/Dt vanishes and if by p we mean only the pressure relative to the hydrostatic pressure distribution, we extract from (2.38b)

$$0 = \nabla \cdot \mathbf{T}. \tag{6.134}$$

In component representation (see Appendix B) and noting $P_{ij}(r)$ we find for the r-component

$$\frac{\partial p}{\partial r} = \frac{1}{r}\left[\frac{\partial}{\partial r}(r\,P_{rr}) - P_{\varphi\varphi}\right], \tag{6.135}$$

for the φ-component

$$\frac{\partial p}{\partial \varphi} = 0, \tag{6.136}$$

and for the z-component

$$\frac{\partial p}{\partial z} = \frac{1}{r}\frac{\partial}{\partial r}(r\,P_{rz}). \tag{6.137}$$

The right-hand sides of (6.135) and (6.137) are functions of r only. From (6.136) and (6.137) we conclude $p = z\,g(r) + h(r)$ and from (6.135) then $g'(r) = 0$. This means that because of the arbitrary function $h(r)$, p is not necessarily independent of r, although $\partial p/\partial z = -K = \Delta p/l$ is a constant. From integration of the Eq. (6.137) we obtain the distribution

$$\tau_{rz} = P_{rz} = -\frac{Kr}{2} + \frac{C}{r}, \tag{6.138}$$

where we set $C = 0$, since the friction stresses in the center of the pipe cannot become infinite. Using

$$\tau_{rz}(R) = -\tau_w = -\frac{KR}{2} \tag{6.139}$$

instead of K we introduce the shear stress at the pipe wall, and write (6.138) in the form

$$\tau_{rz} = -\tau_w \frac{r}{R}, \tag{6.140}$$

from which we find the statement valid for all constitutive relations that the shear stress τ_{rz} is a linear function of r. Now we could have obtained this statement more easily from the balance of momentum in integral form, but it has arisen here from the exemplary application of Cauchy's equation.

We shall now specifically use the power law (3.17), and assume that $\dot\gamma = \mathrm{d}u/\mathrm{d}r$ is everywhere less than zero. This is not exactly true since, for symmetry reasons, $\dot\gamma$ is equal to zero in the center of the tube. Using (3.13) we extract from (6.140) the equation

$$\tau_{rz} = m\left[-\frac{\mathrm{d}u}{\mathrm{d}r}\right]^{n-1}\frac{\mathrm{d}u}{\mathrm{d}r} = -\tau_w\frac{r}{R}. \tag{6.141}$$

We find the velocity distribution to be

$$u = R\int_{r/R}^{1}(\tau_w/m)^{1/n}(r/R)^{1/n}\mathrm{d}(r/R), \tag{6.142}$$

or, after integrating

$$u = \left[\frac{\tau_w}{m}\right]^{\frac{1}{n}}\frac{n}{n+1}R\left[1-\left[\frac{r}{R}\right]^{\frac{n+1}{n}}\right]. \tag{6.143}$$

The volume flux is

$$\dot V = \frac{n}{3n+1}(\tau_w/m)^{1/n}\pi R^3 \tag{6.144}$$

and therefore the average velocity

$$\overline{U} = \dot V/(\pi R^2) = \frac{n}{3n+1}(\tau_w/m)^{1/n}R. \tag{6.145}$$

Finally, from (6.144) and (6.139) the pressure drop follows

$$\Delta p = p_1 - p_2 = 2m\frac{l}{R}\left[\frac{\dot V}{\pi R^3}\frac{3n+1}{n}\right]^n. \tag{6.146}$$

6.3.2 Steady Flow Between a Rotating Disk and a Fixed Wall

We consider the flow with the velocity field

$$u_\varphi = r\,\Omega(z), \quad u_z = u_r = 0, \tag{6.147}$$

of Fig. 6.10 whose form is suggested by the no slip condition on the rotating plate

$$u_\varphi(h) = r\,\Omega_R. \tag{6.148}$$

We shall first ask under which conditions the field satisfies Cauchy's equations. The flow shown in Fig. 6.10 occurs in some forms of viscometers which is why these flows are named *viscometric flows*. The calculation of the rate of deformation tensor (see Appendix B) leads to the matrix representation

$$[\mathbf{E}] = \begin{bmatrix} e_{\varphi\varphi} & e_{\varphi z} & e_{\varphi r} \\ e_{z\varphi} & e_{zz} & e_{zr} \\ e_{r\varphi} & e_{rz} & e_{rr} \end{bmatrix} = \begin{bmatrix} \frac{1}{2}\mathbf{A}_{(1)} \end{bmatrix} = \frac{1}{2}\begin{bmatrix} 0 & \dot\gamma & 0 \\ \dot\gamma & 0 & 0 \\ 0 & 0 & 0 \end{bmatrix}, \tag{6.149}$$

with $\dot\gamma = 2\,e_{\varphi z} = r\,d\Omega/dz$, so that the first Rivlin-Ericksen tensor indeed has the same form as in a unidirectional flow.

Therefore the stress tensor has the form (3.35), where here $\vec e_1$ points in the φ-direction, $\vec e_2$ in the z-direction and $\vec e_3$ in the r-direction. Using this stress tensor and the symmetry condition $\partial/\partial\varphi = 0$, the components of Cauchy's equations in cylindrical coordinates are

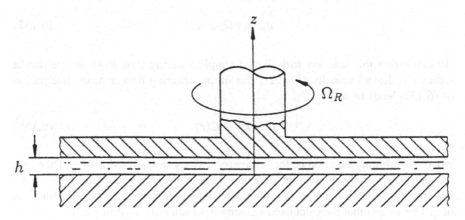

Fig. 6.10 Shearing flow between a rotating disk and a fixed wall

$$r: \quad -\varrho\,\Omega^2(z) = -\frac{1}{r}\frac{\partial p}{\partial r} + \frac{1}{r}\frac{\partial P_{rr}}{\partial r} + \frac{1}{r^2}\left(P_{rr} - P_{\varphi\varphi}\right), \tag{6.150}$$

$$\varphi: \qquad\qquad 0 = \frac{\partial P_{z\varphi}}{\partial z}, \text{ and} \tag{6.151}$$

$$z: \qquad\qquad 0 = -\frac{\partial p}{\partial z} + \frac{\partial P_{zz}}{\partial z}. \tag{6.152}$$

From (3.35) the friction stresses only depend on $\dot{\gamma}$. But from (6.151) we see that $P_{z\varphi} = \tau_{z\varphi}$ is not a function of z and for symmetry reasons not a function of φ either. Therefore the shear stress $\tau_{z\varphi}$ is only a function of r, as is $\dot{\gamma} = r\,d\Omega/dz$

$$r\frac{d\Omega}{dz} = g(r); \tag{6.153}$$

integration of (6.153) gives

$$u_\varphi = r\,\Omega(z) = z\,g(r) + C. \tag{6.154}$$

The no slip condition on the fixed wall implies

$$u_\varphi(0) = r\,\Omega(0) = 0, \tag{6.155}$$

therefore $C = 0$. From (6.148) it follows that

$$u_\varphi(h) = r\,\Omega_R = h\,g(r), \tag{6.156}$$

and therefore $g(r) = \Omega_R\,r/h$, so that the solution is

$$u_\varphi = r\,\Omega_R\,z/h. \tag{6.157}$$

By comparing this solution with that of simple shearing flow (6.4) we see that at radius r with wall velocity $U = r\Omega_R$ the simple shearing flow appears. Integration of (6.152) leads us to

$$p = P_{zz} + C(r), \tag{6.158}$$

where the arbitrary function is, for symmetry reasons, not a function of φ. Therefore the pressure is only a function of r and thus the whole right-hand side of the Eq. (6.150) is only a function of r. On the left-hand side however there is a function of z. This means that the calculated velocity field can only exist in the limit $\varrho \to 0$, that is, by ignoring the inertia terms.

If the inertia of the fluid cannot be ignored, secondary flows form and the form of solution (6.147) is not permissible. As well as the kinematic restriction (class c) in Sect. 4.4), a dynamic restriction also arises (class b) in Sect. 4.4), while no restrictions of any kind were necessary as far as the constitutive relation is concerned. If we introduce (6.158) into (6.150), then $C(r)$ can be expressed through the normal stress differences. Incidentally, by measuring the force on the plate with radius R and the pressure at $r = 0$, the normal stresses of a fluid can be determined by a viscometer which is built according to the principles in Fig. 6.10.

6.3.3   Unsteady Unidirectional Flows of a Second Order Fluid

We extend the velocity field given in (6.147) to the case where the disk carries out a rotational oscillation

$$\varphi_R = \hat{\varphi}_R\, e^{i\omega t}, \tag{6.159}$$

and instead of (6.147) we now write

$$u_\varphi = r\, \hat{\Omega}(z) e^{i\omega t}. \tag{6.160}$$

(As in (6.101) we shall use complex notation and allot physical meaning to the real part only.) The component of Cauchy's equations (6.151) in the φ-direction, with the unsteady flow now considered additionally, contains the inertia term $\varrho\, \partial u_\varphi/\partial t$ on the left-hand side. Since we are ignoring inertia terms, this term also vanishes in the limiting case $\varrho \to 0$. The Eqs. (6.150) to (6.157) are therefore still valid since no restriction has been made relative to the constitutive relation. Since the inertia terms have been ignored, the problem is unsteady only because of the boundary condition. With

$$\Omega_R = \dot{\varphi}_R = i\,\omega\,\hat{\varphi}_R\, e^{i\omega t} \tag{6.161}$$

we extract directly from (6.157) the unsteady (more exactly the quasi-steady) velocity field as

$$u_\varphi = r\, i\,\omega\,\hat{\varphi}_R \frac{z}{h} e^{i\omega t}. \tag{6.162}$$

By comparing this with (6.160) we get

$$\hat{\Omega}(z) = i\,\omega\,\hat{\varphi}_R \frac{z}{h}.$$ (6.163)

We now calculate the torque acting on the oscillating disk with radius R due to the shear stress $\tau_{z\varphi}$

$$M = 2\pi \int_0^R \tau_{z\varphi} r^2 \mathrm{d}r.$$ (6.164)

Since the flow is a simple shearing flow at fixed r (cf. (6.4)), where the z-direction corresponds to the x_2-direction and the φ-direction to the x_1-direction, to calculate $\tau_{z\varphi}$ it is enough to determine τ_{12} in simple shearing flow of a second order fluid; from (3.40) this is

$$\tau_{12} = \eta A_{(1)12} + \beta A_{(1)1j} A_{(1)2j} + \gamma A_{(2)12}.$$ (6.165)

We already have the first Rivlin-Ericksen tensor from Sect. 1.2.4 as

$$A_{(1)12} = 2e_{12} = \frac{\partial u_1(x_2)}{\partial x_2}.$$ (6.166)

In unsteady unidirectional flow $A_{(2)12}$ is not equal to zero and is calculated from (1.69)

$$A_{(2)12} = \frac{\mathrm{D}}{\mathrm{D}t}\left[\frac{\partial u_1(x_2)}{\partial x_2}\right] + A_{(1)j2}\frac{\partial u_j}{\partial x_1} + A_{(1)1j}\frac{\partial u_j}{\partial x_2}.$$ (6.167)

Since $u_2 = u_3 = 0$ and u_1 is only a function of x_2, $A_{(2)12}$ reduces to

$$A_{(2)12} = \frac{\partial^2 u_1}{\partial x_2 \partial t}.$$ (6.168)

With (6.162) and (6.165) we therefore obtain the shear stress

$$\tau_{z\varphi} = \tau_{12} = i\,\omega\,\hat{\varphi}_R \frac{r}{h}(\eta + i\,\omega\,\gamma)e^{i\omega t},$$ (6.169)

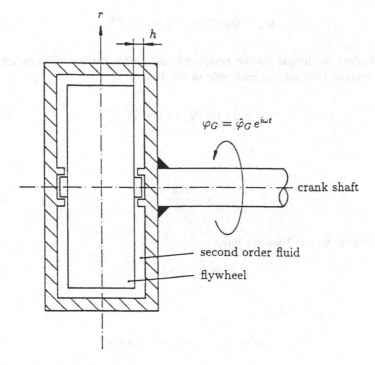

Fig. 6.11 Torsional vibration damper

and finally the torque as

$$M = i\,\omega\,\hat{\varphi}_R(\eta + i\omega\gamma)\mathrm{e}^{i\omega t}2\pi \int\limits_0^R \frac{r^3}{h}\mathrm{d}r = i\,\omega\,\hat{\varphi}_R(\eta + i\omega\gamma)\mathrm{e}^{i\omega t}\frac{\pi R^4}{2h}. \qquad (6.170)$$

This equation can be applied in the *damping of rotational oscillations* of crankshafts. Now the damper consists of a casing attached to the crankshaft with a fulcrumed disk on the inside (shown in Fig. 6.11). When the crankshaft carries out rotational oscillations

$$\varphi_G = \hat{\varphi}_G \mathrm{e}^{i\omega t} \qquad (6.171)$$

the disk inside the casing lags the motion of the casing because of its rotational inertia Θ. The viscoelastic fluid inside the case, which we idealize as a second order fluid, is sheared by the relative motion between the case and the disk. If φ_D describes the rotational oscillation of the disk, the relative motion is

$$\varphi_R = \hat{\varphi}_R\, e^{i\omega t} = (\hat{\varphi}_G - \hat{\varphi}_D)e^{i\omega t}. \tag{6.172}$$

If we neglect the torque on the peripheral surface for reasons of simplicity, the torque from (6.170) acts on each side of the disk

$$M = \frac{1}{2}\chi\, i\,\omega\,\hat{\varphi}_R(\eta + i\,\omega\,\gamma)e^{i\omega t}, \tag{6.173}$$

where

$$\chi = \frac{\pi R^4}{h} \tag{6.174}$$

is a geometric factor. Then we have

$$\Theta\,\ddot{\varphi}_D = 2M \tag{6.175}$$

or

$$-\omega^2\Theta\,\hat{\varphi}_D = i\,\omega\,\hat{\varphi}_R(\eta + i\,\omega\,\gamma)\chi. \tag{6.176}$$

It follows from (6.176) that

$$\hat{\varphi}_D = \left[\frac{\chi\gamma}{\Theta} - i\,\frac{\chi\eta}{\Theta\,\omega}\right]\hat{\varphi}_R. \tag{6.177}$$

Without loss of generality we assume $\hat{\varphi}_R$ to be real. Then the phase angle of $\hat{\varphi}_D$ is given by

$$\tan\alpha = \left[-\frac{\chi\eta}{\Theta\,\omega}\right]\left[\frac{\chi\gamma}{\Theta}\right]^{-1} = -\frac{\eta}{\omega\,\gamma}. \tag{6.178}$$

Since $\gamma < 0$, α varies between $3\pi/2$ and π. It follows from (6.177) that

$$|\hat{\varphi}_D|/|\hat{\varphi}_R| = \frac{\chi\eta}{\Theta\,\omega}\sqrt{1 + (\omega\,\gamma/\eta)^2}, \tag{6.179}$$

or using (6.178)

$$|\hat{\varphi}_D|/|\hat{\varphi}_R| = \frac{\chi\eta}{\Theta\,\omega}\frac{\sqrt{\tan^2\alpha + 1}}{\tan\alpha} = \frac{\chi\eta}{\Theta\,\omega}\frac{1}{|\sin\alpha|}. \tag{6.180}$$

Using (6.172) we further extract the relation

$$\hat{\varphi}_G/\hat{\varphi}_R = 1 + \hat{\varphi}_D/\hat{\varphi}_R, \tag{6.181}$$

and by using (6.177) again we find the equation

$$|\hat{\varphi}_G/\hat{\varphi}_R|^2 = \left[1 + \frac{\chi\gamma}{\Theta}\right]^2 + \left[\frac{\chi\eta}{\Theta\omega}\right]^2, \tag{6.182}$$

which we shall make use of later. We first calculate the work W done by the torque $2\,M$ per period of oscillation $T = 2\pi/\omega$, where we note that only the real part of the quantities has any physical meaning. We obtain the integral

$$W = \int_0^T \Re(2M)\Re(\dot{\varphi}_R)\mathrm{d}t, \tag{6.183}$$

whose integrand we transform using (6.175) to

$$\Re(2M)\Re(\dot{\varphi}_R) = -\omega^2\,\Theta\,\Re(\hat{\varphi}_D\mathrm{e}^{i\omega t})\,\Re(i\,\omega\,\hat{\varphi}_R\mathrm{e}^{i\omega t}). \tag{6.184}$$

Because of (6.178) we shall write the complex angle $\hat{\varphi}_D$ as

$$\hat{\varphi}_D = |\hat{\varphi}_D|\mathrm{e}^{i\alpha}, \tag{6.185}$$

and since $\hat{\varphi}_R$ is purely real

$$\hat{\varphi}_R = |\hat{\varphi}_R|, \tag{6.186}$$

the following expression arises from (6.184)

$$\Re(2M)\,\Re(\dot{\varphi}_R) = \Theta\omega^3|\hat{\varphi}_D||\hat{\varphi}_R|\left(\cos\alpha\cos\omega t\sin\omega t + |\sin\alpha|\sin^2\omega t\right). \tag{6.187}$$

After carrying out the integration, (6.183) furnishes the result

$$W = \pi\,\Theta\,\omega^2|\hat{\varphi}_R||\hat{\varphi}_D||\sin\alpha|, \tag{6.188}$$

which we bring to dimensionless form using the reference work $\frac{\pi}{2}\,\Theta\,\omega^2|\hat{\varphi}_G|^2$

$$W^+ = \frac{2W}{\pi\,\Theta\,\omega^2|\hat{\varphi}_G|^2} = \frac{\frac{2\chi\eta}{\Theta\omega}}{\left[\frac{\chi\eta}{\Theta\omega}\right]^2 + \left[1 + \frac{\chi\gamma}{\Theta}\right]^2}, \tag{6.189}$$

where we have also made use of (6.180) and (6.182). W^+ is a function of the two dimensionless groups $(\chi\eta)/(\Theta\omega)$ and $\chi\gamma/\Theta$, and therefore represents a surface, as is shown in Fig. 6.12 together with the projection in the $(\chi\eta)/(\Theta\omega) - W^+$ plane for negative $\chi\gamma/\Theta$ values.

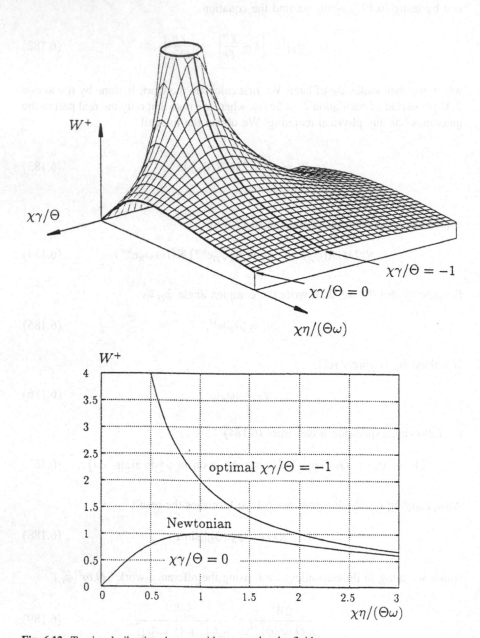

Fig. 6.12 Torsional vibration damper with a second order fluid

The operation points of all possible dampers which use fluid as the dissipating medium lie on the surface W^+. Of particular interest are the two sketched curves $\gamma = 0$, which corresponds to a Newtonian fluid, and $\chi \gamma / \Theta = -1$, which is clearly optimal in the following sense: for a given $(\chi \eta)/(\Theta \omega)$ the greatest possible damping is achieved on this curve. If $(\chi \eta)/(\Theta \omega) = 1$ this curve takes on double the value of the maximum damping possible with a Newtonian fluid.

For second order fluids the "tuning" $\chi \gamma / \Theta = -1$, which can always be reached for a given material constant γ by suitable choice of Θ or χ, is even frequency independent, i.e., the damper achieves the highest damping in the entire frequency domain. Real fluids only obey this law for small enough frequencies (where the memory time of the fluid is small compared to the period), so that γ is more or less strongly dependent on Ω, and so the damper can only be used at optimal tuning within a restricted range of frequencies.

6.4 Unidirectional Flows of a Bingham Material

6.4.1 Channel Flow of a Bingham Material

We shall consider the steady, fully developed flow of a Bingham material through a two-dimensional channel of height h and shall assume that the pressure gradient $\partial p/\partial x = -K$ is negative, that the upper wall $(y = h)$ is moved in the positive x-direction with velocity U and that the wall shear stress on the lower wall $(y = 0)$ is larger than the yield stress. All other cases can be referred back to this case. The x- and y-components of Cauchy's equation simplify to

$$\frac{\partial p}{\partial x} = \frac{\partial \tau'_{xy}}{\partial y} \tag{6.190}$$

and

$$\frac{\partial p}{\partial y} = \frac{\partial \tau'_{yy}}{\partial y}, \tag{6.191}$$

since in established channel flow the components of the stress deviator are not functions of x. From (6.191) we conclude

$$p = \tau'_{yy} + f(x) \tag{6.192}$$

and, using (6.190) we see that $f'(x) = \text{const} = -K$, so that

$$\tau'_{xy} = -Ky + \tau_w, \tag{6.193}$$

where τ_w is the shear stress on the lower wall ($y = 0$). Since $\tau_w > \vartheta$, the material flows near to the wall until the stress falls below the yield stress ϑ at a height

$$y = \kappa_1 h = (\tau_w - \vartheta)/K \qquad (6.194)$$

and the material becomes solid. As y becomes larger the shear stress finally becomes negative, until at

$$y = \kappa_2 h = (\tau_w + \vartheta)/K \qquad (6.195)$$

the negative shear stress $-\tau'_{xy}$ is equal to the yield stress, after which the material flows again. Clearly du/dy is positive in the first flow zone as we conclude from the constitutive relation (3.60) and (3.62). For the unidirectional flow at hand the constitutive relation has the form

$$\tau'_{xy} = \eta_1 \frac{du}{dy} + \vartheta \, \mathrm{sgn}\left(\frac{du}{dy}\right) \qquad (6.196a)$$

$$\tau'_{yy} = 0. \qquad (6.196b)$$

From (6.193) and (6.196a, 6.196b) and using the boundary condition $u(0) = 0$ we obtain the velocity distribution in the first flow zone as

$$\frac{u}{U} = -\frac{K h^2}{2\eta_1 U}\left(\left(\frac{y}{h}\right)^2 - 2\kappa_1 \frac{y}{h}\right), \qquad (6.197)$$

which for $\kappa_1 \geq 1$ is already the distribution in the whole channel. In the second flow zone du/dy is negative and the velocity distribution is

$$\frac{u}{U} = 1 + \frac{K h^2}{2\eta_1 U}\left(1 - \left(\frac{y}{h}\right)^2 - 2\kappa_2\left(1 - \frac{y}{h}\right)\right), \qquad (6.198)$$

where the boundary condition $u(h) = U$ has been used. As expected du/dy vanishes at the yield surfaces. The velocity at the yield surfaces $y = \kappa_1 h$ and $y = \kappa_2 h$ is equal to the solid body velocity which, from (6.197) is

$$U_S = \frac{K h^2 \kappa_1^2}{2\eta_1} \qquad (6.199)$$

and from (6.198) is

$$U_S = U + \frac{Kh^2}{2\eta_1}(1 - \kappa_2)^2. \tag{6.200}$$

It follows from (6.199) and (6.200) that

$$\kappa_1^2 - (1 - \kappa_2)^2 = \frac{2U\eta_1}{Kh^2}; \tag{6.201}$$

together with (6.194) and (6.195), (6.201) uniquely determines the flow at a given pressure gradient $-K$, plate velocity U and material properties η_1 and ϑ.

First we shall consider the case where no solid is formed, therefore (6.197) represents the entire velocity distribution, which with the condition $u(y = h) = U$ gives for the shear stress τ_w

$$\tau_w = \vartheta + \frac{U\eta_1}{h} + \frac{Kh}{2}. \tag{6.202}$$

We can easily convince ourselves that for nonvanishing K the velocity distribution is that of Couette-Poiseuille flow (6.19). From (6.194) and (6.202) we conclude that no solid is formed if $2\eta_1 U/(Kh^2) > 1$.

If the second flow zone in the channel is not formed but a solid does arise this then, by previous assumptions, adheres to the upper wall, and from (6.200) we have $\kappa_2 = 1$. Equation (6.201) furnishes the value $\kappa_1 = \sqrt{2U\eta_1/(Kh^2)}$, which, with (6.194) determines the shear stress at the wall. With (6.197) it also gives the velocity distribution for this case (Fig. 6.13).

As already explained above, in general a solid forms between the two flow zones. With the dimensionless numbers $2U\eta_1/(Kh^2)$ and $2\vartheta/(Kh)$, abbreviated to A and B respectively, and using (6.194), (6.195) and (6.201), we determine the position of the yield surfaces as

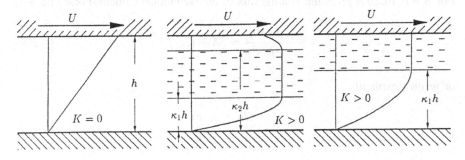

Fig. 6.13 Channel flow of a Bingham material

$$\kappa_1 = \frac{A + (1 - B)^2}{2(1 - B)} \tag{6.203}$$

and

$$\kappa_2 = \frac{A + (1 - B^2)}{2(1 - B)}. \tag{6.204}$$

Since $0 < \kappa_1 < \kappa_2 < 1$, we infer the inequalities

$$1 > \frac{2\vartheta}{Kh} > 0 \tag{6.205}$$

and

$$\left(1 - \frac{2\vartheta}{Kh}\right)^2 > \frac{2U\eta_1}{Kh^2}. \tag{6.206}$$

For pure pressure driven flows ($U = 0$) and $2\vartheta/(Kh) \geq 1$ the solid occupies the whole channel.

Using the quantities A and B the volume flux (per unit depth) is represented by the equation

$$\frac{12\,\dot{V}\eta_1}{Kh^3} = 1 + 3A - \frac{3}{2}B + \frac{1}{2}B^3 + \frac{3A^2}{2(1-B)^2} - \frac{3A^2}{2(1-B)}, \tag{6.207}$$

which for pure pressure driven flow ($A = 0$) reduces to

$$\frac{12\,\dot{V}\eta_1}{Kh^3} = \left(1 - \frac{3}{2}B + \frac{1}{2}B^3\right). \tag{6.208}$$

For $B = 0$, (6.207) gives the volume flux of the Newtonian Couette-Poiseuille flow

$$\frac{12\,\dot{V}\eta_1}{Kh^3} = 3A + 1, \tag{6.209}$$

or written explicitly

$$\dot{V} = \frac{Uh}{2} + \frac{Kh^3}{12\eta_1}. \tag{6.210}$$

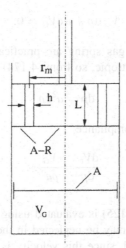

Fig. 6.14 Sketch of shock absorber using electro-rheological fluid

Finally we note that the Eqs. (6.197) to (6.201) and (6.203) to (6.210) are valid for any sign of U and K, as long as the absolute value of B is taken.

We apply the results to shock absorber using electro-rheological (or magneto-rheological) fluids, which under the influence of a strong electric (magnetic) field behave as Bingham media, even if they exhibit Newtonian behavior without field. For the control volume of Fig. 6.14, we have (2.8) in the form

$$\iint\limits_{A-R,R,A} \vec{u} \cdot \vec{n}\, dS = 0, \qquad (6.211)$$

where A is the cross-sectional area of the inner cylinder, R the cross-sectional ring shaped channel between piston and cylinder and $A - R$ the piston area. We assume $A/R \ll 1$ and therefore $A - R \approx A$ and obtain from (6.211)

$$-A\, u_P + \dot{V} + \frac{dV_G}{dp}\frac{dp}{dt} = 0. \qquad (6.212)$$

$-Au_P$ is the volume displaced by the piston per unit time, $\dot{V}$ the volume flux through the channel and the third term represents the change of Volume per unit time by the displacement of the intermediate (mass and frictionless) divider, which forms a separate gas chamber acting as a gas spring. dV_P/dp is the volume compliance and the inverse the volume stiffness. We find an expression for the volume compliance by noting that the mass in the chamber is constant

$$V_G\,d\rho + \rho\,dV_G = 0. \tag{6.213}$$

The changes of state in the gas spring are practically isentropic and since the velocity is small, even homentropic, so from (4.174)

$$dp = a^2 d\rho \tag{6.214}$$

which have for the volume compliance

$$\frac{dV_G}{dp} = \frac{V_G}{\rho a^2}. \tag{6.215}$$

For small volume changes (6.125) is evaluated using the undisturbed state.

The velocity of the piston may be neglected in the computation of the volume flow through the ring channel, since this velocity is much smaller than the flow velocity on account of $R/A \ll 1$. For the same reason the channel height is much smaller than the mean radius and the flow in channel is essentially plane. Then (6.207) provides the expression for the volume flow per unit length of the channel, so here we have

$$\dot{V} = 2\pi\, r_m \Delta p\, \frac{h^3}{12\eta_1 L}\left(1 - \frac{3}{2}B + \frac{1}{2}B^3\right);$$

$$\text{for } B < 1 \quad \text{and} \quad \dot{V} = 0; \quad \text{for } B \geq 1,$$

where $\Delta p = KL$ is the pressure difference $p - p_0$ across the piston and $B = 2\vartheta L/(\Delta p\,h)$. Since p_0 is time independent we obtain from (6.212) a nonlinear differential equation for Δp

$$\frac{d(\Delta p)}{dt} = \frac{A u_P(t) - 2\pi\, r_m \Delta p\, \dfrac{h^3}{12\eta_1\, L}\left(1 - \dfrac{3}{2}\dfrac{2\vartheta L}{\Delta p\, h} + \dfrac{1}{2}\left(\dfrac{2\vartheta L}{\Delta p\, h}\right)^3\right)}{d V_G/dp}; \tag{6.216}$$

$$\text{for} \quad B < 1 \quad \text{and} \quad \frac{d(\Delta p)}{dt} = \frac{A u_P(t)}{d V_G/dp}; \quad \text{for} \quad B \geq 1.$$

For a given piston motion $x_P = x_0 \sin(\omega t)$; $\dot{x}_P = u_P(t)$ say, the equation is integrated numerically giving the force acting on the piston. It is customary to describe the damper characteristic by graphing $F(u_P)$ since the circumscribed area is a measure of the dissipated energy. This graph is displayed in Fig. 6.15 for a Bingham Material with yield stress $\vartheta = 5000$ N/m^2 and $\vartheta = 0$ (Newtonian Fluid).

However the comparison is in so far misleading as damper using Newtonian Fluids are not designed according to the principles outlined in Fig. 6.14. These shocks absorbers have pressure dependent throttle openings. The work done by the

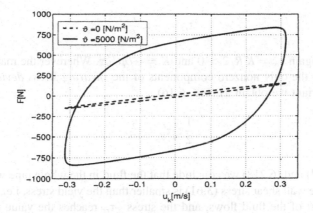

Fig. 6.15 Damper characteristic

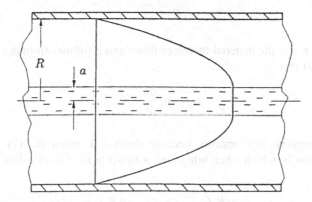

Fig. 6.16 Pipe flow of a Bingham material

piston is here first converted into kinetic energy which is subsequently dissipated. This damper characteristic is nearly independent of viscosity and therefore independent of ambient temperature. (The dissipation itself is of course due to viscosity).

6.4.2 Pipe Flow of a Bingham Material

Because of the kinematic restriction the steady flow of a Bingham material through a circular pipe with radius R is also a unidirectional flow. As explained in Sect. 6.3.1, for any material behavior we obtain a shear stress distribution in the pipe linearly dependent on the distance from the center r

$$\tau_{rz} = -\tau_w \frac{r}{R}, \tag{6.217}$$

where here again $\tau_w = KR/2 > 0$ and $K = -\partial p/\partial z$. Wherever the material flows, τ_{rz} or τ_{zr} are the only nonzero components of the *shearing stress deviator*, whose second invariant we obtain as (Fig. 6.16).

$$\frac{1}{2}\tau'_{ij}\tau'_{ij} = \tau^2_{rz}. \tag{6.218}$$

Using (6.217) and (6.218) we conclude that the fluid in the whole pipe will not flow as long as the wall shear stress (3.61) is smaller than the yield stress, i.e., $\tau_w < \vartheta$. For $\tau_w > \vartheta$ a part of the fluid flows, and the stress $-\tau_{rz}$ reaches the value of the yield stress at the radius $r = a$

$$\frac{a}{R} = \frac{\vartheta}{\tau_w}. \tag{6.219}$$

In the region $r > a$ the material therefore flows and it follows from the constitutive relation (3.60) that

$$\tau_{rz} = \eta_1 \frac{du}{dr} - \vartheta, \tag{6.220}$$

where the negative sign appears because $du/dr < 0$. From (6.217) we find an equation for du/dr, which when integrated with $u(r = R) = 0$ furnishes the velocity distribution

$$u(r) = \frac{\tau_w R}{2\eta_1}\left(1 - \left(\frac{r}{R}\right)^2\right) - \frac{\vartheta R}{\eta_1}\left(1 - \left(\frac{r}{R}\right)\right). \tag{6.221}$$

For $\vartheta = 0$ we recover the well known form for Newtonian fluids. In the region $r < a$ (6.221) yields the constant velocity in the center of the pipe as

$$u_{max} = \frac{\tau_w R}{2\eta_1}\left(1 - \frac{a}{R}\right)^2 = \frac{\tau_w R}{2\eta_1}\left(1 - \frac{\vartheta}{\tau_w}\right)^2, \tag{6.222}$$

and finally we obtain the volume flux as

$$\dot{V} = \frac{\pi \tau_w R^3}{4\eta_1}\left(1 - \frac{4}{3}\frac{\vartheta}{\tau_w} + \frac{1}{3}\left(\frac{\vartheta}{\tau_w}\right)^4\right). \tag{6.223}$$

Chapter 7
Fundamentals of Turbulent Flow

7.1 Stability and the Onset of Turbulence

We shall now follow on from the discussion of laminar pipe flow. There we determined that the pressure drop is proportional to the volume flux, a result which agrees with experiment only for Reynolds' numbers smaller than a critical Reynolds' number. If this critical Reynolds' number is exceeded the pressure drop increases sharply and finally becomes proportional to the square of the flux through the tube. At the same time there is a striking change in the behavior of the flow.

Below the critical Reynolds' number straight particle paths parallel to the pipe wall with a unidirectional or laminar flow motion are seen, so that this flow form has the name *laminar flow*. The particle paths can be observed by using a glass tube, where color is introduced into the fluid at one point, and so a streakline appears, which, for steady flow, coincides with the pathline. In laminar flow a fine thread appears which will only spread out from the very small effect of the molecular diffusion.

If the Reynolds' number is increased sufficiently, the flow becomes very clearly unsteady: the thread waves back and forth and spreads out much faster than would be expected from molecular diffusion. At only a small distance from where the color is introduced, the thread has mixed with the fluid. This form of flow is called *turbulent flow*. A characteristic sign of turbulent flow is the strongly increased diffusion which expresses itself in the rapid spreading out of the color thread. We have already mentioned other characteristics: three-dimensionality and unsteadiness of the always rotational flow, and stochastic behavior of the flow quantities.

Of course the *transition* to turbulence does not only occur in pipe flows but in all laminar flows, particularly laminar boundary layers. In as much as the laminar flows discussed up to now have been exact solutions of the Navier-Stokes equations, these solutions hold in principle for arbitrarily large Reynolds' numbers. For these

© Springer Nature Switzerland AG 2020
J. H. Spurk and N. Aksel, *Fluid Mechanics*,
https://doi.org/10.1007/978-3-030-30259-7_7

solutions to be realized in nature however, not only must the Navier-Stokes equations be satisfied, but the flows must also be stable with respect to small disturbances. However this is no longer the case above the critical Reynolds' number, where even a vanishingly small disturbance is enough to induce the transition to the turbulent flow form.

In most of the laminar flows mentioned, the Reynolds' number below which all small disturbances die away can be theoretically calculated. In others, in particular in Hagen-Poiseuille flows, no critical Reynolds' number has been found: it appears as if this flow is theoretically stable at all Reynolds' numbers. However in nature turbulent flow forms do appear here, as has just been shown in the example above of pipe flow. Historically the investigation into turbulent flows began with pipe flows (Reynolds 1883). It is probable that the instability of pipe flow develops from a disturbance in the pipe entrance where a pipe flow with parabolic velocity profile has not yet developed. The experimentally determined critical Reynolds' number strongly depends on the conditions of the approach flow at the entrance. Critical Reynolds' numbers up to 40,000 have been measured for approach flows which are especially free from disturbances, whereas for the disturbed approach flows typical in technical applications, the critical Reynolds' number drops to 2300. Even if the approach flow is highly disturbed, pipe flow remains laminar when the Reynolds' number is lower than 2000. A valid measure for the critical Reynolds' number under the conditions found in technical applications is

$$Re_{crit} = (\bar{U}d/\nu)_{crit} = 2300. \tag{7.1}$$

For Reynolds' numbers $Re < Re_{crit}$ it is recommended to compute the pressure drop from the laws of laminar pipe flow, while if $Re > Re_{crit}$ the pressure drop follows from the corresponding laws of turbulent pipe flow.

However, what happens in pipe flow makes it clear that the Reynolds' number at which the flow becomes turbulent is generally different from the Reynolds' number at which the flow becomes unstable for the first time. Both Reynolds' numbers are often called the critical Reynolds' number, but the difference between them is important, because the instability of a flow relative to small disturbances does not necessarily and directly imply the transition to turbulent flow. In general a new, more complicated but still laminar flow evolves, which, as the Reynolds' number increases, becomes unstable and possibly develops into a new laminar flow, but could also make the transition to turbulence. The transition from unstable to fully turbulent flow has until now only been accessible by direct numerical simulation.

Experimental investigations are very difficult because the flow is particularly sensitive to unavoidable and often unknown disturbing influences which can still decisively change the transition behavior. Frequently the Reynolds' numbers of stability and that of transition to turbulence lie close to one another, especially if there is a high degree of disturbance in the approach flow.

The laminar flow between two rotating cylinders offers an example of the large difference between the two Reynolds' numbers. In this case the first instability is interesting because it is closely related to the instability of the density stratification: a small parcel of fluid moved up from radius r to radius $r_1 > r$ by a small disturbance brings the angular momentum $L = ru_\varphi$ with it in the absence of friction forces. The velocity of the small quantity of fluid on its new path r_1 is L/r_1 and its centripetal acceleration is L^2/r_1^3. There it is acted on by the surrounding pressure gradient which, from (6.40) is given by

$$\varrho \frac{u_{\varphi 1}^2}{r_1} = \varrho \frac{L_1^2}{r_1^3} = \left.\frac{\partial p}{\partial r}\right|_{r_1}. \tag{7.2}$$

If the quantity of fluid is shifted back to the initial radius r by this pressure gradient then the flow is stable. We are led therefore to the necessary condition for stability

$$\left.\frac{\partial p}{\partial r}\right|_{r_1} = \varrho \frac{L_1^2}{r_1^3} > \varrho \frac{L^2}{r_1^3}, \tag{7.3}$$

i.e.,

$$r_1 u_{\varphi 1} > r u_\varphi. \tag{7.4}$$

The potential vortex with $r u_\varphi = $ const is apparently just the "neutral" velocity distribution. However the velocity distribution is unstable if $r u_\varphi$ is larger at the smaller radius than it is at the bigger one, as for example if the outer cylinder does not move and only the inner cylinder is rotated.

So far these considerations only hold for frictionless fluids. If we take the friction into account we find the critical Reynolds' number to be

$$\Omega_I R_I \frac{h}{\nu} = 41.3\sqrt{\frac{R_I}{h}}, \tag{7.5}$$

where h denotes the gap width. A new laminar flow forms above this Reynolds' number; vortices turning alternately to the left and to the right appear regularly and their axis of symmetry is in the direction of the axis of the cylinder (*Taylor vortices*). The transition to turbulence only takes place at much higher Reynolds' numbers, about 50 times greater than the Reynolds' number at which stability is lost. This flow phenomenon is also of technical interest since it can appear wherever a shaft rotates in a bore, for example in radial bearings.

7.2   Reynolds' Equations

In fully developed turbulent flow, i.e., after the transition has been completed, the flow quantities are *random quantities*. The flow may be considered to be the superposition of a basic or main flow with irregular stochastic fluctuations in the velocity or in other fluid mechanical quantities. The velocity field is therefore represented as follows

$$u_i(x_j, t) = \bar{u}_i(x_j, t) + u_i'(x_j, t). \tag{7.6}$$

This decomposition is particularly appropriate if the fluctuation velocity u_i' is much smaller than the basic velocity $\bar{u}_i$. The basic velocity corresponds to the mean value of the velocity. In even the most general case we can form the *mean value of the velocity* $\bar{u}_i(x_j, t)$, as well as other mean quantities, using

$$\bar{u}_i(x_j, t) = \lim_{n \to \infty} \frac{1}{n} \sum_{k=1}^{n} u_i^{(k)}(x_j, t), \tag{7.7}$$

where the flow is realized n times and each time the velocity $u_i(x_j, t)$ is determined at the same place x_j at the same instant t. In flows where these mean values are independent of time, that is in *statistically steady processes*, we have, instead of the ensemble-mean value calculated from (7.7), the time-mean value calculated from the formula

$$\bar{u}_i(x_j) = \lim_{T \to \infty} \frac{1}{T} \int_{t-T/2}^{t+T/2} u_i(x_j, t) \mathrm{d}t, \tag{7.8}$$

which would require only one experimental realization. In what follows we shall restrict ourselves to incompressible flows which are steady in the mean. We shall now insert (7.6) and the corresponding form for the pressure

$$p = \bar{p} + p' \tag{7.9}$$

in the continuity Eq. (2.5) as well as into the Navier-Stokes equations in the form (4.9a) and shall subject the resulting equation to the averaging according to (7.8). From (7.8) and (7.6) we have the following rules for the calculation of the mean values of two arbitrary random quantities g and f

$$\bar{\bar{g}} = \bar{g}, \tag{7.10a}$$

$$\overline{g+f} = \bar{g}+\bar{f}, \tag{7.10b}$$

$$\overline{\bar{g}f} = \bar{g}\bar{f}, \tag{7.10c}$$

$$\overline{\partial g/\partial s} = \frac{\partial \bar{g}}{\partial s}, \quad \text{and} \tag{7.10d}$$

$$\overline{\int f \mathrm{d}s} = \int \bar{f}\mathrm{d}s, \tag{7.10e}$$

where s is any one of the independent variables x_i or t, and we obtain from the continuity equation

$$\frac{\partial \bar{u}_i}{\partial x_i} = 0, \tag{7.11}$$

since $\overline{u_i'}$ vanishes as a consequence of (7.8) and therefore so does $\overline{\partial u_i'/\partial x_i}$ from (7.10d). It thus follows that for the fluctuating velocities we have

$$\frac{\partial u_i'}{\partial x_i} = 0. \tag{7.12}$$

For the same reason, all terms linear in the fluctuating quantities vanish from the Navier-Stokes equations

$$\frac{\overline{\partial u_i'}}{\partial t} = \frac{\overline{\partial^2 u_i'}}{\partial x_i \partial x_j} = \frac{\overline{\partial p'}}{\partial x_i} = \overline{u_j'\frac{\partial \bar{u}_i}{\partial x_j}} = 0, \tag{7.13}$$

and we obtain the equation

$$\varrho\bar{u}_j\frac{\partial \bar{u}_i}{\partial x_j} + \varrho\overline{u_j'\frac{\partial u_i'}{\partial x_j}} = \varrho\,k_i - \frac{\partial \bar{p}}{\partial x_i} + \eta\frac{\partial^2 \bar{u}_i}{\partial x_j \partial x_j}. \tag{7.14}$$

Along with the following relation from the continuity equation

$$\frac{\partial}{\partial x_j}\left(u_i'u_j'\right) = u_j'\frac{\partial}{\partial x_j}\left(u_i'\right) \tag{7.15}$$

Equation (7.14) can be rewritten in the form first cited by Reynolds

$$\varrho\,\bar{u}_j\frac{\partial\bar{u}_i}{\partial x_j} = \varrho\,k_i - \frac{\partial\bar{p}}{\partial x_i} + \eta\frac{\partial^2\bar{u}_i}{\partial x_j\partial x_j} - \frac{\partial\left(\varrho\,\overline{u_i'u_j'}\right)}{\partial x_j}. \tag{7.16}$$

The vanishing of the mean linear terms means physically that their contributions to the integral (7.8) cancel. In other words: in the mean the fluctuating quantities are just as often positive as negative. This is not the case for the nonlinear terms (e.g. $\overline{u_i'u_j'}$), as is obvious for the main diagonal components of the tensor $\overline{u_i'u_j'}$, that is $\overline{u_1'u_1'}$, $\overline{u_2'u_2'}$, $\overline{u_3'u_3'}$. But even the terms $\overline{u_1'u_2'}$, etc., which are the velocity components in two different directions are in general nonzero. They would only be zero if we were dealing with statistically independent quantities. However the components of the velocity are *correlated*. As a measure of the *correlation* between two fluctuating quantities g' and f' we use the expression

$$R = \frac{\overline{g'f'}}{\sqrt{\overline{g'^2}\,\overline{f'^2}}}, \tag{7.17}$$

here

$$R_{ij}(x_k,t) = \frac{\overline{u_i'u_j'}}{\sqrt{\overline{u_i'^2}\,\overline{u_j'^2}}}, \tag{7.18}$$

or, more generally, for the correlation between two velocity components $u_i'(x_k,t)$ and $u_j'(x_k+r_k,t+\tau)$

$$R_{ij}(x_k,t,r_k,\tau) = \frac{\overline{u_i'(x_k,t)u_j'(x_k+r_k,t+\tau)}}{\sqrt{\overline{u_i'^2}(x_k,t)\overline{u_j'^2}(x_k+r_k,t+\tau)}}. \tag{7.19}$$

(In Eqs. (7.18) and (7.19) we do not sum over the indices i and j.)

The special forms for the spatial and temporal correlation (autocorrelation) arise from (7.19) for $\tau = 0$ and $r_k = 0$ respectively. If the distance $|\vec{r}|$ between $\vec{x}$ and $\vec{x}+\vec{r}$, at which the velocity components in (7.19) are to be taken, tends to infinity the velocity components become statistically independent and the correlation disappears. A measure of the range of correlation between two velocity components in the x_1-direction, taken at a distance r along the x_1-axis and at the same time ($\tau = 0$), is the integral length scale

$$L(\vec{x}, t) = \int_0^\infty R_{11}(\vec{x}, t, r, 0)\mathrm{d}r, \tag{7.20}$$

which represents the spatial extent of *turbulent fluctuation*. For $\tau \to \infty$ the correlation goes to zero as well. An integral time scale analogous to (7.20) can also be introduced.

Of course the steady basic flow $\bar{u}_i$ can be kinematically restricted, for example unidirectional flows or two-dimensional rotationally symmetric flows. However the superimposed fluctuating motion u'_i is always three-dimensional, and of course, unsteady.

The steady basic flow must satisfy the Reynolds' equation (7.16) and the continuity Eq. (7.11). However these equations are not enough to determine the basic flow because the terms $-\varrho \overline{u'_i u'_j}$ introduced by the averaging appear as unknowns. These terms represent averaged momentum fluxes (per unit area) and give rise to a force in the j-direction on a surface whose normal is in the i-direction. They are known as *Reynolds' stresses*. The tensor of these stresses is clearly symmetric since the order of the indices results from the arbitrary order of the factors. We summarize the whole stress tensor in the form

$$T_{ij} = \bar{\tau}_{ij} - \varrho \overline{u'_i u'_j}, \tag{7.21}$$

or for the assumed incompressible flow ($\partial u_k / \partial x_k = 0$) because of (3.1a), also in the form

$$T_{ij} = -\bar{p}\,\delta_{ij} + 2\eta\,\bar{e}_{ij} - \varrho \overline{u'_i u'_j}. \tag{7.22}$$

We then write the Reynolds' equations without volume body forces in the form

$$\varrho\,\bar{u}_j \frac{\partial \bar{u}_i}{\partial x_j} = \frac{\partial T_{ji}}{\partial x_j}. \tag{7.23}$$

The divergence of the Reynolds' stresses (final term on the right-hand side of (7.16)) acts on the basic flow as an additional but unknown force (per unit volume). In turbulent flow this force is in general much larger than the divergence of the viscous stresses, which in the incompressible flow assumed here corresponds to the term $\eta\, \partial^2 \bar{u}_i / (\partial x_j \partial x_j)$. Only in the immediate neighborhood of solid walls do the fluctuation velocities and with them the Reynolds' stresses decrease to zero. This is because the fluctuation velocities, just like the average velocities, have to obey the no slip condition so that the viscous stresses predominate right at the wall, in a region called the *viscous sublayer*.

It would now appear obvious to construct differential equations for the unknown Reynolds' stresses in a systematic manner, which, along with the Reynolds'

Eq. (7.16) and the continuity Eq. (7.11), would form a complete system of differential equations. In order to find the appropriate equations we introduce (7.6) and (7.9) into the Navier-Stokes Eq. (4.9a) and subtract the Reynolds' Eq. (7.16). In this way we are led to the equation for the fluctuating velocity field

$$\varrho\left(\frac{\partial u_i'}{\partial t} + \bar{u}_k\frac{\partial u_i'}{\partial x_k} + u_k'\frac{\partial u_i}{\partial x_k} + u_k'\frac{\partial \bar{u}_i}{\partial x_k}\right) = -\frac{\partial p'}{\partial x_i} + \frac{\partial}{\partial x_k}\left(\varrho\,\overline{u_i'u_k'}\right) + \eta\frac{\partial^2 u_i'}{\partial x_k\partial x_k}. \quad (7.24)$$

We can multiply this equation by u_j' and find a further equation by interchanging i and j. After averaging, we can add these equations to furnish equations for the Reynolds' stresses. We shall not perform this calculation because it is clear that multiplying (7.24) by u_j' introduces terms of the form $\overline{u_j'u_k'u_i'}$ into the problem as new unknowns.

On the other hand if we find new differential equations for these triple correlations they will contain quadruple correlations, and so on. Therefore this process fails to complete the system of equations. The problem of closing this system of equations represents the fundamental problem of fully developed turbulent flow, and until now remains unresolved.

All attempts so far to make the system of equations determinate have been partly based on considerable simplifications and hypotheses. At the lowest level, the closure of the system of equations is accomplished by using relationships between the Reynolds' stresses and the mean velocity field. These semi-empirical relationships represent *turbulence models*, which can take on the form of algebraic relationships or of differential equations, and which are classified according to the number of differential equations. As the name "semi-empirical" implies, they all contain quantities which have to be determined experimentally.

As a consequence of the turbulent fluctuating motion, not only is the momentum flux increased (expressed through the Reynolds' stresses), but so also are the heat and diffusion fluxes. In order to discuss the turbulent heat flux we shall start with the energy Eq. (4.2), where we may not set the material change of the density $D\varrho/Dt$ to zero if external heating of the fluid takes place as is the case in heat transfer problems. While the change in density can be ignored for liquids, this is not so for gases. If we ignore the change in density for liquids from (4.2) and since $de = cdT$, we obtain directly an equation for the temperature field

$$\varrho\,c\frac{DT}{Dt} = \Phi + \frac{\partial}{\partial x_i}\left(\lambda\frac{\partial T}{\partial x_i}\right). \quad (7.25)$$

The remaining simplifications for gases arise from (4.176) if we ignore the density changes which result from changes in the pressure, since the equations necessary for this ((4.182), (4.184) and (4.188)) are satisfied. If there is external heating the entropy change in (4.176) may not be ignored, and gives rise to a change in the density. For the calorically perfect gas, Eq. (4.176) leads, by (4.177), to the expression

$$\frac{1}{\varrho}\frac{D\varrho}{Dt} = -\frac{1}{c_p}\frac{Ds}{Dt},$$

(7.26)

from which, using Gibbs' relation (2.133) the expression

$$\frac{1}{\varrho}\frac{D\varrho}{Dt} = -\frac{1}{T}\frac{DT}{Dt}$$

(7.27)

follows. We can also extract this directly from the thermal equation of state $\varrho = \varrho(p, T)$ if we consider that the change in state of a material particle is isobaric if the change in density as a result of pressure change is ignored. Using (7.27) the energy Eq. (4.2) for gases at low flow velocities ($M \to 0$) takes the form

$$\varrho\, c_p \frac{DT}{Dt} = \Phi + \frac{\partial}{\partial x_i}\left(\lambda \frac{\partial T}{\partial x_i}\right).$$

(7.28)

In accordance with (4.180), the dissipation can be neglected under the assumptions made, or in other words: the work of deformation (per unit time and volume) transformed irreversibly to heat, hardly produces any raise in temperature. We note however that the dissipation plays a decisive role as a loss in the balance of the mechanical energy in turbulent flow, and may on no account be ignored there. (The corresponding balance equation for the kinetic energy of the fluctuating motion is obtained from (7.24) if we multiply this equation by u_i' and then perform the averaging process.)

We shall now insert (7.6) and the corresponding form for the temperature

$$T = \overline{T} + T'$$

(7.29)

into (7.25) (or for gases into (7.28)), where as explained the dissipation function is ignored and in addition λ is taken to be constant

$$\varrho\, c\left(\frac{\partial(\overline{T} + T')}{\partial t} + (\bar{u}_i + u_i')\frac{\partial(\overline{T} + T')}{\partial x_i}\right) = \lambda \frac{\partial^2(\overline{T} + T')}{\partial x_i \partial x_i}.$$

(7.30)

Noting the rules (7.10), averaging leads us to the equation for the mean temperature

$$\varrho\, c\, \bar{u}_i \frac{\partial \overline{T}}{\partial x_i} = -\varrho\, c\, \overline{u_i' \frac{\partial T'}{\partial x_i}} + \lambda \frac{\partial^2 \overline{T}}{\partial x_i \partial x_i},$$

(7.31)

which, because of (7.12), can also be expressed in the form

$$\varrho\, c\, \bar{u}_i \frac{\partial \overline{T}}{\partial x_i} = -\varrho\, c \frac{\partial}{\partial x_i} \left(\overline{u_i' T'} \right) + \lambda \frac{\partial^2 \overline{T}}{\partial x_i\, \partial x_i}. \tag{7.32}$$

By analogy to the Reynolds' stresses, a *"turbulent heat flux vector"* appears here

$$q_i = \varrho\, c\, \overline{u_i' T'}, \tag{7.33}$$

which is unknown, just like the Reynolds' stresses, and which prevents the solution of Eq. (7.32). What has been said in connection with the Reynolds' equation is also valid here: the closure of the system of equations takes place by a semi-empirical relation between the turbulent heat flux vector and the average velocity and temperature fields.

7.3　Turbulent Shear Flow Near a Wall

Turbulent shear flows play an important role in technical applications because they are met in channel and pipe flows as well as in turbulent boundary layer flows. The emphasis here is on the profiles of the mean velocity and on the resistance laws. We can already obtain important insights into the behavior of turbulent shear flows if we consider the simplest case of a unidirectional flow with a vanishing pressure gradient along a smooth flat wall.

In laminar flow with a vanishing pressure gradient and with the basic assumptions of unidirectional flow ($u_1 = f(x_2)$, $u_2 = u_3 = 0$) the Navier-Stokes equations simplify to

$$0 = \eta \frac{d^2 u_1}{dx_2^2}, \tag{7.34}$$

from which we infer the constant shear stress $\tau_{21} = P_{21} = \eta\, du_1/dx_2$ and the known linear velocity distribution of the simple shearing flow. Using the same assumption that the mean quantities, i.e., the nonvanishing velocity components $\bar{u}_1$ and the Reynolds' stresses only dependent on x_2, we still obtain quite generally from the Reynolds' Eqs. (7.23)

$$0 = \frac{\partial T_{ji}}{\partial x_j}. \tag{7.35}$$

Using the Cartesian coordinate notation x, y, z and Cartesian velocity components u, v and w, we extract from the first of these equations under the assumption of vanishing x component of the pressure gradient, the equation

$$0 = \frac{d}{dy}\left(\eta \frac{d\bar{u}}{dy} - \varrho \overline{u'v'}\right);$$ (7.36)

we shall not be interested in the other two component equations just now. Integrating (7.36)

$$\text{const} = \tau_w = \eta \frac{d\bar{u}}{dy} - \varrho \overline{u'v'}$$ (7.37)

furnishes the statement that the total shear stress T_{21}, i.e., the sum of the viscous stress $P_{21} = \tau_{21} = \eta\, d\bar{u}/dy$ and the Reynolds' stress $-\varrho\,\overline{u'v'}$, is constant and therefore independent of y. We have already identified the constant of integration as the shear stress τ_w at the wall, since for $y = 0$ the Reynolds' stress vanishes as a consequence of the no slip condition. Because of the (unknown) Reynolds' stress appearing in (7.36), the distribution of the mean velocity $\bar{u} = f(y)$ is now no longer a linear function.

In recalling technical applications, in particular those of established turbulent channel and pipe flows, the question of the practical importance of the result (7.37) arises. In these flows (as well as in most boundary layer flows) the pressure gradient does not vanish, but in channel and pipe flow is the only source of motion. Just as with laminar flows, the shear stress is then not constant, but is a linear function of y (channel flow) or r (pipe flow). For a nonvanishing pressure gradient, it follows from the first component of Eq. (7.35) that

$$0 = -\frac{\partial \bar{p}}{\partial x} + \frac{d}{dy}\left(\eta \frac{d\bar{u}}{dy} - \varrho \overline{u'v'}\right),$$ (7.38)

and from the second component of Eq. (7.35)

$$0 = -\frac{\partial \bar{p}}{\partial y} + \frac{d}{dy}\left(-\varrho \overline{v'^2}\right),$$ (7.39)

while the third leads us to $\partial \bar{p}/\partial z = 0$. We conclude from (7.39) that the sum of $\bar{p}$ and $\varrho\,\overline{v'^2}$ is only a function of x, and therefore that $\partial \bar{p}/\partial x$ is only a function of x since the Reynolds' stress $-\varrho\,\overline{v'^2}$ by assumption only depends on y. Since the second term in (7.38) does not depend on x, it also follows that $\partial \bar{p}/\partial x$ is a constant. The entire shear stress, which we now abbreviate to

$$\tau = \eta \frac{d\bar{u}}{dy} - \varrho \, \overline{u' v'} \tag{7.40}$$

is therefore (as in the laminar case) a linear function of y

$$\tau = \frac{d\bar{p}}{dx} y + \text{const.} \tag{7.41}$$

We determine the constant of integration by noting that in the middle of the channel $(y = h)$ the shear stress vanishes because for symmetry reasons $d\bar{u}/dy$ and $\overline{u' v'}$ are zero there. Therefore we write (7.41) in the form

$$\tau = -\frac{\partial \bar{p}}{\partial x} h \left(1 - \frac{y}{h}\right) = \tau_w \left(1 - \frac{y}{h}\right), \tag{7.42}$$

where we have denoted the shear stress on the lower wall as τ_w.

For turbulent pipe flow a linear shear stress distribution also arises, as can be shown by considerations analogous to the laminar case (6.138) or (6.140). (Since the results of this section are not only valid for pipe flows, we depart from (6.140) and denote the coordinate in the axial direction as x.) We infer from (7.42) that close to the wall $(y/h \ll 1)$ the entire shear stress is virtually constant and so a layer exists there in which the effect of the pressure gradients can be neglected; the simple Eq. (7.37) is here applicable. This does not only hold for the channel and pipe flows already mentioned, but also for turbulent boundary layer flows. In all these flows, a layer close to the wall exists where the outer boundaries of the flow, e.g., the height of the channel or the thickness of the boundary layer, have no effect, and in which layer the flow is independent of these quantities. We recognize the consequences if we bring (7.37) to the form

$$\frac{\tau_w}{\varrho} = \nu \frac{d\bar{u}}{dy} - \overline{u' v'}, \tag{7.43}$$

from which we see directly that $\tau_w \varrho$ has the dimension of the square of a velocity. Thus we introduce the *friction velocity* as a reference velocity

$$u_* = \sqrt{\frac{\tau_w}{\varrho}}, \tag{7.44}$$

which is also physically significant since it provides a measure for the turbulent fluctuation velocity. Equation (7.43) may now be written in the form

$$1 = \frac{d(\bar{u}/u_*)}{d(y u_*/\nu)} - \frac{\overline{u' v'}}{u_*^2}. \tag{7.45}$$

Thus the mean velocity $\bar{u}$ when referred to u_* is only a function of the dimensionless coordinate $y u_*/\nu$, in which the *friction length* ν/u_* appears as reference length. (If h were also introduced as a reference length, $\bar{u}/u_*$ would have to depend additionally on the dimensionless quantity $h u_*/\nu$; in other words, the processes taking place in the layer near the wall would depend on the distance to the opposite wall.) From (7.45) we infer the so-called *law of the wall*

$$\frac{\bar{u}}{u_*} = f\left(y\frac{u_*}{\nu}\right), \tag{7.46}$$

and the corresponding equation

$$\frac{\overline{u'v'}}{u_*^2} = g\left(y\frac{u_*}{\nu}\right). \tag{7.47}$$

The law of the wall was formulated by Prandtl (1925) and is one of the most important results from turbulence theory. It is clear from what was said earlier that the functions $f(y u_*/\nu)$ and $g(y u_*/\nu)$ are universal functions, and so are the same for all turbulent flows. Equation (7.45) on its own is not enough to find the form of the universal function f, since this function contains the unknown Reynolds' stresses. As already noted many times, the Reynolds' stresses tend to zero directly at the wall, and close to the wall we can express the universal function f in a Taylor expansion about $y = 0$. For simplicity we introduce

$$y_* = y\frac{u_*}{\nu}, \tag{7.48}$$

and since $\bar{u}(0) = 0$ write

$$\frac{\bar{u}(y_*)}{u_*} = \left.\frac{\mathrm{d}(\bar{u}/u_*)}{\mathrm{d}y_*}\right|_0 y_* + \left.\frac{\mathrm{d}^2(\bar{u}/u_*)}{\mathrm{d}y_*^2}\right|_0 \frac{1}{2}y_*^2 + \cdots. \tag{7.49}$$

From (7.45) the first coefficient follows as

$$\left.\frac{\mathrm{d}(\bar{u}/u_*)}{\mathrm{d}y_*}\right|_0 = 1, \tag{7.50}$$

since $\overline{u'v'}\big|_0 = 0$. We find the other coefficients through repeated differentiation of the Eq. (7.45) and evaluation at $y = 0$ and get

$$\frac{d^2(\bar{u}/u_*)}{dy_*^2}\bigg|_0 = \frac{1}{u_*^2}\frac{d(\overline{u'v'})}{dy}\frac{dy}{dy_*} = \frac{1}{u_*^2}\frac{\nu}{u_*}\left[\overline{\frac{\partial u'}{\partial y}v'} + \overline{\frac{\partial v'}{\partial y}u'}\right]_0 = 0, \qquad (7.51)$$

where the zero arises on the right-hand side because u' and v' always vanish at $y = 0$. The third derivative follows as

$$\frac{d^3(\bar{u}/u_*)}{dy_*^3}\bigg|_0 = \frac{1}{u_*^2}\left(\frac{\nu}{u_*}\right)^2\left[\overline{\frac{\partial^2 u'}{\partial y^2}v'} + 2\overline{\frac{\partial u'}{\partial y}\frac{\partial v'}{\partial y}} + \overline{\frac{\partial^2 v'}{\partial y^2}u'}\right]_0 = 0. \qquad (7.52)$$

The first and the last terms in the brackets vanish because the fluctuation velocities at the wall are zero. Since then their derivatives in the x- and z-directions are also zero, it follows from the continuity Eq. (7.12) that $\partial v'/\partial y$ vanishes at the wall. If we differentiate (7.52) again using the same line of reasoning we find for the fourth derivative

$$\frac{d^4(\bar{u}/u_*)}{dy_*^4}\bigg|_0 = \frac{1}{u_*^2}\left(\frac{\nu}{u_*}\right)^3\left[3\overline{\frac{\partial u'}{\partial y}\frac{\partial^2 v'}{\partial y^2}}\right]_0. \qquad (7.53)$$

In order to evaluate this expression we would need to know the field of the fluctuation velocities. Since the expression does not have to vanish from pure kinematic reasons, we assume that it is in general nonzero. Thus we conclude from the Taylor expansion (7.49) that

$$\frac{\bar{u}}{u_*} = y_* + O(y_*^4). \qquad (7.54)$$

Therefore the fluctuating motion influences the velocity profile only in the terms of order y_*^4. Accordingly there exists a layer in which, although the fluctuating motion itself is not zero, the distribution of the average velocity is mainly influenced by the viscous shear stresses. Thus the name *viscous sublayer* is justified. For dimensional reasons the thickness of this layer must be of the order of magnitude of the friction length ν/u_*. Since there is no other typical length available, we set $\delta_v = \beta\nu/u_*$ for the thickness of the viscous sublayer, where β is a pure number to be determined by experiment.

Of course there is no sudden transition from the viscous layer to the region where the Reynolds' stress $-\varrho\,\overline{u'v'}$ is important. As the distance from the wall increases, the effect of the viscosity eventually completely disappears (as far as momentum transfer is concerned) and the velocity distribution is fully determined by the Reynolds' stresses.

It is this fact which allows us to state the universal function f in this region. Firstly it follows from (7.40) that the shear stress in this region is

$$\tau_w = \tau_t = -\varrho \, \overline{u' v'}, \tag{7.55}$$

where we write τ_t to express that we are only dealing with the turbulent shear stress. The velocity distribution cannot be calculated from (7.55) since there is no relation between the Reynolds' stress and the mean velocity. It is clear that we should find this relation using some turbulence model. One such model is the *Boussinesq formulation* of the Reynolds' stresses

$$-\varrho \, \overline{u' v'} = A \frac{\partial \bar{u}}{\partial y}, \tag{7.56}$$

where A is the *turbulent transport coefficient*, or $\nu_t = A/\varrho$ is the so-called *eddy viscosity*. Clearly the Boussinesq formulation follows the formulation of the viscous shear stress $\tau = \eta \, \partial \bar{u}/\partial y$ and only shifts the problem of the unknown Reynolds' stresses to the problem of the unknown eddy viscosity. The most simple assumption we can make is that A is constant; however we cannot do so in wall bounded flows since the Reynolds' stresses must vanish near the wall. For so-called "free turbulent shear flow", i.e., for turbulent flows not bounded by walls, as typically encountered in jets and *wakes*, the assumption of constant eddy viscosity can be useful.

Using the concept known as the "mixing length", Prandtl found a relationship between the eddy viscosity ν_t and the mean velocity field. The basic idea is that the turbulent stresses arise through macroscopic momentum exchange in the same manner as the viscous stresses arise from molecular exchange of momentum. Molecular momentum exchange occurs when a molecule at position y with a velocity u in the x-direction moves to position $y - l$ under thermal motion where its velocity is $u - du$. Therefore the molecule carries over the velocity difference $du = l \, du/dy$, where l is the distance in the y-direction between two molecular collisions. These motions proceed in both directions, and therefore momentum is carried over from the faster layer to the slower and vice versa. The number of molecules (per unit volume) moving parallel to the y-axis in the $\pm \, y$-direction is one third of the total number, and one third also moves parallel to the x and z axes, respectively. The molecules move with thermal velocity v and the mass flux per unit volume is thus $(1/3) \, \varrho v$. The molecular momentum flux which manifests itself as the viscous shear stress $\tau_{21} = \eta \, du/dy$ is therefore proportional to $(1/3) \, \varrho v \, du/dy$. Although in this extremely simplified derivation, where all molecules have the same thermal velocity v, they do not affect each other except during collisions and should only move parallel to the coordinate axes. This formula leads to a very good value for the viscosity ($\eta = (1/3) \, l \, \varrho \, v$) of dilute gases.

In carrying these ideas over to turbulent exchange motion it is assumed that *turbulent fluid parcels*, i.e., fluid masses which move more or less as a whole, behave like molecules, thus moving over the distance l unaffected by their surroundings, "mixing" with their new surroundings and so losing their identity. The fluctuation u' in the velocity is, from the above point of view, proportional to $l \, d\bar{u}/dy$. The "mixing" of two turbulent fluid parcels goes along with a displacement of the fluid, which gives rise to the transverse velocity v' whose magnitude is therefore proportional to $l\,d\bar{u}/dy$. (This is in contrast to the molecular momentum exchange where the thermal velocity is independent of du/dy.) Thus, if we absorb the proportionality factor into the unknown *mixing length* l, the turbulent shear stress τ_t becomes

$$\tau_t = -\varrho\,\overline{u'\,v'} = \varrho\,l^2\left(\frac{d\bar{u}}{dy}\right)^2. \tag{7.57}$$

The change of sign in (7.57) comes from the fact that a parcel of fluid coming from above (v' negative) generally carries a positive u' with it. If we also take account of the fact that the sign of τ_t is the same as that of $d\bar{u}/dy$, we can write down *Prandtl's mixing length formula*

$$\tau_t = \varrho\,l^2\left|\frac{d\bar{u}}{dy}\right|\frac{d\bar{u}}{dy}, \tag{7.58}$$

and the eddy viscosity

$$\nu_t = l^2\left|\frac{d\bar{u}}{dy}\right|, \tag{7.59}$$

which, from Prandtl's mixing length model, therefore depends on the velocity gradient.

At first sight the unknown eddy viscosity in (7.59) has only been replaced by the unknown mixing length. However the latter is more accessible to physical insight, so that "rational" assumptions for the mixing length are likely to be more easy to make. Yet no generally valid representation for the mixing length has been found so far. The mixing length concept is based on the unrealistic assumption that turbulent fluid parcels whose typical diameters are of the order of magnitude of the mixing length l, traverse this distance l without influence from the surrounding fluid. The mixing length formula serves only as a very rough description of shear turbulence. Like all algebraic turbulence models it has the disadvantage that the Reynolds' stresses only depend on the local mean velocity field, while in general the Reynolds' stresses depend on the history of the velocity field and require a formulation more in line with the constitutive relations of non-Newtonian, viscoelastic fluids.

Although there are typical experiments which clearly contradict the mixing length idea, it is certainly a model which is very useful and simple to apply, and the predictions of this model compare favorably to models which take the history of the velocity field into account. The model can also be tensorially generalized; we shall not go into this but shall now turn towards the application of the law of the wall.

Since the Reynolds' stresses must vanish at the wall, we choose l to be proportional to y

$$l = \kappa y. \tag{7.60}$$

This choice is also inspired by dimensional reasons, since close to the wall where the law of the wall holds, but outside the viscous sublayer, there is no typical length we can use and all physically relevant lengths must be proportional to y. Since the shear stress is constant and thus equal to the wall shear stress ($\tau = \text{const} = \tau_w = \varrho\, u_*^2$), we find the following relation from the mixing length formula

$$u_* = \kappa y \frac{d\bar{u}}{dy}, \tag{7.61}$$

whose integration leads to the desired universal velocity distribution, the so-called *logarithmic law of the wall*

$$\frac{\bar{u}}{u_*} = \frac{1}{\kappa}\ln y + C. \tag{7.62}$$

We may reach this important result purely from dimensional considerations without having to refer to Prandtl's mixing length formula. Since viscosity has no effect in the region in which we are interested, the fluid properties are only described by the density ϱ. In the relation between the constant shear stress and the velocity distribution $\bar{u} = f(y)$, aside from the density, only the change in $\bar{u}$ with y may appear, since the momentum flux connected with the shear stress is only present if the velocity changes in the y-direction. Thus $\bar{u}$ itself may not appear in the relation we are looking for. The only quantities which occur are therefore the shear stress $\tau_t = \tau_w$, ϱ and the derivatives $d^n\bar{u}/dy^n$ of the velocity distribution, among which the functional relation

$$f\left(\tau_t,\ \varrho,\ \frac{d^n\bar{u}}{dy^n}\right) = 0 \tag{7.63}$$

exists. This relation must be reducible to a relation between dimensionless quantities. If we assume that the first two derivatives $d\bar{u}/dy$ and $d^2\bar{u}/dy^2$ characterize the velocity distribution, we can form only one dimensionless quantity, namely

$$\Pi_1 = \frac{(d^2\bar{u}/dy^2)^2}{(d\bar{u}/dy)^4} \frac{\tau_t}{\varrho}. \tag{7.64}$$

Restricting ourselves to the first two derivatives, the relation we require is

$$f(\Pi_1) = 0, \quad \text{or} \quad \Pi_1 = \text{const.} \tag{7.65}$$

We denote the absolute constant Π_1 as κ^2 and find from (7.64)

$$\tau_t = \varrho\, \kappa^2 \left(\frac{d\bar{u}/dy}{d^2\bar{u}/dy^2}\right)^2 \left(\frac{d\bar{u}}{dy}\right)^2. \tag{7.66}$$

Comparing this with the mixing length formula (7.57), we deduce a formula for the mixing length from dimensional analysis

$$l = \left|\kappa \frac{d\bar{u}/dy}{d^2\bar{u}/dy^2}\right|. \tag{7.67}$$

We make no further use of this here. Using $\tau_t = \tau_w = \varrho\, u_*^2$ in (7.66) we obtain a differential equation for the distribution of the mean velocity

$$\frac{d^2\bar{u}}{dy^2} + \frac{\kappa}{u_*}\left(\frac{d\bar{u}}{dy}\right)^2 = 0, \tag{7.68}$$

where the sign of the second term has been chosen positive since the curvature of the velocity profile is negative for flow in the positive x-direction. The solution of (7.68) is the logarithmic law of the wall (7.62)

$$\frac{\bar{u}}{u_*} = \frac{1}{\kappa}\ln y + C.$$

This velocity profile does not hold for $y \to 0$, but only to the edge of a layer near the wall which we divide up into the already mentioned viscous sublayer and an *intermediate* or *buffer layer* where the Reynolds' stresses decrease as the wall is approached, while the viscous stresses increase. The velocity at the edge of this layer therefore depends on the viscosity. The constant in (7.62) serves to fit the velocity in the logarithmic part of the law of the wall to this velocity and so depends on the viscosity too. The second constant of integration which appears when we solve (7.68) is fixed so that $d\bar{u}/dy$ tends to infinity when y tends to zero. Now (7.62) does not hold in this region (as just stated), but taking the limit $y \to 0$ in (7.62) corresponds to a very thin viscous sublayer where $d\bar{u}/dy$ must become correspondingly large. We then set the constant C to

$$C = B + \frac{1}{\kappa}\ln\frac{u_*}{\nu} \tag{7.69}$$

and obtain (7.62) in the dimensionally homogeneous form of the logarithmic law of the wall

$$\frac{\bar{u}}{u_*} = \frac{1}{\kappa}\ln\left(y\frac{u_*}{\nu}\right) + B. \tag{7.70}$$

This important velocity distribution is met in every turbulent flow near a smooth wall, in channel and pipe flows, as well as in all turbulent boundary layer flows. Equation (7.70) is valid in a domain described by the inequality

$$\frac{\nu}{u_*} \ll y \ll \delta, \tag{7.71}$$

where δ stands for either the boundary layer thickness, or else half the channel height or the pipe radius. The constants κ and B are independent of the viscosity and therefore also independent of the Reynolds' number $(u_*\delta/\nu)$. They are absolute constants for flow bounded by a smooth wall and are found experimentally. Different measurements show a certain amount of scatter in these values, which is in part explained by the fact that fully turbulent flow was not realized in the experiments, or that the shear stresses were not constant because of very large pressure gradients (see (7.41)). In the region

$$30 \leq y\frac{u_*}{\nu} \leq 1000 \tag{7.72}$$

we find reasonably good agreement for $\kappa \approx 0.4$ and $B \approx 5$. (The values given in the literature for κ vary between 0.36 and 0.41, and for B between 4.4 and 5.85. In applications it is sufficient to round κ off to 0.4 ($1/\kappa = 2.5$) and B to 5.)

From measurements, the entire law of the wall may be divided into three regions, where of course there is no sudden transition from one region to the next:

viscous sublayer (linear region)	$0 < y\,u_*/\nu < 5$
buffer layer	$5 < y\,u_*/\nu < 30$
logarithmic layer	$y\,u_*/\nu > 30$

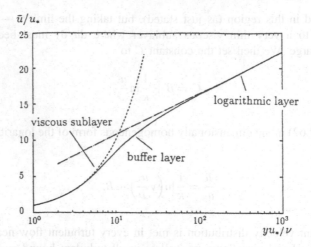

Fig. 7.1 Universal velocity distribution on the logarithmic scale

The velocity profile in the viscous sublayer and the logarithmic layer are sketched in Fig. 7.1 on the logarithmic scale. Figure 7.2 shows the same profiles on the linear scale.

A series of analytic expressions has been given for the buffer layer, which have the character of interpolation formulae between the linear and logarithmic laws, but there are also closed form expressions which describe the entire wall region. We shall not go any further into these, because the resistance laws which we will discuss shortly do not require the exact distribution of the mean velocity in the buffer layer. The exact distribution can however be important in heat transfer problems.

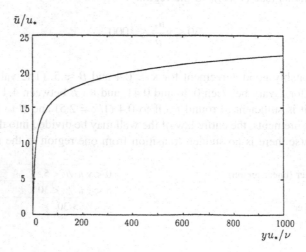

Fig. 7.2 Universal velocity distribution on the linear scale

7.4 Turbulent Flow in Smooth Pipes and Channels

In the last section it was shown that the universal law of the wall holds for all turbulent flows, but is restricted to a distance which is small in comparison to half of the channel height h or to the pipe radius. At large distances from the wall the effect of the opposite wall becomes noticeable, and as already mentioned, the velocity distribution then also depends on the Reynolds' number $u_* R / \nu$, where R stands for one of the above typical lengths, so that the corresponding distribution for the law of the wall takes on the form

$$\frac{\bar{u}}{u_*} = F \left(u_* \frac{R}{\nu}, u_* \frac{y}{\nu} \right), \tag{7.73}$$

where y is measured from the wall and so in a circular cross-section we have $y = R - r$. If, for fixed u_* / ν, we consider the limit $u_* R / \nu \to \infty$, then with $u_* R / \nu$, R itself vanishes from the relation (7.73), and we return to the law of the wall. Now if we take the limit $u_* R / \nu \to \infty$ for fixed R it also means that y would no longer appear in the relation. In order not to lose relevant information in these limits, we form the entirely equivalent form

$$\frac{\bar{u}}{u_*} = F \left(u_* \frac{R}{\nu}, \frac{y}{R} \right). \tag{7.74}$$

Now taking the limit $u_* / \nu \to \infty$ at fixed R, the effect of the Reynolds' number vanishes because $u_* R / \nu \to \infty$ and therefore so does the effect of the viscosity on the distribution of the mean velocity outside the wall region

$$\frac{\bar{u}}{u_*} = F \left(\frac{y}{R} \right). \tag{7.75}$$

In this equation the viscosity appears only indirectly through u_*, i.e., through the shear stress on the wall and through the condition that (7.75) must be fitted to the value of the velocity which is given by the law of the wall. In order to find the unknown function F the same considerations hold as those which led to (7.70), except that the shear stress τ_t now depends on y. Instead of following the reasoning which led to the logarithmic law of the wall, we shall determine the function $F(y/R)$ so that it agrees with the law of the wall $f(u_* y / \nu)$ in that region where both distributions must coincide with one another, i.e., for $y/R \ll 1$ and simultaneously $u_* y / \nu \gg 1$. Since the magnitude of the velocity, in contrast to the velocity distribution, is directly dependent on the Reynolds' number we require that the derivatives of the velocity distributions agree in the *overlap region* $y_* \gg 1$, $y/R \ll 1$

$$\frac{d\bar{u}}{dy} = \frac{u_*}{R}\frac{dF}{d\eta} = \frac{u_*^2}{\nu}\frac{df}{dy_*}, \tag{7.76}$$

where $y_* = yu_*/\nu$ and $\eta = y/R$. These variables are entirely independent of each other, since changing R will for example leave y_* unaffected. Multiplying by y/u_* leads to

$$\eta\frac{dF}{d\eta} = y_*\frac{df}{dy_*} = const, \tag{7.77}$$

since the equation can only hold if both sides are equal to a constant. By integrating (7.77) as before, we obtain in a completely different manner the logarithmic law of the wall (7.70)

$$f = \frac{\bar{u}}{u_*} = \frac{1}{\kappa}\ln\frac{yu_*}{\nu} + B,$$

and equally find a logarithmic law for the region where the influence of R is felt

$$F = \frac{\bar{u}}{u_*} = \frac{1}{\kappa}\ln\frac{y}{R} + const. \tag{7.78}$$

If in (7.78) we set $y = R$, with $\bar{u}(R) = U_{max}$ it follows that

$$\frac{\bar{u} - U_{max}}{u_*} = \frac{1}{\kappa}\ln\frac{y}{R}, \tag{7.79}$$

where we note that (7.78) is really being applied outside the region in which it is valid, which from the derivation is restricted to $y/R \ll 1$. The more general form of (7.79)

$$\frac{\bar{u} - U_{max}}{u_*} = f\left(\frac{y}{R}\right) \tag{7.80}$$

is known as the *velocity defect law*. By subtracting (7.79) from the law of the wall, we acquire the expression

$$\frac{U_{max}}{u_*} = \frac{1}{\kappa}\ln\left(u_*\frac{R}{\nu}\right) + B, \tag{7.81}$$

which shows explicitly how the maximum velocity depends on the Reynolds' number u_*R/ν. For given U_{max} and R, (7.81) is an implicit function for u_* or for the shear stress and therefore for the pressure gradient $K(K = -\partial p/\partial x)$. Therefore (7.81) is already a resistance law. We express it in the form (6.60) and use the velocity averaged over the cross-section of the pipe, denoted by $\bar{U}$, as a reference velocity

$$\pi R^2 \bar{U} = 2\pi \int_0^R \bar{u}(R - y)dy. \tag{7.82}$$

Using the distribution of the mean velocity $\bar{u}$ given by (7.79) (which already represents a good description of the whole velocity distribution over the entire pipe cross-section), we find

$$\bar{U} = U_{max} - 3.75u_*, \tag{7.83}$$

and thus with (7.81)

$$\frac{\bar{U}}{u_*} = \frac{1}{\kappa} \ln \frac{u_* R}{\nu} + B - 3.75, \tag{7.84}$$

which relates the velocity $\bar{U}$ and the wall shear stress. With

$$\tau_w = \varrho\, u_*^2 = \frac{K R}{2} \tag{7.85}$$

we find

$$\zeta = \frac{l}{d}\lambda = \frac{p_1 - p_2}{\bar{U}^2 \varrho/2} = \frac{K l}{\bar{U}^2 \varrho/2} = 4\frac{u_*^2 l}{\bar{U}^2 R}, \tag{7.86}$$

or

$$\lambda = 8\frac{u_*^2}{\bar{U}^2}, \tag{7.87}$$

where $d = 2R$. Using these we write the Eq. (7.84) in the form

$$2\sqrt{\frac{2}{\lambda}} = \frac{1}{\kappa} \ln\left(\frac{1}{4}\frac{\bar{U} d}{\nu}\sqrt{\frac{\lambda}{2}}\right) + B - 3.75. \tag{7.88}$$

If instead of the natural logarithm we use the logarithm to the base ten (Brigg's logarithm), we finally obtain

$$\frac{1}{\sqrt{\lambda}} = 2.03 \lg\left(Re\sqrt{\lambda}\right) - 0.8 \tag{7.89}$$

with the Reynolds' number $Re = \bar{U}d/\nu$. The constant -0.8 does not exactly correspond to the calculated value $-\left[\ln(4\sqrt{2})/\kappa - B + 3.75\right]/(2\sqrt{2})$, but is adjusted to fit experimental results.

We can easily see that for a plane channel of height $2\,h$ we reach the same form as in (7.89), by writing the Reynolds' number using the hydraulic diameter introduced in (6.67) (here $d_h = 4\,h$). However the constant in the relation corresponding to (7.83) has a somewhat different value

$$\bar{U} = U_{max} - 2.5u_* \tag{7.90}$$

Experiments show that the formula for the circular pipe also describes the resistance for noncircular cross-sections if the Reynolds' number is formed with the hydraulic diameter, as we have already remarked. In reality only the turbulent flow in circular pipes and in annular conduits is unidirectional. Contrary to the laminar flow of Newtonian fluids, a fully formed turbulent flow through a pipe with general cross-sectional shape is no longer unidirectional. It forms a *secondary flow* with a velocity component perpendicular to the axial direction.

This secondary flow transports momentum into the "corners" (Fig. 7.3), which also gives rise to large velocities there. The result is that the shear stress along the

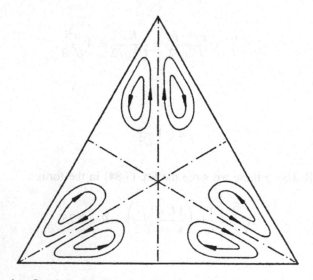

Fig. 7.3 Secondary flow in a pipe with triangular cross-section

entire wetted surface is almost constant, which is to be considered as the assumption needed to apply the concept of hydraulic diameter. Therefore we do not expect that the formula for the circular pipe still applies if, for example, the angles of a triangular cross-section are too small to allow an effective secondary flow.

7.5 Turbulent Flow in Rough Pipes

Pipes used in applications are always more or less "rough". While in laminar flow the wall roughness hardly affects the resistance, its influence in the turbulent case is quite considerable if the mean *protrusion height* k is greater than the thickness of the viscous sublayer. (Here we assume that the roughness is fully characterized by k or k/R, as is the case in closely spaced protrusions.) An essential parameter is the ratio of the protrusion height k to the viscous length ν/u_*. If the protrusion height lies in the linear region of the velocity profile, that is

$$u_* \frac{k}{\nu} \leq 5, \qquad (7.91)$$

the effect on the resistance is negligible; in this case we speak of *hydraulically smooth* surfaces. If the protrusion height is considerably greater than the thickness of the buffer region we speak of a *dynamically completely rough* surface, which is characterized by the inequality

$$u_* \frac{k}{\nu} \geq 70. \qquad (7.92)$$

Experiments show that the viscosity, that is the Reynolds' number, then no longer has an effect on the friction factor. As we have seen in connection with (7.69) and (7.70), the friction influence appears in the logarithmic law only through the constant of integration C. In the case of a completely rough wall the viscous sublayer no longer exists. The constant C is then to be fixed so that a dimensionally homogeneous form of the velocity distribution arises in which the viscosity no longer appears. Therefore we set

$$C = B - \frac{1}{\kappa} \ln k \qquad (7.93)$$

and are led to the logarithmic law of the wall for completely rough walls

$$\frac{\bar{u}}{u_*} = \frac{1}{\kappa} \ln \frac{y}{k} + B. \qquad (7.94)$$

From measurements the constant B is found to be

$$B = 8.5. \tag{7.95}$$

The velocity defect law (7.79) is not affected by the wall roughness; this equation continues to be valid, as is Eq. (7.83). Thus we find the equation corresponding to (7.84) is

$$\frac{\bar{U}}{u_*} = \frac{1}{\kappa}\ln\frac{R}{k} + 8.5 - 3.75. \tag{7.96}$$

Using (7.87) we obtain the *resistance law* of the completely rough pipe as

$$\lambda = 8\left(2.5\ln\frac{R}{k} + 4.75\right)^{-2}, \tag{7.97}$$

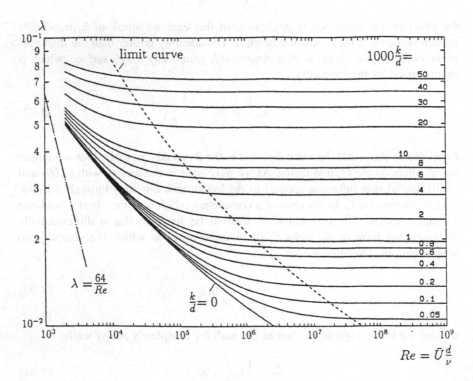

Fig. 7.4 Friction factor for circular pipes

or using the logarithm to the base ten as before

$$\lambda = \left(2 \lg \frac{R}{k} + 1.74\right)^{-2}, \tag{7.98}$$

where the constant found from calculations (1.68) is replaced with the value 1.74, which is in better agreement with the experiments.

Finally we note Colebrook's formula

$$\frac{1}{\sqrt{\lambda}} = 1.74 - 2 \lg \left(\frac{k}{R} + \frac{18.7}{Re\sqrt{\lambda}}\right), \tag{7.99}$$

which interpolates the entire spectrum from "hydraulically smooth" to "completely rough" very well. We see that for $Re \rightarrow \infty$ (vanishing viscosity effects) this becomes Eq. (7.98), and for $k/R \rightarrow 0$ the resistance formula for the smooth pipe appears. For practical purposes Colebrook's formula is graphed in Fig. 7.4.

With increasing Reynolds' number the viscous length ν/u_* and therefore the protrusion height become continuously smaller from where the pipe may be considered as being completely rough (i.e. (7.89) is valid). If we set (7.92) with the equality sign into (7.98), and then using (7.87) to replace u_* with $\bar{U}(\lambda/8)^{1/2}$, we obtain the limit curve $\lambda_l = f(Re)$ (dashed line in Fig. 7.4).

Chapter 8
Hydrodynamic Lubrication

8.1 Reynolds' Equation of Lubrication Theory

The geometric characteristics of the unidirectional flows discussed in Chap. 6 are their infinite extension in the flow direction and the fact that the flow cross-section does not change in the flow direction. Because of these kinematic restrictions the nonlinear terms in the equations of motion vanish, simplifying the mathematical treatment considerably. Now unidirectional flows do not really occur in nature, but they are suitable models for the flows often met in applications whose extension in the flow direction is much larger than their lateral extension. Frequently the cross-section is not constant, but varies, even if only weakly, in the flow direction. As well as the channel and pipe flows with slowly varying cross-section, a typical example is the flow in a *journal bearing* (Fig. 6.3), where a flow channel with slightly varying cross-section is formed due to the displacement of the journal.

We now search for a criterion to neglect the convective terms in the Navier–Stokes equations and consider the lubrication gap shown in Fig. 8.1. This arises from the flow channel of a simple shearing flow if the upper wall is inclined to the *x*-axis at an angle α. Since the fluid adheres to the wall it is pulled into the narrowing gap so that a pressure builds up in the gap; this pressure is quite substantial for $h/L \ll 1$ and can, for example, support a load which acts on the upper wall.

Further basic considerations regarding neglecting the convective terms are based on plane two-dimensional and steady flow. On the lower wall the normal velocity component (here the *y* component) vanishes as a consequence of the kinematic boundary condition. Exactly the same holds at the upper wall; because of the no slip condition the component of velocity in the *y*-direction is $v = -\alpha U$ and is at most $O(\alpha U)$ anywhere in the fluid film. Then, from the continuity equation

$$\frac{\partial u}{\partial x} + \frac{\partial v}{\partial y} = 0$$

© Springer Nature Switzerland AG 2020
J. H. Spurk and N. Aksel, *Fluid Mechanics*,
https://doi.org/10.1007/978-3-030-30259-7_8

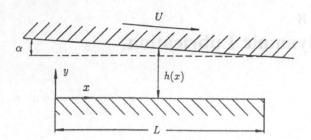

Fig. 8.1 Lubrication gap

we find the following estimate for plane two-dimensional incompressible flow

$$\frac{\partial u}{\partial x} \sim \alpha \frac{U}{\bar{h}},$$ (8.1)

so that the first component of the Navier–Stokes Eq. (4.9a), ignoring the volume body force, leads to the order of magnitude equation

$$\varrho \left(\alpha \frac{U^2}{\bar{h}} + \alpha \frac{U^2}{\bar{h}} \right) \sim -\frac{\partial p}{\partial x} + \eta \left(\alpha^2 \frac{U}{\bar{h}^2} + \frac{U}{\bar{h}^2} \right).$$ (8.2)

Here $\bar{h}$ is a mean distance between the upper and lower walls, which in bearings is typically of the order

$$\bar{h} \sim \alpha L.$$ (8.3)

For $\alpha \ll 1$ we ignore the first term in brackets on the right-hand side and for the ratio of the convective terms to the remaining friction term we obtain the expression

$$\frac{\varrho \, \alpha U^2 / \bar{h}}{\eta U / \bar{h}^2} = \alpha \, Re,$$ (8.4)

where

$$Re = \frac{\varrho \, U \bar{h}}{\eta}$$ (8.5)

is the Reynolds' number formed with mean wall distance and wall velocity. Consequently we can ignore the convective terms, and in steady flow therefore all the inertia terms, if

$$\alpha\, Re \ll 1 \tag{8.6}$$

holds. We emphasize that a small Reynolds' number is sufficient but not necessary for (8.6) to be true. In reality, such high Reynolds' numbers can be reached in bearings that the flow becomes turbulent. However in this chapter we shall restrict ourselves to laminar flow. The criterion (8.6) is also valid for unsteady flows if the typical time τ is of the order L/U or $\bar{h}/(\alpha U)$, since then the local acceleration is of the same order of magnitude as the convective.

Under the condition (8.6), the terms in (8.2) which do not contain α must all balance, and the x component of the Navier–Stokes equations reduces to

$$\frac{\partial p}{\partial x} = \eta \frac{\partial^2 u}{\partial y^2}. \tag{8.7}$$

Using (8.1) and since $v \sim \alpha U$, the component of the Navier–Stokes equations in the y-direction leads to the order of magnitude equation

$$\varrho\left(\alpha^2 \frac{U^2}{\bar{h}} + \alpha^2 \frac{U^2}{\bar{h}}\right) \sim -\frac{\partial p}{\partial y} + \eta\left(\alpha^3 \frac{U}{\bar{h}^2} + \alpha \frac{U}{\bar{h}^2}\right), \tag{8.8}$$

from which we infer the equation

$$0 = \frac{\partial p}{\partial y}. \tag{8.9}$$

However (8.7) and (8.9) correspond exactly to the differential equations of the Couette-Poiseuille flow (6.13) and (6.14). Therefore we can immediately state the solution (where, because $\alpha \ll 1$, the x component of the wall velocity is equal to U)

$$\frac{u}{U} = \frac{y}{h(x)} - \frac{\partial p}{\partial x}\frac{h^2(x)}{2\eta U}\left(1 - \frac{y}{h(x)}\right)\frac{y}{h(x)}. \tag{8.10}$$

Since the gap height h depends on the coordinate x, the flow is only "locally" a Couette-Poiseuille flow.

We now calculate the volume flux in the x-direction per unit depth (that is, per unit length in the z-direction)

$$\dot{V}_x = \int_0^{h(x)} \vec{u}\cdot\vec{e}_x\,\mathrm{d}y = \int_0^{h(x)} u\,\mathrm{d}y, \tag{8.11}$$

which must be independent of x for the plane two-dimensional flow considered here. From (8.11) it follows as in the case of channel flow that

$$\dot{V}_x = \frac{1}{2} U h(x) - \frac{\partial p}{\partial x} \frac{h^3(x)}{12\eta}, \tag{8.12}$$

and differentiating with respect to x leads to a differential equation for the pressure distribution in the fluid film

$$\frac{\partial}{\partial x} \left(\frac{h^3}{\eta} \frac{\partial p}{\partial x} \right) = 6U \frac{\partial h}{\partial x}. \tag{8.13}$$

This equation is the two-dimensional form of a general equation which we shall now develop, and which is called the *Reynolds' equation*, but is obviously not to be confused with Eq. (7.16) of the same name.

If there also exists a flow in the z-direction, as well as (8.7) we have the equation

$$\frac{\partial p}{\partial z} = \eta \frac{\partial^2 w}{\partial y^2}, \tag{8.14}$$

which has the same form as (8.7). In order to calculate the volume flux per unit depth $\dot{V}_z$ in the z-direction, it suffices to replace $\partial p / \partial x$ by $\partial p / \partial z$ in (8.12), and to write the wall velocity in the z-direction (W) thus

$$\dot{V}_z = \frac{1}{2} W h(x, z) - \frac{\partial p}{\partial z} \frac{h^3(x, z)}{12\eta}, \tag{8.15}$$

where we have allowed for the fact that the gap height may also depend on z.

In the general case we also allow h to depend on z as well as x in (8.12). We combine both the volume fluxes $\dot{V}_x$ and $\dot{V}_z$ vectorially as

$$\vec{V} = \dot{V}_x \vec{e}_x + \dot{V}_z \vec{e}_z. \tag{8.16}$$

Now this plane two-dimensional field must satisfy the continuity equation

$$\frac{\partial \dot{V}_x}{\partial x} + \frac{\partial \dot{V}_z}{\partial z} = 0, \tag{8.17}$$

a result which is easily seen if we apply the continuity equation in the integral form (2.8) to a cylindrical control volume of base area $dx\,dz$. Using (8.12) and (8.15) the Reynolds' equation arises directly from Eq. (8.17).

$$\frac{\partial}{\partial x}\left(\frac{h^3}{\eta}\frac{\partial p}{\partial x}\right) + \frac{\partial}{\partial z}\left(\frac{h^3}{\eta}\frac{\partial p}{\partial z}\right) = 6\left(\frac{\partial(hU)}{\partial x} + \frac{\partial(hW)}{\partial z}\right). \tag{8.18}$$

If the plates are rigid bodies the derivatives $\partial U/\partial x$ and $\partial W/\partial z$ on the right-hand side vanish. Further the plate velocity W in the z-direction is often zero.

8.2 Statically Loaded Bearing

8.2.1 Infinitely Long Journal Bearing

To discuss the journal bearing extending to infinity in the z-direction in Fig. 8.2, we use (8.12). The radius of the bearing shell is

$$R_S = R + \bar{h} = R\left(1 + \frac{\bar{h}}{R}\right), \tag{8.19}$$

if $\bar{h}$ is the average height of the lubrication gap (radial clearance) and R is the radius of the journal. Typical values of the *relative bearing clearance*

$$\psi = \frac{R_S - R}{R} = \frac{\bar{h}}{R} \tag{8.20}$$

lie in the region of 10^{-3}. If the center of the journal is displaced by distance e on the line $\varphi = 0$, the distance to the surface of the journal measured from the center of the bearing shell is

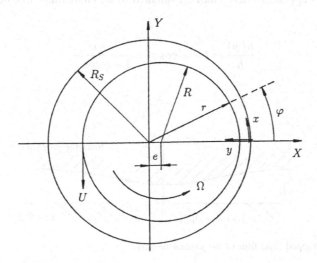

Fig. 8.2 Geometry of the journal bearing

$$r = R + e \cos \varphi = R\left(1 + \frac{e}{R}\cos \varphi\right) \tag{8.21}$$

for $e/R \ll 1$, and because $\psi \ll 1$, we have for the distance between the surface of
the journal and the surface of the bearing shell

$$h(\varphi) = R_S - r = \bar{h}(1 - \epsilon \cos \varphi), \tag{8.22}$$

where

$$\epsilon = \frac{e}{\bar{h}} \tag{8.23}$$

is the *relative eccentricity*. Since ψ is very small the fact that the lubrication gap or
the fluid film is curved is not important; let us consider the fluid film to be
unwrapped (Fig. 8.3) and set $dx = R d\varphi$.

With the notation we have introduced, we write (8.12) in the form

$$\frac{\dot{V}_x}{h^3(\varphi)} = \frac{\Omega R}{2h^2(\varphi)} - \frac{1}{12\eta R}\frac{\partial p}{\partial \varphi}. \tag{8.24}$$

We integrate this equation from 0 to 2π and because of course $p(0) = p(2\pi)$, we find
the (constant) volume flux $\dot{V}_x$ to be

$$\dot{V}_x = \frac{\Omega R \bar{h}}{2} \int_0^{2\pi} \left(\frac{\bar{h}}{h(\varphi)}\right)^2 d\varphi \left(\int_0^{2\pi} \left(\frac{\bar{h}}{h(\varphi)}\right)^3 d\varphi\right)^{-1}. \tag{8.25}$$

The integrals appearing here can be evaluated in an elementary manner using the
substitution

$$\frac{h(\varphi)}{\bar{h}} = 1 - \epsilon \cos \varphi = \frac{1 - \epsilon^2}{1 + \epsilon \cos \chi}, \tag{8.26}$$

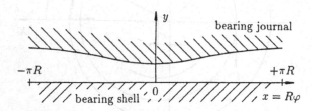

Fig. 8.3 Unwrapped fluid film of the journal bearing

but for the time being we shall abbreviate them as I_2 and I_3

$$\dot{V}_x = \frac{1}{2}\Omega R^2 \psi \frac{I_2}{I_3}. \tag{8.27}$$

Thus the pressure gradient follows from (8.12) as

$$\frac{\partial p}{\partial x} = \frac{1}{R}\frac{\partial p}{\partial \varphi} = 6\frac{\eta \Omega R}{h^2(\varphi)}\left(1 - \frac{\bar{h}}{h(\varphi)}\frac{I_2}{I_3}\right). \tag{8.28}$$

Of particular technical interest is the force exerted on the journal by the fluid, or the "load-bearing capacity" of the bearing, which corresponds to the negative of this force per unit length

$$\vec{F} = \int_0^{2\pi} \vec{t}R\,d\varphi, \tag{8.29}$$

where $\vec{t}$ is the stress vector with the components t_X and t_Y in the X-Y coordinate system of Fig. 8.2. In order to calculate this stress vector we first consider the component of the stress tensor in the x-y system of the lubrication gap. Here the flow is locally a unidirectional flow and therefore the stress tensor has the components $\tau_{xx} = \tau_{yy} = -p$ and $\tau_{xy} = \tau_{yx} = \eta\partial u/\partial y$, and because of (8.28), we write the order of magnitude equation for their ratio

$$\frac{\tau_{xx}}{\tau_{xy}} \sim \frac{\eta\,UR/\bar{h}^2}{\eta U/\bar{h}} = \frac{R}{\bar{h}} = \frac{1}{\psi}. \tag{8.30}$$

Therefore it is sufficient to consider only the normal stress $-p$, and so the stress vector at the journal has the form $\vec{t} = -p\vec{n}$, where $\vec{n}$ has the components $n_X = \cos\varphi$ and $n_Y = \sin\varphi$. Thus we have

$$F_X = -\int_0^{2\pi} p\,\cos\varphi\,R\,d\varphi \tag{8.31}$$

and

$$F_Y = -\int_0^{2\pi} p\,\sin\varphi\,R\,d\varphi. \tag{8.32}$$

Since $\cos\varphi$ is an even function, so too are $h(\varphi)$ and all powers of $h(\varphi)$. From (8.28), $\partial p/\partial\varphi$ is then also an even function, and the pressure itself must be an odd function of φ. The X component of the force then vanishes. The Y component balances the load acting on the journal and the journal shifts perpendicular to the direction in which the force acts. Partial integration of Eq. (8.32) leads to

$$F_Y = Rp \cos \varphi \big|_0^{2\pi} - R \int_0^{2\pi} \frac{\partial p}{\partial \varphi} \cos \varphi \, d\varphi. \tag{8.33}$$

The first term on the right-hand side vanishes, and we obtain

$$F_Y = -6 \frac{\eta \Omega R}{\psi^2} \int_0^{2\pi} \left(\left(\frac{\bar{h}}{h(\varphi)} \right)^2 - \frac{I_2}{I_3} \left(\frac{\bar{h}}{h(\varphi)} \right)^3 \right) \cos \varphi \, d\varphi. \tag{8.34}$$

We denote the first part of the integral with I_4 and the second with I_5 and bring (8.34) to the form

$$So = F_Y \frac{\psi^2}{\eta \, \Omega R} = 6 \frac{I_2 I_5 - I_3 I_4}{I_3}. \tag{8.35}$$

On the left is now a dimensionless force which is called the *Sommerfeld number*. Often $2R$ is used in the definition instead of R and so $So = F_Y \psi^2 / (2\eta \Omega R)$; using this definition, the Sommerfeld number S frequently found in American literature is given by the relation $S = 1/(2\pi So)$.

Finally we shall consider the friction torque exerted on the journal through the shear stress. Let us take the shear stress from (8.10) as

$$\tau_{xy} = \eta \frac{\partial u}{\partial y}\bigg|_h = \eta \, U \left(\frac{1}{h} + \frac{\partial p}{\partial x} \frac{h}{2\eta \, U} \right), \tag{8.36}$$

and then with (8.28)

$$\tau_{xy} = \eta \frac{\Omega}{\psi} \left(4 \frac{\bar{h}}{h} - 3 \frac{I_2}{I_3} \frac{\bar{h}^2}{h^2} \right), \tag{8.37}$$

and using this the friction torque becomes

$$T = R^2 \int_0^{2\pi} \tau_{xy} d\varphi = \frac{\eta \Omega R^2}{\psi} \left(4I_1 - 3 \frac{I_2^2}{I_3} \right), \tag{8.38}$$

that is

$$T \frac{\psi}{\eta \Omega R^2} = \frac{4I_1 I_3 - 3I_2^2}{I_3}. \tag{8.39}$$

Here we state the integrals used

$$I_1 = \int_0^{2\pi} (1 - \epsilon \cos\varphi)^{-1} d\varphi = \frac{2\pi}{(1 - \epsilon^2)^{1/2}};$$ (8.40)

$$I_2 = \int_0^{2\pi} (1 - \epsilon \cos\varphi)^{-2} d\varphi = \frac{2\pi}{(1 - \epsilon^2)^{3/2}};$$ (8.41)

$$I_3 = \int_0^{2\pi} (1 - \epsilon \cos\varphi)^{-3} d\varphi = \frac{\pi(2 + \epsilon^2)}{(1 - \epsilon^2)^{5/2}};$$ (8.42)

$$I_4 = \int_0^{2\pi} \cos\varphi (1 - \epsilon \cos\varphi)^{-2} d\varphi = \frac{I_2 - I_1}{\epsilon}; \quad \text{and}$$ (8.43)

$$I_5 = \int_0^{2\pi} \cos\varphi (1 - \epsilon \cos\varphi)^{-3} d\varphi = \frac{I_3 - I_2}{\epsilon}.$$ (8.44)

Equations (8.27), (8.35) and (8.39) can now also be expressed explicitly as functions of the relative eccentricity ϵ

$$\dot{V}_x (\Omega R^2 \psi)^{-1} = \frac{1 - \epsilon^2}{2 + \epsilon^2},$$ (8.45)

$$So = F_Y \psi^2 (\eta \Omega R)^{-1} = \frac{12\pi\epsilon}{\sqrt{1 - \epsilon^2}(2 + \epsilon^2)}, \quad \text{and}$$ (8.46)

$$T\psi (\eta \Omega R^2)^{-1} = \frac{4\pi(1 + 2\epsilon^2)}{\sqrt{1 - \epsilon^2}(2 + \epsilon^2)}.$$ (8.47)

We conclude from (8.46) that the eccentricity becomes very small ($\epsilon \to 0$) either if the bearing is lightly loaded or else if the number of revolutions ($\sim \Omega$) is very large, and we then speak of "fast running bearings". In the limit $\epsilon \to 0$ we have for the friction torque

$$T = 2\pi \eta \Omega \frac{R^2}{\psi};$$ (8.48)

a result already obtained and known as *Petroff's formula* (cf. Sect. 6.1.2). We further infer from these equations that if the viscosity η becomes smaller, both the load-bearing capacity and the friction torque decrease.

The pressure distribution can be determined from (8.28) with the substitution (8.26), and we finally obtain

$$p = C - 6\frac{\eta\Omega}{\psi^2}\epsilon\frac{\sin\varphi(2 - \epsilon\cos\varphi)}{(2+\epsilon^2)(1 - \epsilon\cos\varphi)^2}\,. \tag{8.49}$$

Since up until now we have only applied boundary conditions to the velocity, the pressure can only be determined up to a constant (as is always the case in incompressible flow). This constant can be ascertained physically if the pressure p_A is given at some position $\varphi = \varphi_A$ (usually $\varphi_A = \pi$) using an axial oil groove, in which the pressure is maintained, for example, by an oil pump pressure, or into which oil is introduced at ambient pressure.

If this pressure is too low, the pressure in the bearing becomes theoretically negative (Fig. 8.4). However fluids in thermodynamic equilibrium cannot maintain negative pressure. The fluid begins to vaporize if the pressure drops below the *vapor pressure* $p_v(T)$, and we say that the fluid "cavitates", i.e., bubbles filled with vapor (or air) form. (Of course this does not only occur in bearings, but whenever the pressure in a liquid drops below the vapor pressure.) The ensuing *two-phase-flow* is so difficult that a solution for this *cavitation region* is still unknown. It is seen experimentally that when the cavitation limit is reached, the fluid film ruptures and "fluid filaments" form, while the hollow space is filled with vapor, or with air which was either dissolved in the lubricant or has penetrated if the ends of the bearing are exposed to the atmosphere. In any case it is to be stressed that the fluid no longer fills up the diverging region of the gap and therefore the continuity

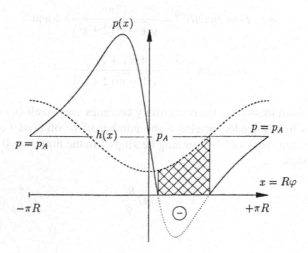

Fig. 8.4 Pressure distribution in the fluid film

equation in the form (8.17) is not satisfied in this region. It follows that all our conclusions which were based on the continuity equation are no longer valid, in particular the asymmetry of the pressure distribution and the vanishing of the X-component of the force. As experiments show, the pressure in this region is essentially constant, and for bearings which are "ventilated" it is equal to the atmospheric pressure. Because of this experimental fact it is recommended to set the pressure equal to the ambient pressure, that is, to set the pressure difference to the surroundings to zero. (It is only this pressure difference that determines the load bearing capacity.) However the extent of the cavitation region is still unknown and has to be calculated simultaneously with the pressure distribution.

We are led to a somewhat simpler problem if we lay down the so-called *Reynolds' boundary condition*. To do this we assume that the cavitation region always ends at the widest part of the lubricating film ($\varphi = \pi$), so that the pressure build up begins there and the appropriate boundary condition reads

$$p(\pi) = 0. \tag{8.50}$$

In general we shall only satisfy this boundary condition if pressureless oil (i.e. at ambient pressure) is supplied to this position by an oil groove. The end of the pressure distribution, and therefore the start of the cavitation region, is determined by simultaneously satisfying the two boundary conditions

$$p(\varphi_E) = 0, \quad \left.\frac{dp}{d\varphi}\right|_{\varphi_E} = 0. \tag{8.51}$$

Measured pressure distributions agree well with calculations based on these boundary conditions, which also show that the position of the start of the pressure is not critical.

8.2.2 Infinitely Short Journal Bearing

The other limiting case of considerable interest is the infinitely short journal bearing, whose width B is much smaller that the diameter of the journal. In this bearing the volume flux as a consequence of the pressure gradient in the x-direction can be neglected but not the Couette flow as a consequence of the wall velocity. This means that the term $\partial p/\partial x$ in (8.18) drops out and integrating over z leads us to the equation

$$p = 6\frac{\eta\Omega R}{h^3}\frac{dh}{dx}\frac{z^2}{2} + C_1 z + C_2 \,. \tag{8.52}$$

The constants of integration are determined from the boundary condition

$$p\left(z = +\frac{B}{2}\right) = p\left(z = -\frac{B}{2}\right) = 0 \,, \tag{8.53}$$

so that the pressure distribution becomes

$$p = -3\frac{\eta\Omega}{R^2\psi^2}\left(\frac{B^2}{4} - z^2\right)\frac{\epsilon\sin\varphi}{(1 - \epsilon\cos\varphi)^3} \,. \tag{8.54}$$

Again this is antisymmetric, and is negative for $0 < \varphi < \pi$. In practice the pressure in this region is often set to zero. This step of eliminating the negative pressure, called the *half Sommerfeld boundary condition* is occasionally applied to infinitely long journal bearings, but there it leads to results which do not agree with experimental results as well as the theoretical results based on the Reynolds' boundary conditions.

If we also integrate over the width of the bearing, we use the substitution (8.26) to determine the force components explicitly from (8.31) and (8.32)

$$F_X = -\frac{\eta\Omega B^3}{\psi^2 R}\frac{\epsilon^2}{(1 - \epsilon^2)^2} \,, \tag{8.55}$$

$$F_Y = \frac{\eta\Omega B^3}{4\psi^2 R}\frac{\pi\epsilon}{(1 - \epsilon^2)^{3/2}} \,. \tag{8.56}$$

8.2.3 Journal Bearing of Finite Length

It is worth noting that an analytic solution can be found for a finite journal bearing, based on the Sommerfeld boundary condition, but it leads to an antisymmetric pressure distribution with negative pressures, which are not realized in the bearing. The calculation of the bearing under the realistic Reynolds' boundary conditions demands that we use numerical methods, since the outflow boundary, that is the curve where $p = dp/d\varphi = 0$ is met, is unknown.

If no oil groove is available at the position $\varphi = \pi$ to fix the pressure there, the start of the pressure distribution follows first along an unknown curve, which is determined by the boundary conditions on the pressure ($p = 0$) and on the pressure gradient ($\partial p/\partial n = 0$). However experimental results show that these boundary conditions do not predict the start or end of the pressure very precisely (although for some applications the prediction is precise enough). The actual boundary conditions demand that we deal with the "pressureless" region, where the flow is very complicated and where the surface tension also plays a considerable role.

8.3 Dynamically Loaded Bearings

A dynamic bearing load occurs if the center of the journal carries out motion. The resulting forces can, in certain circumstances, increase the motion of the journal. We then talk about *hydrodynamic instability*, which typically occurs with a frequency corresponding to half the rotational frequency of the shaft. The assumptions discussed earlier $\left(\tau \sim \Omega^{-1} \sim R/U\right)$ for neglecting the acceleration $\partial u/\partial t$ are then valid. The effect of the journal motion needs only be considered in the continuity Eq. (8.17). The motion of the journal gives rise to a volume flux (per unit area) in the y-direction, given by $\vec{u} \cdot \vec{n}$, where $\vec{u}$ is the flow velocity at the journal, i.e., at the upper wall of the lubrication gap. The gap height h is now a function of time, given in the most general case by the equation

$$y = h(x, z, t), \tag{8.57}$$

or else implicitly

$$F(x, y, z, t) = y - h(x, z, t) = 0. \tag{8.58}$$

The kinematic boundary condition (4.170) $DF/Dt = 0$ directly furnishes

$$\vec{u} \cdot \vec{n} = -\frac{\partial F/\partial t}{|\nabla F|}, \tag{8.59}$$

or since

$$|\nabla F| = \sqrt{1 + \left(\frac{\partial h}{\partial x}\right)^2 + \left(\frac{\partial h}{\partial z}\right)^2} \approx 1,$$

also

$$\vec{u} \cdot \vec{n} = \frac{\partial h}{\partial t}. \tag{8.60}$$

This term is to be added to the left-hand side of Eq. (8.17), so that we now have the Reynolds' equation in the form

$$\frac{\partial}{\partial x}\left(\frac{h^3}{\eta}\frac{\partial p}{\partial x}\right) + \frac{\partial}{\partial z}\left(\frac{h^3}{\eta}\frac{\partial p}{\partial z}\right) = 6\left(\frac{\partial(hU)}{\partial x} + \frac{\partial(hW)}{\partial z} + 2\frac{\partial h}{\partial t}\right). \qquad (8.61)$$

8.3.1 Infinitely Long Journal Bearing

We use (8.61) for the infinitely long bearing, but shall now only calculate the contribution of the pressure field stemming from the journal motion along the X-axis. The film thickness (8.22) now assumes the form

$$h(\varphi, t) = \bar{h}[1 - \epsilon(t)\cos\varphi] \qquad (8.62)$$

(see Fig. 8.2), from which we see that the change in the film thickness is

$$\frac{\partial h}{\partial t} = -\bar{h}\dot{\epsilon}\cos\varphi, \qquad (8.63)$$

with $\dot{\epsilon} = d\epsilon/dt$. As before, we set $dx = R\,d\varphi$ and by integrating (8.61) obtain

$$\frac{h^3}{\eta R}\frac{\partial p}{\partial \varphi} = -12\int_0^\varphi \bar{h}\dot{\epsilon}\cos\varphi\, R\,d\varphi = -12R\,\bar{h}\dot{\epsilon}\sin\varphi, \qquad (8.64)$$

since, for symmetry reasons

$$\left.\frac{\partial p}{\partial \varphi}\right|_{\varphi=0} = 0.$$

Integrating again first leads us to

$$p = -12\eta\,\bar{h}\dot{\epsilon}\,R^2 \int \frac{\sin\varphi}{h^3}\,d\varphi + \text{const}, \qquad (8.65)$$

and with $dh = \bar{h}\epsilon\sin\varphi\,d\varphi$ then immediately to

$$p = 12\frac{\eta\,\dot{\epsilon}\,R^2}{\bar{h}^2}\left(\frac{1}{2\epsilon(1 - \epsilon\cos\varphi)^2} + C\right). \qquad (8.66)$$

We note that p is an even function of φ here, so that (8.32) implies that the Y-component of the force vanishes. Equation (8.31) combined with (8.43) furnishes the X-component of the force (per unit depth) exerted by the fluid on the journal as

$$F_X = -\int_0^{2\pi} p \cos\varphi\, Rd\varphi = -12\pi\eta R^3 \frac{\dot{\epsilon}}{\bar{h}^2(1-\epsilon^2)^{3/2}}\,. \qquad (8.67)$$

8.3.2 Dynamically Loaded Slider Bearing

In applying (8.61) to the plane "slider bearing" shown in Fig. 8.5, we obtain from integrating twice over x

$$p(x,t) = 6\eta\, U\left[\int_0^x \frac{1}{h^2}dx + \frac{2}{U}\int_0^x \frac{1}{h^3}\left(\int_0^x \frac{\partial h}{\partial t}dx\right)dx + \frac{C}{6U}\int_0^x \frac{1}{h^3}dx\right]. \qquad (8.68)$$

One of the constants of integration appearing has already been determined by the boundary condition

$$p(x=0) = p(h_1) = 0, \qquad (8.69)$$

while we fix the constant C by the second boundary condition

$$p(x=L) = p(h_2) = 0\,. \qquad (8.70)$$

If we wish to make further progress we need to know the film thickness. If the walls forming the gap are straight and rigid, that is

$$h(x,t) = h_1(t) - \alpha x = h_1(t) - \frac{h_1(t)-h_2(t)}{L}x, \qquad (8.71)$$

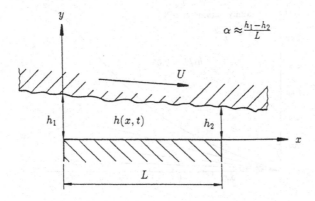

Fig. 8.5 Geometry of the slider bearing

the integration over h can be carried out. After determining the constants of integration, and using the abbreviation

$$\dot{h} = \frac{\partial h}{\partial t}$$

Equation (8.68) assumes the form

$$p(x,t) = 3\frac{\eta U}{\alpha h_0}\left(1 - \frac{2\dot{h}}{\alpha U}\right)\left(\left(\frac{h_0}{h_1} - 1\right)^2 - \left(\frac{h_0}{h(x,t)} - 1\right)^2\right). \qquad (8.72)$$

Note that the pressure remains zero for positive squeeze motion $\dot{h} = (1/2)\,\alpha\,U$.
    At the film thickness

$$h_0 = 2\frac{h_1 h_2}{h_1 + h_2} \qquad (8.73)$$

the extremum of the pressure distribution is met. The velocity distribution is linear over the gap height at $h = h_0$. For $\dot{h} = 0$ we obtain the pressure distribution of the plain "slider" (Fig. 8.6), while for $U = 0$ the formula for pure squeeze flow arises (Fig. 8.7), which, for the special case $\alpha = 0$, that is $h = h(t)$ reduces to

$$p(x,t) = -6\frac{\eta \dot{h} L^2}{h^3}\left(1 - \frac{x}{L}\right)\frac{x}{L}. \qquad (8.74)$$

Integrating the pressure distribution (8.72) leads to the load-bearing capacity of the slider bearing (per unit depth)

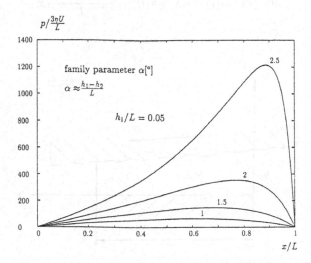

Fig. 8.6 Pressure distribution in the lubrication gap fluid film $(\dot{h} = 0)$, for various angles of inclination α

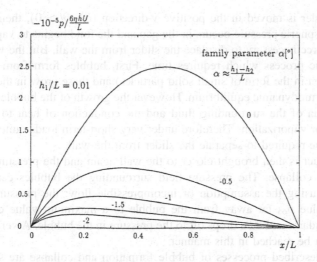

Fig. 8.7 Pressure distribution in the squeeze gap fluid film ($U = 0$) for various angles of inclination α

$$F_y = 6\frac{\eta\,U}{\alpha^2}\left(1 - \frac{2\dot{h}}{\alpha U}\right)\left(\ln\frac{h_1}{h_2} + 2\frac{1 - h_1/h_2}{1 + h_1/h_2}\right). \tag{8.75}$$

From (8.75), for $U = 0$ and in the limit $\alpha \to 0$ we obtain the formula

$$F_y = -\frac{\eta\,\dot{h}\,L^3}{h^3}, \tag{8.76}$$

which also follows from direct integration of (8.74). In the limit $h(t) \to 0$, (8.76) renders an infinitely large force. For a given force, (8.76) represents a differential equation for the motion $h(t)$ of the slider, whose solution is

$$t = \frac{\eta L^3}{2F_y}\frac{1}{h^2} + \text{const.} \tag{8.77}$$

We dispose of the constant of integration using the initial condition

$$h(t = 0) = h_A \tag{8.78}$$

and obtain

$$t = \frac{\eta L^3}{2F_y}\left(\frac{1}{h^2} - \frac{1}{h_A^2}\right); \tag{8.79}$$

this means that under a finite force, the slider cannot reach the wall within a finite time.

If the slider is moved in the positive y-direction ($\partial h/\partial t > 0$), then a pressure below atmospheric pressure occurs in the gap and the fluid begins to vaporize. This limits the force necessary to displace the slider from the wall. But the vaporization is a dynamic process which requires time. First, bubbles form from "cavitation nuclei" (often in the form of small solid particles) and these grow in the attempt to produce thermodynamic equilibrium. However the growth of the bubbles is affected by the inertia of the surrounding fluid and the conduction of heat to the bubble necessary for vaporization. Therefore under very short-term loads quite substantial forces can be required to separate the slider from the wall.

If the slider is then brought closer to the wall again and the pressure increases, the bubbles collapse. The pressure field surrounding the bubbles can be easily determined using the assumption of incompressible flow: the pressure increases from the value p_∞ far away from the bubble to a maximum value close to the bubble boundary, and then drops off to the pressure in the bubble. A very high peak pressure can be reached in this manner.

All the described processes of bubble formation and collapse are summarized within the concept of *cavitation*. As already mentioned, this phenomenon does not only occur in the fluid film, but in fact anywhere that the pressure drops below vapor pressure, for example in the flow past a body where the pressure drops below atmospheric pressure close to the thickest part of the body (see Fig. 10.14), and can therefore reach values below vapor pressure. Bubbles can then form in the low pressure region, and be carried into the higher pressure region where they collapse, so that the surface of the body is acted on by pulsating pressure loads, eventually leading to the destruction of the surface. The collapse of the bubble goes hand in hand with a cracking noise, giving the first indication of cavitation in, for example, hydraulic machines. Cavitation in the fluid film of the finger joints is probably the origin of "knuckle cracking": by tugging a finger, low pressure is produced in the fluid in the joint and a bubble forms. The collapse of this bubble causes a pressure wave which is perceived as the cracking noise. The squeeze flow mentioned also forms in valve seats, and this can lead to the phenomenon of cavitation if the valve is opened too quickly.

We also wish to point out the mathematical relationship between the pure squeeze flow between parallel walls and the steady pressure driven unidirectional flow. Since for $\alpha = 0$ the gap height is not a function of position, the Reynolds' equation appears in the form of Poisson's equation for the pressure

$$\nabla \cdot \nabla p = \Delta p = \frac{12\eta}{h^3}\frac{\partial h}{\partial t}, \tag{8.80}$$

where the right-hand side is to be taken as a constant, since time only appears parametrically. Equation (8.80) is of the same form as (6.72)

$$\Delta u = -\frac{K}{\eta},$$

and we can take its solution directly from Sect. 6.1, replacing u by p, and $-K/\eta$ by $12\eta\dot{h}/h^3$. (With regard to the results we note that the coordinates y, x perpendicular to the flow direction in Sect. 6.1 are to be replaced by the coordinates x, z here, and so the channel height h corresponds to the length of the slider L here.) In this analogy, the volume flux $\dot{V}$ clearly corresponds to the force and the average velocity $\bar{U}$ to the pressure averaged over the cross-section of the slider. In this manner we carry over the velocity distribution originating from the pressure gradient in (6.19) to the pressure distribution (8.74), and the volume flux in (6.21) to the load-bearing capacity (8.76).

In the fluid film of a cylindrical slider with circular cross-section $A = \pi R^2$ we find from the analogy with (6.53) the pressure distribution

$$p(r,t) = -\frac{3\eta\,\dot{h}}{h^3}\left(R^2 - r^2\right), \tag{8.81}$$

and with (6.55) and (6.58) the load-bearing capacity

$$F_y = -\frac{3\,\pi\,\eta\,\dot{h}}{2h^3}\,R^4. \tag{8.82}$$

Similarly, from the pressure driven part of the velocity distribution (6.65), we obtain the pressure distribution for a slider with circular ring cross-section $A = \pi\left(R_O^2 - R_I^2\right)$

$$p(r,t) = -\frac{3\,\eta\,\dot{h}}{h^3}\left(R_O^2 - r^2 - \left(R_O^2 - R_I^2\right)\frac{\ln r/R_O}{\ln R_I/R_O}\right), \tag{8.83}$$

and finally from (6.66) the load-bearing capacity

$$F_y = -\frac{3\pi\,\eta\,\dot{h}}{2h^3}\left(R_O^4 - R_I^4 + \frac{\left(R_O^2 - R_I^2\right)^2}{\ln R_I/R_O}\right). \tag{8.84}$$

We refrain from carrying over the results for channel flow through rectangular, triangular and elliptical cross-sections to the pressure distribution and load-bearing capacity of sliders with corresponding cross-sections, but do note that solutions known from the theory of elasticity can be used here too.

8.3.3 Squeeze Flow of a Bingham Material

As we have already shown, the assumptions of lubrication theory imply that squeeze flows can be locally considered as unidirectional flows. Therefore the equations of motion are valid in the form (6.190) and (6.191), which reduce to (8.7) and (8.9) wherever the material flows, since the material then behaves like a generalized Newtonian fluid. However the wall shear stress τ_w now depends parametrically on x, and, because of (6.192) and (6.193), necessarily the pressure gradient $-K$ too, as well as the dimensionless positions κ_1 and κ_2 of the yield surfaces in the velocity distributions (6.197) and (6.198). As was explained in connection with the Eqs. (8.17) and (8.61), the continuity equation in the integral form for plane two-dimensional unsteady squeeze flow reads

$$\frac{\partial \dot{V}}{\partial x} + \frac{\partial h}{\partial t} = 0. \tag{8.85}$$

The volume flux vanishes at the position $x = L/2$ of the gap for symmetry reasons, so that integrating (8.85) leads to the relation

$$\dot{V} = -\dot{h}\left(x - \frac{L}{2}\right). \tag{8.86}$$

Since the local volume flux $\dot{V}(x)$ is equal to the volume flux (6.208) of pure pressure driven flow, if $-K$ is replaced by the local pressure gradient, a nonlinear equation for the pressure gradient arises directly from (8.86)

$$\frac{h^3}{12\eta_1}\frac{\partial p}{\partial x}\left\{1 + 3\frac{\vartheta}{h}\left(\frac{\partial p}{\partial x}\right)^{-1} - 4\left(\frac{\vartheta}{h}\left(\frac{\partial p}{\partial x}\right)^{-1}\right)^3\right\} = \dot{h}\left(x - \frac{L}{2}\right), \tag{8.87}$$

or

$$\left(\frac{h}{\vartheta}\frac{\partial p}{\partial x}\right)^3 + \left(3 - \frac{12\eta_1\dot{h}L}{\vartheta h^2}\left(\frac{x}{L} - \frac{1}{2}\right)\right)\left(\frac{h}{\vartheta}\frac{\partial p}{\partial x}\right)^2 - 4 = 0, \tag{8.88}$$

where, because of the symmetry already mentioned, we have restricted ourselves to the region $L/2 \leq x \leq L$.

To calculate the pressure distribution from this differential equation we would first be required to solve the cubic equation for the pressure gradient and then decide which of the three roots makes physical sense. The solution may not be found analytically for arbitrary values of the squeeze velocity and the channel dimensions, and in these cases we are directed towards a numerical solution for a given x. We gain an overview by looking for approximate solutions for large and small values of

$$A := -\frac{12\eta_1\dot{h}L}{\vartheta h^2} \quad \text{with} \quad A > 0. \tag{8.89}$$

It is directly obvious that $(h/\vartheta)\partial p/\partial x = -2$ is a root of the Eq. (8.88) for $A = 0$. The material does not yet flow for this value, that is, the load given by this value does not yet lower the plate. For small values of A we obtain an asymptotic expansion for the pressure gradient, by setting

$$\frac{h}{\vartheta}\frac{\partial p}{\partial x} = -2 + \varepsilon \tag{8.90}$$

and insert this into (8.88). Comparing terms of the same order of magnitude leads us to the equation

$$\varepsilon = \pm \frac{2}{\sqrt{3}} A^{1/2} \left(\frac{x}{L} - \frac{1}{2}\right)^{1/2}, \tag{8.91}$$

and then with this to

$$\frac{h}{\vartheta}\frac{\partial p}{\partial x} = -2 \left(1 + \left(\frac{A}{3}\right)^{1/2} \left(\frac{x}{L} - \frac{1}{2}\right)^{1/2}\right) \quad \text{for} \quad A \to 0, \tag{8.92}$$

where we choose the sign in (8.91) so that the load, or the magnitude of the pressure gradient, increases as A becomes larger. For very large A we directly infer from (8.88) the pressure gradient

$$\frac{h}{\vartheta}\frac{\partial p}{\partial x} = -A \left(\frac{x}{L} - \frac{1}{2}\right), \tag{8.93}$$

which corresponds to the Newtonian limiting case from (8.74). As before we look for an asymptotic expansion and set

$$\frac{h}{\vartheta}\frac{\partial p}{\partial x} = -A \left(\frac{x}{L} - \frac{1}{2}\right) + \varepsilon. \tag{8.94}$$

From (8.88) we use the assumption $A(x/L - 1/2) \gg \varepsilon$ and thus obtain the equation

$$\frac{h}{\vartheta}\frac{\partial p}{\partial x} = -\left(3 + A\left(\frac{x}{L} - \frac{1}{2}\right)\right), \tag{8.95}$$

which, however, does not hold near $x = L/2$. Integrating Eqs. (8.92) and (8.95) with the boundary condition $p(x = L) = 0$ leads to the pressure distribution (relative to the ambient pressure)

$$p = \frac{2\vartheta L}{h} \left(1 - \frac{x}{L} + \frac{1}{3}\left(\frac{A}{6}\right)^{1/2} \left(1 - \left(\frac{x}{L} - \frac{1}{2}\right)^{3/2} 2\sqrt{2}\right)\right) \quad \text{for} \quad A \to 0, \tag{8.96}$$

or

$$p = \frac{3\vartheta L}{h}\left(1 - \frac{x}{L} - \frac{A}{6}\left(\left(\frac{x}{L}\right)^2 - \frac{x}{L}\right)\right) \quad \text{for} \quad A \to \infty, \tag{8.97}$$

and thus to the load-bearing capacity (per unit depth)

$$F = 2\int_{L/2}^{L} p\,\mathrm{d}x = \frac{\vartheta L^2}{2h}\left(1 + \frac{4}{5}\sqrt{\frac{A}{6}}\right) \quad \text{for} \quad A \to 0, \tag{8.98}$$

and

$$F = \frac{3\vartheta L^2}{4h}\left(1 + \frac{A}{9}\right) \quad \text{for} \quad A \to \infty. \tag{8.99}$$

To conclude, we refer to a kinematic contradiction in this solution: since the pressure gradient and thus the positions of the yield surfaces depend parametrically on x, the velocities at the yield surfaces are functions of x. The contradiction becomes clear if we use the Bingham constitutive relation (3.63) and (3.64). Since the rigid solid body here only carries out a translation, the velocity at the solid body side of the yield surfaces is independent of x and thus the no slip condition (4.159) is violated. Numerical calculations (for the rotationally symmetric case) show that although the pressure distribution, the load-bearing capacity and the velocity distributions are essentially correctly predicted by lubrication theory, the yield surfaces are indeed not predicted correctly. However the yield surfaces from lubrication theory do resemble surfaces of constant value of the stress invariants, if these assume a value slightly different from $\vartheta : \left(\tau'_{ij}\tau'_{ij}/2\right)^{1/2} \sim 1.05\vartheta$. Because of this, the solution on the basis of lubrication theory is of sufficient accuracy for most engineering applications.

8.4 Thin-Film Flow on a Semi-Infinite Wall

The assumptions which underlie hydrodynamic lubrication theory are frequently found to be valid bases of other technically important flows which on a superficial glance have nothing in common with lubrication theory. A typical feature of these flows is the gradual thinning of the film flow, which creates a locally-valid film of constant thickness. As an example of this type of film flow we now consider the steady plane flow on a semi-infinite wall and tie it up with the corresponding film flow on an infinitely long wall (see Sect. 6.1.3). We retain the notation of the cited section, and place the origin of coordinates, whose position is arbitrary on an infinite wall, at the leading edge with the negative x-direction along the surface. For

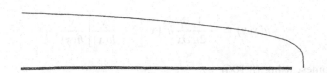

Fig. 8.8 Film flow over a horizontal plate

the sake of simplicity we assume the plate to be horizontal, so that β (in the notation of Sect. 6.1.3) is zero.

A prescribed volume flux is maintained. The flow must go over the leading edge, and we expect that as it does so the surface of the moving film will drop. The form of the surface of the film is unknown and must be determined as part of the solution. The differential equations which the solution must satisfy can be taken directly from Sect. 6.1.3 (see (6.26) and (6.27)), and involve putting $\beta = 0$ (Fig. 8.8)

$$\frac{\partial p}{\partial x} = \eta \frac{\partial^2 u}{\partial y^2}, \tag{8.100}$$

$$\frac{\partial p}{\partial y} = -\rho g. \tag{8.101}$$

Using the no-slip condition (6.28) and the continuity of the stress vector on the *free surface* (6.29) we obtain

$$u(0) = 0, \tag{8.102}$$

$$n_j \tau_{ji(1)} = n_j \tau_{ji(2)}. \tag{8.103}$$

Using formula (4.164) we find the normal vector to the free surface, $y = h(x)$, i.e.,
$n_j = \left(-h'(x) \big/ \sqrt{1 + h'^2(x)}, 1 \big/ \sqrt{1 + h'^2(x)}, 0 \right)$.

Where lubrication theory is valid the variation of $h'(x)$ is very small, so we write the normal vector as $n_j = (0, 1, 0)$, which is exactly the same as in Sect. 6.1.3 Accordingly the boundary conditions (6.31), (6.33) hold on $y = h(x)$, i.e.,

$$p_1 = p_2 = p_0, \tag{8.104}$$

$$\frac{\partial u}{\partial y} = 0. \tag{8.105}$$

The pressure term, given by (6.35), can be substituted here

$$p(x, y) = p_0 + \rho g(h(x) - y), \tag{8.106}$$

where (unlike the film flow on an infinite plate) the film height is an as yet unknown function of x. Integrating (8.100) and taking into account the boundary conditions (8.102) and (8.105), we obtain the result

$$u(x,y) = -\frac{1}{2\eta}\frac{\partial p}{\partial x}h^2(x)\left[2 - \frac{y}{h(x)}\right]\frac{y}{h(x)},\tag{8.107}$$

which becomes, using (8.106)

$$u(x,y) = -\frac{g}{2\nu}h'(x)h^2(x)\left[2 - \frac{y}{h(x)}\right]\frac{y}{h(x)}.\tag{8.108}$$

To calculate the film height, we use the kinematic free surface boundary condition (4.170), which becomes in steady flow

$$\vec{u}\cdot\nabla F = 0 = -v(x,y) + h'(x)u(x,y),\tag{8.109}$$

and on $y = h(x)$ this reduces to

$$h'(x) = \frac{v(x,y)}{u(x,y)}.\tag{8.110}$$

We can obtain the v-component of velocity on the free surface using the two-dimensional continuity equation $\partial u/\partial x + \partial v/\partial y = 0$, which yields

$$v(x,h(x)) = -\int_0^{h(x)}\frac{\partial u}{\partial x}dy = \frac{g}{2\nu}h^2(x)h'^2(x) + \frac{g}{3\nu}h^3h''(x)\tag{8.111}$$

and it follows, using $u(x, h(x))$ from (8.107), either that

$$h'(x) = -\frac{h(x)h''(x)}{3h'(x)},\tag{8.112}$$

or that

$$\frac{d}{dx}\left[h'(x)h^3(x)\right] = 0.\tag{8.113}$$

It is important to notice also that the differential equation for the film thickness can be immediately derived from the Reynolds' lubrication equation. We first obtain from (8.106) the value $\rho g h'(x)$ for $\partial p/\partial x$; then when this value is substituted in (8.13), using $U = 0$ and constant η, we obtain (8.113) as before.

The starting-point for the derivation of the Reynolds' lubrication equation is the expression for constant volume flux. For film flows this takes the form

$$\dot{V}_x = \int_0^{h(x)} u\,dy = -\frac{g}{3\nu}h^3(x)h'(x),\tag{8.114}$$

which is clearly the first integral of (8.113). The expression $\dot{V}_x 3\nu/g$ has dimension $(length)^3$, so that a characteristic length L for this problem can be defined as

$$L = \left(\frac{\nu \dot{V}_x}{g}\right)^{\frac{1}{3}}.$$

Integrating (8.113) once again, we get

$$\frac{h^4}{L^4} = 12\left(-\frac{x}{L} + c\right), \tag{8.115}$$

which, for a given value of L does not depend on special values of volume flux, density, viscosity or indeed gravitational force. The integration constant c cannot be determined from the present theory. This shows once more that the flow as we approach the edge is not known, the reason for which is that the rate of fall in the free surface, namely $h'(x)$, is no longer small. Obviously one could construct a solution in the neighborhood of the edge, but this would in general be dependent on the Reynolds' number. It is clear from the form of the solution that the thickness of the film far from the edge does not depend on the thickness at the edge, which obviously increases rapidly as we move inwards, and so the constant c may be set to zero, which corresponds to the assumption $h(0) = 0$. Even though the solution is not correct at the edge, far from the edge relatively small errors in height arise from this, as has been confirmed by experiment. If greater accuracy is required, the constant c may be found from measured values.

When the local Reynolds' number is defined by

$$Re = u(x, h)\frac{h}{\nu},$$

we find that

$$\alpha\, Re = h'(x)Re = \frac{9}{2}\frac{\dot{V}_x^2}{h^3 g},$$

which is independent of viscosity.

8.5   Flow Through Particle Filters

Particle filters are installed to remove soot particles from the exhausts of diesel engines, since such particles may present a health hazard. Often the filters are made out of long rectangular tubes bounded by ceramic walls: each entry tube has four sidewalls, each of which adjoins an exit tube, and in the same way each exit tube is bounded by four entry tubes. A cross-section of the filter would thus resemble a

chess-board, the black squares representing the cross-sections of (say) the entry
tubes, the white those of the exit tubes. The entry and exit tubes are closed at the
outer and inner ends of the filter respectively.

Exhaust gas, laden with soot particles, flows into an entry tube, and its gaseous
phase then flows into the adjoining exit tubes through the pores in the walls.
Particles which are too large to penetrate the walls remain inside the entry tube and
are deposited on its walls. Such deposits naturally increase the thickness of the wall,
so that an increased pressure difference is required to maintain the same volume flux
through the wall. The filter must be renewed when this pressure loss impairs engine
performance. The renewal is brought about by burning the soot particles at a high
temperature, and (after a relatively long time interval) removing the accumulated
ash from the sites of the deposits.

Typical ratios of tube thickness to length are about $c/L \approx 4 \cdot 10^{-3}$. The volume
flux entering the tube, namely $\bar{U} c^2$, must flow out through the four side walls, so that
$\bar{U} c^2 = 4\bar{V} Lc$. From this it follows that the streamlines are inclined to the axis of the
tube at an angle roughly $\bar{V}/\bar{U} \approx 10^{-3}$. We can therefore assume that the flow is
locally parallel. The topic of parallel flow in a square tube (or tube of triangular
cross-section, which also is used in filters) has already been discussed in Sect. 6.1.6,
where (6.89) gives the mean velocity in the form of a rapidly-convergent series,
summation of which yields

$$\bar{U} = -\frac{\partial p(x)}{\partial x} \frac{c^2}{4\eta} \cdot 0.4217 , \qquad (8.116)$$

where the pressure gradient is an as yet undetermined function of x. The outflow
through the side walls of the entry tube is, over a distance dx, $-4\bar{V}\, c\mathrm{d}x$; this equals
the change in volume flux in the entry tube, which is $\mathrm{d}\bar{U}_{in}\, c^2$, and it follows that

$$\frac{\mathrm{d}\bar{U}_{in}}{\mathrm{d}x} = -\frac{4\bar{V}}{c} . \qquad (8.117)$$

Similarly the change in mean velocity at the corresponding position in the exit
tube is

$$\frac{\mathrm{d}\bar{U}_{out}}{\mathrm{d}x} = +\frac{4\bar{V}}{c} . \qquad (8.118)$$

The local volume flux across unit surface area, namely $\bar{V}$, through the wall at x is
related to the pressure difference $p_{in}(x) - p_{out}(x)$ by

$$\bar{V} = \frac{k}{\eta} \frac{p_{in} - p_{out}}{s} , \qquad (8.119)$$

where s is the thickness of the porous layer.

The formula (8.119) was originally derived on the basis of experimental evidence by Darcy in 1856. We will discuss this eponymous law at a later stage. Here we content ourselves with observing that the permeability constant k is an empirical constant, which is characterised by the number, size and shape of the pores.

Inserting (8.116) into (8.117) and (8.118) and using Darcy's law from (8.119) results in two second-order coupled linear differential equations in the variables $p_{in}(x)$ and $p_{out}(x)$

$$\frac{\partial^2 p_{in}}{\partial x^2} = 16k \frac{p_{in} - p_{out}}{0.4217s\,c^3},$$
$$\frac{\partial^2 p_{out}}{\partial x^2} = -16k \frac{p_{in} - p_{out}}{0.4217s\,c^3}. \tag{8.120}$$

It follows immediately that the sum of the pressures in the entry and exit tubes is a linear function of x. The system is in fact a fourth-order boundary value problem, with prescribed boundary conditions at $x = 0$ and $x = L$. As the volume flux $\dot{v}$ on entry to an entry tube may be found by dividing the total entry flux by the number of entry tubes, with (8.116) we have the boundary condition

$$\frac{\partial p_{in}(0)}{\partial x} = \frac{4\eta\dot{v}}{0.4217c^4}. \tag{8.121}$$

All the gas flowing through an entry tube will have drained through the side walls on reaching the end $x = L$, and accordingly

$$\frac{\partial p_{in}(L)}{\partial x} = 0. \tag{8.122}$$

Since the exit tubes are closed at $x = 0$, it follows that the mean velocity is zero at this position, so that

$$\frac{\partial p_{out}(0)}{\partial x} = 0, \tag{8.123}$$

whilst at $x = L$ we have

$$p_{out}(L) = p_0, \tag{8.124}$$

where p_0 is atmospheric pressure omitting pressure losses in the exhaust pipe. Obviously the linear system of equations may be solved using the well-known change of variables $p_{in,out} = A_{in,out}\, e^{\lambda x}$. This leads to a boundary value problem with a complicated analytical solution, details of which are omitted here; numerical

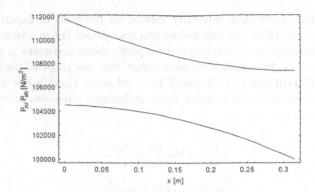

Fig. 8.9 Pressure distribution in a particle filter

values may be readily obtained with the aid of widely-available computer program. (Special algorithms are required for purely numerical integration.)

Figure 8.9 shows a typical example of such a solution. Obviously there is a length of filter for which the pressure loss is a minimum for some given filter geometry and mass flux: a longer filter will naturally lead to a smaller velocity $\bar{V}$ and thus to a reduced pressure drop across the wall, while on the other hand the overall pressure loss from filter entry to exit will increase because of the greater length of tube over which the shear stress will now operate. In Fig. 8.9, the length is so chosen as to minimise the overall pressure loss.

Because of the finite (though small) thickness of the tube walls, the cross-sectional area of the oncoming stream is reduced. This leads to pressure losses, which will be discussed in Sect. 9.1.4. The fact, however, that the velocity is suddenly increased on entry even reduces those pressure losses necessary for the flow to attain its fully-formed velocity profile.

8.6 Flow Through a Porous Medium

The individual stream tubes in the porous medium are made up of more or less rapidly varying channels, so that the typical angle of declination of the streamlines is appreciable. Thus the neglect of the convective terms in the equations of motion on the basis of sufficiently small αRe cannot be justified. Therefore we require that the Reynolds' number itself (in a sense to be defined later) is sufficiently small. The limiting case $Re \to 0$, whose technical meaning is given in Sect. 4.1.3, leads to Eq. (4.35), which, together with the continuity equation, defines the creeping flow equations to be dealt with in Chap. 13.

For reasons to be discussed, the calculation of flow in a porous medium cannot take place on the basis of (4.35) and the continuity equation. The equations of motion applicable in this instance are, however, closely related to those of rectilinear flow or, more precisely those of lubrication flow. The discussion will proceed on this basis although the flow is formally a creeping flow.

A porous medium is a structure which is often made up of granular or fibrous materials. The cross sections of the pores are as a rule so small, that the Reynolds' number which is formed with typical length d and typical velocity u of the fluid phase, is small in the pores: thus $u\,d\,\rho/\eta \ll 1$. The precise geometry of the pores is of course unknown; however even for a given geometry a calculation of the velocity field or the pressure field is not justified, owing to the complex geometry of the numerous pores. We must therefore restrict ourselves to treating mean values over numerous pores. Thus, volume flux over an element of area is interpreted as the local velocity in a porous medium. The corresponding velocity components in the coordinate directions are given by volume fluxes through surface elements whose normals are in the coordinate directions. The linear dimensions of the surface elements are large compared to d, however they must be small compared to the linear dimensions of the greater region of interest, so that the mean value is valid as the local velocity. In the same way we can also treat the mean value of the pressure $\bar{p}$ in a volume element whose height is large compared to d and has as its base a surface element considered above. As a consequence of the small Reynolds' number and the linearity of Eq. (4.35) the inertia force is small and one can expect that the pressure gradient is proportional to the mean velocity, as is the case for unidirectional flows. It may be remarked that inertia forces do not come into play in some other flows; thus: for laminar rectilinear flows the inertia terms vanish on kinematic grounds irrespective of the Reynolds' number; for locally rectilinear flows inertia terms may be neglected because the product of Reynolds' number times the deviation angle of the streamlines is small; in the present case inertia terms can be ignored because the Reynolds' number itself is small. Viscosity must enter into the relation between pressure gradient and mean velocity. This is because the only forces remaining available to balance the pressure gradients are due to viscosity. Furthermore, a quantity whose dimension is the square of a length must enter into this relation. In rectilinear flows, for example, this quantity is the square of the transverse thickness; say the square of the channel thickness. We now proceed to consider this relationship from a more general point of view.

One can easily show on dimensional grounds, that

$$\bar{U}_i = -\frac{k_{ij}}{\eta}\frac{\partial \bar{p}}{\partial x_j}\,. \tag{8.125}$$

The tensor k_{ij} is constant, provided the properties of the porous medium are homogenous, i.e., independent of position. For an isotropic medium it takes the form

$$k_{ij} = k\delta_{ij}\,, \tag{8.126}$$

then it follows that

$$\bar{U}_i = \frac{k}{\eta}\frac{\partial \bar{p}}{\partial x_i} \qquad (8.127)$$

and so

$$\vec{\bar{U}} = -\frac{k}{\eta}\nabla \bar{p}. \qquad (8.128)$$

It is therefore the same relationship between mean pressure gradient and mean velocity, which also arises in lubrication theory (8.12) and in rectilinear flow (6.58). In sedimentary solids it is often a tensor; in consequence the resistance depends on the direction of flow and is in general greater in the direction normal to the sedimentation. This appears to be the case in the deposition of soot in particle filters which we have already discussed. In this instance a thin layer of soot markedly increases the pressure loss. The ceramic material of the filter is of course isotropic. For a clean filter therefore, Eq. (8.127) comes into play, and integration of Eq. (8.119) can be carried out with $p(x_2 = 0) = p_{in}$; $p(x_2 = s) = p_{out}$, where as usual $\bar{p}$, $\bar{U}_2$, are replaced by p, $\bar{V}$, respectively and from symmetry $\bar{V} = \bar{W}$.

There is a number of models for the structure of a porous medium from which the permeability can be calculated. The simplest example is that of a solid through which a bundle of cylindrical holes is bored. Then the mean velocity over a surface F (in the above sense) is

$$\bar{U} = \frac{\dot{V}_{tot}}{F} = \frac{\pi R^4}{8\eta}\frac{N}{F}\frac{\partial \bar{p}}{\partial x}, \qquad (8.129)$$

where N is the number of holes crossing the surface F and in which (6.63) gives the volume flux through a single hole. The ratio N/F is at the same time the ratio of the voids in the material $N\pi R^2 L$ to the entire volume, namely FL where L is the length of the hole. This ratio is called the porosity, namely n of the medium.

Therefore we can also write (8.127) in the form

$$\bar{U} = n\frac{R^2}{8\eta}\frac{\partial \bar{p}}{\partial x} = n\frac{d^2}{32\eta}\frac{\partial \bar{p}}{\partial x} \qquad (8.130)$$

and identify the permeability as

$$k = \frac{n d^2}{32}. \qquad (8.131)$$

The factor 1/32 is valid only for bundles of straight holes and this model is somewhat unrealistic. When we replace it by a form factor $f(s)$ and the porosity by a function of n we obtain a more general expression

$$k = f(s)f(n)\,d^2\,,\tag{8.132}$$

where d then is a typical transverse measure of the hole, or grain size in the case of granular solids; the form factor and the porosity factor may then be found experimentally.

Measurements show that Darcy's law is valid for Reynolds' numbers $Re = \bar{U}\,d\,\rho/\eta$ up to about 10. This may appear surprising since we have applied the law under the express proviso that inertia forces are very small. This condition (just as with the condition $\alpha\,Re \ll 1$, which is the basis of locally parallel flow) was based solely on the equations of motion without taking into account either the complex flow paths or the boundary conditions.

For those cross sections, which are relatively small compared to path lengths, the spread of vorticity, into the centre of the channel, takes place by diffusion and is practically uninfluenced by convection. (The vorticity is produced at the wall where the no-slip condition is enforced by the viscosity on the adhering fluid). The formula (4.11), for the viscous force per unit volume, show that it is of great importance whenever the vorticity is large.

In spatially restricted regions of flow, diffusion produces the vorticity field, and this explains why we see viscous influences still operating at Reynolds' numbers greater than anticipated. It is known from the discussion in Sect. 4.1.3, that in the absence of convection the flow is determined only by friction, and is completely independent of Reynolds' number. The results calculated on the assumptions that $Re \ll 1$, or $\alpha\,Re \ll 1$ are then valid even when these characteristic quantities are markedly greater than unity. The precise limits of validity vary from case to case. However, in spatially restricted regions discrepancy between theory and experiment is still acceptable up to a Reynolds' number of about 10. As the influence of the inertia terms increases, the question of the stability of the flow arises and the flow enters the transition phase to turbulence. As a rule in technical applications one seeks a sufficient distance from turbulence so as to keep pressure losses low.

From a mathematical point of view we now remark that for homogenous and isotropic permeability, that is when (8.127) is valid, these equations lead to an important and far reaching conclusion, namely that the mean velocity field is a potential flow. As has been explained in Sect. 1.2.4 when the velocity field is irrotational then curl $\vec{U} = 0$. The continuity equation for incompressible flows remains unchanged for the mean flow

$$\frac{\partial \bar{U}_i}{\partial x_i} = 0\,.\tag{8.133}$$

The Laplace equation for the mean pressure follows then from (8.127)

$$\frac{\partial^2 \bar{p}}{\partial x_i \partial x_i} = 0 . \tag{8.134}$$

It should be emphasised, that only the mean velocity $\vec{U}(\vec{x})$ is irrotational. The actual velocity field $\vec{u}(\vec{x})$ is of course not irrotational, on the contrary: the diffusion of vorticity inside the holes or cracks gives rise to the precise character of these flows. Potential flows, where the velocity field is universally irrotational, will be discussed in Chap. 10, where a detailed account of potential theory is provided as it lies at the foundation of classical fluid mechanics. Suffice it to say that the methods of potential theory are applicable here in so far as the assumptions of permeability are fulfilled. Flows through porous media occur frequently in nature, for example ground water flows or flows of oil or gas through sand or rock, such as sandstone or limestone.

8.7 Hele-Shaw Flows

A flow which is closely related to the solutions discussed already, is that between two parallel plates, separated by a narrow gap h; between the plates a section of a cylinder is inserted, whose characteristic cross section is d. The undisturbed flow consists of a Poiseuille flow (Sect. 6.1.2) with mean velocity components $\bar{U}$, $\bar{W}$ in the x- and z-directions respectively. The x-component follows from (6.22); thus

$$\bar{U} = -\frac{h^2}{12\eta} \frac{\partial p}{\partial x} , \tag{8.135}$$

while the z-component is obtained by replacing $\partial p/\partial x$, $\bar{U}$ by $\partial p/\partial z$, $\bar{W}$ respectively; thus

$$\bar{W} = -\frac{h^2}{12\eta} \frac{\partial p}{\partial z} . \tag{8.136}$$

It is well known that the pressure gradient in the undisturbed flow is constant. In the presence of a cylinder, however, this is no longer the case.

The connection between pressure gradient and mean velocity is locally valid provided $(h/d)Re$, $(Re = \bar{U}h/\nu)$ is sufficiently small; it should be noted that (h/d) plays the role of the inclination angle α here. It then follows from the continuity equation, namely

$$\frac{\partial \bar{U}}{\partial x} + \frac{\partial \bar{W}}{\partial z} = 0 , \tag{8.137}$$

that the pressure satisfies the Laplace equation

$$\frac{\partial^2 p}{\partial x^2} + \frac{\partial^2 p}{\partial z^2} = 0. \tag{8.138}$$

The boundary conditions for the pressure on the contour of the body, namely

$$F(x,z) = 0 = -z + f(x) = 0,$$

may be obtained by substitution (8.135), (8.136) into (4.169), thus

$$\vec{U} \cdot \nabla F = \nabla p \cdot \nabla F = 0. \tag{8.139}$$

From $\vec{U} \cdot \nabla F = 0$, it follows that

$$f'(x) = \bar{W}/\bar{U} \quad \text{on} \quad z = f(x), \tag{8.140}$$

which essentially shows that the body contour is a streamline. Equation (8.140) also provides the streamlines of the flow when the velocity field is given. These streamlines may also be found by eliminating the curve parameter s in Eq. (1.11a). Since the mean velocity field is irrotational the streamlines are those of a potential flow. The same differential equations remain valid for all streamlines of the local velocity field $\vec{u}(x, y, z)$ in the plane $y = const.$ This becomes evident when the ratio w/u is formed by means of Eq. (6.19) and the corresponding equation for w (where the wall velocity is taken to be zero) and is then substituted in (8.140). It follows that the streamlines in all planes $y = const.$ are congruent to one another.

An experimental setup, based on the above theory, was used by Hele-Shaw in 1889 to visualise the streamlines of potential flows around a variety of cylindrical bodies, especially bluff bodies. (Potential flows at greater Reynolds' numbers around bluff bodies are not realised otherwise because of flow separation). Obviously the kinematic boundary condition on the body is satisfied, whereas this is not the case for the no-slip condition. Since the fluid must adhere to the cylinder, Eqs. (8.139) and (8.140) are no longer valid in that region next to the cylinder whose thickness is of order h. This error may be reduced by making the gap between the plates arbitrarily small. But this leads to even smaller Reynolds' numbers, so that due to the neglect of the no-slip condition the validity of the solution is restricted to Reynolds' numbers less than unity. In fact noticeable deviations from the theoretical predictions become evident when $(h/d)Re \approx 4$.

Chapter 9
Stream Filament Theory

9.1 Incompressible Flow

We shall now follow on with our earlier statement that in many technically interesting problems the entire flow region can by represented as a single streamtube,
and the behavior of the flow is then characterized by its behavior at a median
streamline. Within the framework of this assumption, the flow quantities are only
functions of the arc length s along the streamline, and possibly of the time t. Thus
the flow quantities are assumed constant over the cross-section of the streamtube.
Now this assumption does not have to be satisfied for the entire streamtube (at least
not in steady flow), but only in those sections of the streamtube where we wish to
calculate the flow in this *quasi-one-dimensional* approximation. Therefore the flow
must be at least piecewise *uniform*, i.e., essentially constant over the cross-section,
and also may not change too strongly in the flow direction: this assumes that the
cross-section is a slowly varying function of the arc length s. In between these
uniform regions the flow can exhibit a three-dimensional character, but cannot be
computed there using stream filament methods.

The assumption of constant flow variables over the cross-section requires that
the friction effect is negligible, because we know from Chap. 6 that the flow
quantities vary considerably over the cross-section of streamtubes bounded by walls
if the flow is dominated by frictional effects, as is the case in fully developed pipe
flow. Even in these flows, the concept of stream filament theory can be applied if
the distribution of the flow quantities over the cross-section is known, or else it
must be possible to make reasonable assumptions about these distributions. In
particular attention must be paid in the calculation of quantities averaged over the
cross-section: the averaged velocity calculated from the continuity equation, which
we used as the typical velocity in the resistance laws cannot be used in the balances
of energy and momentum. This is because, for example, the momentum flux $\varrho \overline{U}^2 A$
in a circular pipe formed with this averaged velocity constitutes only 75% of the
actual momentum flux through the circular cross-section in laminar flow.

© Springer Nature Switzerland AG 2020
J. H. Spurk and N. Aksel, *Fluid Mechanics*,
https://doi.org/10.1007/978-3-030-30259-7_9

In turbulent flow the velocity distributions are flatter and the difference between the maximum and the average velocities is therefore much smaller. The assumption of constant velocity over the cross-section is therefore a much better approximation in turbulent flow than it is in laminar.

9.1.1   Continuity Equation

We first bring the continuity equation to a form useful in the context of stream filament theory. For this we assume that the cross-sectional area of the streamtube is given in the form $A = A(s, t)$ and that all flow quantities only depend on the arc length s and the time t. For the section of the streamtube in Fig. 9.1 the continuity equation is

$$\int_0^L \frac{\partial \varrho}{\partial t} A \, ds - \varrho_1 u_1 A_1 + \varrho_2 u_2 A_2 + \iint_{(S_w)} \varrho \vec{u} \cdot \vec{n} \, dS = 0. \tag{9.1}$$

If the cross-section of the tube does not change in time the integral over the wall S_w vanishes. Otherwise we take the surface S_w as given by the equation

$$r = R(t, \varphi, s), \tag{9.2}$$

or in its implicit form

$$F(t, \varphi, s, r) = r - R(t, \varphi, s) = 0. \tag{9.3}$$

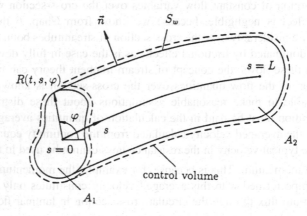

Fig. 9.1 The continuity equation in stream filament theory

From the kinematic boundary condition (4.170) we get the normal component of the flow velocity at the moving wall as

$$\vec{u} \cdot \vec{n} = \vec{u} \cdot \frac{\nabla F}{|\nabla F|} = \frac{1}{|\nabla F|} \frac{\partial R}{\partial t}, \tag{9.4}$$

and we note that $1/|\nabla F|$ is the component n_r of the normal to the surface in the r-direction. Using this we write the integral over S_w in the form

$$\iint\limits_{(S_w)} \varrho \vec{u} \cdot \vec{n} \, dS = \iint\limits_{(S_w)} \varrho \frac{\partial R}{\partial t} n_r \, dS = \int\limits_0^L \int\limits_0^{2\pi} \varrho \frac{\partial R}{\partial t} R \, d\varphi \, ds, \tag{9.5}$$

since $n_r \, dS = R \, d\varphi \, ds$ is the projection of the surface element dS in the radial direction. From

$$A = \int\limits_0^{2\pi} \int\limits_0^R r \, dr \, d\varphi, \tag{9.6}$$

it follows that

$$\frac{\partial A}{\partial t} = \int\limits_0^{2\pi} R \frac{\partial R}{\partial t} d\varphi, \tag{9.7}$$

and finally the continuity equation is

$$\int\limits_0^L \frac{\partial \varrho}{\partial t} A \, ds + \int\limits_0^L \varrho \frac{\partial A}{\partial t} \, ds - \varrho_1 u_1 A_1 + \varrho_2 u_2 A_2 = 0. \tag{9.8}$$

In stream filament theory this equation holds quite generally. However in most technical applications the streamtube cross-section does not change in time, so that the second integral is equal to zero.

In incompressible flow the first integral vanishes if we again assume that incompressibility implies constant density (see discussion in Sect. 4.1.1). Therefore, for incompressible steady or unsteady flow, if A does not vary in time, the relation

$$u_1 A_1 = u_2 A_2 \qquad (9.9)$$

holds. In compressible flow, the first integral in (9.8) only vanishes if the flow is steady.

9.1.2   Inviscid Flow

Incompressible inviscid flows can already be dealt with using Bernoulli's equation (4.61) or (4.62) and the continuity equation. We show how these are applied in an example of the steady discharge from a vessel (Fig. 9.2) and consider the entire flow space as a streamtube. Figure 9.2 shows clearly that the only region where the assumptions of stream filament theory are not satisfied is in the transition between the large cross-section A_1 and the smaller cross-section A_2. We assume that the depth h does not vary with time: this occurs if the ratio A_1/A_2 is large enough or if there is an appropriate influx into the vessel.

The flow is then steady and it follows from Bernoulli's equation (4.62) that

$$\frac{u_1^2}{2} + g h = \frac{u_2^2}{2}, \qquad (9.10)$$

where we have already made use of the fact that $p_1 = p_2 = p_0$. Using the continuity equation (9.9) to solve for u_2 furnishes the discharge velocity

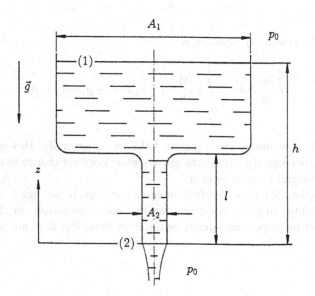

Fig. 9.2 Discharge from a vessel

$$u_2 = \sqrt{\frac{2g\,h}{1 - (A_2/A_1)^2}}. \tag{9.11}$$

For $A_2/A_1 \to 0$ we obtain the famous Torricelli's formula

$$u_2 = \sqrt{2g\,h}. \tag{9.12}$$

For the limit $A_2/A_1 \to 1$ we find $u_2 \to \infty$, which, from (9.9) and also (9.10) would imply $u_1 \to \infty$. This unrealistic result comes from the fact that for $A_2/A_1 \to 1$, u_1 and u_2 cannot satisfy the Eqs. (9.9) and (9.10) simultaneously. In this case the fluid cannot completely fill the cross-section A_2, so that the assumptions which led to (9.9) break down. For given u_1, (9.10) leads to

$$u_2 = \sqrt{u_1^2 + 2g\,h} \tag{9.13}$$

and thus, from the continuity equation, furnishes the largest possible cross-section A_2 which will produce a uniform flow

$$A_2 = \frac{A_1}{\sqrt{1 + 2g\,h/u_1^2}}. \tag{9.14}$$

If the exit cross-section is any larger, the fluid separates from the wall, and we see an unsteady and asymmetric exit flow. What has already been said furnishes the explanation of why the exit tube in funnels is of conical shape. If we consider A_2 as a function of z, the cross-sectional shape at which separation is just prevented along the whole tube is

$$A_2(z) = \frac{A_1}{\sqrt{1 + 2g(h - z)/u_1^2}}. \tag{9.15}$$

For this cross-sectional shape the pressure p in the fluid as a function of z is equal to the ambient pressure p_0. A jet issuing from a circular tube at the height h takes on this cross-sectional distribution, since its velocity increases with increasing $(h - z)$ under the effect of gravity.

We shall now consider unsteady discharge and for simplicity assume the ratio $A_1/A_2 \to \infty$. For $t < 0$ the cross-section of (2) is closed; it is suddenly opened at $t = 0$. At every time t, Bernoulli's equation for unsteady flow (4.61) holds

$$\varrho \int \frac{\partial u}{\partial t}\, ds + \varrho \frac{u^2}{2} + p + \varrho g z = C. \tag{9.16}$$

Here the integral along the streamline (which is fixed in space) is to be taken from the height h to the exit area. However in the transition region the flow is three-dimensional and cannot be described by stream filament theory. For $A_1/A_2 \to \infty$ the section of the tube contributes the greatest amount to the integral and we shall only take this into account. The contribution of the integration path in the container is negligible, since there $u \approx 0$ for all times, and therefore $\partial u/\partial t \approx 0$. Then it follows from (9.16), since again we have $p_1 = p_2 = p_0$, that

$$\int_0^l \frac{\partial u}{\partial t}\, ds + \frac{u_2^2}{2} = g h + \frac{u_1^2}{2}, \tag{9.17}$$

and also, since u in the pipe is not a function of s ($u = u_2$), that

$$l \frac{du_2}{dt} = g h - \frac{u_2^2}{2}, \tag{9.18}$$

where we have neglected the term $u_1^2/2$ because $A_1/A_2 \to \infty$. Integrating (9.18) with the initial condition $u_2(0) = 0$ leads to the solution

$$u_2(t) = \sqrt{2 g h}\, \tanh\!\left(\frac{\sqrt{2 g h}}{2l} t\right), \tag{9.19}$$

which shows that the maximum discharge velocity is reached for $t \to \infty$ and is equal to the steady velocity of Torricelli's formula. A more precise account of the flow in the transition region between container and exit tube would have led to a slightly different "effective" length l, and there only the time constant

$$\tau = \frac{2l}{\sqrt{2 g h}} \tag{9.20}$$

would have been affected. For $t = 3\tau$ the steady velocity is effectively reached, but during this time, for a large but finite A_1/A_2, the height of the water has barely dropped. The discharge after this is quasi-steady: the exit velocity can be calculated from (9.12) using the water height $h(t)$ at the time t. With this assumption, we determine the time required for the height to drop from h_0 to the actual height $h(t)$. From the continuity equation and Torricelli's formula for $A_1/A_2 \to \infty$, we obtain the differential equation for the water height

$$u_1 = -\frac{dh}{dt} = \frac{A_2}{A_1}\sqrt{2g\,h(t)}, \tag{9.21}$$

whose solution with $h(0) = h_0$ reads

$$t = \frac{A_1}{A_2}\sqrt{\frac{2}{g}}\left(\sqrt{h_0} - \sqrt{h(t)}\right). \tag{9.22}$$

The other limiting case $A_1/A_2 \to 1$, i.e., $l \to h$ in (9.17) leads to the result $du/dt = g$ (free fall), which is as expected since the bounding walls of the container exerts no force on the fluid.

9.1.3 Viscous Flow

While friction is negligible for pipe lengths l which are not too large, compared to the diameter, say, friction losses are noticeable for long pipes. Within the framework of stream filament theory, these losses can only be discussed phenomenologically and are introduced as additional pressure drops according to Eq. (6.60)

$$\Delta p_l = \varrho\frac{u^2}{2}\lambda\frac{l}{d_h}, \tag{9.23a}$$

or

$$\Delta p_l = \zeta\varrho\frac{u^2}{2}, \quad \text{with} \quad \zeta = \lambda\frac{l}{d_h}. \tag{9.23b}$$

The formulae (9.23a, 9.23b) correspond to the pressure loss in pipes of constant cross-section. If the cross-section is not constant we can consider these formulae to apply locally

$$d(\Delta p_l) = \varrho\frac{u^2(s)}{2}\frac{\lambda(s)}{d_h(s)}ds, \tag{9.24}$$

so that the equation

$$\Delta p_l = \varrho\frac{u_1^2}{2}\int_1^2\left(\frac{A_1}{A(s)}\right)^2\frac{\lambda(s)}{d_h(s)}ds \tag{9.25}$$

holds for the pressure loss between positions (1) and (2). We rewrite this equation using the loss factor ζ

$$\Delta p_l = \varrho \frac{u_1^2}{2} \zeta, \tag{9.26}$$

where

$$\zeta = \int\limits_{1}^{2} \left(\frac{A_1}{A(s)}\right)^2 \frac{\lambda(s)}{d_h(s)} \, ds. \tag{9.27}$$

In doing this we always refer the loss coefficient to the dynamic pressure $\varrho u_1^2/2$ at position before the loss has occurred. (In literature ζ is often referred to the dynamic pressure behind the position of loss.)

For long enough pipes, the pipe flow friction coefficients (cf. Chaps. 6 and 7) can be used. However we must recall that fully developed pipe flow only begins at a certain distance after the pipe entrance. A boundary layer forms at the pipe entrance. Its thickness increases with increasing distance from the entrance, until the boundary layer finally grows together and fills the whole cross-section. Only somewhat after this position do we perceive fully developed pipe flow, whose velocity profile no longer changes as the flow progresses down the pipe (cf. Fig. 9.3). Since the volume flux $\dot{V}$ is independent of s, the fluid not yet affected by the friction is accelerated. In steady flow the pressure drop over the entrance length l_E may be calculated from Bernoulli's equation for loss free flow, since the streamline in the center of the pipe is not yet affected by the friction

$$p_1 - p_2 = \frac{\varrho}{2}\left(4\overline{U}^2 - \overline{U}^2\right) = 3\frac{\varrho}{2}\overline{U}^2. \tag{9.28}$$

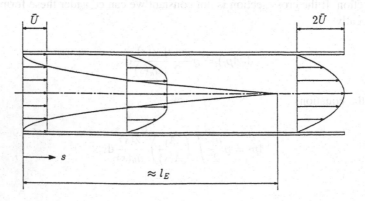

Fig. 9.3 Laminar entrance flow

Even if we assume that the contribution to the pressure drop as a consequence of the friction stresses on the pipe wall at the entrance section is the same as in fully developed pipe flow, we obtain a larger pressure drop, because the flux of the kinetic energy at the entrance is smaller than in the region of fully developed pipe flow. We estimate this additional work from the energy equation, where we neglect the dissipated energy. In incompressible and adiabatic flow we then have $De/Dt = 0$, and it follows that

$$-\pi R^2 \varrho \frac{\overline{U}^3}{2} + \varrho\pi \int\limits_0^R u^3(r)\, r\, \mathrm{d}r = \pi R^2 (p_1 - p_2)_{kin}\overline{U}. \tag{9.29}$$

After carrying out the integration we obtain the pressure drop due to the increase in kinetic energy

$$(p_1 - p_2)_{kin} = \frac{\varrho}{2}\overline{U}^2. \tag{9.30}$$

The pressure drop which results from the wall shear stresses has to be added on to this. We estimate this pressure drop as if the formula for fully developed pipe flow were also to hold in the entrance section, so that the total pressure drop becomes

$$\Delta p_{tot} = (p_1 - p_2)_{kin} + \zeta\frac{\varrho}{2}\overline{U}^2, \tag{9.31}$$

or else, using (6.61)

$$\Delta p_{tot} = \frac{\varrho}{2}\overline{U}^2\left(1 + \frac{l_E}{d}\frac{64}{Re}\right). \tag{9.32}$$

The total pressure drop corresponds to the pressure drop along the streamline in (9.28). Using Eqs. (9.28) and (9.32) we obtain an estimate for the entrance length in the laminar case

$$l_{E(laminar)} = \frac{Re}{32}d. \tag{9.33}$$

Here we are only dealing with a rough estimate. In reality the transition proceeds asymptotically: numerical solutions of the Navier-Stokes equations, in agreement with measurements, show that the velocity in the center of the pipe has reached about 90% of its maximum value at the entrance length given above (99% of the maximum velocity is finally reached when $l/d = 0.056\,Re$). In turbulent flow the velocity profile is flatter and the maximum velocity is only about 20% greater than the average velocity (cf. (7.83), (7.87), and (7.89) for $Re \approx 10^5$). Therefore the work required to increase the kinetic energy is almost negligible and the drag formula (7.89) for fully developed pipe flow can also be applied in the entrance region. The entrance length can be obtained from

$$l_{E(turbulent)} = 0.39 \, Re^{1/4} d, \tag{9.34}$$

which is much smaller than that of the laminar flow.

We shall now extend Bernoulli's equation (4.62) phenomenologically to include the pressure losses

$$\varrho \frac{u_1^2}{2} + p_1 + \varrho \, g \, z_1 - \Delta p_l = \varrho \frac{u_2^2}{2} + p_2 + \varrho \, g \, z_2, \tag{9.35}$$

where instead of $\overline{U}$ we write u, since in stream filament theory it is always the average velocity that is meant. The pressure loss in unsteady flow is only known for a few special cases, and in general it is not allowed to apply (9.35) to unsteady flows as well, while retaining the steady loss coefficients. It follows from (9.35) that the discharge velocity in the example in Fig. 9.2 is not (9.11) but rather

$$u_2 = \sqrt{\frac{2(\varrho \, gh - \Delta p_l)}{\varrho \left(1 - (A_2/A_1)^2\right)}}. \tag{9.36}$$

However, since the loss occurs essentially only in the pipe with the cross-sectional area A_2 where the entrance velocity is also u_2, we write

$$\Delta p_l = \zeta \, \varrho \frac{u_2^2}{2}, \tag{9.37}$$

and with this

$$u_2 = \sqrt{\frac{2g \, h}{1 + \zeta - (A_2/A_1)^2}}. \tag{9.38}$$

Here we recall that ζ in general depends on the Reynolds number and hence on u_2, so that (9.38) still does not give the exit velocity explicitly. If we assume, for example, that fully developed laminar pipe flow occurs over the entire length, i.e., that

$$\zeta = \frac{64 \, l}{Re \, d}$$

holds, and neglect $(A_2/A_1)^2$, the explicit formula follows

$$u_2 = 8\frac{\eta l}{R^2 \varrho}\left(\sqrt{1 + \frac{2g\,h}{[(8\eta l)/(R^2\varrho)]^2}} - 1\right). \tag{9.39}$$

If the effect of the losses in the pipe is large, i.e., for large ζ, we also get by expanding the square root

$$u_2 = \frac{\varrho\,gh}{8\eta\,l}R^2; \tag{9.40}$$

this result can also be obtained directly from (9.38).

In order to calculate the force exerted on the fluid by the vessel, we use the balance of momentum in the integral form (2.40) and apply it to the section of the streamline in Fig. 9.1. If $\vec{\tau}$ is the unit tangent vector of the average streamline fixed in space, using the assumptions of stream filament theory we get the equation

$$\int_0^L \frac{\partial(\varrho u)}{\partial t}\vec{\tau}A\,ds - \varrho_1 u_1^2 A_1 \vec{\tau}_1 + \varrho_2 u_2^2 A_2 \vec{\tau}_2 + \iint\limits_{(S_w)} \varrho\,u\,\vec{\tau}(\vec{u}\cdot\vec{n})\,dS$$

$$= p_1 A_1 \vec{\tau}_1 - p_2 A_2 \vec{\tau}_2 + \iint\limits_{(S_w)} \vec{t}\,dS. \tag{9.41}$$

We assume the flow to be uniform only at positions (1) and (2) so that the friction stresses vanish only there. The last integral represents the force exerted on the flow by the walls. Therefore the force exerted on the walls by the flow is exactly the negative of this integral. The surface integral on the left-hand side of (9.41) vanishes if the cross-section A does not vary in time. Otherwise we calculate the normal components $\vec{u}\cdot\vec{n}$ on S_w by (9.4) and, by a consideration completely analogous to (9.5) and (9.6) obtain the equation

$$\iint\limits_{(S_w)} \varrho\,u\vec{\tau}(\vec{u}\cdot\vec{n})dS = \int_0^L \varrho\,u\vec{\tau}\frac{\partial A}{\partial t}\,ds, \tag{9.42}$$

so that the balance of momentum appears in the form

$$\int_0^L \frac{\partial(\varrho u)}{\partial t}\vec{\tau}A\,ds + \int_0^L \varrho\,u\vec{\tau}\frac{\partial A}{\partial t}\,ds - \varrho_1 u_1^2 A_1 \vec{\tau}_1 + \varrho_2 u_2^2 A_2 \vec{\tau}_2$$

$$= p_1 A_1 \vec{\tau}_1 - p_2 A_2 \vec{\tau}_2 - \vec{F}, \tag{9.43}$$

which is generally valid within the framework of stream filament theory.

In applying (9.43) to the discharge vessel for unsteady flow, we again come up against the known difficulty that in order to work out the integral we need to know the flow quantities along the streamline. However, in the transition region between the large cross-section A_1 and the smaller one A_2 the quantities are unknown. For $A_1/A_2 \to \infty$ the section of the pipe again gives the greatest contribution. The second integral drops out since the cross-section is not a function of time. Further, both u and $\vec{\tau}$ are constant along the pipe, and because $\varrho_1 = \varrho_2 = \varrho$, we finally obtain

$$\vec{F} = \vec{\tau} \left(-\varrho A_2 l \frac{du_2}{dt} + \varrho\, u_1^2 A_1 - \varrho\, u_2^2 A_2 + p_1 A_1 - p_2 A_2 \right), \tag{9.44}$$

where we have not yet used $p_1 = p_2 = p_0$. In steady flow the first term in the brackets also drops out. Because $A_2/A_1 \to 0$, the momentum flux

$$\varrho\, u_1^2 A_1 = \varrho\, u_2^2 A_1 \frac{A_2^2}{A_1^2}$$

can be neglected. (If the velocities over the cross-section are not constant, as in fully developed laminar flow, the momentum fluxes are to be determined by integrating over the actual distribution.)

9.1.4   Application to Flows with Variable Cross-Section

The results are generally applicable to flow through pipes whose cross-sections narrow in the s-direction, such as often appear in applications in the form of *nozzles*. Nozzles serve to transform the pressure energy into kinetic energy, e.g., the blade rows in turbomachines often act as nozzles. Now the pressure decreases in the flow direction in a nozzle, and in addition they are almost always very short, so that fully developed flow cannot form. Both of these facts mean that the effect of friction is reduced. If necessary the effect of friction can be accounted for through a separate boundary layer calculation. No free surface appears in these applications and if we take the pressure relative to the hydrostatic pressure, Bernoulli's equation in steady flow reads

$$\varrho \frac{u_1^2}{2} + p_1 = \varrho \frac{u_2^2}{2} + p_2. \tag{9.45}$$

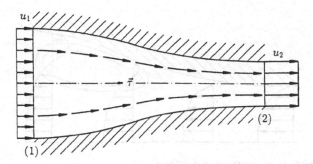

Fig. 9.4 Nozzle flow

Instead of (9.11) we then obtain

$$u_2 = \sqrt{\frac{2\Delta p/\varrho}{1 - (A_2/A_1)^2}} \qquad (9.46)$$

for the velocity at position (2), which shows that the "driving force" of the flow is the pressure difference $\Delta p = p_1 - p_2$. The magnitude of the force on the nozzle in Fig. 9.4 can be expressed in quantities at the position (1) using (9.44) and (9.45)

$$F = \varrho\, u_1^2 A_1 - \varrho\, u_1^2 \left(\frac{A_1}{A_2}\right)^2 A_2 + p_1 A_1 - A_2 \left\{ \varrho\, \frac{u_1^2}{2} \left[1 - \left(\frac{A_1}{A_2}\right)^2 \right] + p_1 \right\}. \qquad (9.47)$$

The flow processes within widening pipes are considerably more complicated. These find uses as *diffusers* and serve to transform kinetic energy into pressure energy. Since u_2 becomes smaller, the pressure here increases in the flow direction and even if the section of pipe is short (in fact, especially if it is short) boundary layer separation can occur at the wall, thus affecting the entire flow if the surface ratio A_2/A_1 is very large. In a diffuser the fluid particles must advance into regions of higher pressure, which they are only able to do because of their kinetic energy. If the Reynolds number is large a boundary layer forms close to the wall, where the particle velocity is smaller than the average velocity. The particles in the boundary layer have lost some of their kinetic energy through dissipation. Now the remaining kinetic energy is no longer enough to overcome the increasing pressure and the particles come to a standstill and finally, under the influence of the pressure gradient, are driven back opposite to their original direction of motion. All these events constitute the phenomenon of *boundary layer separation*, which we will discuss in detail in Sect. 12.1.4. Vortices form in the separated region and are kept in motion by the friction stresses and by turbulent stresses exerted by the unseparated flow. The separated flow is usually unsteady. A typical flow form is sketched in Fig. 9.5.

Because of the displacement action of the separated boundary layer the still unaffected core of the flow experiences a smaller cross-section increase than that which corresponds to the actual channel geometry. As a result the pressure build up

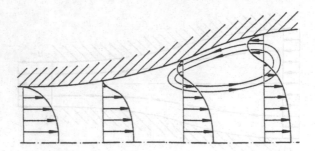

Fig. 9.5 Boundary layer separation in a diffuser

is smaller than expected. Usually the flow is evened out again downstream from the separation point by the transfer of momentum from the core flow to the separated flow. However, the work dissipated by the friction stresses leads to an additional pressure loss. The ratio of the actual pressure increase reached in the diffuser to that theoretically obtainable (i.e. according to the loss free Bernoulli equation) is called the *diffuser efficiency*

$$\eta_D = \frac{(p_2 - p_1)_{real}}{(p_2 - p_1)_{ideal}} = \frac{\varrho/2(u_1^2 - u_2^2) - \Delta p_l}{\varrho/2(u_1^2 - u_2^2)},\qquad(9.48)$$

where here too we set Δp_l for the pressure loss in the diffuser

$$\Delta p_l = \zeta\,\varrho\frac{u_1^2}{2},$$

so that we obtain the equation

$$\eta_D = 1 - \zeta\frac{1}{1 - (A_1/A_2)^2}.\qquad(9.49)$$

Here we have also made use of the continuity Eq. (9.9). The efficiency depends on the *opening angle* δ of the diffuser (Fig. 9.6). The highest efficiencies are reached for opening angles of

$$5° < \delta < 10°\qquad(9.50)$$

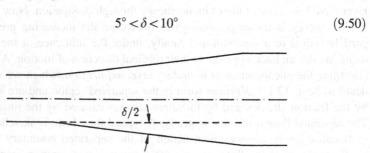

Fig. 9.6 Diffuser opening angle

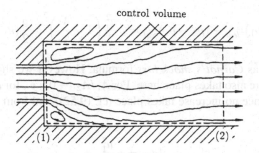

Fig. 9.7 Step expansion of the cross-section

and then amount to about 85%. If, for a given area ratio, the opening angles are smaller, the diffuser becomes so long that the friction losses from the walls become important, while for larger opening angles boundary layer separation occurs.

An abrupt change of cross-section can serve as an "ideally poor" diffuser (Fig. 9.7). Here the separation position is at the point where the cross-section changes.

The pressure over the cross-section is practically constant at the position (1), since the streamline curvature is very small (cf. (4.44) $\partial p/\partial n \approx 0$). In the subsonic flow considered here, the pressure in the jet is then generally equal to the surrounding pressure. (We shall see later that in compressible flow, waves can occur in the jet and as a consequence the pressure in the jet can be different from the surrounding pressure.) Thus the same pressure as in the jet also acts on the face of the cross-section widening. At position (2), the flow is uniform again and the pressure is p_2. Applying the balance of momentum to the control volume sketched, we find from (9.44) that

$$F = \varrho\, u_1^2 A_1 - \varrho\, u_2^2 A_2 + p_1 A_1 - p_2 A_2. \tag{9.51}$$

In doing this we have not made any of the simplifications beyond those associated with stream filament theory. If we neglect the contribution of the shear stress at the pipe wall to the force F, it then is simply the product $-p_1(A_2 - A_1)$ of the pressure and the face area, and we have for the pressure difference

$$(p_2 - p_1)_{real} = \varrho\, u_1^2 \frac{A_1}{A_2}\left(1 - \frac{A_1}{A_2}\right). \tag{9.52}$$

The pressure difference in loss free flow is found from Bernoulli's equation (9.45) as

$$(p_2 - p_1)_{ideal} = \varrho\, \frac{u_1^2}{2}\left(1 - \frac{A_1^2}{A_2^2}\right), \tag{9.53}$$

and thus the pressure loss reads

$$\Delta p_{IC} = (p_2 - p_1)_{ideal} - (p_2 - p_1)_{real} = \varrho \frac{u_1^2}{2} \left(1 - \frac{A_1}{A_2} \right)^2 = \frac{\varrho}{2} (u_1 - u_2)^2, \quad (9.54)$$

a relation known as *Carnot's shock loss* formula. However we should note that an increase in pressure also takes place here. For $A_1/A_2 \to 0$, i.e., for discharge into an infinitely large space no increase takes place (cf. (9.52)) and from (9.54) we obtain the *exit loss*

$$\Delta p_{IE} = \varrho \frac{u_1^2}{2}. \tag{9.55}$$

This is precisely the kinetic energy necessary to maintain the flow through the pipe. This exit loss can be reduced with a diffuser at the exit.

A similar loss to that of sudden area expansion also occurs for abrupt area contraction (Fig. 9.8). The reason for this can be seen in the separation of the flow at the sharp convex edge of the channel narrowing which the flow cannot follow. The separated stream then contracts to the cross-section $A_3 = \alpha A_2$, where α is dependent on the cross-section ratio A_1/A_2, and is called the *contraction coefficient*. The losses mainly arise during jet spreadingand can therefore by estimated using Carnot's shock loss formula

$$\Delta p_{IC} = \varrho \frac{u_3^2}{2} \left(1 - \frac{A_3}{A_2} \right)^2 = \varrho \frac{u_2^2}{2} \left(\frac{1 - \alpha}{\alpha} \right)^2. \tag{9.56}$$

The contraction coefficient α can be determined theoretically for $A_1/A_2 \to \infty$. For a plane two-dimensional orifice, the methods of function theory (Sect. 10.4.7) lead to $\alpha = 0.61$. For a circular aperture we find $\alpha = 0.58$ by numerical methods. Jet contraction and the losses associated with it can be minimized by rounding off the corner at the transition in cross-section. In bends or elbows, flow separation occurs usually at the inner side of the bend, and with it contraction of the main jet. The flow is smoothed out sufficiently far from the bend and we again have uniform flow. Contraction also occurs in pipe branches and valves and losses are associated with

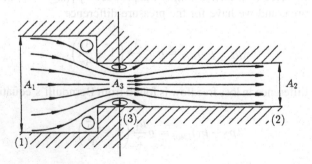

Fig. 9.8 Step contraction of the cross-section

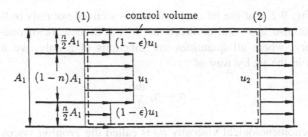

Fig. 9.9 Mixing

the subsequent smoothing out of the velocity. The flow patterns are usually so complicated that the losses cannot be estimated and we are nearly always referred to empirical data. Because of the large number of geometrical shapes, we refer to manufacturers' data and handbooks for the corresponding loss coefficients.

Associated with the sudden cross-section expansion is the mixing process of Fig. 9.9. Applying the balance of momentum in integral form and neglecting the shear stresses at the wall we obtain the increase in pressure due to the mixing process as

$$p_2 - p_1 = \varrho u_1^2 (1 - n) + \varrho u_1^2 (1 - \epsilon)^2 n - \varrho u_2^2. \tag{9.57}$$

From the continuity equation it follows that

$$u_2 = u_1 (1 - n\epsilon), \tag{9.58}$$

and therefore the pressure increase is

$$p_2 - p_1 = n(1 - n)\epsilon^2 \varrho u_1^2, \tag{9.59}$$

which is always positive because $n \leq 1$. For $\epsilon = 1$ we obtain the result (9.52) by replacing A_1 by $(1 - n) A_2$.

9.1.5 Viscous Jet

The discussion of viscosity effects in flow filament theory was so far restricted to flows in pipes and ducts, were the no-slip condition gives rise to shear stresses. We now turn to the effect normal viscous stresses have on the flow. Normal viscous stresses are important in free jets of highly viscous liquids as they occur, e.g., in the discharge from a vessel under the influence of gravity. For inviscid fluids this case has been treated in Sect. 9.1.2. However it would in general be quite wrong to compute the normal stresses a posteriori from this solution.

In case of viscous jets we neglect air friction at the free surface, as was already done in (6.33) so that only normal stresses occur in the jet. We know from

Sect. 9.1.2, Fig. 9.2, that the jet contracts and is strained not only in the (negative) z-direction but also in r- and φ-directions. In the framework of one-dimensional filament theory, where all quantities are functions of z only, we introduce the normal stress in the jet by way of

$$\sigma = \eta_T \frac{du}{dz}, \qquad (9.60)$$

where the phenomenological viscosity η_T is called the *Trouton* viscosity.

We determine the *Trouton* viscosity by requiring that the energy dissipated per unit time and volume $\sigma\, du/dz$ be equal to the energy $P_{zz}e_{zz} + P_{rr}e_{rr} + P_{\varphi\varphi}e_{\varphi\varphi}$ dissipated by the actual stretchings in z-, r-, φ-directions. From Appendix B 2 we have

$$e_{zz} = \frac{du_z}{dz} = \frac{du}{dz} \qquad (9.61)$$

and find from the continuity equation $A(z) = \pi r^2(z) = \text{const}$

$$\frac{du}{dz} = -\frac{u}{A}\frac{dA}{dz} = -\frac{2}{r}u\frac{dr}{dz} = -\frac{2}{r}u_r, \qquad (9.62)$$

where u_r is the material derivative of $r(z)$. From (9.62) and Appendix B 2 follows

$$e_{\varphi\varphi} = \frac{1}{r}u_r = -\frac{1}{2}\frac{du}{dz}, \quad \text{and} \qquad (9.63)$$

$$e_{rr} = \frac{\partial u_r}{\partial r} = -\frac{1}{2}\frac{du}{dz}. \qquad (9.64)$$

Since the stretchings are the same, so are the normal friction stresses irrespective of constitutive relations

$$P_{rr} = P_{\varphi\varphi}. \qquad (9.65)$$

Using this the above claim leads to

$$\eta_T = \frac{P_{zz} - P_{rr}}{du/dz}, \qquad (9.66)$$

which is occasionally cited as a "definition" of the *Trouton* viscosity. With $P_{zz} = 2\eta\, du/dz$ and $P_{rr} = -\eta\, du/dz$ the *Trouton* viscosity is three times the shear viscosity for Newtonian fluids

$$\eta_T = \frac{2\eta\, du/dz + \eta\, du/dz}{du/dz} = 3\,\eta. \qquad (9.67)$$

Starting point for the equation of motion of the jet is (2.18). Choosing $A dz$ as the infinitesimal integration domain for the volume integrals, where A is the cross-sectional area of the jet, the integrand may be considered constant in the domain, so that the right side of (2.18) becomes $\rho(D\vec{u}/Dt)A\,dz$ and the first integral on the left side $\rho\vec{k}A dz$. The stress vector $(-p+\sigma)\vec{n}$ varies over the surface and the integration for the surface force in (2.18) must be carried out. The integration of the pressure over the closed surface $A + dA$, A, dM vanishes as the pressure on the outer covering dM is the same as on the cross sections $A + dA$ and A. According to Fig. 9.2 z-direction is anti-parallel to the body force $\vec{g}$, so on $A + dA$, $\vec{n} = \vec{e}_z$ and the viscous normal stress there is $\sigma + d\sigma$. On A $\vec{n} = -\vec{e}_z$ and the viscous normal stress is σ. Viscous normal stresses are obviously not present on dM. The surface force is therefore $((A + dA)(\sigma + d\sigma) - A\sigma)\vec{e}_z$ which brings (2.18) to the form

$$\rho \frac{D\vec{u}}{Dt} A\,dz = d(A\sigma)\vec{e}_z + \rho \vec{k} A\,dz. \tag{9.68}$$

Since $\vec{u} = -u\,\vec{e}_z$ and $\vec{k} = -g\,\vec{e}_z$, and using (9.60) and (9.61) this can be rewritten as

$$\frac{d^2u}{dz^2} - \frac{1}{u}\left(\frac{du}{dz}\right)^2 + \frac{\rho}{\eta_T}\left(u\frac{du}{dz} - g\right) = 0. \tag{9.69}$$

Contrary to the corresponding equation for the inviscid flow, which follows from (9.69) in the limit $\eta \to 0$, (9.69) is a second order differential equation requiring two boundary (or initial) conditions. The nonlinear equation is best integrated numerically as an initial value problem and (9.69) is now integrated from the jet exit to the jet length L. It is then expedient to choose the positive z-direction parallel to the flow direction, i.e., replace $\vec{e}_z$ by $-\vec{e}_z$ in the above equations. (This does not change (9.69)). One initial condition is the exit velocity of the jet from the mouth, and a second initial condition (for du/dz) may be found from the momentum equation in the form (2.40). For a jet of length L exiting at $x = 0$ and a control volume consisting of the areas $A(0)$ and $A(L)$, and the outer covering M we find the momentum equation in the form

$$-\rho\,u(0)A(0)\vec{u}(0) + \rho\,u(L)A(L)\vec{u}(L) = -\sigma(0)A(0)\vec{e}_z + \sigma(L)A(L)\vec{e}_z + \rho\vec{g}V, \tag{9.70}$$

where $V = \int_0^L A(\zeta)\,d\zeta$ is the jet volume and $\vec{e}_z$ now points in flow direction. With the mass flow $\dot{m} = \rho\,u(0)A(0)$ we also have the scalar form

$$\dot{m}(u(0) - u(L)) - \eta_T(u'(0)A(0) - u'(L)A(L)) + \rho\,g\,V = 0, \tag{9.71}$$

where the prime indicates differentiation with respect to z. Integration proceeds using the known value $u(0) = U$ say and an estimated value of $u'(0)$. If the

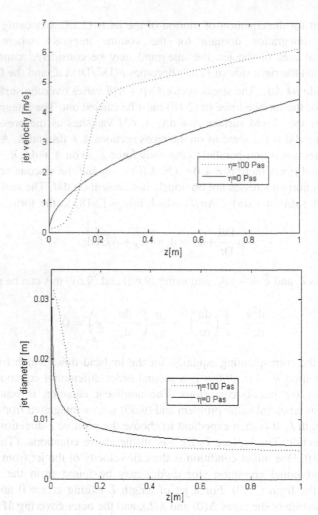

Fig. 9.10 Diameter and velocity distribution

momentum Eq. (9.71) is not satisfied by the values found at $x = L$, $u'(0)$ is varied until it is. Figure 9.10 shows the diameter and velocity distribution so computed for an inviscid jet and a viscous jet having a shear viscosity of 100 *Pas*, a value typically for molten glass.

Simpler forms of (9.70) emerge if the weight and the momentum are neglected and find application in fiber spinning.

9.2 Steady Compressible Flow

9.2.1 Flow Through Pipes and Ducts with Varying Cross-Section

We shall first restrict ourselves to those steady flows where, from the estimate (4.184), compressibility effects are to be expected, and therefore the inequality $M^2 \ll 1$ is no longer satisfied. A number of flow attributes then arise which do not occur in incompressible flow.

In steady, homentropic flow, which is barotropic, we can still calculate the flow quantities at the position (2) from Bernoulli's equation (4.64) and the continuity equation

$$\varrho_1 u_1 A_1 = \varrho_2 u_2 A_2, \tag{9.72}$$

which follows from (9.8), given the quantities at the position (1). Instead of Bernoulli's equation, the energy equation together with the condition that the entropy is constant along the streamline, can be used. While the energy which is dissipated into heat is lost in the case of incompressible flow and can therefore no longer be transformed into mechanical energy, the energy transformed into heat is still usable in compressible flow. We see from the energy Eq. (4.2) for adiabatic incompressible flow,

$$\frac{De}{Dt} = \frac{\Phi}{\varrho}, \tag{9.73}$$

that all the dissipated energy flows into the increase of the internal energy, which incidentally does not depend on the density, since density is a constant rather than a state variable in incompressible flow. The corresponding equation for compressible flow reads

$$\frac{De}{Dt} + p\frac{Dv}{Dt} = \frac{\Phi}{\varrho}, \tag{9.74}$$

showing that part of the dissipated energy can be converted to work of expansion. However the irreversible process of dissipation increases the entropy, so that Bernoulli's equation is no longer applicable. Its place is taken by the energy Eq. (2.114), which we shall first bring it to a form suitable for stream filament theory. We further assume that the flow at positions (1) and (2) is uniform, and so the friction stresses and temperature gradients vanish here, but as before we allow friction and heat conduction processes between these positions. We take the streamtube wall to be at rest, but allow moving surfaces S_f within the streamtube (e.g., moving blades of turbomachines). Further, for reasons already explained, we neglect the work of the volume body force, and for the section of the streamtube in Fig. 9.1 we obtain

$$
\int_0^L \frac{\partial}{\partial t}\left(\varrho\frac{u^2}{2}+\varrho e\right)A\,ds - \left(\frac{u_1^2}{2}+e_1\right)\varrho_1 u_1 A_1 + \left(\frac{u_2^2}{2}+e_2\right)\varrho_2 u_2 A_2 +
$$

$$
- p_1 u_1 A_1 + p_2 u_2 A_2 = \iint\limits_{(S_f)} u_i t_i\,dS - \iint\limits_{(S_w)} q_i n_i\,dS. \tag{9.75}
$$

We write $\dot{Q}$ for the heat supplied through the wall and P for the power supplied from the moving surfaces. We specialize the equation to steady flow and use the continuity equation (9.72)

$$
\frac{u_2^2}{2}+e_2+\frac{p_2}{\varrho_2}=\frac{u_1^2}{2}+e_1+\frac{p_1}{\varrho_1}+\frac{\dot{Q}+P}{\varrho_1 u_1 A_1}. \tag{9.76}
$$

With the definition of enthalpy (2.117) we write

$$
\frac{u_2^2}{2}+h_2=\frac{u_1^2}{2}+h_1+q+w, \tag{9.77}
$$

where for short we have set

$$
q=\frac{\dot{Q}}{\varrho_1 u_1 A_1} \tag{9.78}
$$

and

$$
w=\frac{P}{\varrho_1 u_1 A_1}. \tag{9.79}
$$

For adiabatic flow ($q = 0$) where no work is supplied ($w = 0$), the energy equation assumes that same form as the Eq. (4.150) for inviscid flow

$$
\frac{u_1^2}{2}+h_1=\frac{u_2^2}{2}+h_2=h_t. \tag{9.80}
$$

However we should note that (9.80) only holds between two positions which are in equilibrium states, i.e., where there are no temperature and velocity gradients. On the other hand the energy equation for isentropic flow holds for every point on the streamline. Since every point on the streamline is in equilibrium in isentropic flow, (9.80) reduces to (4.150). The result (9.80) also becomes obvious if we recall that, although the viscous stresses perform work within the control volume, this is dissipated into heat and therefore implies no net change in the energy.

The influence of the *Mach number* gives rise to further differences between compressible and incompressible flow. We shall see that surfaces of discontinuity, over which flow quantities change discontinuously, are possible also for steady supersonic flow ($M > 1$). The most important of these surfaces has already been discussed in connection with Fig. 4.28. We shall first however examine the effect that the Mach number has on the relation between the cross-sectional area A and the velocity u in isentropic flow. In incompressible flow this relation is directly obvious from the continuity equation

$$u A = \text{const} \tag{9.81}$$

as A becomes large u must decrease, and vice versa. However the continuity equation for compressible flow

$$\varrho u A = \text{const} \tag{9.82}$$

contains the additional variable ϱ, so that we should expect differing behavior. If we call the arc-length along the streamline x (to distinguish it from the entropy s), by logarithmic differentiation of (9.82) with respect to x we obtain the expression

$$\frac{1}{u}\frac{du}{dx} + \frac{1}{A}\frac{dA}{dx} + \frac{1}{\varrho}\frac{d\varrho}{dx} = 0. \tag{9.83}$$

For isentropic flow, thus $p = p(\varrho)$, we have from the definition of the *speed of sound*

$$a^2 = \left(\frac{\partial p}{\partial \varrho}\right)_s, \tag{9.84}$$

in particular $dp/d\varrho = a^2$ and therefore from (9.83)

$$\frac{1}{u}\frac{du}{dx} + \frac{1}{A}\frac{dA}{dx} + \frac{1}{a^2\varrho}\frac{dp}{dx} = 0. \tag{9.85}$$

Using the component of Euler's equation in the direction of the streamline,

$$\varrho u \frac{\partial u}{\partial x} = -\frac{\partial p}{\partial x}, \tag{9.86}$$

we then obtain the equation

$$\frac{1}{u}\frac{du}{dx} + \frac{1}{A}\frac{dA}{dx} = \frac{u}{a^2}\frac{du}{dx}, \tag{9.87}$$

which we reduce to

$$\frac{1}{u}\frac{du}{dx}(1 - M^2) = -\frac{1}{A}\frac{dA}{dx}. \tag{9.88}$$

For $M < 1$ we obtain qualitatively the same behavior as in incompressible flow: increasing cross-sectional area $(dA/dx > 0)$ corresponds to decreasing velocity $(du/dx < 0)$ and vice versa. However, for $M > 1$ (9.88) shows that if the cross-sectional area $(dA/dx > 0)$ increases the velocity must also increase $(du/dx > 0)$, or if the area decreases so does the velocity. If dA/dx vanishes, i.e., if the cross-sectional area has an extremum, then either $M = 1$ or $u(x)$ also has an extremum. Since du/dx must remain finite, the Mach number $M = 1$ is only reached at the position where the cross-sectional area has an extremum, that being a minimum.

If the Mach number at this section of minimum area, also called the *throat*, is not one, the velocity has an extremum there. The possible flows in converging-diverging channels are sketched in Fig. 9.11. These flow forms are only realized if the pressure ratio across the entire converging-diverging channel is properly adjusted.

For the nozzle flows occurring in applications, say, in turbomachines or in jet engines, one of the following questions mostly arises: either the cross-section $A(x)$ of the nozzle is given, and the flow quantities are required as a function of

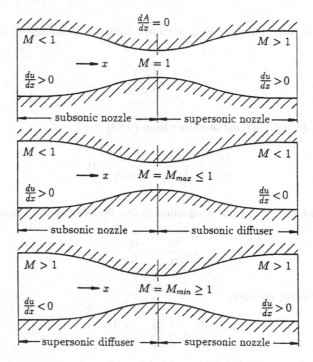

Fig. 9.11 Possible flow forms in converging-diverging channels

x (*direct problems*), or else the velocity $u(x)$ is known, and it is the associated cross-section that is required (*indirect problems*). Closed formulae can be given in the case of the calorically perfect gas and isentropic flow.

However, to begin with we shall discuss the more general solution for real gases, and shall assume that the equations of state are given in the well-known form of the *Mollier diagram*. We characterize the thermodynamic state of the gas by the *reservoir values* or *total values*: if the gas discharges from a large reservoir, the gas is in its *state of rest* inside the reservoir, and it is therefore said to be in its *reservoir state* or *total state*. In particular in the case of the calorically perfect gas, this state is used as a convenient reference state which can be defined at every point in the flow field as the state which would be attained were the gas brought to rest isentropically.

We infer from the energy equation (9.80) that the *reservoir enthalpy* h_t has the same value in adiabatic flow whether the gas has been brought to rest isentropically or not. We call h_t a *conserved quantity*. In the calorically perfect gas, because $h = c_p T$, the same also holds for the *reservoir* or *total temperature* T_t. On the other hand, the pressure depends on how the gas has been brought to rest, i.e., on the particular change of state. The *reservoir* or *total pressure* is only reached again if this change of state is isentropic. Therefore in this sense the total pressure is not a conserved quantity. It changes when the entropy changes, for example, if the gas passes through a shock wave.

As the governing equation for the design of the nozzle, besides the continuity equation

$$\varrho u A = \dot{m}, \tag{9.89}$$

we use the energy equation which holds for every point along the streamline in isentropic flow

$$\frac{u^2}{2} + h = h_t. \tag{9.90}$$

Besides h_t and p_t, the pressure drop across the nozzle $p_1 - p_2$ and the mass flux $\dot{m}$ must be given in the problem. For the direct problem we first form the variable

$$\varrho u = \frac{\dot{m}}{A(x)}, \tag{9.91}$$

whose right-hand side is a given function of x. We then note the values of h and ϱ along the isentrope s_t which is fixed by h_t and p_t, and insert (9.91) as well as the relation found for $h(\varrho)$ into the energy equation and obtain an equation for $\varrho(x)$

$$h(\varrho) + \frac{1}{2\varrho^2}\left(\frac{\dot{m}}{A(x)}\right)^2 = h_t. \tag{9.92}$$

We then solve this equation graphically or numerically for a given $A(x)$. Using the known $\varrho(x)$, we find the remaining variables of state $h(x)$, $T(x)$ and $p(x)$ along the isentrope s_t. In this manner the entire flow process can be found. The speed of sound a is determined by noting the values of p and ϱ along the isentrope s_t and forming the derivative $dp/d\varrho = a^2$, graphically if necessary. Now the Mach number distribution $M(x)$ can be found.

In indirect problems, we first calculate $h(x)$ from the given distribution $u(x)$ and find all the associated variables of state along the isentrope s_t. With $\varrho(x)$ now known, the distribution of the cross-section $A(x)$ follows from the continuity Eq. (9.89). For calorically perfect gas, closed form solutions can be given for the flow quantities. To do this we proceed by first giving the flow quantities as functions of the Mach number and then the cross-sectional area of the nozzle as a function of the Mach number. However first we introduce Bernoulli's equation for calorically perfect gas. From the isentropic relation for calorically perfect gas

$$p = C \varrho^{\gamma}, \tag{9.93}$$

we calculate the pressure function P as

$$P = \int \frac{dp}{\varrho} = C^{1/\gamma} \frac{\gamma}{\gamma - 1} p^{(\gamma-1)/\gamma}. \tag{9.94}$$

Replacing C by (9.93) evaluated at the reference state, i.e.,

$$C = p_1 \varrho_1^{-\gamma},$$

we extract

$$P(p) = \frac{\gamma}{\gamma - 1} \frac{p_1}{\varrho_1} \left(\frac{p}{p_1}\right)^{(\gamma-1)/\gamma}, \tag{9.95}$$

or by directly applying the isentropic relation (9.93),

$$P(p, \rho) = \frac{\gamma}{\gamma - 1} \frac{p}{\varrho}. \tag{9.96}$$

By doing this, Bernoulli's equation assumes the same form as the energy equation

$$\frac{u^2}{2} + \frac{\gamma}{\gamma - 1} \frac{p}{\varrho} = \text{const}, \tag{9.97}$$

while (9.95) leads to

$$\frac{u^2}{2} + \frac{\gamma}{\gamma-1}\frac{p_1}{\varrho_1}\left(\frac{p}{p_1}\right)^{(\gamma-1)/\gamma} = \text{const} \qquad (9.98)$$

or

$$\frac{u_1^2}{2} + \frac{\gamma}{\gamma-1}\frac{p_1}{\varrho_1} = \frac{u_2^2}{2} + \frac{\gamma}{\gamma-1}\frac{p_1}{\varrho_1}\left(\frac{p_2}{p_1}\right)^{(\gamma-1)/\gamma}. \qquad (9.99)$$

In particular we refer to the last form as Bernoulli's equation for compressible flow of a calorically perfect gas.

We now obtain the discharge velocity from a large reservoir as

$$u_2 = \sqrt{2\frac{\gamma}{\gamma-1}\frac{p_1}{\varrho_1}\left(1 - \left(\frac{p_2}{p_1}\right)^{(\gamma-1)/\gamma}\right)}. \qquad (9.100)$$

Equation (9.100) corresponds to Torricelli's formula for incompressible flow and is called the *Saint-Venant-Wantzel formula*. The greatest velocity in steady flow is reached for $p_2 = 0$, i.e., for expansion into a vacuum

$$u_{max} = \sqrt{2\frac{\gamma}{\gamma-1}\frac{p_1}{\varrho_1}}. \qquad (9.101)$$

If air under normal conditions expands into a vacuum, we obtain a maximum velocity of about

$$u_{max} \approx 735\,\text{m/s}. \qquad (9.102)$$

In order to represent the thermodynamic variables as functions of the Mach number, we rewrite Bernoulli's equation (9.97) with the expression from (9.93),

$$a^2 = \gamma\frac{p}{\varrho}, \qquad (9.103)$$

and obtain

$$\frac{u^2}{2} + \frac{1}{\gamma-1}a^2 = \frac{1}{\gamma-1}a_t^2, \qquad (9.104)$$

or for the ratio of the *total* to the *local temperature*,

$$\frac{T_t}{T} = \left(\frac{a_t}{a}\right)^2 = \frac{\gamma - 1}{2}M^2 + 1. \tag{9.105}$$

Using the isentropic relation (9.93) and the equation of state for the thermally perfect gas $p = \varrho RT$ we then obtain

$$\frac{p_t}{p} = \left(\frac{T_t}{T}\right)^{\gamma/(\gamma-1)} = \left(\frac{\gamma - 1}{2}M^2 + 1\right)^{\gamma/(\gamma-1)} \tag{9.106}$$

and

$$\frac{\varrho_t}{\varrho} = \left(\frac{T_t}{T}\right)^{1/(\gamma-1)} = \left(\frac{\gamma - 1}{2}M^2 + 1\right)^{1/(\gamma-1)}. \tag{9.107}$$

We call the flow variables encountered at $M = 1$ *critical* or *sonic* and denote them with the superscript $*$. These values differ from the total values only by constant factors and are therefore often used as reference values. In particular for diatomic gases ($\gamma = 1.4$), we find

$$\frac{a^*}{a_t} = \left(\frac{2}{\gamma + 1}\right)^{1/2} = 0.913, \tag{9.108}$$

$$\frac{p^*}{p_t} = \left(\frac{2}{\gamma + 1}\right)^{\gamma/(\gamma-1)} = 0.528, \quad \text{and} \tag{9.109}$$

$$\frac{\varrho^*}{\varrho_t} = \left(\frac{2}{\gamma + 1}\right)^{1/(\gamma-1)} = 0.634. \tag{9.110}$$

We shall now obtain the relation between the Mach number and the cross-sectional area. It follows from the continuity equation that

$$\dot{m} = \varrho u A = \varrho^* u^* A^* = \varrho^* a^* A^*, \tag{9.111}$$

in which A^* is the cross-section where $M = 1$ is reached. We also use this cross-section as the reference cross-section, even if the Mach number $M = 1$ is not reached in the nozzle, and define it with the given mass flux $\dot{m}$ as

$$A^* = \frac{\dot{m}}{\varrho^* a^*}. \tag{9.112}$$

In

$$\frac{A}{A^*} = \frac{\varrho^* \, \varrho_t a^*}{\varrho_t \varrho u},$$

(9.113)

we replace a^*/u using the energy Eq. (9.104)

$$u^2 + \frac{2}{\gamma - 1}a^2 = u^{*2} + \frac{2}{\gamma - 1}a^{*2} = \frac{\gamma + 1}{\gamma - 1}a^{*2},$$

(9.114)

then replace ϱ_t/ϱ and ϱ^*/ϱ_t using (9.107) and (9.110), respectively, and in this manner finally obtain the desired relation

$$\left(\frac{A}{A^*}\right)^2 = \frac{1}{M^2}\left[\frac{2}{\gamma + 1}\left(1 + \frac{\gamma - 1}{2}M^2\right)\right]^{(\gamma + 1)/(\gamma - 1)}.$$

(9.115)

If the mass flux, total state and cross-section are given, the Mach number distribution in the nozzle is known from (9.115). Using (9.105), (9.106) and (9.107) we then know the distribution of the temperature, pressure and density in the nozzle. The velocity then follows from (9.113). The relations mentioned are tabulated for $\gamma = 1.4$ in Appendix C and depicted in Fig. 9.12. In agreement with the qualitative considerations, Fig. 9.12 shows that in order to reach *supersonic velocities* the cross-section must increase again. Converging-diverging nozzles were first used in steam turbines and are known as *Laval nozzles*, but they find many other applications in, for example, rocket engines, nozzles in supersonic wind tunnels, etc.

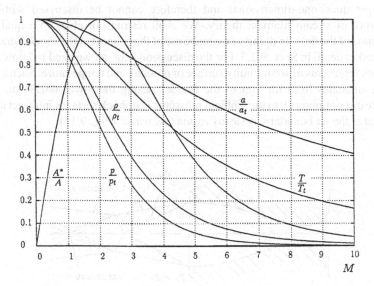

Fig. 9.12 Area ratio and variables of state as functions of the Mach number for steady flow of a perfect diatomic gas ($\gamma = 1.4$)

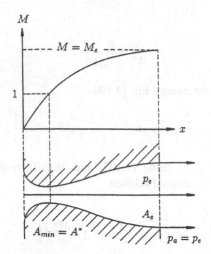

Fig. 9.13 Correctly expanding nozzle

However, in order to produce supersonic velocities a large enough pressure drop is required over the nozzle. We shall discuss the possible operational states of the Laval nozzle, starting with the normal case where the ambient pressure p_a is so chosen that it is the same as the pressure p_e at the nozzle exit given by the area ratio A^*/A_e (Fig. 9.13).

If the ambient pressure is increased, we then talk of an over expanded jet, because the gas in the nozzle expands to a lower than the ambient pressure: $p_e < P_a$. At first the flow in the nozzle does not change (curve 1 in Fig. 9.15). Outside the nozzle the flow is no longer quasi-one-dimensional and therefore cannot be discussed within the framework of stream filament theory. We shall restrict ourselves to a qualitative description of the flow. In doing so we make use of the concept of the *shock*, which will be treated in detail in Sect. 9.2.3. For the discussion here, all we need to know is that the shock represents a discontinuity surface of pressure and temperature. Such a shock surface emanates from the rim of the nozzle, raising the lower nozzle discharge pressure discontinuously to the ambient pressure. The shock surfaces intersect and are reflected at the jet boundary as steady *expansion waves* (Fig. 9.14).

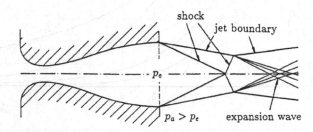

Fig. 9.14 Overexpanded jet

A rhombic pattern characteristic of supersonic jets arises and this is sometimes visible to the naked eye in exhaust jets of rocket engines, because the temperature of the fluid particles is raised by passing through the shock and then lowered again by passing through the expansion waves, where the intrinsic luminosity of the exhaust is altered in a corresponding manner.

If the ambient pressure is further raised, the shock moves into the nozzle and forms a *normal shock wave* in the nozzle. This discontinuous pressure increase positions itself in the nozzle just so that the required ambient pressure is reached. Behind the shock the flow is subsonic, as we will show later. The section of the nozzle behind the shock then works as a subsonic diffuser which theoretically raises the pressure behind the shock to the ambient pressure. However, in practice a flow separation occurs and the actual gain in pressure is so small that the pressure behind the shock is actually about the same as the ambient pressure. The subsonic jet cannot sustain steady waves, and for (almost) parallel discharge, the pressure in the jet must be the same as the ambient pressure (curve 2 in Fig. 9.15).

If the ambient pressure is raised even further, the shock migrates further into the nozzle and it becomes weaker, since the Mach number in front of the shock becomes smaller. If the ambient pressure is so increased that the shock finally reaches the throat of the nozzle, the shock strength drops to zero and the whole nozzle contains subsonic flow (curve 3 in Fig. 9.15). If we increase p_a even further, the Mach number has a maximum at the throat, but $M = 1$ is no longer reached

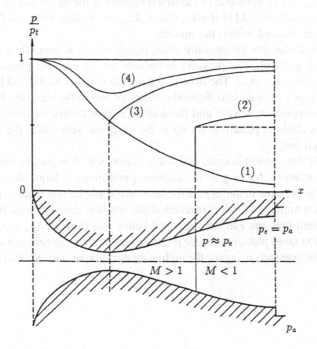

Fig. 9.15 Overexpanding nozzle

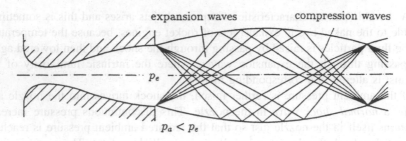

Fig. 9.16 Under expanded jet

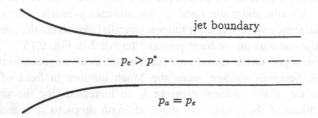

Fig. 9.17 Subsonic nozzle and subsonic jet

(curve 4 in Fig. 9.15); the value of the Mach number at the throat can be determined from the area relation (9.115) if we set $A = A_{min}$. A^* is then only a reference area which is never realized within the nozzle.

In underexpanded jets the pressure at the nozzle exit p_e is larger than the ambient pressure p_a (Fig. 9.16). The pressure is reduced to the ambient pressure through stationary expansion waves. The flow in the nozzle remains unaffected by this. The expansion waves penetrate into themselves and are then reflected at the boundary of the jet as "compression waves" and these often reform themselves into a shock. In this manner a rhombic pattern is set up in the jet again, very much the same as for over-expanded jets.

In a purely convergent nozzle, no steady supersonic flow can be formed in the above stated manner. As long as the ambient pressure p_a is larger than the critical pressure p^*, the pressure in the jet p_e is the same as the ambient pressure p_a (Fig. 9.17).

If the Mach number $M = 1$ is reached at the smallest cross-section, then $p_e = p^*$ and the ambient pressure can be decreased below this pressure $(p_a < p_e)$. Then an after-expansion takes place in the free jet: the pressure at the nozzle exit is expanded to the ambient pressure p_a again through stationary expansion waves (Fig. 9.18).

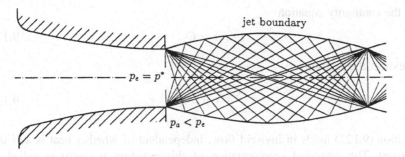

Fig. 9.18 Subsonic nozzle with after expansion in the jet

9.2.2 Constant Area Flow

As a further application of stream filament theory we shall consider the flow in a
duct where the cross-sectional area remains the same, without moving internal
surfaces or friction, but with heat transfer through the pipe wall. Then (9.77) holds

$$\frac{u_2^2}{2} + h_2 = \frac{u_1^2}{2} + h_1 + q. \tag{9.116}$$

In the application of the balance of momentum, we assume here that there is no
friction at the wall. Because of $\vec{F} = 0$ and $A_1 = A_2$ we obtain from (9.43)

$$\varrho_2 u_2^2 + p_2 = \varrho_1 u_1^2 + p_1. \tag{9.117}$$

With the continuity equation

$$\varrho_2 u_2 = \varrho_1 u_1, \tag{9.118}$$

and the equation of state $h = h(p, \varrho)$, e.g., for the calorically perfect gas

$$h = \frac{\gamma}{\gamma - 1} \frac{p}{\varrho}, \tag{9.119}$$

four equations with four unknowns are available. For a real gas this system of
equations can be solved iteratively, but for the perfect gas the solution can be stated
explicitly. However, here we only want to demonstrate an important property of this
flow. From the balance of momentum

$$\varrho u^2 + p = C_1, \tag{9.120}$$

and the continuity equation

$$\varrho u = C_2,\qquad\qquad(9.121)$$

we extract the relation

$$\frac{C_2^2}{\varrho}+p=C_1.\qquad\qquad(9.122)$$

Equation (9.122) holds in inviscid flow, independent of whether heat is added or removed. The graphical representation of this equation $p = p(\varrho)$ is called the *Rayleigh curve*. In general we can state the equations of state for the enthalpy and entropy of a substance, $h = h(p, \varrho)$ and $s = s(p, \varrho)$, often in the form of a diagram. For the perfect gas these are the Eqs. (9.119) and

$$s = s_0 + c_v \ln\left(\frac{p}{p_0}\left(\frac{\varrho}{\varrho_0}\right)^{-\gamma}\right).\qquad\qquad(9.123)$$

Using these two equations of state the Rayleigh curve can be transformed into an h-s-diagram (Fig. 9.19). If we heat the gas we raise its entropy and move along the curve from left to right. We obtain the velocity in the pipe by differentiating (9.122) and inserting (9.121) to get

$$u^2\,\mathrm{d}\varrho = \mathrm{d}p\qquad\qquad(9.124)$$

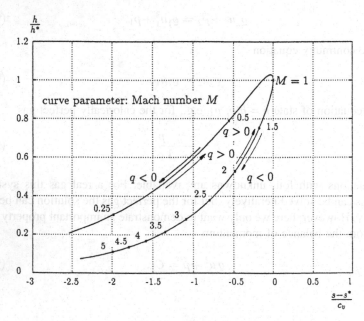

Fig. 9.19 Rayleigh curve for the perfect diatomic gas ($\gamma = 1.4$)

or

$$u^2 = \left(\frac{\mathrm{d}p}{\mathrm{d}\varrho}\right)_R, \tag{9.125}$$

where the index R means that the change of pressure with density is to be taken along the Rayleigh curve. If the heating is sufficiently high, we reach a point where $(\mathrm{d}s/\mathrm{d}h)_R = 0$ and which lies on the isentrope $s = $ const. Therefore for this point we have

$$u^2 = \left(\frac{\mathrm{d}p}{\mathrm{d}\varrho}\right)_R = \left(\frac{\partial p}{\partial \varrho}\right)_s = a^2, \tag{9.126}$$

and we see that this point corresponds to $M = 1$. If we cool the gas we decrease its entropy and we move along the curve from right to left. On the upper part of the curve (the subsonic branch) the Mach number is increased by heating as a consequence of the increase in entropy, and we see that there is a region where the enthalpy decreases with increasing entropy.

For a perfect gas this means that the temperature decreases there while the entropy increases. Clearly we cannot move through the point $M = 1$ from either the subsonic or the supersonic branch by heating, since the entropy would then have to decrease under heating. Of course, starting for example from the subsonic branch, we can apply heat until $M = 1$ is reached, and then remove heat to move back along the supersonic branch. If the Mach number $M = 1$ is reached at the exit (2) in the duct flow in Fig. 9.20, the greatest possible heat is thus added for a given mass flux.

If, in spite of this, we increase the heating further, the flow conditions change at position (1): the mass flux and with this the Mach number are reduced, so that increased heating again leads to $M = 1$ at the position (2).

We shall now consider the case where no heat is supplied in a duct of constant cross-section, but where friction may occur. From the continuity Eq. (9.121) and the energy equation

$$\frac{u^2}{2} + h = C_3, \tag{9.127}$$

which we derive from (2.114) in the same manner that led to (9.77), we obtain the *Fanno curve* $h = h(\varrho)$

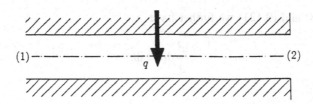

Fig. 9.20 Pipe flow with addition of heat

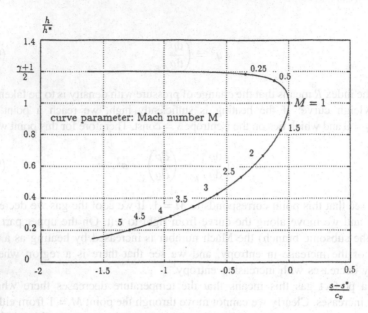

Fig. 9.21 Fanno curve for the perfect diatomic gas ($\gamma = 1.4$)

$$\frac{1}{2}\left(\frac{C_2}{\varrho}\right)^2 + h = C_3. \tag{9.128}$$

This can also be transformed into an h-s-diagram using the equation of state $s = s(\varrho, h)$ (Fig. 9.21). The Fanno curve is valid for a duct flow without heating, independent of the size of the wall friction. On this curve there is again a point where $(\mathrm{d}s/\mathrm{d}h)_F = 0$ and through which the isentrope goes. From Gibbs' relation

$$T\,\mathrm{d}s = \mathrm{d}h - \frac{\mathrm{d}p}{\varrho}, \tag{9.129}$$

it follows that for this point

$$\left(\frac{\mathrm{d}p}{\mathrm{d}h}\right)_F = \varrho = \left(\frac{\partial p}{\partial h}\right)_s. \tag{9.130}$$

Using (9.128) and (9.121) we further have

$$\frac{u^2}{\varrho}\,\mathrm{d}\varrho = \mathrm{d}h \tag{9.131}$$

or

$$\left(\frac{dh}{d\varrho}\right)_F = \frac{u^2}{\varrho} = \left(\frac{\partial h}{\partial \varrho}\right)_s \tag{9.132}$$

and, because of $\varrho = (\partial p/\partial h)_s$ finally

$$u^2 = \left(\frac{\partial p}{\partial h}\right)_s \left(\frac{\partial h}{\partial \varrho}\right)_s = \left(\frac{\partial p}{\partial \varrho}\right)_s = a^2. \tag{9.133}$$

The velocity associated with the point is again the velocity of sound. The upper part of the curve is the subsonic branch and the lower is the supersonic branch. Since in a flow where there is friction the entropy can only increase, the Mach number always increases in the subsonic branch until $M = 1$, but always decreases in the supersonic branch until the Mach number $M = 1$ is reached. Again the velocity of sound is reached at the end of the pipe. If we increase the effect of friction, for example by lengthening the pipe in the subsonic region, then the mass flux must decrease. In the supersonic region, if the length of the duct is greater than that where $M = 1$ is reached at the exit, a shock wave occurs and this brings the flow to subsonic velocity.

9.2.3 The Normal Shock Wave Relations

The shock wave mentioned in connection with the nozzle flow, i.e., the discontinuous transition from supersonic to subsonic velocity, occurs very often in supersonic flows. Here we shall discuss the *normal shock wave*, in which the shock surface is perpendicular to the velocity. However the more general relations of the *oblique shock wave* can be obtained from the results.

For most purposes it is enough to consider the shock wave as a surface of discontinuity across which the flow variables suddenly change. In what follows we shall derive relations from the conservation laws from which the quantities behind the shock can be determined knowing the corresponding ones in front of the shock. Strictly speaking the shock is not a surface of discontinuity. The quantities actually change continuously over a distance which is of the order of magnitude of the mean free path, and thus can be taken as infinitesimally small in almost all technical problems. Inside the shock the heat conduction and friction effects play a decisive role and the structure can be determined from, among other things, the Navier-Stokes equations. The theoretical and experimental results agree well for small supersonic Mach numbers. However we shall not go into the calculation of the shock structure here since in practice it is usually enough to know the change in flow quantities across the shock.

We assume that changes in velocity and temperature in front of and behind the actual shock vanish, or are at least much smaller than the changes within the shock itself. Since the thickness of the shock is very small, we neglect all volume integrals in the application of the conservation laws to the shock. (In particular, this is also the case for the unsteady flow to be discussed later.) In addition we neglect the external heating, since the surface of integration S_w in the balance of energy (9.75) tends to zero. From the continuity Eq. (9.8), the balance of momentum (9.41) and the balance of energy (9.75), we then obtain

$$\varrho_1 u_1 = \varrho_2 u_2, \tag{9.134}$$

$$\varrho_1 u_1^2 + p_1 = \varrho_2 u_2^2 + p_2, \quad \text{and} \tag{9.135}$$

$$\frac{u_1^2}{2} + h_1 = \frac{u_2^2}{2} + h_2, \tag{9.136}$$

where the index 1 denotes the position just in front of the shock and the index 2 the position just behind (Fig. 9.22).

Since the thickness of the shock is assumed to be infinitesimally small, the areas A_1 and A_2 are the same, even if the cross-section of the duct varies. The balance laws provide three equations for the four unknowns u_2, ϱ_2, p_2 and h_2. The system is made determinate by the addition of the equation of state

$$p = p(\varrho, h) \tag{9.137}$$

in the form of a Mollier chart, or else for the perfect gas

$$p = \varrho h \frac{\gamma - 1}{\gamma}. \tag{9.138}$$

With these, knowing the state in front of the shock, the state behind the shock can be determined, and the shock structure itself does not need to be known.

In general, only compression shock waves occur where $\varrho_2 > \varrho_1$, but expansion shock waves are also possible, according to the second law of thermodynamics if the inequality $(\partial^2 p/\partial v^2)_s < 0$ holds, as is possible, for example, near the critical point.

In what follows we shall only deal with compression shocks and first shall discuss the application of the conservation laws for a real gas whose Mollier chart is

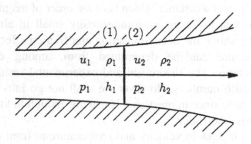

Fig. 9.22 Normal shock wave

given. If we insert the continuity Eq. (9.134) into the balances of momentum (9.135) and of energy (9.136) we obtain

$$p_2 - p_1 = \varrho_1 u_1^2 \left(1 - \frac{\varrho_1}{\varrho_2} \right) \tag{9.139}$$

and

$$h_2 - h_1 = \frac{u_1^2}{2} \left[1 - \left(\frac{\varrho_1}{\varrho_2} \right)^2 \right]. \tag{9.140}$$

The following calculation proceeds best if, for a given state in front of the shock, we estimate the density ratio ϱ_1/ϱ_2 across the shock, since it, contrary to the pressure and temperature ratios, remains finite even for a very strong shock. From (9.139) and (9.140) we directly obtain the pair of values (h_2, p_2), and using these find a new ϱ_2 from the Mollier diagram, from which a more precise estimate of the density ratio ϱ_1/ϱ_2 is obtained. Usually a few iterations are enough to determine the state behind the shock to the required precision.

Again for the calorically perfect gas closed relations can be given. We first of all eliminate the velocity u_1 from (9.139), (9.140), and obtain a relation solely between thermodynamic quantities, the so-called *Hugoniot relation*

$$h_2 - h_1 = \frac{1}{2} (p_2 - p_1) \left(\frac{1}{\varrho_1} + \frac{1}{\varrho_2} \right), \tag{9.141}$$

which still holds in general. Using (9.138) we find for the perfect gas the relation

$$\frac{p_2}{p_1} = \frac{(\gamma + 1)\varrho_2/\varrho_1 - (\gamma - 1)}{(\gamma + 1) - (\gamma - 1)\varrho_2/\varrho_1}, \tag{9.142}$$

between the pressure and the density ratios, from which we infer, for $p_2/p_1 \to \infty$, the maximum density ratio

$$\left(\frac{\varrho_2}{\varrho_1} \right)_{max} = \frac{\gamma + 1}{\gamma - 1}. \tag{9.143}$$

Contrary to this *Hugoniot change of state* (Fig. 9.23), we have for the isentropic change of state

$$\frac{p_2}{p_1} = \left(\frac{\varrho_2}{\varrho_1} \right)^{\gamma}, \tag{9.144}$$

and in the limit $p_2/p_1 \to \infty$ we obtain an infinitely large density ratio ϱ_2/ϱ_1. The maximum density ratio across a shock for diatomic gases with $\gamma = c_p/c_v = 7/5$ is then $\varrho_2/\varrho_1 = 6$, for fully excited internal degrees of freedom of the molecular

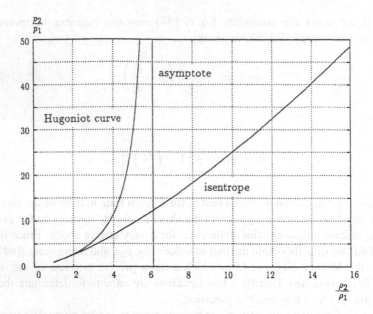

Fig. 9.23 Hugoniot curve for the perfect diatomic gas ($\gamma = 1.4$)

vibration ($\gamma = 9/7$) we find $\varrho_2/\varrho_1 = 8$, and for monatomic gases ($\gamma = 5/3$) then $\varrho_2/\varrho_1 = 4$.

We note that because $p = \varrho\, RT$,

$$\frac{p_2}{p_1} = \frac{\varrho_2}{\varrho_1}\frac{T_2}{T_1} \tag{9.145}$$

holds, and so for the limit $p_2/p_1 \to \infty$, T_2/T_1 also tends to infinity. If we solve (9.139) for the velocity, it follows that

$$u_1^2 = \frac{p_1}{\varrho_1}\left(\frac{p_2}{p_1} - 1\right)\left(1 - \frac{\varrho_1}{\varrho_2}\right)^{-1}, \tag{9.146}$$

and with $a^2 = \gamma p/\varrho$ for the calorically perfect gas also

$$\left(\frac{u_1}{a_1}\right)^2 = M_1^2 = \frac{1}{\gamma}\left(\frac{p_2}{p_1} - 1\right)\left(1 - \frac{\varrho_1}{\varrho_2}\right)^{-1}, \tag{9.147}$$

from which we can eliminate ϱ_1/ϱ_2 using the Hugoniot relation (9.30).

In this manner we obtain an equation for the pressure ratio

$$\left(\frac{p_2}{p_1} - 1\right)^2 - 2\frac{\gamma}{\gamma+1}(M_1^2 - 1)\left(\frac{p_2}{p_1} - 1\right) = 0, \qquad (9.148)$$

which, besides the trivial solution $p_2/p_1 = 1$ (no shock), also has the solution

$$\frac{p_2}{p_1} = 1 + 2\frac{\gamma}{\gamma+1}(M_1^2 - 1). \qquad (9.149)$$

This is an explicit relation between the pressure ratio across the shock and the Mach number M_1 in front of the shock. For $M_1 = 1$ both solutions merge into one another and the shock becomes a sound wave. Equation (9.149) shows that, for a shock wave $(p_2/p_1 > 1)$ the Mach number M_1 must be greater than one, and that for a very strong shock $(M_1 \to \infty)$ the pressure ratio tends to infinity. If we replace p_2/p_1 in (9.149) using the Hugoniot relation (9.142), we acquire the equation for the jump in density

$$\frac{\varrho_2}{\varrho_1} = \frac{(\gamma+1)M_1^2}{2+(\gamma-1)M_1^2}, \qquad (9.150)$$

which leads us again to the result (9.143) for $M_1 \to \infty$. Because of (9.145) we can now use (9.149) and (9.150) to obtain the jump in temperature as

$$\frac{T_2}{T_1} = \frac{p_2}{p_1}\frac{\varrho_1}{\varrho_2} = \frac{[2\gamma M_1^2 - (\gamma-1)][2+(\gamma-1)M_1^2]}{(\gamma+1)^2 M_1^2}. \qquad (9.151)$$

To find the Mach number behind the shock, we use the continuity Eq. (9.134) and $a^2 = \gamma p/\varrho$ to get

$$M_2^2 = \left(\frac{u_2}{a_2}\right)^2 = u_1^2\left(\frac{\varrho_1}{\varrho_2}\right)^2\frac{\varrho_2}{\gamma p_2} = M_1^2\frac{p_1\varrho_1}{p_2\varrho_2}, \qquad (9.152)$$

from which, using (9.149) and (9.150), we finally find

$$M_2^2 = \frac{\gamma+1+(\gamma-1)(M_1^2 - 1)}{\gamma+1+2\gamma(M_1^2 - 1)}. \qquad (9.153)$$

We infer from this equation that in a normal shock wave, because $M_1 > 1$, the Mach number behind the shock is always lower than 1. In the case of a very strong shock $(M_1 \to \infty)$, M_2 takes on the limiting value

$$M_2\big|_{(M_1 \to \infty)} = \sqrt{\frac{1}{2}\frac{\gamma-1}{\gamma}}. \qquad (9.154)$$

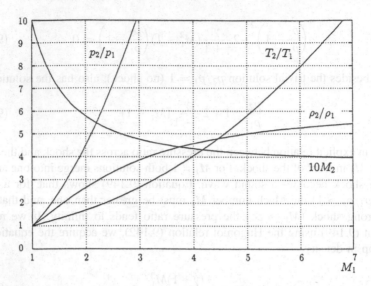

Fig. 9.24 Mach number and variables of state behind a shock as functions of the Mach number in front of the shock

The shock relations are tabulated in Appendix C for $\gamma = 1.4$ and are graphed in Fig. 9.24.

Because of the irreversible processes (friction, heat conduction), the entropy increases through the shock. We apply Reynolds' transport theorem (1.96) to (2.143) and it follows that for an infinitesimally thin shock,

$$\iint\limits_{(S)} \varrho \, s(\vec{u} \cdot \vec{n}) \, dS > 0, \tag{9.155}$$

or using the continuity equation (9.134),

$$s_2 - s_1 > 0. \tag{9.156}$$

We can confirm this explicitly for the calorically perfect gas if we use the equation arising from (9.123)

$$s_2 - s_1 = c_v \ln \left[\frac{p_2}{p_1} \left(\frac{\varrho_2}{\varrho_1} \right)^{-\gamma} \right] \tag{9.157}$$

and eliminate the density ratio using the Hugoniot relation (9.142)

$$s_2 - s_1 = c_v \ln \left[\frac{p_2}{p_1} \left(\frac{(\gamma - 1)p_2/p_1 + \gamma + 1}{(\gamma + 1)p_2/p_1 + \gamma - 1} \right)^{\gamma} \right]. \tag{9.158}$$

For $p_2/p_1 \to \infty$ the entropy difference tends logarithmically to infinity. For a weak shock we set $p_2/p_1 = 1 + \alpha$ and write out the following relation by expanding the right-hand side for small α

$$\frac{s_2 - s_1}{c_v} = \frac{\gamma^2 - 1}{12\gamma^2} \left(\frac{p_2 - p_1}{p_1} \right)^3 . \tag{9.159}$$

This shows directly that for calorically perfect gases, $p_2 - p_1$ must always be greater than zero so that only compression shock waves can occur, since otherwise the entropy would have to decrease across the shock.

9.3 Unsteady Compressible Flow

Just as in steady compressible flow, shocks also occur in unsteady compressible flow as surfaces separating different flow regions across which the shock relations must be satisfied. The shock relations therefore play the role of boundary conditions. First we shall discuss the shock relations for a shock wave which is in motion. As we have already noted, the volume integrals drop out of the balance laws even for unsteady shocks (shocks moving with varying velocity) as long as the shock thickness is taken to be infinitesimally thin. Therefore the balance Eqs. (9.134) to (9.136) and all the relations derived from these are still valid; we only have to pay attention to choose the correct velocities in front of and behind the shock.

To do this we shall consider a shock moving with velocity u_s (t) in a duct (not necessarily with constant cross-section) (Fig. 9.25).

Let the flow in front of the shock have the velocity u'_1 and the thermodynamic quantities p_1, ϱ_1 and h_1. We characterize the gas velocity in this system with a dash and call the reference system in which the duct is at rest and the shock is moving the *laboratory frame*, because this is the system in which experimental measurements are often made.

We distinguish this reference system from the *moving reference frame* in which the shock is at rest and so the steady shock relations (9.149), (9.150), (9.151) and (9.153) are valid. The moving reference frame is obtained by superimposing the shock velocity to all velocities in such a way that the shock itself is at rest (Fig. 9.26). Doing this we obtain the transformation equations

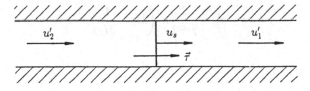

Fig. 9.25 Shock in the laboratory frame

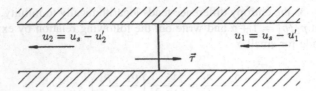

Fig. 9.26 Shock in the moving reference frame

$$u_1 = u_s - u_1', \tag{9.160}$$

and

$$u_2 = u_s - u_2', \tag{9.161}$$

where we take the velocities to be positive if they are opposite to $\vec{\tau}$, as denoted by the arrows in Fig. 9.26. The results from the steady shock relations can thus be carried over to the laboratory frame. Often the velocity in front of the shock is zero in the laboratory system, therefore,

$$u_1' = 0; \quad \text{since} \quad u_1 = u_s, \tag{9.162}$$

it is sufficient to replace M_1 in the shock relations for perfect gas by the shock Mach number

$$M_s = \frac{u_s}{a_1} \tag{9.163}$$

in order to obtain the shock relations of the moving shock. Then for the velocity u_2 in the moving frame we use (9.161) and the continuity equation in this frame

$$\varrho_1 u_s = \varrho_2 u_2 \tag{9.164}$$

to obtain the relation

$$u_2' = u_s \left(1 - \frac{\varrho_1}{\varrho_2}\right). \tag{9.165}$$

In this equation we can again replace ϱ_1/ϱ_2 using the shock relation (9.150), so that

$$u_2' = \frac{2}{\gamma + 1} a_1 \left(M_s - \frac{1}{M_s}\right). \tag{9.166}$$

For very high shock Mach numbers we find that

$$u_2'\big|_{(M_2\to\infty)} = \frac{2}{\gamma+1} u_s, \tag{9.167}$$

and we see that gas flows with high velocities can be produced behind the shock. However, further consideration shows that while the gas reaches supersonic velocity, the Mach number M_2' remains bounded as a consequence of the strong heating of the gas. From (9.161) we have for the Mach number

$$M_2' = \frac{u_2'}{a_2} = M_s \frac{a_1}{a_2} - M_2, \tag{9.168}$$

where we replace a_1/a_2 by $\sqrt{T_1/T_2}$ and introduce the resulting expression in the shock relation (9.151). We then extract the limit

$$M_2'\big|_{(M_s\to\infty)} = \sqrt{\frac{2}{\gamma(\gamma-1)}} \approx 1.89 \quad (\text{for } \gamma = 1.4) \tag{9.169}$$

for $M_s \to \infty$. (At high Mach numbers, air shows real gas effects and consequently higher values of M_2' are reached.)

To calculate the unsteady flow in the framework of stream filament theory we begin with the differential form of the balance equations. We obtain the differential form of the continuity equation from its integral form (9.8) if we integrate there only over the differential length dx and replace the quantities at position (2) by a Taylor expansion about the position (1). This leads to the equation

$$A\frac{\partial\varrho}{\partial t}dx + \varrho\frac{\partial A}{\partial t}dx - \varrho\,u\,A + \left(\varrho + \frac{\partial\varrho}{\partial x}dx\right)\left(u + \frac{\partial u}{\partial x}dx\right)\left(A + \frac{\partial A}{\partial x}dx\right) = 0, \tag{9.170}$$

which reduces to

$$\frac{\partial(\varrho A)}{\partial t} + \frac{\partial(\varrho\,u\,A)}{\partial x} = 0, \tag{9.171}$$

where the terms quadratic in dx drop out in the limit $dx \to 0$. For the differential form of the equations of motion we begin directly with (4.56) and neglect the volume body forces

$$\frac{\partial u}{\partial t} + u\frac{\partial u}{\partial x} = -\frac{1}{\varrho}\frac{\partial p}{\partial x}. \tag{9.172}$$

In (9.172) the friction effects can also be accounted for phenomenologically, by adding the additional pressure gradients according to (9.26). As already noted, however, the friction coefficients are mostly unknown for unsteady flow. Therefore in what follows we shall restrict ourselves to loss free adiabatic flow, which is then isentropic. From the general equation of state $\varrho = \varrho(p, s)$, and using $Ds/Dt = 0$ it follows that

$$\frac{D\varrho}{Dt} = \left[\frac{\partial \varrho}{\partial p}\right]_s \frac{Dp}{Dt} = a^{-2}\frac{Dp}{Dt}. \tag{9.173}$$

In addition we restrict ourselves to flows through ducts of constant cross-section. Then the continuity Eq. (9.171) takes on the form (2.3a), which reads here

$$\frac{D\varrho}{Dt} + \varrho\frac{\partial u}{\partial x} = 0. \tag{9.174}$$

If we also insert (9.173), after multiplying with a/ϱ we obtain

$$\frac{1}{\varrho a}\frac{\partial p}{\partial t} + \frac{u}{\varrho a}\frac{\partial p}{\partial x} + a\frac{\partial u}{\partial x} = 0. \tag{9.175}$$

Adding this equation to the equation of motion (9.172) we extract the interesting relation

$$\frac{\partial u}{\partial t} + (u + a)\frac{\partial u}{\partial x} + \frac{1}{\varrho a}\left(\frac{\partial p}{\partial t} + (u + a)\frac{\partial p}{\partial x}\right) = 0. \tag{9.176}$$

If we view this equation in connection with the general time derivative (1.19) (there applied to the temperature) we come to the following interpretation: along the path of an observer described by $dx/dt = u + a$, the change du/dt is equal to the change dp/dt multiplied by $-(\varrho a)^{-1}$. Instead of the partial differential equation (9.176) two coupled ordinary differential equations then appear

$$du + \frac{1}{\varrho a}dp = 0 \quad \text{along} \quad dx = (u + a)dt. \tag{9.177}$$

If we subtract (9.175) from (9.172) we extract the equation

$$\frac{\partial u}{\partial t} + (u - a)\frac{\partial u}{\partial x} - \frac{1}{\varrho a}\left(\frac{\partial p}{\partial t} + (u - a)\frac{\partial p}{\partial x}\right) = 0, \tag{9.178}$$

from which follow the two ordinary differential equations

$$du - \frac{1}{\varrho a}dp = 0 \quad \text{along} \quad dx = (u - a)dt. \tag{9.179}$$

$Ds/Dt = 0$ (cf. (4.48)) clearly implies that the change in the entropy of a material particle vanishes; expressed otherwise, the change in entropy along a particle path is zero

$$ds = 0 \quad \text{along} \quad dx = u\,dt. \tag{9.180}$$

The rearrangement and interpretation described have allowed us to reduce the three nonlinear partial differential equations (9.172), (9.174), and (4.48) to a system of six ordinary differential equations. From a mathematical point of view, we note that this equivalence represents the fundamental content of the *theory of characteristics*, which is a theory for the solution of systems of *hyperbolic* differential equations. The system of Eqs. (9.172), (9.174), and (4.48) is of this hyperbolic kind. The method of solution can also be carried over to steady supersonic flow because the differential equations describing supersonic flow are hyperbolic. We call the solution curves of the differential equations

$$\frac{dx}{dt} = u \pm a \quad \text{and} \quad \frac{dx}{dt} = u \tag{9.181}$$

in the *x-t*-plane *characteristics*; the path of a particle is therefore also a characteristic. The differential equations which are valid along these characteristics are called *characteristic* or sometimes *compatibility relations*.

As an example of their application we shall consider homentropic flow, for which

$$\frac{\partial s}{\partial x} = 0 \tag{9.182}$$

is valid and because of $Ds/Dt = 0$ also

$$\frac{\partial s}{\partial t} = 0. \tag{9.183}$$

Therefore, the entropy is constant in the entire *x-t*-plane, in particular along the characteristic lines. The Eq. (9.180) which determine the distribution of the entropy thus drop out. From (9.93),

$$p = C\varrho^\gamma,$$

where C is an absolute constant due to the constancy of the entropy, it follows that

$$\frac{dp}{d\varrho} = a^2 = C\gamma\varrho^{\gamma-1}. \tag{9.184}$$

Using this, the compatibility relations (9.177) and (9.179),

$$\mathrm{d}u \pm \frac{1}{\varrho a}\mathrm{d}p = \mathrm{d}u \pm \frac{a}{\varrho}\mathrm{d}\varrho = \mathrm{d}u \pm \sqrt{\gamma C}\,\varrho^{(\gamma-1)/2}\frac{\mathrm{d}\varrho}{\varrho} = 0 \qquad (9.185)$$

can be directly integrated

$$u + \sqrt{\gamma C}\,\frac{2}{\gamma - 1}\varrho^{(\gamma-1)/2} = u + \frac{2}{\gamma - 1}a = 2r, \qquad (9.186)$$

$$u - \sqrt{\gamma C}\,\frac{2}{\gamma - 1}\varrho^{(\gamma-1)/2} = u - \frac{2}{\gamma - 1}a = -2s, \qquad (9.187)$$

The constant of integration $2r$ is constant along the characteristic described by $\mathrm{d}x/\mathrm{d}t = u + a;\ -2s$ is constant along the characteristic $\mathrm{d}x/\mathrm{d}t = u - a$. We call these constants of integration *Riemann invariants*.

We now use these equations to calculate the flow in an infinitely long duct. Since the duct has no ends we are dealing with a pure *initial value problem*. At time $t = 0$ the initial condition in the duct is given by $u(x, 0)$ and $a(x, 0)$ (Fig. 9.27). We are looking for the state of the flow at a later instant in time t_0 at the place x_0, which is denoted in the x-t-plane as the point $P_0 = P(x_0, t_0)$ (Fig. 9.28). The quantities $2r$ and $-2s$ are constant along the characteristics and are given by the initial conditions. Therefore, we must have

$$2r = u(x_A, 0) + \frac{2}{\gamma - 1}a(x_A, 0) = u(x_0, t_0) + \frac{2}{\gamma - 1}a(x_0, t_0), \qquad (9.188)$$

and

$$-2s = u(x_B, 0) - \frac{2}{\gamma - 1}a(x_B, 0) = u(x_0, t_0) - \frac{2}{\gamma - 1}a(x_0, t_0), \qquad (9.189)$$

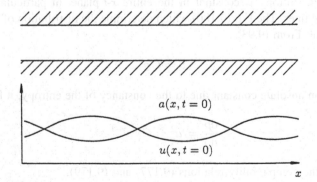

Fig. 9.27 Initial distributions

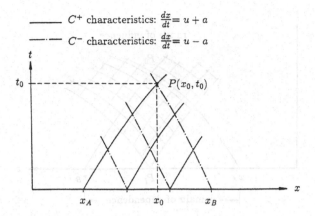

Fig. 9.28 Characteristics in the x–t plane

where x_A and x_B are points through which the characteristics pass. With these we now know u and a at the point P_0

$$u(x_0, t_0) = r - s,$$ (9.190)

$$a(x_0, t_0) = \frac{\gamma - 1}{2}(r + s).$$ (9.191)

The characteristics which run through the point P_0 and points x_A and x_B are not yet known. They can be determined by approximation: we fix a series of points along the x-axis, and at these we know the directions of the characteristics from the initial values. We approximate the characteristics by their tangents at these points. At the point where the tangents cross we can determine the values of u and a by the above method; but by doing this we again know the directions of the characteristics in these points and can approximate again. This process must be carried out until we reach the point P_0 that we want. The state of the flow at the point P_0 only depends on the initial data in the interval between x_A and x_B. We call this interval the *domain of dependence* of the point P_0 (Fig. 9.29). On the other hand the initial conditions at a point P_I only act in a bounded region, the *range of influence* of the point P_I.

Another example is the so-called *piston problem*, an *initial-boundary value problem*: in an infinitely long tube at position $x = 0$ there is a piston which is accelerated suddenly to a constant velocity $-|u_p|$ at time $t = 0$. The state in the tube before the piston is set in motion is given by $u = 0, a = a_4$. Therefore, the initial conditions are

$$u(x > 0, t = 0) = 0; \quad a(x > 0, t = 0) = a_4$$ (9.192)

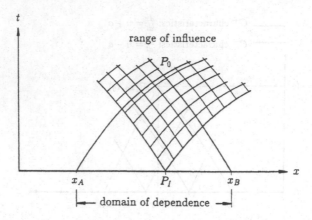

Fig. 9.29 Domain of dependence and range of influence in the x–t plane

and

$$-|u_p| \leq u(x = 0, t = 0) \leq 0. \tag{9.193}$$

The initial condition (9.193) comes from the fact that as the piston is set suddenly in motion at the position $x = 0$ at time $t = 0$ (thus in an infinitesimally short time), the gas must pass through the whole velocity interval, from the undisturbed velocity $u = 0$ up to the velocity given by the kinematic boundary condition

$$u(x = x_p, t) = -|u_p| \tag{9.194}$$

at the piston (piston path $x_p = -|u_p|t$). Therefore, the point $P_s = P(0, 0)$ is a *singular point* in the x-t-plane. To solve the problem we have available the equations for the characteristics

$$\frac{dx}{dt} = u \pm a \tag{9.195}$$

as well as the Eqs. (9.190) and (9.191), which read in general

$$u = r - s \tag{9.196}$$

and

$$a = \frac{\gamma - 1}{2}(r + s), \tag{9.197}$$

where the *Riemann constants* r and s are given by (9.186) and (9.187). As before we determine their values from the initial conditions. First it follows from (9.192) that

$$2r = \frac{2}{\gamma - 1} a_4 \quad \text{and} \quad -2s = -\frac{2}{\gamma - 1} a_4, \tag{9.198}$$

and then we find from (9.196) $u = 0$ and from (9.197) $a = a_4$ in a solution domain outside the range of influence of the singular point P_s, where the initial condition (9.193) holds. From

$$\frac{dx}{dt} = +a_4 \quad \text{it follows that} \quad x = +a_4 t + \text{const} \tag{9.199}$$

and from

$$\frac{dx}{dt} = -a_4 \quad \text{it follows that} \quad x = -a_4 t + \text{const.} \tag{9.200}$$

The characteristics with the positive sign on "a" point to the right and are therefore called *forward-facing characteristics* for short, although in general they could point to the left also; we shall denote them less ambiguously as C^+ *characteristics*. The characteristics with the negative sign on "a" are called *backward-facing* or C^- *characteristics*. The constants of integration are determined by the x value, the characteristics take at $t = 0$.

The range of influence of the singular point P_s is bounded on the right by the C^+ characteristic through P_s, for which $u = 0$ still holds. Between this characteristic $x = a_4 t$ and the x-axis the flow velocity is $u = 0$ and the velocity of sound is $a = a_4$. Physically this characteristic can be interpreted as a wave which reports the first effect of the piston motion to the gas at rest in the tube. In compressible media such a report can only propagate at a finite velocity, namely the velocity of sound. However a whole bundle of forward-facing characteristics whose slopes $dx/dt = u + a$ take on all the values between a_4 and $-|u_p| + a_3$ run through the singular point P_s. These characteristics are already drawn as straight lines in Fig. 9.30, since we shall presently show that both u and a are constant along these C^+ characteristics.

We calculate u at the point P_1 in Fig. 9.30 and using (9.186) and (9.187) obtain from (9.196)

$$u = \frac{1}{2}\left(u + \frac{2}{\gamma - 1} a\right) - \frac{1}{2}\left(\frac{2}{\gamma - 1} a_4\right), \tag{9.201}$$

where $-2s$ is fixed by the initial condition (9.192) and $2r$ follows from the initial condition (9.193).

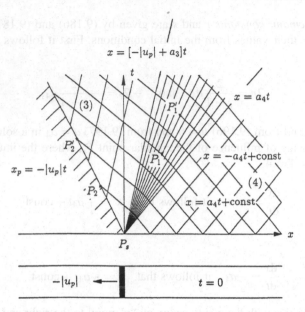

Fig. 9.30 x-t diagram of the piston problem

If we now calculate the velocity u at the point P_1' on the same C^+ characteristic, we are led to exactly the same equation, because the value of the Riemann invariant $2r$ is the same along the same characteristic, and the value of $-2s$ is also the same on all C^- characteristics, since they come from a region of homogeneous flow conditions. The velocity of sound a at the point P_1 follows from (9.197) as

$$a = \frac{\gamma - 1}{2}\left[\frac{1}{2}\left(u + \frac{2}{\gamma - 1}a\right) + \frac{1}{2}\left(\frac{2}{\gamma - 1}a_4\right)\right], \qquad (9.202)$$

and just as before we show that a has the same value at the point P_1' as at the point P_1. Thus u and a are constant on forward-facing characteristics, and the equation of the characteristics through the origin P_s reads

$$x = (u + a)t, \quad -|u_p| \le u \le 0. \qquad (9.203)$$

We insert this equation into (9.201), solve for u and in this way obtain u explicitly as a function of x and t

$$u = \frac{2}{\gamma + 1}\left(\frac{x}{t} - a_4\right). \qquad (9.204)$$

We insert this u into (9.203) and obtain

$$a = \frac{\gamma - 1}{\gamma + 1}\frac{x}{t} + \frac{2}{\gamma + 1}\, a_4 \,. \tag{9.205}$$

We could have reached exactly the same results using (9.202). We obtain the last characteristic belonging to the "fan" from (9.203) if we set $u = |u_p|$ there. We call the velocity of sound met on this characteristic a_3, and calculate it from (9.205) by inserting $x = (-|u_p| + a_3)t$

$$a_3 = -\frac{\gamma - 1}{2}|u_p| + a_4. \tag{9.206}$$

In order to calculate the velocity of sound at the piston (point P_2) we can use Eq. (9.202) by setting $u = -|u_p|$ and $a = a_p$,

$$a_p = -\frac{\gamma - 1}{2}|u_p| + a_4, \tag{9.207}$$

and comparing this with (9.206) shows

$$a_p = a_3. \tag{9.208}$$

Since the same result is found for every point P_2 on the path of the piston we conclude that the velocity of sound $a = a_3$ and the velocity $u = -|u_p|$ prevail in the region between the piston path $x_p = -|u_p|t$ and the last C^+ characteristic $x = (-|u_p| + a_3)t$.

Therefore, we find three different solution regions. The first region, labeled (4) in Fig. 9.30, is between the positive x-axis and the initial characteristic $x = a_4 t$ of the fan. The flow velocity is $u = 0$ and the velocity of sound is a_4 there. All characteristics in this region are parallel lines, Next to this is the solution region between the initial characteristic $x = a_4 t$ and the end characteristic $x = (-|u_p| + a_3)t$, where u and a are given by (9.204) and (9.205). This region represents the so-called *expansion wave* which widens as it moves into the positive x-direction. The C^+ characteristics there are straight lines, spacing out to form a fan; the C^- characteristics in this region are no longer straight lines. We call this region the *expansion fan*. Next to this is the region labeled (3) between the end characteristic and the path of the piston, in which all the characteristics are again straight lines. The flow is homentropic, i.e., (9.93) or else

$$\frac{p}{p_4} = \left(\frac{\varrho}{\varrho_4}\right)^{\gamma} \tag{9.209}$$

is valid everywhere; therefore

$$\frac{p_3}{p_4} = \left(\frac{T_3}{T_4}\right)^{\gamma/(\gamma-1)} = \left(\frac{a_3}{a_4}\right)^{2\gamma/(\gamma-1)} = \left(1 - \frac{\gamma-1}{2}\frac{|u_p|}{a_4}\right)^{2\gamma/(\gamma-1)}.$$ (9.210)

A vacuum is produced at the piston base if

$$|u_p| = \frac{2}{\gamma-1}a_4.$$ (9.211)

Since u_p is equal to the gas velocity at the bottom of the piston, Eq. (9.211) represents the maximum possible velocity which can be reached in unsteady expansion of a calorically perfect gas. This is considerably larger than the maximum velocity in steady flow (cf. (9.101)). Of course the result (9.211) does not contradict the energy equation. If the piston is moved even faster, a region in which vacuum prevails is generated between the piston and the gas.

Just as in (9.201) we obtain the pressure distribution in the expansion fan as

$$\frac{p}{p_4} = \left(1 + \frac{\gamma-1}{2}\frac{u}{a_4}\right)^{2\gamma/(\gamma-1)}$$ (9.212)

or written explicitly in x and t

$$\frac{p}{p_4} = \left(\frac{\gamma-1}{\gamma+1}\frac{x}{a_4 t} + \frac{2}{\gamma+1}\right)^{2\gamma/(\gamma-1)}$$ (9.213)

The density distribution in the expansion fan is calculated from

$$\frac{\varrho}{\varrho_4} = \left(\frac{p}{p_4}\right)^{1/\gamma}.$$ (9.214)

In Fig. 9.31 we see the distribution of velocity u and pressure p for fixed t. It is clear from the figure that the flow quantities may have discontinuities in their derivatives. This is typical for the solution of hyperbolic equations. (Discontinuities in the derivatives also propagate along the characteristic lines.) To complete the picture we state the particle path in the expansion fan

$$x = -\frac{2}{\gamma-1}a_4 t + \frac{\gamma+1}{\gamma-1}a_4 t_0\left(\frac{t}{t_0}\right)^{2/(\gamma+1)},$$ (9.215)

which is obtained as a solution of the linear differential equation

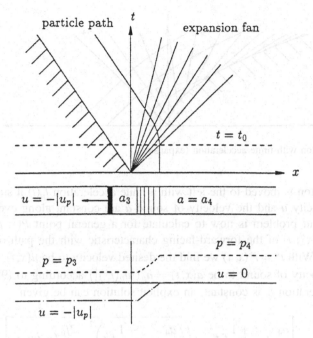

Fig. 9.31 Distribution of the velocity and the pressure inside an expansion fan

$$\frac{dx}{dt} = u = \frac{2}{\gamma+1}\left(\frac{x}{t} - a_4\right) \tag{9.216}$$

with the initial condition $x(t_0) = a_4 t_0$ using standard methods. The equation of the C^-characteristic is found from the solution of the differential equation

$$\frac{dx}{dt} = u - a = \frac{3-\gamma}{\gamma+1}\frac{x}{t} - \frac{4}{\gamma+1}a_4 \tag{9.217}$$

to be

$$x = -\frac{2}{\gamma-1}a_4 t + \frac{\gamma+1}{\gamma-1}a_4 t_0 \left(\frac{t}{t_0}\right)^{(3-\gamma)/(\gamma+1)}. \tag{9.218}$$

The solution of the initial-boundary value problem we have discussed is one of the very few exact and closed solutions for the nonlinear system (9.172), (9.174) and (4.48). Basically this is due to the fact that no typical lengths enter into the problem. Since no typical time occurs, the independent variables can also only appear in combinations of x/t. Therefore the problem only depends on one *similarity variable x/t*.

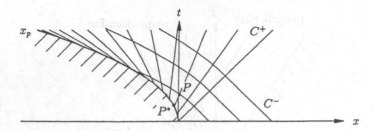

Fig. 9.32 Piston with finite acceleration (expansion)

If the piston is moved to the left with a finite acceleration $f_p(t)$ it still holds that the gas velocity u and the velocity of sound a are constant along every C^+ characteristic. Our problem is now to calculate for a general point $P(x, t)$, the intersection P^* $(x^*,\ t)$ of the forward-facing characteristic with the path of the piston (Fig. 9.32). With $t^* = t^*(x,\ t)$ we find the desired velocity to be $u(x, t) = -\left|u_p(t^*)\right|$ and the velocity of sound to be $a(x, t) = a_p\left(\left|u_p(t^*)\right|\right)$ according to (9.207). If the piston acceleration f_p is constant, an explicit solution can be given

$$u(x,t) = -\left[\frac{a_4}{\gamma} + \frac{\gamma+1}{2\gamma}f_p t - \sqrt{\left(\frac{a_4}{\gamma} + \frac{\gamma+1}{2\gamma}f_p t\right)^2 - \frac{2f_p}{\gamma}(a_4 t - x)}\right], \quad (9.219)$$

$$a(x,t) = a_4 + \frac{\gamma-1}{2}u(x,t), \quad (9.220)$$

which is valid for $x \leq a_4 t$. To the right of the first C^+ characteristic, thus for $x > a_4 t$, we again have $u = 0$ and $a = a_4$.

If the piston moves with finite acceleration in the positive x-direction, then compression waves occur, which satisfy exactly the same equations as the expansion waves. However now the characteristics of the same family (C^+) can intersect each other. Yet at the point of intersection of the characteristics, the solution is no longer unique since different values of the Riemann invariants r hold along the different characteristics. Since, for example, the velocity is $u = r - s$, different velocities occur at the same point, something that is, of course, physically impossible. The characteristics form an envelope, and in the region enclosed within the envelope the solutions are no longer unique. Experiments show that in these cases a shock wave forms (Fig. 9.33).

The shock begins at the cusp P of the envelope, that is, at the position where the solution ceases to be unique. The appearance of the shock is to be expected physically because in the region where the characteristics crowd together (just before they cross) the flow quantities change rapidly and the changes in the velocity and temperature in the x-direction become so large that friction and heat conduction can no longer be neglected. The equation for the envelope can be stated in a closed form under certain assumptions (i.e. for certain piston paths). However we can

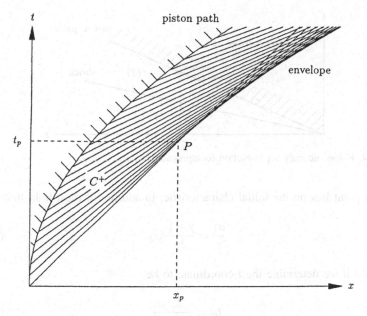

Fig. 9.33 Piston with finite acceleration (compression)

determine the starting point of the envelope (and thus the position x_P in the tube at which the shock wave appears at time t_P) generally if the acceleration of the piston is at no time larger than the initial acceleration. Then it is enough to consider a constantly accelerated piston. Under this assumption we obtain the velocity field immediately from (9.219) if we replace f_p there by $-f_p$

$$u(x,t) = \sqrt{\left(\frac{a_4}{\gamma} - \frac{\gamma+1}{2\gamma}f_p t\right)^2 + \frac{2f_p}{\gamma}(a_4 t - x)} - \left(\frac{a_4}{\gamma} - \frac{\gamma+1}{2\gamma}f_p t\right) \quad (9.221)$$

and thus determine $\partial u/\partial x$ to be

$$\frac{\partial u}{\partial x} = -\left[\left(\frac{a_4}{\gamma} - \frac{\gamma+1}{2\gamma}f_p t\right)^2 + \frac{2f_p}{\gamma}(a_4 t - x)\right]^{-1/2}\frac{f_p}{\gamma}. \quad (9.222)$$

Since $\partial u/\partial x$ tends to infinity at the cusp of the envelope, we obtain this point by setting the expression in brackets in (9.222) to zero. Because of $x \le a_4 t$ the bracket only vanishes for

$$x_P = a_4 t_P, \quad (9.223)$$

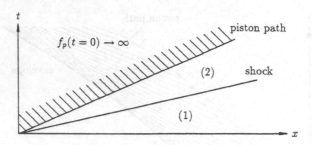

Fig. 9.34 Piston suddenly set in motion (compression)

i.e., the point lies on the initial characteristic. In addition it must hold that

$$\frac{a_4}{\gamma} = \frac{\gamma+1}{2\gamma} f_p t_P,$$
(9.224)

from which we determine the t-coordinate to be

$$t_P = \frac{2}{\gamma+1} \frac{a_4}{f_p}.$$
(9.225)

For $f_p(0) \to \infty$ the starting point lies at the origin. If the piston is set in motion suddenly with constant velocity, the shock forms at the origin of the x-t-plane (Fig. 9.34), i.e., the shock immediately forms which moves ahead of the piston at constant velocity.

If we apply the Eqs. (9.177) and (9.179) to liquids, the velocity u is often much smaller than the velocity of sound a. Then the density and the velocity of sound in the flow change very little from their undisturbed values a_4 and ϱ_4, so that instead of (9.177) and (9.179) we can write

$$du + \frac{1}{\varrho_4 a_4} dp = 0 \quad \text{for} \quad C^+ : x = +a_4 t + \text{const}$$
(9.226)

and

$$du - \frac{1}{\varrho_4 a_4} dp = 0 \quad \text{for} \quad C^- : x = -a_4 t + \text{const.}$$
(9.227)

Here then the characteristics are always straight lines in the x–t diagram. These equations are the starting point for the numerical calculation of *pressure waves* (*hydraulic shock* or also called *water hammer*) in hydraulic pipes (hydroelectric power stations, fuel injection systems, water mains, etc.), as they may occur if valves are suddenly opened or closed. Since in liquids the sound speed and the density is comparatively high, the pressure changes can become so large, even for small changes in velocity, that conduits may be damaged structurally.

If valves are shut suddenly, the pressure downstream can drop below the vapor pressure, so that the fluid cavitates. As the cavity is refilled, a very high pressure occurs again. A velocity change of, for example, $\Delta u = 2\,\mathrm{m/s}$ originating with a closure of valve leads to a pressure wave which propagates with the velocity of sound $a_4 = 1400$ m/s (for water) upstream, with

$$\Delta p = 2\,\mathrm{m/s} \cdot 1400\,\mathrm{m/s} \cdot 1000\,\mathrm{kg/m^3} = 28\,\mathrm{bar}.$$

However, the effective velocity of sound is often smaller, on the one hand because the elasticity of the pipe wall lowers the propagation velocity, and on the other hand because small air bubbles are often found in the fluid and these also lower the effective speed of sound.

Chapter 10
Potential Flows

As the discussions in Sects. 4.1 and 4.3 have already shown, solid walls and discontinuities in the tangential velocity represent surfaces from which angular velocity $(\vec{\omega} = \operatorname{curl} \vec{u}/2)$ diffuses into the flow field. Since the widths of the developing regions (boundary layers) tend to zero in the limit $Re \to \infty$, the flow can be treated within the framework of potential theory. Because of the kinematic restriction of irrotationality, only the kinematic boundary condition, but not the no slip condition, can be satisfied. Therefore potential flows, although they are exact solutions of the Navier–Stokes solutions in the incompressible case, can in general only describe the flow field of an inviscid fluid (with exceptions, like the potential vortex for the flow around a rotating cylinder). However, the results of a calculation for inviscid fluid can be carried over to real flows as long as the flow does not separate. If separation does occur, the boundaries of the separation region are generally not known. In cases where these boundaries are known or can be reasonably estimated, a theory based on inviscid flow can also be useful.

Besides ignoring the viscosity, the overriding simplifications of the theory of potential flows stem from the introduction of a *velocity potential* and the use of Bernoulli's equation (with Bernoulli's constant the same value everywhere in the flow field). Therefore the flow is described by the continuity equation (2.3) and Bernoulli's equation (4.73). We introduce the velocity potential from (1.50)

$$u_i = \frac{\partial \Phi}{\partial x_i}$$

into the continuity equation. To do this we use the assumption of barotropy, which is already within Bernoulli's equation

$$dP = \frac{1}{\varrho} dp, \tag{10.1}$$

© Springer Nature Switzerland AG 2020
J. H. Spurk and N. Aksel, *Fluid Mechanics*,
https://doi.org/10.1007/978-3-030-30259-7_10

or, by (9.184)

$$dp = a^2 d\varrho, \tag{10.2}$$

also

$$dP = \frac{a^2}{\varrho} d\varrho, \tag{10.3}$$

and express $\partial\varrho/\partial t$ and $\partial\varrho/\partial x_i$ by $\varrho\, a^{-2}\partial P/\partial t$ and $\varrho\, a^{-2}\partial P/\partial x_i$, respectively. Thus we obtain (2.3a) in the form

$$a^{-2}\frac{\partial P}{\partial t} + a^{-2}\frac{\partial \Phi}{\partial x_i}\frac{\partial P}{\partial x_i} + \frac{\partial^2 \Phi}{\partial x_i \partial x_i} = 0, \tag{10.4}$$

which together with Bernoulli's equation (4.73)

$$\frac{\partial \Phi}{\partial t} + \frac{1}{2}\frac{\partial \Phi}{\partial x_i}\frac{\partial \Phi}{\partial x_i} + P + \psi = C(t)$$

furnishes two coupled equations for the two unknowns P and Φ. In applying potential theory to compressible flow, ψ can usually be ignored. However it is seldom necessary to solve these nonlinear equations, which almost always require numerical methods.

10.1   One-Dimensional Propagation of Sound

We shall consider the case where $u_i = \partial\Phi/\partial x_i$ and $\partial P/\partial x_i$ are so small that all the nonlinear terms can be neglected, and ϱ and a can be approximated by the unperturbed quantities ϱ_0 and a_0. Within the scheme of Sect. 4.4, besides the simplifications of type (a) in the constitutive relation (zero viscosity) and of type (c) in the kinematics (potential flow), another simplification of type (b) in the dynamics (neglecting the convective terms) appears. In spite of this it is clear from the derivation that we are still dealing with compressible flows ($D\varrho / Dt \neq 0.$). Under this assumption, the continuity equation reads

$$\frac{\partial P}{\partial t} + a_0^2 \frac{\partial^2 \Phi}{\partial x_i \partial x_i} = 0, \tag{10.5}$$

while Bernoulli's equation assumes the form

$$\frac{\partial \Phi}{\partial t} + P = 0, \tag{10.6}$$

where the constant has been absorbed into the potential. Equation (10.6) corresponds to the linearized form of Euler's equation $\varrho \partial u_i / \partial t = -\partial p / \partial x_i$. If we differentiate (10.6) with respect to t and subtract (10.5), we obtain

$$\frac{\partial^2 \Phi}{\partial t^2} - a_0^2 \frac{\partial^2 \Phi}{\partial x_i \partial x_i} = 0. \tag{10.7}$$

This is the *wave equation*: it is the most important special case of a *hyperbolic* partial differential equation. In (10.7) it describes the velocity potential Φ of sound, in electrodynamics it describes the propagation of electromagnetic waves and in the theory of oscillations, the transverse oscillations of strings and membranes, or the longitudinal oscillations in elastic bodies.

For the one-dimensional propagation of sound, for example in tubes, we obtain (10.7) in the form

$$\frac{\partial^2 \Phi}{\partial t^2} = a_0^2 \frac{\partial^2 \Phi}{\partial x^2}, \tag{10.8}$$

whose general solution is known as *d'Alembert's solution*

$$\Phi = h(x - a_0 t) + g(x + a_0 t). \tag{10.9}$$

This solution can be directly verified by insertion. The unknown functions h and g are determined by the initial and boundary conditions of a specific problem. From (10.9) we obtain the velocity u as

$$u = \frac{\partial \Phi}{\partial x} = h'(x - a_0 t) + g'(x + a_0 t), \tag{10.10}$$

where the dash denotes the derivatives of the functions with respect to their arguments. Then from (10.6) we extract the pressure function as

$$P = -\frac{\partial \Phi}{\partial t} = a_0 h'(x - a_0 t) - a_0 g'(x + a_0 t). \tag{10.11}$$

For $x = a_0 t + \text{const}$, that is, along the C^+ characteristics introduced in Chap. 9, (10.10) furnishes

$$u = g'(x + a_0 t) + \text{const},\qquad (10.12)$$

and (10.11)

$$P = -a_0 g'(x + a_0 t) + \text{const}.\qquad (10.13)$$

Within the framework of the assumptions being used, in (10.1) we replace ϱ by ϱ_0, and compare (10.12) and (10.13) to obtain

$$dp + \varrho_0 a_0 du = 0 \quad \text{along} \quad x = a_0 t + \text{const}.\qquad (10.14)$$

In the same manner we find

$$dp - \varrho_0 a_0 du = 0 \quad \text{along} \quad x = -a_0 t + \text{const}.\qquad (10.15)$$

These are again the Eqs. (9.226) and (9.227). In Chap. 9 we dealt with the nonlinear propagation of waves, but the assumptions which led to the Eqs. (9.226) and (9.227) reduced the general problem of nonlinear waves to the problem of *acoustics*. We note that d'Alembert's solution is a special application of the theory of characteristics described in Chap. 9.

We shall first consider the application of d'Alembert's solution to the initial value problem, where the distributions of u and $P = (p - p_0)/\varrho_0$ are given for the time $t = 0$

$$u(x,0) = u_I(x), \quad P(x,0) = P_I(x).\qquad (10.16)$$

Using this it follows from (10.10) that

$$u_I(x) = h'(x) + g'(x),\qquad (10.17)$$

and from (10.11)

$$P_I(x) = a_0 h'(x) - a_0 g'(x).\qquad (10.18)$$

With these we express the unknown functions $h'(x)$ and $g'(x)$ in terms of the initial distributions

$$h'(x) = \frac{1}{2}\left[u_I(x) + a_0^{-1} P_I(x)\right], \quad \text{and}\qquad (10.19)$$

$$g'(x) = \frac{1}{2}\left[u_I(x) - a_0^{-1} P_I(x)\right].\qquad (10.20)$$

We insert these now known functions into, for example, the formula for velocity (10.10)

$$u(x,t) = \frac{1}{2}[u_I(x - a_0t) + u_I(x + a_0t)] + \frac{1}{2}a_0^{-1}[P_I(x - a_0t) - P_I(x + a_0t)].$$

$$(10.21)$$

For simplicity let us take $P_I(x) \equiv 0$ in the following example: for u we set the initial condition

$$u(x,0) = u_I(x) = \begin{cases} 0 & \text{for } x > b \\ 1 & \text{for } |x| \leq b \\ 0 & \text{for } x < -b \end{cases}, \qquad (10.22)$$

and we infer from (10.21) that the initial rectangular distribution (10.22) resolves into two rectangular waves each of half the initial amplitude, of which one moves to the right and the other to the left. From

$$u(x,t) = \frac{1}{2}u_I(x - a_0t) + \frac{1}{2}u_I(x + a_0t), \qquad (10.23)$$

the initial distribution

$$u_I(x) = \frac{1}{2}u_I(x) + \frac{1}{2}u_I(x) \qquad (10.24)$$

is generated for $t = 0$. For $t = t_1$, we obtain for the first wave $1/2u_I(x - a_0t_1)$, that is, the same rectangular function, now displaced a_0t_1 to the right. For the second wave we find $1/2u_I(x + a_0t_1)$, i.e., again the same rectangle but now displaced by $-a_0t_1$ (therefore to the left), as is also clear in Fig. 10.1. Along the characteristics $x = a_0t + \text{const}$ and $x = -a_0t + \text{const}$ the value of the amplitudes remains the same.

We now consider the *initial-boundary value problem* with a rigid wall at the position $x = 0$, and so the kinematic boundary initial condition implies that the velocity u vanishes there. The initial condition u_I is the function shown in Fig. 10.2a, where again we set $P_I \equiv 0$. We look for a solution in the semi-infinitely long tube ($x \geq 0$) with the initial condition

$$u(x,0) = u_I(x), \quad x \geq 0, \qquad (10.25)$$

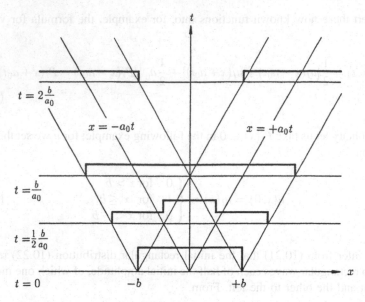

Fig. 10.1 Propagation of a rectangular disturbance

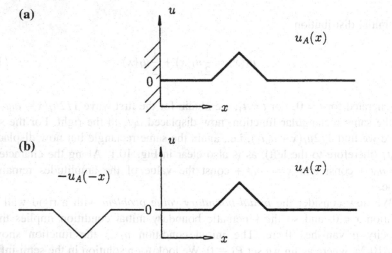

Fig. 10.2 Initial distributions of **a** the initial-boundary value problem, and **b** the equivalent initial value problem

and the boundary condition

$$u(0, t) = 0, \quad t \geq 0. \tag{10.26}$$

This initial-boundary value problem is equivalent to the pure initial value problem of the pipe extending to infinity in both directions, with the initial distribution sketched in Fig. 10.2b

$$u(x, 0) = \begin{cases} + u_I(+x) & \text{for } x \geq 0 \\ - u_I(-x) & \text{for } x < 0 \end{cases}. \tag{10.27}$$

With (10.21) the solution for u reads

$$u(x, t) = \frac{1}{2}[u_I(x - a_0 t) + u_I(x + a_0 t)]. \tag{10.28}$$

For $x \geq a_0 t$ the argument is $x - a_0 t \geq 0$, and with (10.27) we write

$$u(x, t) = \frac{1}{2}[u_I(x - a_0 t) + u_I(x + a_0 t)], \quad x \geq a_0 t. \tag{10.29}$$

For $x < a_0 t$ the argument is $x - a_0 t < 0$, and from (10.28) and (10.27) we then obtain

$$u(x, t) = \frac{1}{2}[-u_I(-x + a_0 t) + u_I(x + a_0 t)], \quad x < a_0 t. \tag{10.30}$$

Because of the properties of the function u_I given in (10.27), $u(x, t)$ satisfies the initial condition (10.25) and the boundary condition (10.26), so that (10.29) and (10.30) together represent the solution of the initial-boundary value problem. However it is more perceptive to view the graphical solution of the equivalent initial value problem. The initial distribution shown in Fig. 10.2b is again resolved into two waves, one moving to the right and one to the left, each with the initial velocity a_0. At the position $x = 0$ the superimposed waves cancel each other out, so that the boundary condition $u(0, t)$ is always satisfied. The graphical solution is shown in Fig. 10.3. The solution has physical meaning only for $x \geq 0$.

In addition to d'Alembert's solution, the method of separation is also applicable to the linear wave equation (10.8). We start out directly with the differential equation for the velocity u, which also satisfies the wave equation

$$\frac{\partial^2 u}{\partial t^2} = a_0^2 \frac{\partial^2 u}{\partial x^2}. \tag{10.31}$$

Now we shall treat the problem where there are rigid walls at the positions $x = 0$ and $x = l$, and so the boundary conditions read

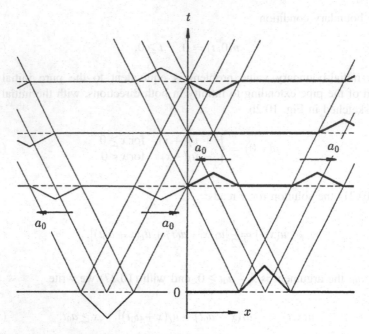

Fig. 10.3 Propagation of a wave in the semi-infinite pipe

$$u(0, t) = u(l, t) = 0. \tag{10.32}$$

As initial conditions we take

$$u(x, 0) = u_I(x), \tag{10.33}$$

and again

$$P(x, 0) = P_I(x) = 0. \tag{10.34}$$

From (10.34) we obtain a second initial condition for u

$$\left.\frac{\partial u}{\partial t}\right|_{t=0} = 0, \tag{10.35}$$

which is a consequence of the linearized Euler's equation $\varrho \partial u / \partial t = -\partial p / \partial x$. The separation form

$$u(x, t) = T(t)X(x) \tag{10.36}$$

leads to

$$\frac{T''}{T} = a_0^2 \frac{X''}{X} = \text{const} = -\omega^2, \tag{10.37}$$

with the solutions

$$T = C_1 \cos(\omega t) + C_2 \sin(\omega t), \quad \text{and} \tag{10.38a}$$

$$X = C_3 \cos\left(\frac{\omega x}{a_0}\right) + C_4 \sin\left(\frac{\omega x}{a_0}\right). \tag{10.38b}$$

The initial condition (10.35) requires that $C_2 = 0$; C_3 vanishes because of the boundary condition $u(0, t) = 0$, so that for the velocity we obtain

$$u(x, t) = A \cos(\omega t) \sin\left(\frac{\omega x}{a_0}\right), \tag{10.39}$$

where $A = C_1 C_4$. For

$$\omega_k = k\, a_0 \frac{\pi}{l}, \quad k = 1, 2, 3, \ldots \tag{10.40}$$

Eq. (10.39) now also satisfies the boundary condition $u(l, t) = 0$. The $\omega_k/2\pi$ are the *eigenfrequencies* of the column of fluid in the tube of length l. (Note the following, if one of these eigenfrequencies $\omega_k/2\pi$ lies close to the eigenfrequency of an elastic element (such as a valve) which comes into contact with the fluid, self excited oscillations can occur.)

With (10.40) we obtain the solutions

$$u_k = A_k \cos\left(\frac{k\pi a_0 t}{l}\right) \sin\left(\frac{k\pi x}{l}\right), \tag{10.41}$$

whose sum, because of the linearity of (10.31) is also a solution. Therefore the general solution reads

$$u = \sum_{k=1}^{\infty} A_k \cos\left(\frac{k\pi a_0 t}{l}\right) \sin\left(\frac{k\pi x}{l}\right). \tag{10.42}$$

The initial condition (10.33) leads to the equation

$$u(x,0) = u_I(x) = \sum_{k=1}^{\infty} A_k \sin\left(\frac{k\pi x}{l}\right), \quad 0 \leq x \leq l. \tag{10.43}$$

This is an instruction to expand the initial distribution $u_I(x)$ into a sine series, whose coefficients are determined from

$$A_k = \frac{2}{l} \int_0^l u_I(x) \sin\left(\frac{k\pi x}{l}\right) dx. \tag{10.44}$$

With this the velocity field is known. We calculate the pressure field from (10.5)

$$\frac{\partial P}{\partial t} = -a_0^2 \frac{\partial^2 \Phi}{\partial x^2} = -a_0^2 \frac{\partial u}{\partial x} = -a_0^2 \sum_{k=1}^{\infty} \frac{k\pi}{l} A_k \cos\left(\frac{k\pi a_0 t}{l}\right) \cos\left(\frac{k\pi x}{l}\right) \tag{10.45}$$

as

$$P = -a_0 \sum_{k=1}^{\infty} A_k \sin\left(\frac{k\pi a_0 t}{l}\right) \cos\left(\frac{k\pi x}{l}\right), \tag{10.46}$$

where the constants of integration appearing are set to zero because of the initial condition (10.34).

10.2 Steady Compressible Potential Flow

As a further case of compressible potential flow which emerges from the simplifications in the general equations (10.4) and (4.73), we shall consider steady flow. From the continuity equation (10.4) we then extract

$$a^{-2} \frac{\partial \Phi}{\partial x_i} \frac{\partial P}{\partial x_i} + \frac{\partial^2 \Phi}{\partial x_i \partial x_i} = 0, \tag{10.47}$$

and from Bernoulli's equation (4.73) neglecting the volume body forces,

$$\frac{1}{2} \frac{\partial \Phi}{\partial x_j} \frac{\partial \Phi}{\partial x_j} + P = C. \tag{10.48}$$

With the help of (10.48) we eliminate P from (10.47) and bring the resulting equation to the form

$$a^{-2} \frac{\partial \Phi}{\partial x_i} \frac{\partial}{\partial x_i} \left(\frac{1}{2} \frac{\partial \Phi}{\partial x_j} \frac{\partial \Phi}{\partial x_j} \right) = \frac{\partial^2 \Phi}{\partial x_i \partial x_i}, \tag{10.49}$$

from which a nonlinear partial differential equation for the velocity potential Φ follows

$$a^{-2} \frac{\partial \Phi}{\partial x_i} \frac{\partial \Phi}{\partial x_j} \frac{\partial^2 \Phi}{\partial x_i \partial x_j} = \frac{\partial^2 \Phi}{\partial x_i \partial x_i}. \tag{10.50}$$

This equation holds without restrictions for steady subsonic ($M < 1$), transonic ($M \approx 1$), and supersonic flows ($M > 1$). The steady homenergic hypersonic flow ($M \gg 1$) is generally not a potential flow, as Crocco's law (4.157) implies, so that (10.50) is not used in that case.

Equation (10.50) is the starting point of classical *aerodynamics*. The analytic method of the solution of (10.50) exploits simplifications which arise from the Mach number range, and/or from "linearizations". An example of this is the flow past slender bodies, although in practice numerical methods are more often called into play. With the known potential Φ the velocity field is then also known: $\vec{u} = \nabla \Phi$. From Bernoulli's equation (10.48) the pressure function P then follows, from this the pressure, and finally the density. From (9.95) we can compute the pressure of the calorically prefect gas

$$\frac{p}{p_0} = \left(\frac{\gamma - 1}{\gamma} \frac{\varrho_0}{p_0} P \right)^{\gamma/(\gamma-1)}, \tag{10.51}$$

and with (9.93) then the density

$$\frac{\varrho}{\varrho_0} = \left(\frac{\gamma - 1}{\gamma} \frac{\varrho_0}{p_0} P \right)^{1/(\gamma-1)}. \tag{10.52}$$

10.3 Incompressible Potential Flow

The simplifications arising from the assumption of incompressibility have already been explained: incompressibility can be seen as a particular form of the constitutive relation ($D\varrho/Dt = 0$) or as a kinematic restriction (div $\vec{u} = 0$). Besides this kinematic restriction of divergence free flow (solenoidal flow), irrotationality (curl $\vec{u} = 0$) appears in addition in incompressible potential flow. From (2.5)

$$\frac{\partial u_i}{\partial x_i} = 0$$

together with (1.50)

$$u_i = \frac{\partial \Phi}{\partial x_i}$$

the already known linear potential equation (*Laplace's equation*) follows

$$\frac{\partial^2 \Phi}{\partial x_i \partial x_i} = 0. \tag{10.53}$$

Laplace's equation is the most important form of a partial differential equation of the *elliptic type,* which occurs here as the differential equation for the velocity potential of volume preserving fluid motion. (As already mentioned, Laplace's equation is, together with Poisson's equation, the subject of *potential theory.* It occurs in many branches of physics, and describes, for example, the gravitational potential, from which we can calculate the mass body force of gravity $\vec{k} = -\nabla \psi$. In electrostatics it determines the potential of the electric field, and in magneto-statics the potential of the magnetic field. The temperature distribution in a solid body with steady heat conduction also obeys this differential equation.)

From the derivation, it is clear that (10.53) holds both for steady and unsteady flows. The unsteadiness of the incompressible potential flow exhibits itself only in Bernoulli's equation (4.61) or (4.73), in which now $P = p/\varrho$. We also obtain Laplace's equation (10.53) from the potential equation (10.50), or directly from (10.4) by taking the limit $a^2 \to \infty$ there. Taking this limit actually corresponds to $D\varrho/Dt = 0$, because it follows from $dp/d\varrho = a^2$ that then

$$\frac{D\varrho}{Dt} = a^{-2}\frac{Dp}{Dt} \to 0. \tag{10.54}$$

The treatment of incompressible flow is not exhausted by solving Laplace's equation for given boundary conditions and then computing the pressure distribution from Bernoulli's equation. As we have seen, associated with the lift is a circulation around a body. The changes in time and space of the circulation are subject to Thomson's and Helmholtz's vortex theorems, which also must be satisfied in the solution of the flow past a body. The changes in circulation give rise to surfaces of discontinuity and vortex filaments where the vorticity does not vanish, as sketched in Figs. 4.6, 4.18, 4.20 and 4.21. In incompressible flow the solenoidal term $\vec{u}_R$ from (4.111) or (4.123), whose calculation requires knowledge of the vorticity distribution, is added to the velocity field $\nabla\Phi$. For these reasons, computing the flow past a body turns out to be more difficult than just the classical solution of Laplace's equation.

In problems of flow past a body where there is no lift, discontinuity surfaces and vortex filaments do not occur. Then the flow field only depends on the instantaneous boundary conditions, i.e., on the instantaneous position and velocity of the body. Physically this is explained by the infinitely large velocity of sound which imposes the time varying boundary conditions on the entire flow field instantaneously. In lift problems, the discontinuity surface develops behind the body, and its position and extension, and with this the lift itself, depend on the history of the motion of the body. Now this problem is easier in steady flow, but even there it is necessary to make some assumptions about the position of the discontinuity surface. Here we shall only deal with flow without lift, and with steady flow where lift occurs but where no discontinuities appear in the velocity.

In problems involving flow past a body, the domain of flow reaches to infinity. As well as the boundary conditions at the body already mentioned, conditions at infinity must then also be given (we have already made use of these in Sect. 4.2). We only state these conditions, which are based on the existence of the integrals occurring in Green's formulae (e.g. (4.114)), from potential theory.

If $U_{\infty i}$ is the velocity at infinity, we then have

(a) for a three-dimensional rigid body

$$u_i \sim U_{\infty i} + O(r^{-3}) \quad \text{for} \quad r \to \infty, \tag{10.55}$$

or

$$\Phi \sim U_{\infty i} x_i + O(r^{-2}) \quad \text{for} \quad r \to \infty, \tag{10.56}$$

i.e., the perturbation in the velocity originating from the body must die away as r^{-3};

(b) for a plane rigid body without circulation

$$u_i \sim U_{\infty i} + O(r^{-2}) \quad \text{for} \quad r \to \infty; \quad \text{or} \tag{10.57}$$

(c) for a plane rigid body with circulation

$$u_i \sim U_{\infty i} + O(r^{-1}) \quad \text{for} \quad r \to \infty. \tag{10.58}$$

If the body experiences a change in volume, we then have in the three-dimensional case

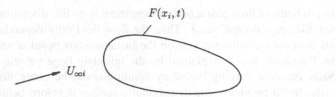

$$F(x_i, t)$$

$$U_{\infty i}$$

Fig. 10.4 Flow past a body

$$u_i \sim U_{\infty i} + O\left(r^{-2}\right) \quad \text{for} \quad r \to \infty,$$

and for the plane two-dimensional case

$$u_i \sim U_{\infty i} + O\left(r^{-1}\right) \quad \text{for} \quad r \to \infty.$$

The "direct problem" of potential theory is mathematically represented as follows: the surface of the body (Fig. 10.4) is given in the most general case as $F(\vec{x}, t) = 0$. Then Laplace's equation is to be solved under the boundary condition (4.170) and the condition at infinity (10.56). With Φ known, we determine the velocity field from $u_i = \partial\Phi/\partial x_i$ and the pressure from Bernoulli's equation

$$\frac{\partial\Phi}{\partial t} + \frac{1}{2}\frac{\partial\Phi}{\partial x_i}\frac{\partial\Phi}{\partial x_i} + \frac{p}{\varrho} = C, \tag{10.59}$$

where we have assumed that the pressure is constant at infinity and only the pressure differences caused by the motion are of interest, so that the mass body force does not appear explicitly in (10.59). The *direct problem* can, with reasonable effort, only be solved in a closed form for a few bodies which are geometrically very simple, like rectangles, spheres, cylinders and, ellipsoids. For the body shapes met in practice we are reduced to using numerical methods.

Therefore in what follows we shall deal with the *indirect problem*, where we examine known solutions of Laplace's equation to see if they represent flows of practical interest. In doing this, solutions from electrostatics, in particular, can be carried over to fluid mechanics.

10.3.1 Simple Examples of Potential Flows

In these indirect problems we shall first examine solutions in the form of polynomials. In this manner we are led to three solutions of particular importance: translation flows, plane two-dimensional, and rotationally symmetric stagnation point flows.

The potential of the *translational flow* is given by

$$\Phi = U_{\infty i} x_i = U_\infty x + V_\infty y + W_\infty z; \qquad (10.60)$$

we have already used this in (10.56). Equation (10.60) clearly satisfies Laplace's equation. The potential of the translational flow is part of every flow past a finite body. The particular form, for which the velocity vector

$$\vec{u} = \nabla \Phi = U_\infty \vec{e}_x + V_\infty \vec{e}_y + W_\infty \vec{e}_z$$

is parallel to the *x*-axis, is

$$\Phi = U_\infty x, \qquad (10.61)$$

and is called *parallel flow*.

The polynomial

$$\Phi = \frac{1}{2} \left(a x^2 + b y^2 + c z^2 \right) \qquad (10.62)$$

satisfies Laplace's equation

$$\frac{\partial^2 \Phi}{\partial x^2} + \frac{\partial^2 \Phi}{\partial y^2} + \frac{\partial^2 \Phi}{\partial z^2} = 0, \qquad (10.63)$$

assuming the coefficients satisfy the condition

$$a + b + c = 0. \qquad (10.64)$$

The choice $c = 0$, that is $a = -b$, leads to steady *plane stagnation point flow*

$$\Phi = \frac{a}{2} \left(x^2 - y^2 \right), \qquad (10.65)$$

with the velocity components

$$\begin{aligned} u &= ax, \\ v &= -ay, \\ w &= 0. \end{aligned} \qquad (10.66)$$

This represents the inviscid flow against a flat wall (Fig. 10.5). From (1.11) we obtain the equation of the streamlines as

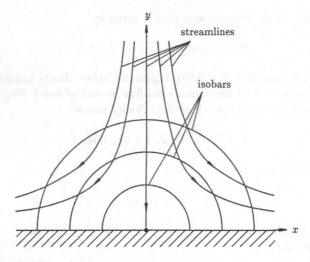

Fig. 10.5 Plane stagnation point flow

$$\frac{dy}{dx} = -\frac{y}{x},$$ (10.67)

and integrating this leads to hyperbolae

$$xy = x_0 y_0,$$ (10.68)

where x_0, y_0 is the position through which the streamline runs. From Bernoulli's equation (10.59) and with (10.65) it follows that the pressure distribution is

$$\frac{a^2}{2}(x^2 + y^2) + \frac{p}{\varrho} = \frac{p_g}{\varrho},$$ (10.69)

where Bernoulli's constant C is determined by the pressure at the *stagnation point*. The stagnation point is the point on the body at which the velocity vanishes ($\vec{u} = 0$); the pressure there is p_g, and, from Bernoulli's equation, this is the highest pressure occurring at the body. The lines of equal pressure (isobars) are circles. The pressure at the wall decreases in the direction of flow, so that no boundary layer separation occurs even for viscous flow. Contrary to a flow where the pressure increases in the direction of flow, fluid particles in the boundary layer here do not come to a standstill. As we shall show in Sect. 12.1, the boundary layer in the present case has constant thickness, which tends to zero as $\nu \to 0$.

Plane stagnation point flow is met close to the stagnation point (or more exactly, the stagnation line) of a plane body (Fig. 10.6), but it is only realized locally there, as we see from the fact that the incident flow velocity tends to infinity as $y \to \infty$.

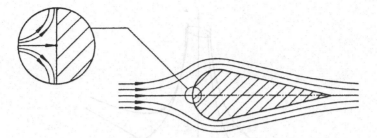

Fig. 10.6 Flow past a plane airfoil close to the forward stagnation point

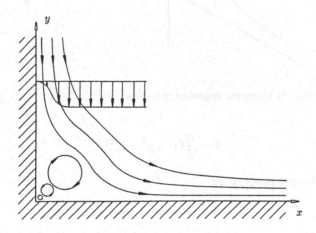

Fig. 10.7 Flow in a right-angled corner

Within the framework of the inviscid theory, each streamline can be viewed as a wall, in particular the streamline $x = 0$, i.e., the y-axis, but we see here that this potential flow would not occur in a real (viscous) fluid.

Now along the y-axis the pressure in the direction of flow increases. In the boundary layer formed along the y-axis the fluid particles have lost kinetic energy, and their remaining kinetic energy is not enough to advance them into the region of increasing pressure. The motion comes to a standstill and thus to boundary layer separation, as sketched in Fig. 10.7. We examine the *boundary layer separation* and the eddy creation in creeping flows in Chaps. 12 and 13.

If we consider now the flow in all four quadrants, no solid wall appears on which the no slip boundary conditions would be enforced physically and so the potential flow satisfies all boundary conditions. By (4.11) it is then an exact solution of Navier–Stokes equations.

The choice $b = a$, that is $c = -2a$, in (10.62) leads to the potential of *rotationally symmetric stagnation point flow* (Fig. 10.8)

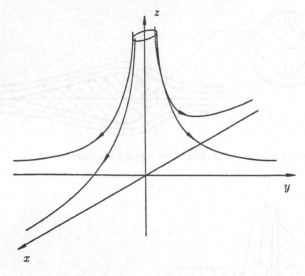

Fig. 10.8 Rotationally symmetric stagnation point flow

$$\Phi = \frac{a}{2}\left(x^2 + y^2 - 2z^2\right), \tag{10.70}$$

whose velocity components are

$$u = ax, \quad v = ay, \quad w = -2az. \tag{10.71}$$

The equations for the streamlines can be brought to the form

$$\frac{dx}{dy} = \frac{u}{v} = \frac{x}{y}, \quad \frac{dx}{dz} = \frac{u}{w} = -\frac{x}{2z}, \quad \frac{dy}{dz} = \frac{v}{w} = -\frac{y}{2z}. \tag{10.72}$$

The integral curves of the first equation in (10.72) represent the projection of the streamlines into the x-y-plane. These are the straight lines

$$x = C_1 y \tag{10.73}$$

through the origin. The integral curves of the two other differential equations are the projections into the x-z-plane

$$x^2 z = C_2, \tag{10.74}$$

and into the y-z-plane

$$y^2 z = C_3. \tag{10.75}$$

From Bernoulli's equation we extract the pressure field as

$$\frac{a^2}{2} \left(x^2 + y^2 + 4z^2 \right) + \frac{p}{\varrho} = \frac{p_g}{\varrho}, \tag{10.76}$$

where p_g is again the pressure in the stagnation point.

We are led to *unsteady stagnation point flow* if the coefficient a depends on the time: $a = a(t)$. The directions of the associated velocity fields (10.66) and (10.71) are clearly steady, and so they are of the form (1.13). The streamlines are fixed in space for unsteady stagnation point flow too, as we see directly from the fact that a does not arise in the equations for the streamlines. In order to determine the pressure field we now have to apply Bernoulli's equation for unsteady flow, leading to

$$\frac{1}{2}\frac{da}{dt} \left(x^2 + y^2 - 2z^2 \right) + \frac{a^2}{2} \left(x^2 + y^2 + 4z^2 \right) + \frac{p}{\varrho} = \frac{p_g}{\varrho}. \tag{10.77}$$

Of particular importance in potential theory are *singular* or *fundamental solutions*. With the help of these fundamental solutions, solutions to the direct problem can also be formed using, for example, integration processes. As a typical example we consider the potential of a *point source*

$$\Phi = \frac{A}{r}, \tag{10.78}$$

which we met in Eq. (4.115) as a Green's function. Just as was the case there, r is the distance from the position $\vec{x}'$ of the source to the position $\vec{x}$ where the potential Φ is given by (10.78). Therefore in Cartesian coordinates

$$r = \sqrt{(x - x')^2 + (y - y')^2 + (z - z')^2}, \tag{10.79}$$

and for the source at the origin $\vec{x}' = 0$ then

$$r = \sqrt{x^2 + y^2 + z^2} = \sqrt{x_j x_j}. \tag{10.80}$$

Equation (4.111) shows the importance of this singular solution in potential theory. Here the intuitive interpretation should stand in the foreground, but first we show that (10.78) satisfies Laplace's equation. In index notation it follows that

$$\frac{\partial \Phi}{\partial x_i} = -\frac{A}{r^2}\frac{\partial r}{\partial x_i} = -\frac{A}{r^3}x_i,$$

(10.81)

and further

$$\frac{\partial^2 \Phi}{\partial x_i \partial x_i} = -\frac{A}{r^3}\frac{\partial x_i}{\partial x_i} + 3\frac{A}{r^4}\frac{x_i x_i}{r} = \frac{A}{r^3}(-3 + 3) = 0.$$

(10.82)

Therefore Laplace's equation is satisfied where $r \neq 0$. In order to evaluate the behavior at the singular point we calculate the volume flux (for simplicity) through the surface of a sphere with radius r, which we call the *strength m* of the source

$$m = \iint\limits_{(S_{sph})} \vec{u} \cdot \vec{n}\, dS = \iint\limits_{(S_{sph})} \frac{\partial \Phi}{\partial x_i}n_i\, dS = \iint\limits_{(S_{sph})} \frac{\partial \Phi}{\partial r}\, dS.$$

(10.83)

With the surface element $d\Omega$ of the unit sphere we find

$$m = \iint\limits_{(S_{sph})} -A\, d\Omega = -4\pi A.$$

(10.84)

The strength is independent of the radius of the sphere, and we write for the potential of the source

$$\Phi = -\frac{m}{4\pi}\frac{1}{r}.$$

(10.85)

We take a *source* to be the fundamental solution (10.85) with positive strength $m > 0$, and a *sink* to be the same solution with negative strength $m < 0$. The sink flow is realized physically when the volume flux m is withdrawn by suction almost at a point, for example by a very thin tube (Fig. 10.9), but note that the *source flow* cannot be realized in this manner.

The volume flux m violates the continuity equation at the singularity $r = 0$, where $\operatorname{div} \vec{u} = \Delta \Phi$ tends to infinity. This fact can be described with the help of the *Dirac delta function* $\delta(\vec{x} - \vec{x}')$. The delta function is a generalized function with the properties

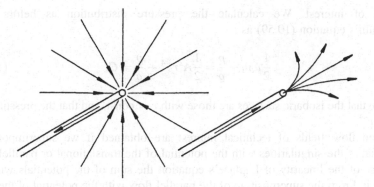

Fig. 10.9 Realizing a point sink and a point source

$$\delta(\vec{x} - \vec{x}') = 0 \quad \text{for} \quad \vec{x} \neq \vec{x}' \tag{10.86}$$

and

$$\iiint\limits_{(V')} f(\vec{x}')\delta(\vec{x} - \vec{x}')dV' = f(\vec{x}'), \tag{10.87}$$

if $\vec{x}$ lies in the domain of integration (V'), otherwise the integral vanishes. Using this we write

$$\text{div}\,\vec{u} = \frac{\partial^2 \Phi}{\partial x_i \partial x_i} = m\,\delta(\vec{x} - \vec{x}') \tag{10.88}$$

and we see that the continuity equation is satisfied everywhere with the exception of the singular point $\vec{x} = \vec{x}'$. With $f(\vec{x}') = m$, (10.87) reads

$$\iiint\limits_{(V')} m\,\delta(\vec{x} - \vec{x}')dV' = m. \tag{10.89}$$

If we now consider (10.88) as Poisson's equation (4.102), from its solution (4.103) it follows that the potential of the source is

$$\Phi = -\frac{m}{4\pi}\iiint\limits_{(\infty)} \frac{\delta(\vec{x}'' - \vec{x}')}{|\vec{x} - \vec{x}''|}dV'' = -\frac{m}{4\pi}\frac{1}{|\vec{x} - \vec{x}'|}, \tag{10.90}$$

from which with $\vec{x}' = 0$ we again obtain (10.85). The breakdown of the continuity equation at the singular point is not troublesome if this point is excluded from the region of interest. We calculate the pressure distribution as before from Bernoulli's equation (10.59) as

$$\frac{1}{2}u_iu_i + \frac{p}{\varrho} = \frac{1}{2}A^2r^{-4} + \frac{p}{\varrho} = C, \tag{10.91}$$

and see that the isobaric surfaces are those with $r = $ const, and that the pressure falls off as r^{-4}.

Often flow fields of technical interest are obtained if we superimpose the potential of the singularities with the potential of the translational or parallel flow. Because of the linearity of Laplace's equation the sum of the potentials will also satisfy it. From the superposition of the parallel flow with the potential of the point source at the origin, for example, we obtain the potential

$$\Phi = U_\infty x - \frac{m}{4\pi}\frac{1}{\sqrt{x^2 + y^2 + z^2}}, \tag{10.92}$$

which, in spherical coordinates r, ϑ, φ, (Appendix B) reads

$$\Phi = U_\infty r \cos\vartheta - \frac{m}{4\pi}\frac{1}{r}. \tag{10.93}$$

We first determine whether this flow has a stagnation point, that is, we look for the place where $u_i = 0$. Using (10.81), we get the velocity in index notation as

$$u_i = U_{\infty 1}\delta_{1i} + \frac{m}{4\pi}\frac{1}{r^3}x_i, \tag{10.94}$$

since $U_{\infty i}$ has only one component in the x_1-direction. From the requirement that $u_2 = u_3 = 0$ we conclude that the stagnation point must lie on the x_1-axis. There $x_2 = x_3 = 0$, and $r = |x_1|$, therefore

$$u_1 = U_\infty + \frac{m}{4\pi}\frac{x_1}{|x_1|^3}. \tag{10.95}$$

The equation $u_1 = 0$ only has a real solution on the negative x-axis ($x_1 = x = -|x|$), and therefore the stagnation point lies at

$$x = -\sqrt{\frac{m}{4\pi U_\infty}}. \tag{10.96}$$

At this place the velocity from the source is exactly equal to the incident flow velocity at infinity. The streamline through the stagnation point divides the fluid of the outer flow from the source fluid (Fig. 10.10). This streamline can be viewed as

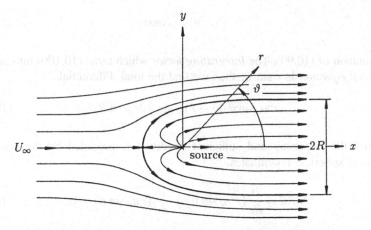

Fig. 10.10 Flow past a semi-infinite, rotationally symmetric body from the superposition of the source potential with the potential of the parallel flow

the wall of a semi-infinite rotationally symmetric body, so that the outer flow represents the flow past such a body. For $r \to \infty$ we again obtain U_∞ for the velocity of the outer flow, just as for the source flow. The fluid coming out of the source flows through the cross-section πR^2, so that

$$m = U_\infty \pi R^2, \tag{10.97}$$

from which we calculate the radius of the body as

$$R = \sqrt{\frac{m}{\pi U_\infty}}. \tag{10.98}$$

Since we are dealing with a flow which is rotationally symmetric about the x-axis, we use spherical coordinates. With the vector element $\mathrm{d}\vec{x}$ in spherical coordinates (Appendix B) we obtain the equation of the streamline as

$$\frac{\mathrm{d}r}{\mathrm{d}\vartheta} = r\frac{u_r}{u_\vartheta}, \tag{10.99}$$

and bring this equation into the form

$$r\,u_r\mathrm{d}\vartheta - u_\vartheta\mathrm{d}r = 0. \tag{10.100}$$

If the left-hand side of (10.100) is a total differential of a function $\Psi(r, \vartheta)$, then

$$\Psi(r, \vartheta) = \text{const} \tag{10.101}$$

is the solution of (10.99). The *integrating factor* which turns (10.100) into an *exact differential equation* is $r \sin \vartheta$; thus we find the total differential

$$r^2 u_r \sin \vartheta \, d\vartheta - r u_\vartheta \sin \vartheta \, dr = d\Psi, \tag{10.102}$$

for which the necessary and sufficient condition is provided by the continuity equation in spherical coordinates

$$\nabla \cdot \vec{u} = \frac{\partial}{\partial r} \left(r^2 u_r \sin \vartheta \right) + \frac{\partial}{\partial \vartheta} \left(r u_\vartheta \sin \vartheta \right) = 0. \tag{10.103}$$

We call Ψ the *stream function*, and in this case of rotationally symmetric flow, *Stokes' stream function*, and we note that this result is independent of the requirement that curl $\vec{u} = 0$, and so it also holds for rotational and viscous flows. From (10.102) we now infer the equations

$$-r u_\vartheta \sin \vartheta = \frac{\partial \Psi}{\partial r} \tag{10.104}$$

and

$$r^2 u_r \sin \vartheta = \frac{\partial \Psi}{\partial \vartheta}, \tag{10.105}$$

from which we see that in rotationally symmetric flow the velocity components can be calculated also from the stream function Ψ. From the condition curl $\vec{u} = 0$, we obtain, with (10.104) and (10.105), the following differential equation for rotationally symmetric flows

$$\frac{\partial}{\partial \vartheta} \left(\frac{1}{r^2 \sin \vartheta} \frac{\partial \Psi}{\partial \vartheta} \right) + \frac{\partial}{\partial r} \left(\frac{1}{\sin \vartheta} \frac{\partial \Psi}{\partial r} \right) = 0, \tag{10.106}$$

to be satisfied by Ψ. We note that, contrary to plane two-dimensional flow, Ψ here does not satisfy Laplace's equation. (A stream function can be introduced in the same manner in cylindrical coordinates for rotationally symmetric flow. The same also holds for plane two-dimensional flows.)

From (10.104) and (10.105) we now calculate the stream function of the point source and parallel flow: with $u_r = \partial\Phi/\partial r$ it follows that

$$u_r = U_\infty \cos \vartheta + \frac{m}{4\pi} \frac{1}{r^2} = \frac{1}{r^2 \sin \vartheta} \frac{\partial \Psi}{\partial \vartheta}, \tag{10.107}$$

and therefore for the stream function

$$\Psi = U_\infty \frac{r^2}{2} \sin^2 \vartheta - \frac{m}{4\pi} \cos \vartheta + f(r). \tag{10.108}$$

By inserting this result into (10.104) and using

$$u_\vartheta = \frac{1}{r} \frac{\partial \Phi}{\partial \vartheta} = -U_\infty \sin \vartheta$$

we obtain the condition $df/dr = 0$, i.e., $f(r) = $ const. Up to a constant the stream function then reads

$$\Psi = U_\infty \frac{r^2}{2} \sin^2 \vartheta - \frac{m}{4\pi} \cos \vartheta, \tag{10.109}$$

from which we read off the stream function of a source in the origin

$$\Psi = -\frac{m}{4\pi} \frac{x}{r}, \tag{10.110}$$

or of a source at position (x', y', z')

$$\Psi = -\frac{m}{4\pi} \frac{x - x'}{\sqrt{(x-x')^2 + (y-y')^2 + (z-z')^2}}. \tag{10.111}$$

With this we extract the equation of the streamlines from (10.101)

$$\Psi = \text{const} = U_\infty \frac{r^2}{2} \sin^2 \vartheta - \frac{m}{4\pi} \cos \vartheta. \tag{10.112}$$

According to (10.96) we have at the stagnation point $\vartheta = \pi$ and therefore

$$\text{const} = \frac{m}{4\pi}, \tag{10.113}$$

from which, with (10.98) we finally obtain the equation of the stagnation streamline as

$$r = \frac{R}{\sin\vartheta}\sqrt{\frac{1+\cos\vartheta}{2}}. \tag{10.114}$$

From Bernoulli's equation

$$\frac{\varrho}{2}U_\infty^2 + p_\infty = p + \frac{\varrho}{2}u_i u_i \tag{10.115}$$

we calculate the pressure at the stagnation point as

$$p_g = \frac{\varrho}{2}U_\infty^2 + p_\infty, \tag{10.116}$$

where the pressure p_g is the *stagnation pressure*, and $\varrho\,U_\infty^2/2$ the *dynamic pressure*. If we were to place a pressure tap at the stagnation point of the body in Fig. 10.10, we would measure the stagnation pressure, as in (10.116). At a pressure tap on the (almost) cylindrical part of the body some distance behind the stagnation point, we would measure the *static pressure* prevailing there. For the dynamic pressure $\varrho\,u_i u_i\,/\,2$, which can be defined at every point in the flow field, we find, from (10.94) and (10.98) the asymptotic behavior

$$\frac{\varrho}{2}u_i u_i \sim \frac{\varrho}{2}U_\infty^2\left(1 + \frac{1}{2}(R/r)^2 + O(R/r)^4\right), \tag{10.117}$$

which, together with Bernoulli's equation, shows that at a point on the surface of the body whose distance from the stagnation point is large compared with R, the static pressure p is practically equal to the static pressure p_∞ at infinity. This fact is exploited by the *Prandtl tube*, with which the dynamic pressure and therefore the velocity can be measured (Fig. 10.11). In doing this it is not necessary that the form (10.114) be realized, but a well rounded nose of the Prandtl tube is sufficient.

In the same manner, sources and sinks can be arranged on the x-axis to generate the flow past a spindle shaped body, as sketched in Fig. 10.12. The contours of the body are calculated using the same methods which led to (10.114). For closed bodies the sum of the strengths of sources and sinks must vanish (*closure condition*)

$$\sum m_i = 0. \tag{10.118}$$

In an obvious manner we shall generalize the methods discussed to a continuous distribution of sources, and shall consider as the simplest case a *source distribution* on a line l along the x-axis (Fig. 10.13). Let $q(x')$ be the *source intensity* (strength per unit length), which can be positive (source) or negative (sink), then the closure condition reads

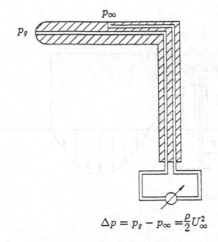

$$\Delta p = p_g - p_\infty = \frac{\rho}{2}U_\infty^2$$

Fig. 10.11 Prandtl tube

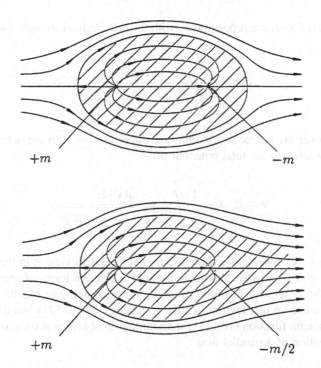

Fig. 10.12 Body generated by sources and sinks

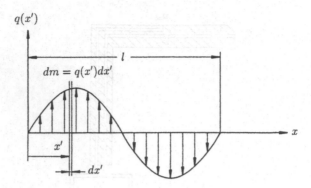

Fig. 10.13 Source distribution

$$\int_0^l q(x')dx' = 0. \tag{10.119}$$

The potential of a source at position x', with the infinitesimal strength $dm = q(x')dx'$ is

$$d\Phi = -\frac{q(x')dx'}{4\pi\sqrt{(x-x')^2+y^2+z^2}}. \tag{10.120}$$

Integration over the source distribution and superposition with the potential of the parallel flow leads to the total potential of

$$\Phi = U_\infty x - \frac{1}{4\pi}\int_0^l \frac{q(x')dx'}{\sqrt{(x-x')^2+y^2+z^2}}. \tag{10.121}$$

Since the flow is rotationally symmetric it is sufficient to view it in the x-y-plane, thus to set $z = 0$. By suitable distribution of $q(x')$, different forms of spindle-shaped bodies can be generated. In order to calculate the contour we require the stream function of a source distribution, which, in analogy to (10.121), we find by integrating the stream function (10.111) for an infinitesimal source at the position x' and the superposition of a parallel flow

$$\Psi = U_\infty \frac{y^2}{2} - \frac{1}{4\pi}\int_0^l \frac{q(x')(x-x')dx'}{\sqrt{(x-x')^2+y^2}}. \tag{10.122}$$

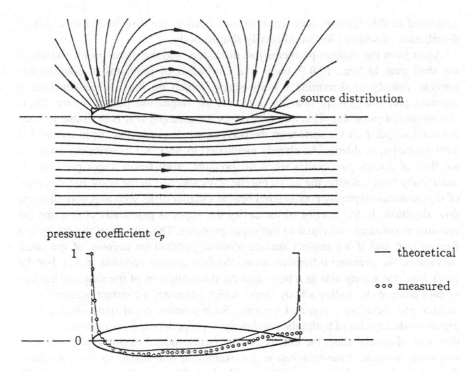

Fig. 10.14 Spindle-shaped airfoil with theoretical and measured pressure coefficient (after Prandtl)

Figure 10.14 shows a body generated in the above manner, and a comparison of the theoretical with the measured *pressure coefficient*. This comparison is also interesting for historical reasons, because it was one of the first systematic pieces of work in the field of aerodynamics. The theoretical pressure coefficient follows from Bernoulli's equation as

$$c_p = \frac{p - p_\infty}{(\varrho/2)U_\infty^2} = 1 - \frac{u^2 + v^2}{U_\infty^2}, \tag{10.123}$$

where the Cartesian velocity components u and v are found from the gradient of the potential (10.121).

The direct problem can also be computed with the help of the singularity distribution. Here the source distribution is to be determined for a given body contour, which leads to an *integral equation*. We are also led to an integral equation if the function $\Phi(\vec{x})$ in (10.121) is given and we are now looking for the source intensity $q(x')$. Incidentally not all rotationally symmetric bodies can be represented by a source distribution on the x-axis. For example, no body, blunter than the body found from the superposition of a translational flow and a point source can be generated in this manner, and we must turn from a line distribution to a surface distribution of sources for an arbitrarily shaped body.

Apart from the indirect problems just discussed, and the direct problems, which we shall treat in Sect. 10.4.9, there is a third class of problems of considerable interest, namely to determine a body shape, which on the surface generates a constant pressure (except near the fore and aft stagnation points, where this is obviously not possible). This question arises for example if it is desirable to keep the surplus speed on the surface of the body as small as possible, or in connection with cavitation problems. As already mentioned in Sect. 8.3.2 cavitation occurs in the flow of liquids past bodies when the pressure drops below vapor pressure. At sufficiently high velocity, the vapor bubble stays attached to the body and we speak of *supercavition* typically encountered behind circular disks with axis normal to the flow direction. In the interior of the cavity the vapor is practically at rest and the pressure is constant and equal to the vapor pressure. The surface of the cavity is a free surface and if we neglect surface tension, justified on account of the small curvatures, the pressure difference across the free surface vanishes (6.31). For the outer flow, the cavity acts as a body and the determination of the shape of the free surface amounts to finding a body shape, which generates a constant pressure on its surface and therefore a constant velocity. Such cavities occur quite often e.g. on highly loaded blades of hydraulic machinery or propellers and are undesirable since they can adversely affect the efficiency. But under special circumstances one tries purposely to achieve supercavitation, for example if one wishes to move a vehicle through the liquid with the smallest possible drag. Such vehicles are equipped with a cavitator at the bow, which generates the cavity that encloses the remaining body. This part of the body has practically no drag quite independent of its shape (The cavitator of course experiences a drag, but it is much smaller than the drag which the fully wetted body would have).

Using the methods of the theory of functions, two-dimensional plane flow problems with free surfaces can still be handled with relative ease (see Sect. 10.4.7), but in general these problems are difficult, even in rotationally symmetric flows that we are treating here. The difficulty is in part due to the fact, that the boundary conditions (4.169) and (4.173) must be satisfied on a not yet known boundary. The determination of the cavity shape is a difficult numerical problem, which we cannot go into here. But the *constant pressure body* is already a very good description of the actual cavity.

In fact the first computations of the cavity shape use this idea (Reichardt 1944) and his most convenient approximate formula for the shape is still in use.

We start from (10.122) and render the coordinates in the x, y-plane of the rotationally symmetric flow dimensionless, using half of the source distribution length $l/2$ Fig. 10.13. We introduce with $U_\infty(l/2) q(x')$ the dimensionless source distribution and refer the stream function to $U_\infty l^2/4$ which brings (10.122) to the form

$$\psi = \frac{y^2}{2} - \frac{1}{4\pi} \int_0^2 \frac{q(x')(x - x')dx'}{\sqrt{(x - x')^2 + y^2}}. \tag{10.124}$$

It will be shown, that the superposition of only three source distribution leads to an almost constant pressure distribution. These are: a linear source distribution $q_1 = -2(x' - 1)$, a cubic distribution $q_2 = -4(x' - 1)^3$, a source of strength $+m$ at $x = 0$ together with a source of strength $-m$ (sink) at $x = 2$. Each of these distributions satisfy the closure condition

$$\int_0^2 q(x')dx' = 0 \tag{10.125}$$

so the body will be closed. For the whole source distribution we set $q(x') = A(q_1 + bq_2 + cq_3)$, with still unknown constants A, b, c, and write for the stream function

$$\psi = \frac{y^2}{2} - \frac{A}{4\pi} \left(\int_0^2 \frac{q_1(x')(x - x')dx'}{\sqrt{(x - x')^2 + y^2}} + b \int_0^2 \frac{q_2(x')(x - x')dx'}{\sqrt{(x - x')^2 + y^2}} \right.$$
$$\left. + c \left(\frac{x}{\sqrt{x^2 + y^2}} - \frac{(x - 2)}{\sqrt{(x - 2)^2 + y^2}} \right) \right), \tag{10.126}$$

where the stream function of the source-sink has been taken directly from (10.111). The other integrals can be solved in closed form, but we prefer a numerical integration and label them ψ_1 and ψ_2. The body contour is given by $\psi = 0$ which provides an implicit relation for the contour $y = y_B(x)$. At first, estimated values for the constants b, c should be used. Then we determine the constant A by choosing the dimensionless radius R. This radius appears at $x = 1$ (from symmetry) and fixes the aspect ratio of the cavity. Evaluating (10.126) with the estimated values we find

$$A = \frac{2\pi R^2}{\psi_1(1, R) + b\psi_2(1, R) + 2c/\sqrt{R^2 + 1}}. \tag{10.127}$$

For a list of x-coordinates the equations $\psi = 0$ are then solved numerically for the y-coordinates. These are of course the coordinates of the contour $y = y_B(x)$. Introducing the body contour into (10.123) leads to $c_p(x, y_B(x)) = c_p(x)$. The constants b, c are now found by trial and error. However if after each trial the function $c_p(x)$ is plotted, the correct choice may be found rather quickly. For the choice $R = 0.16$ the values $b = 0.145$ and $c = 0.214$ are thus found. The resulting

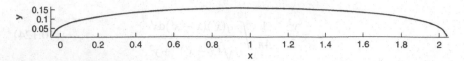

Fig. 10.15 Constant pressure body

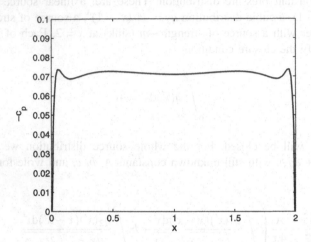

Fig. 10.16 Pressure coefficient of constant pressure body

contour is shown in Fig. 10.15 and the (negative) pressure coefficient $-c_p(x)$ in Fig. 10.16. (Note that Fig. 10.16 is only a section of the whole graph $-c_p(x)$, thus the coefficient at the stagnation points is out of bounds.) The evaluation of (10.123) requires the velocity components u, v. For the rotationally symmetric flow here, these may be determined using the stream function, thus avoiding the use of the potential function as was done using spherical coordinates with (10.104) and (10.105). The continuity equation $\nabla \cdot \vec{u} = 0$ in cylindrical coordinates (from B2) with $\partial/\partial\varphi = 0$ is

$$\frac{\partial(u_r r)}{\partial r} + \frac{\partial(u_z r)}{\partial z} = 0, \tag{10.128}$$

where we have temporarily reverted to dimensional coordinates. Equation (10.128) is necessary and sufficient for the differential

$$d\psi = u_z r\, dr - u_r\, r\, dz \tag{10.129}$$

to be a total differential. Thus

$$u_z = \frac{1}{r}\frac{\partial \psi}{\partial r}, \quad u_r = -\frac{1}{r}\frac{\partial \psi}{\partial z}. \tag{10.130}$$

Since Cartesian coordinates in the plane $z = 0$ correspond to z, r and the velocity components u, v to u_z, u_r we write for (10.130)

$$u = \frac{1}{y}\frac{\partial \psi}{\partial y}, \quad v = -\frac{1}{y}\frac{\partial \psi}{\partial x} \tag{10.131}$$

and in dimensionless form

$$\frac{u}{U_\infty} = \frac{1}{y}\frac{\partial \psi}{\partial y}, \quad \frac{v}{U_\infty} = -\frac{1}{y}\frac{\partial \psi}{\partial x}, \tag{10.132}$$

which may also be used in computing the pressure coefficient.

That part of the body where the pressure is very nearly constant may be considered as the cavity contour. The part of the body near the front stagnation point is taken up by the cavitator so the change in pressure there is of no consequence in this application. The rear stagnation point cannot be part of a bubble since the pressure there would have to be equal to the vapor pressure, while on the other hand it should be the stagnation pressure. There are closure models to circumvent this problem, but they are not free of contradictions. It is very doubtful, if the rear part of the cavity can be properly handled within potential theory. Experimentally one observes unsteady flow at rear part of the cavity with intermittent ejection of vapor from the cavity.

In connection with cavitation we point out that the negative pressure coefficient is here called *Cavitation Number* σ

$$\sigma = \frac{p_\infty - p_v}{(\varrho/2)U_\infty^2}, \tag{10.133}$$

and is the most important characteristic number in cavitation problems. It is noteworthy, that only the difference between the pressure at infinity and the pressure in the bubble is important. The pressure in the bubble can be increased (within limits) be bleeding foreign gas into the cavity and thereby control the cavitation number. This fact increases the usable range of supercavitating vehicles considerably. We also mention that an approximate solution for the cavity shape may be found for $\sigma \ll 1$.

We shall now consider the potential of a source (strength $+m$) and a sink (strength $-m$) on the x-axis at distance Δx (Fig. 10.17)

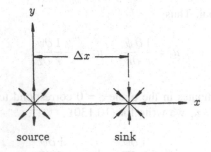

Fig. 10.17 Source-sink pair

$$\Phi = \frac{m}{4\pi}\Delta x \frac{\left[(x-\Delta x)^2 + y^2 + z^2\right]^{-1/2} - (x^2 + y^2 + z^2)^{-1/2}}{\Delta x}. \tag{10.134}$$

We now let the distance Δx shrink to zero and simultaneously raise the strength, so that

$$\lim_{\Delta x \to 0, m \to \infty} m\,\Delta x = M. \tag{10.135}$$

Because of

$$\lim_{\Delta x \to 0} \frac{\left[(x-\Delta x)^2 + y^2 + z^2\right]^{-1/2} - (x^2 + y^2 + z^2)^{-1/2}}{\Delta x} = \frac{\partial(-r^{-1})}{\partial x} = \frac{x}{r^3} \tag{10.136}$$

the potential

$$\Phi = \frac{M}{4\pi}\frac{x}{r^3} \tag{10.137}$$

of a *dipole* at the origin results, which in spherical coordinates reads

$$\Phi = \frac{M}{4\pi}\frac{\cos\vartheta}{r^2}. \tag{10.138}$$

The direction from the sink to the source is the direction of the dipole, and we call M the magnitude of the dipole moment. For this reason the dipole moment is a vector; for the orientation chosen here we have

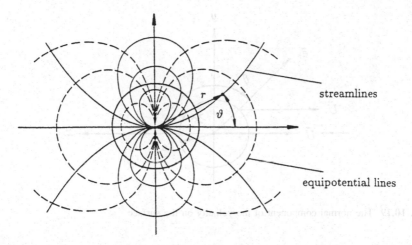

Fig. 10.18 Streamlines and equipotential lines of a three-dimensional dipole

$$\vec{M} = -M\vec{e}_x, \tag{10.139}$$

and in general we obtain the potential of a dipole at the coordinate origin to be (Fig. 10.18)

$$\Phi = -\frac{\vec{M} \cdot \vec{x}}{4\pi|\vec{x}|^3}. \tag{10.140}$$

For the velocity in the radial direction we find

$$u_r = \frac{\partial \Phi}{\partial r} = -\frac{M}{2\pi}\frac{\cos\vartheta}{r^3}, \tag{10.141}$$

and therefore for $r = r_0$

$$u_r(r = r_0) = -\cos\vartheta \cdot \text{const.} \tag{10.142}$$

If we now consider a sphere which moves with velocity

$$\vec{U} = -U\,\vec{e}_x \tag{10.143}$$

(Fig. 10.19), then of course every point on its surface moves with the velocity $-U$ in the x-direction. The component of the velocity normal to the surface of the sphere is

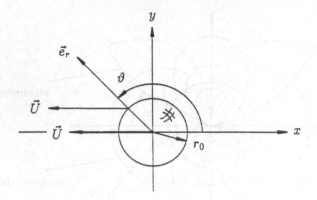

Fig. 10.19 The normal component of the velocity on the surface

$$\vec{U} \cdot \vec{n} = -U\vec{e}_x \cdot \vec{e}_r = -U\cos\vartheta. \tag{10.144}$$

For

$$\vec{U} = \frac{\vec{M}}{2\pi r_0^3}, \tag{10.145}$$

the normal component of the velocity of the dipole on a spherical surface $r = r_0$ is then equal to the normal component of the sphere velocity. But then the value of the velocity is uniquely determined everywhere because the solution of Laplace's equation is unique.

Therefore the dipole field is identical to the velocity field caused by a sphere instantaneously at the origin moving with the velocity according to (10.145). In this instant the flow has the potential

$$\Phi = -\frac{r_0^3}{2}\frac{\vec{U}\cdot\vec{x}}{|\vec{x}|^3}. \tag{10.146}$$

At another instant, when the sphere is at the position $\vec{a}$ (Fig. 10.20) we obtain the potential

$$\Phi = -\frac{r_0^3}{2}\frac{\vec{U}\cdot(\vec{x}-\vec{a})}{|\vec{x}-\vec{a}|^3} = -\frac{r_0^3}{2}\frac{\vec{U}\cdot\vec{r}}{r^3}. \tag{10.147}$$

If we superimpose the potential (10.146) with the potential of the parallel flow, whose velocity corresponds to the negative velocity of the sphere

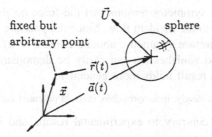

Fig. 10.20 Potential of the sphere

$$-\vec{U} = U_\infty \vec{e}_x,$$

we bring the sphere to rest and obtain the potential of steady flow past a sphere at the origin. In Cartesian coordinates it reads

$$\Phi = U_\infty x + \frac{r_0^3}{2} U_\infty \frac{x}{r^3}, \tag{10.148}$$

and in spherical coordinates

$$\Phi = U_\infty \left(r + \frac{r_0^3}{2r^2} \right) \cos \vartheta. \tag{10.149}$$

We shall now compute the velocity on the surface of the sphere $r = r_0$: for the radial component we obtain

$$u_r = \frac{\partial \Phi}{\partial r} \bigg|_{r=r_0} = U_\infty \left(1 - \frac{r_0^3}{r^3} \right)_{r=r_0} \cos \vartheta = 0, \tag{10.150}$$

as indeed has to be the case in order to satisfy the kinematic boundary condition for a sphere at rest. The velocity component in the ϑ-direction is

$$u_\vartheta = \frac{1}{r} \frac{\partial \Phi}{\partial \vartheta} \bigg|_{r=r_0} = -U_\infty \left(1 + \frac{r_0^3}{2r^3} \right)_{r=r_0} \sin \vartheta = -\frac{3}{2} U_\infty \sin \vartheta. \tag{10.151}$$

The magnitude of this velocity component reaches a maximum at $\vartheta = \pi/2$ and at $\vartheta = 3\pi/2$. We obtain the pressure coefficient from Bernoulli's equation as

$$c_p = \frac{p - p_\infty}{(\varrho/2)U_\infty^2} = 1 - \frac{9}{4} \sin^2 \vartheta. \tag{10.152}$$

It is obvious from symmetry reasons that the force on the sphere has no component perpendicular to the incident flow. Since the pressure distribution is an even function of ϑ, and therefore symmetric about the line $\vartheta = \pi/2, \vartheta = 3\pi/2$, the force in the x-direction also vanishes (as can easily be demonstrated by direct computation). However this result holds more generally:

A body with no lift in steady, incompressible, inviscid potential flow experiences no drag.

This statement is contrary to experimental results and is called *d'Alembert's paradox*.

In potential flow the kinetic energy of the fluid particles increases starting from the forward stagnation point ($\vartheta = \pi$) on the body to reach a maximum at $\vartheta = \pi/2$. This kinetic energy is just enough to carry the fluid particles against the increasing pressure to the rear stagnation point ($\vartheta = 0$). A force towards the front is produced on the rear half of the sphere which is exactly canceled out by the force on the front half of the sphere. In viscous flow the particles have lost kinetic energy in the boundary layer. Their "impetus" is not enough to bring them to the rear stagnation point against the increasing pressure. The fluid particles come to a standstill, and the flow separates from the body. With this any further increase in pressure is inhibited, with the result that the force on the back hemisphere is smaller than that on the front. Thus a drag is produced even if we disregard the friction drag due to the shear stress at the wall. We call this contribution to the drag the *pressure drag*. (This drag can be reduced by replacing the rear hemisphere with a streamlined body shape to prevent separation and again we are led to the spindle-shaped bodies discussed earlier.)

If we consider the flow past a body at small Reynolds' numbers, where the inertia forces (and therefore the kinetic energy) are small compared to the friction forces, then the fluid particles close to the wall are pulled along by the surrounding fluid and carried to the rear stagnation point by the then strong friction forces. Separation then does not occur, and we see a streamline pattern resembling superficially the streamline pattern of a potential flow.

As the Reynolds' number increases, a separation with a steady vortex-ring is formed behind the sphere. The streamlines still close behind the sphere and the vortex. As the Reynolds' number increases even further the vortex becomes larger until it finally becomes unstable and an unsteady *wake* forms. Periodic vortices then separate from the body and are carried away in the wake. The flow past a cylinder in a cross-flow is similar, and is easier to observe. The vortices arrange themselves into a *vortex street* behind the body, since this is again a stable configuration.

At even higher Reynolds' numbers the flow in the wake becomes turbulent, but even then large, ordered vortex-like structures are visible. It is clear that the drag changes greatly with the different flow forms, but however complicated the flow may be, in incompressible flow the drag coefficient c_D is only a function of the Reynolds' number. The function $c_D = c_D(Re)$ for the sphere is shown in Fig. 10.21, together with sketches of the flow configurations at the corresponding Reynolds' numbers $Re = U\,d\,/\,\nu$.

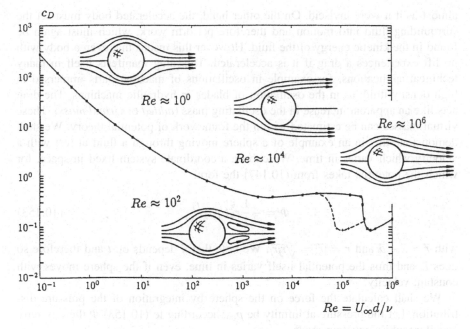

Fig. 10.21 Drag coefficient of the sphere

The sharp drop off of the drag coefficient at $Re \approx 3 \times 10^5$ is due to the transition of the previous laminar boundary layer on the sphere to a turbulent boundary layer. The shear stresses in the turbulent boundary layer are larger, and the outer fluid is able to drag the fluid close to the wall towards the axis. The flow then separates later and the wake becomes narrower. The flow which has not separated acts on a larger part of the back of the sphere, so that a larger force in the forward direction arises and the drag thus becomes smaller. The *transition point* can be shifted towards smaller Reynolds' numbers by roughness protrusions on the surface (as we know from our discussion of turbulent transition in pipe flow) and then the lower drag can be realized at lower Reynolds' numbers (dashed line in Fig. 10.21). An example of this can be found with *golf balls*, whose surfaces are artificially "roughened" *dimples* by indentations.

10.3.2   Virtual Masses

The previous discussion has shown that steady potential flow past a sphere or other blunt body is not found in nature because of the boundary layer separation. However if we suddenly accelerate a body from rest, the flow is described well by potential theory within a certain period of time $\tau \sim O(d/u)$. If the acceleration is large, the inertial forces are larger than the viscous forces and the flow behaves almost as if it were inviscid. On the other hand, the accelerated body must set the surrounding fluid into motion and therefore perform work, which must again be found in the kinetic energy of the fluid. However this means that even a body with no lift experiences a drag if it is accelerated. This drag manifests itself in many technical applications, for example in oscillations of machine parts immersed in high density fluid, as in the oscillation of blades in hydraulic machines. The drag acts like an apparent increase in the oscillating mass (*added* or *virtual mass*). These virtual masses can be estimated within the framework of potential theory. We shall demonstrate this in an example of a sphere moving through a fluid at rest with a velocity which varies in time. We choose a coordinate system fixed in space, for which the potential takes from (10.147) the form

$$\Phi = -\frac{1}{2} r_0^3 U_i(t) \frac{r_i}{r^3}, \tag{10.153}$$

with $\vec{r} = \vec{x} - \vec{a}$ and $r = |\vec{r}| = \sqrt{r_j r_j}$. We note that $\vec{a}$ depends on t and therefore so does $\vec{r}$, and thus the potential itself varies in time, even if the sphere moves with constant velocity.

We shall calculate the force on the sphere by integration of the pressure distribution. Let the pressure at infinity be p_∞, according to (10.153) Φ there is zero, and Bernoulli's equation reads

$$\frac{\partial \Phi}{\partial t} + \frac{1}{2} u_j u_j + \frac{p}{\varrho} = \frac{p_\infty}{\varrho}. \tag{10.154}$$

As we already know, the terms $\frac{1}{2} u_j u_j$, which also occur in the steady case, must cancel out in the integration for the force; therefore we shall ignore them right away. With (10.153) we obtain for $\partial \Phi / \partial t$

$$\frac{\partial \Phi}{\partial t} = -\frac{1}{2} \left(\frac{r_0}{r} \right)^3 \left[r_j \frac{dU_j}{dt} - U_j U_j + \frac{3}{r^2} U_i U_j r_i r_j \right]. \tag{10.155}$$

According to d'Alembert's paradox, only the term with dU_j/dt can provide a contribution to the force, as we can convince ourselves by explicit calculation.

The other term results from the fact that Φ is time dependent even for constant sphere velocity. The pressure on the surface $r = r_0$ arising from this term, because of $r_j / r_0 = n_j$ is

$$\frac{p - p_\infty}{\varrho} = \frac{1}{2} r_0 n_j \frac{dU_j}{dt}. \tag{10.156}$$

Since incompressible flow without circulation react immediately to the instantaneous boundary conditions, it is sufficient to compute the force at the instant in which the center of the sphere passes the coordinate origin. For the sphere moving with U in the positive x-direction, we compute the non-vanishing x-component of the force as

$$F_x = -\iint\limits_{(S)} p \cos \vartheta \, dS,$$

with $dS = r_0^2 \sin \vartheta \, d\vartheta \, d\varphi$:

$$F_x = -\frac{\varrho}{2} \frac{dU}{dt} r_0^3 \int\limits_0^{2\pi} \int\limits_0^{\pi} \cos^2 \vartheta \, \sin \vartheta \, d\vartheta \, d\varphi = -\frac{2}{3} \pi r_0^3 \varrho \frac{dU}{dt}. \tag{10.157}$$

Therefore the sphere experiences a force acting against the acceleration. This statement is valid independent of the chosen coordinate system. If an external force X_x acts on the sphere of mass M, then, taking the drag force F_x into account, it follows from Newton's second law that

$$M \frac{dU}{dt} = X_x + F_x, \tag{10.158}$$

or

$$X_x = \left(M + \frac{2}{3} \pi r_0^3 \varrho \right) \frac{dU}{dt}. \tag{10.159}$$

Therefore if we wish to calculate the acceleration of a sphere in fluid due to an external force, to the actual mass of the sphere M must be added the added or virtual mass

$$M' = \frac{2}{3} \pi r_0^3 \varrho. \tag{10.160}$$

This mass is due to the fact that both the sphere and the surrounding fluid must be accelerated.

The virtual mass of the sphere is precisely half of the fluid mass displaced by the sphere. The additional work per unit time which is performed during acceleration as a consequence of the virtual mass must then be equal to the change in the kinetic energy of the fluid. The kinetic energy in the volume V of fluid is given by

$$K = \iiint\limits_{(V)} \frac{\varrho}{2} u_i u_i \, dV = \frac{\varrho}{2} \iiint\limits_{(V)} \frac{\partial \Phi}{\partial x_i} \frac{\partial \Phi}{\partial x_i} \, dV. \tag{10.161}$$

With

$$\frac{\partial}{\partial x_i}\left(\Phi \frac{\partial \Phi}{\partial x_i}\right) = \frac{\partial \Phi}{\partial x_i}\frac{\partial \Phi}{\partial x_i} + \Phi \frac{\partial^2 \Phi}{\partial x_i \partial x_i} \tag{10.162}$$

and

$$\frac{\partial^2 \Phi}{\partial x_i \partial x_i} = 0,$$

it follows that

$$K = \frac{\varrho}{2}\iiint\limits_{(V)} \frac{\partial}{\partial x_i}\left(\Phi \frac{\partial \Phi}{\partial x_i}\right) dV, \tag{10.163}$$

and further from Gauss' theorem

$$K = \frac{\varrho}{2}\iint\limits_{(S)} \Phi \frac{\partial \Phi}{\partial x_i} n_i \, dS = \frac{\varrho}{2}\iint\limits_{(S)} \Phi \frac{\partial \Phi}{\partial n} \, dS. \tag{10.164}$$

The total kinetic energy of the fluid contained between the surface of the sphere S_s and a surface S_∞ surrounding the whole fluid and extending to infinity ($r \to \infty$, Fig. 10.22) is

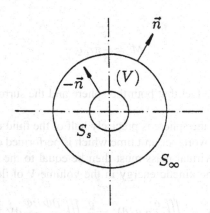

Fig. 10.22 Integration domain

$$K = -\frac{\varrho}{2} \iint\limits_{(S_s)} \Phi \frac{\partial \Phi}{\partial r} \, \mathrm{d}S + \frac{\varrho}{2} \iint\limits_{(S_\infty)} \Phi \frac{\partial \Phi}{\partial r} \, \mathrm{d}S. \tag{10.165}$$

For a sphere instantaneously at the origin, the potential in spherical coordinates is

$$\Phi = -\frac{r_0^3}{2r^2} U \cos \vartheta. \tag{10.166}$$

From this we calculate

$$\frac{\partial \Phi}{\partial r} = U \frac{r_0^3}{r^3} \cos \vartheta, \tag{10.167}$$

and obtain from (10.165)

$$K = -\frac{1}{4} r_0^6 U^2 \varrho \left(-\iint\limits_{(S_s)} r^{-5} \cos^2 \vartheta \, \mathrm{d}S + \iint\limits_{(S_\infty)} r^{-5} \cos^2 \vartheta \, \mathrm{d}S \right). \tag{10.168}$$

The second integral vanishes in the limit $r \to \infty$, and therefore the kinetic energy in the fluid is

$$K = \frac{2}{3} \pi r_0^3 \varrho \frac{U^2}{2}. \tag{10.169}$$

The change in the kinetic energy is

$$\frac{\mathrm{d}K}{\mathrm{d}t} = \frac{2}{3} \pi r_0^3 \varrho U \frac{\mathrm{d}U}{\mathrm{d}t}, \tag{10.170}$$

and equal to the work per unit time of the virtual mass

$$M' \frac{\mathrm{d}U}{\mathrm{d}t} U = \frac{2}{3} \pi r_0^3 \varrho \frac{\mathrm{d}U}{\mathrm{d}t} U. \tag{10.171}$$

As an example we shall consider a sphere of mass M under the influence of gravity in an infinitely extending fluid. The force of gravity $M g$ acts on the sphere. In addition the sphere experiences a hydrostatic lift which is equal to the weight of the displaced fluid (with (10.160) then $2M'g$). The drag due to the virtual mass acts against the acceleration. Therefore the equation of motion reads

388 10 Potential Flows

$$(M + M') \frac{dU}{dt} = M g - 2M'g, \tag{10.172}$$

or

$$\frac{dU}{dt} = \frac{M - 2M'}{M + M'} g. \tag{10.173}$$

with $M = \varrho_s V$ and $M' = \frac{1}{2} \varrho V$ we also write the acceleration as

$$\frac{dU}{dt} = g \frac{\varrho_s - \varrho}{\varrho_s + \varrho/2} = g \frac{\varrho_s/\varrho - 1}{\varrho_s/\varrho + 1/2}. \tag{10.174}$$

If the density of the sphere is much larger than that of the fluid then the acceleration is essentially equal to the gravitational acceleration (as, for example, for a heavy body falling through the atmosphere). If, on the contrary, the density of the fluid is much larger than the density of the sphere, then the sphere moves upwards with an acceleration of 2g (as, for example, a gas bubble in a liquid).

We shall now sketch the manner of computation for the virtual mass of a general body which carries out a pure translational motion: we obtain the velocity field from the solution of Laplace's equation under the boundary conditions

$$\Phi = \text{const} \quad \text{for} \quad r \to \infty \tag{10.175}$$

and

$$u_i n_i = \frac{\partial \Phi}{\partial x_i} n_i = \frac{\partial \Phi}{\partial n} = U_i n_i \quad \text{for} \quad F(x_i, t) = 0. \tag{10.176}$$

Since both the differential equation and the boundary conditions are linear, and the velocity of the body U_i also only appears linearly, U_i can only appear linearly in the solution, which must therefore have the form

$$\Phi = U_i \varphi_i. \tag{10.177}$$

From (10.176) it follows that

$$\frac{\partial \varphi_i}{\partial n} = n_i \quad \text{for} \quad F(x_i, t) = 0, \tag{10.178}$$

where the vector function φ_i thus only depends on the shape of the body. By (10.177) and (10.176) the kinetic energy is

$$K = -\frac{1}{2} U_i U_j \iint\limits_{(S_s)} \varrho \varphi_i n_j dS, \qquad (10.179)$$

where the negative sign appears since now n_j is taken relative to the surface of the body S_s. The integral is a symmetric second order tensor whose six independent components are required in the general case to compute the kinetic energy of the flow generated by the translational motion of the body. (If the body carries out a rotation in addition to a translation, three of these tensors are required.) The components of the tensor have the dimension of mass, and the tensor is called the *virtual* or *added mass tensor*

$$m_{ij} = -\iint\limits_{(S_s)} \varrho \varphi_i n_j dS. \qquad (10.180)$$

With it we can write down the kinetic energy of the fluid

$$K = \frac{1}{2} U_i U_j m_{ij}. \qquad (10.181)$$

If the vector function φ_i is known, then the m_{ij} can be calculated. For the case of the sphere at the origin we have

$$\varphi_i = -\frac{r_0^3}{2r^3} x_i, \qquad (10.182)$$

and, because of $n_j = x_j/r_0$, the virtual mass tensor is computed from

$$m_{ij} = \frac{\varrho}{2r_0} \iint\limits_{(S_s)} x_i x_j \, dS. \qquad (10.183)$$

It is easily shown that the tensor mj in this case is spherically symmetric

$$m_{11} = m_{22} = m_{33} = M'. \qquad (10.184)$$

Taking into account the virtual mass tensor, the equation of motion for a body of mass M acted on by an external force X_i is

$$M \frac{\mathrm{d}U_i}{\mathrm{d}t} + m_{ij} \frac{\mathrm{d}U_j}{\mathrm{d}t} = X_i. \qquad (10.185)$$

From this equation in the form

$$(M\delta_{ij} + m_{ij}) \frac{\mathrm{d}U_j}{\mathrm{d}t} = X_i \qquad (10.186)$$

we see that in general the direction of the acceleration is not the same as the direction of the force. This becomes evident if one tries to accelerate a submerged, asymmetric body in a certain direction.

10.4   Plane Potential Flow

For plane flows a Cartesian coordinate system can always be found where the flow in all planes z = const is the same and the velocity component in the z-direction vanishes. It is often useful, instead of the Cartesian coordinates x, y, to use the polar coordinates r, φ, which we obtain from cylindrical coordinates (Appendix B) by setting z = const there.

10.4.1   Examples of Incompressible, Plane Potential Flows

Here we first have the potential of a source available as a fundamental solution (*line source*, Fig. 10.23), which we met in (4.122) as Green's function

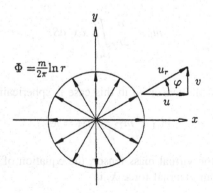

Fig. 10.23 Plane source in the origin

$$\Phi = \frac{m}{2\pi}\ln r, \tag{10.187}$$

with $r^2 = (x - x')^2 + (y - y')^2$ or $r^2 = x^2 + y^2$ for a source at the origin. The velocity components in polar coordinates are then

$$u_r = \frac{\partial \Phi}{\partial r} = \frac{m}{2\pi}\frac{1}{r}, \quad \text{and} \tag{10.188}$$

$$u_\varphi = \frac{1}{r}\frac{\partial \Phi}{\partial \varphi} = 0. \tag{10.189}$$

In Cartesian coordinates, the components read

$$u = \frac{\partial \Phi}{\partial x} = \frac{m}{2\pi}\frac{x}{x^2 + y^2}, \quad \text{and} \tag{10.190}$$

$$v = \frac{\partial \Phi}{\partial y} = \frac{m}{2\pi}\frac{y}{x^2 + y^2}. \tag{10.191}$$

By superimposing a source with the parallel flow in the same manner as before, the flow past a semi-infinite body is generated (Fig. 10.24)

$$\Phi = U_\infty x + \frac{m}{2\pi}\ln\sqrt{x^2 + y^2} = U_\infty r\cos\varphi + \frac{m}{2\pi}\ln r. \tag{10.192}$$

As in the rotationally symmetric case, the superposition of parallel flow with line distributions of sources and sinks gives rise to flows past cylindrical bodies of various shapes. By differentiation of the source potential we obtain the fundamental

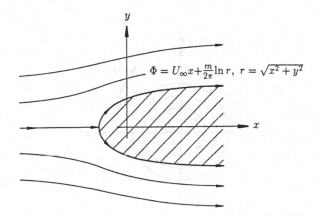

Fig. 10.24 Plane semi-infinite body

solution of the dipole in the plane. The potential of a dipole at the origin orientated in the negative x-direction reads

$$\Phi = \frac{M}{2\pi} \frac{x}{x^2 + y^2} = \frac{M}{2\pi} \frac{\cos \varphi}{r}. \tag{10.193}$$

The velocity potential (10.193) also describes the flow past a circular cylinder with axis in the z-direction moving to the left with velocity

$$U_\infty = \frac{M}{2\pi} \frac{1}{r_0^2}.$$

The superposition of a dipole with a parallel flow generates the flow past a circular cylinder at rest. The associated potential is

$$\Phi = U_\infty x + \frac{M}{2\pi} \frac{x}{x^2 + y^2} = U_\infty \left(r + \frac{r_0^2}{r} \right) \cos \varphi. \tag{10.194}$$

Another important singular solution of Laplace's equation is the *potential vortex* or straight vortex filament which we have already met. The potential of the vortex filament coinciding with the z-axis is given by

$$\Phi = \frac{\Gamma}{2\pi} \varphi = \frac{\Gamma}{2\pi} \arctan \frac{y}{x}. \tag{10.195}$$

For the velocity components in the r- and φ-directions we find (Fig. 10.25)

$$u_r = \frac{\partial \Phi}{\partial r} = 0, \text{ and} \tag{10.196}$$

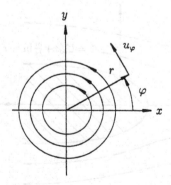

Fig. 10.25 Potential vortex

$$u_\varphi = \frac{1}{r}\frac{\partial \Phi}{\partial \varphi} = \frac{\Gamma}{2\pi}\frac{1}{r}. \tag{10.197}$$

The origin is a singular point: the velocity becomes infinite there. The streamlines are circles. The flow is irrotational with the exception of the singular point. In Sect. 6.1 we also met the velocity field (10.197) as an exact solution of the Navier–Stokes equations, and we showed there that this potential flow arises as a limiting case of viscous fluid between two circular cylinders, if the inner one (radius R_I) rotates and the radius of the outer one becomes infinitely large. The rotating cylinder exerts a friction torque (per unit length) on the fluid, which, because of

$$\tau_w = -\tau_{\varphi r}\bigg|_{R_I} = -\eta\left(\frac{\partial u_\varphi}{\partial r} - \frac{u_\varphi}{r}\right)\bigg|_{R_I} = \frac{\eta \Gamma}{\pi}\frac{1}{R_I^2} \tag{10.198}$$

(see Appendix B) is found to be

$$T = \tau_w 2\pi R_I^2 = 2\Gamma\eta. \tag{10.199}$$

Therefore the torque is independent of the radius, and as a consequence every cylinder of fluid with radius $r \geq R_I$ transmits the same torque. The ring of fluid between R_I and r is not accelerated, in accordance with the fact that the divergence of the friction stresses in incompressible potential flow vanishes. However the power of the friction torque on the cylinder $r = R_I$ is

$$P_I = T\left(\frac{u_\varphi}{r}\right)\bigg|_{R_I} = 2\Gamma\frac{\eta \Gamma}{2\pi}\frac{1}{R_I^2} \tag{10.200}$$

and at the position r

$$P = \frac{\eta \Gamma^2}{\pi r^2}. \tag{10.201}$$

The difference

$$\Delta P = \frac{\eta \Gamma^2}{\pi}\left(\frac{1}{R_I^2} - \frac{1}{r^2}\right) \tag{10.202}$$

is dissipated into heat. This result also shows that an isolated potential vortex without a supply of energy cannot maintain the velocity distribution (10.197). In addition we note that the kinetic energy of this distribution is infinitely large, and therefore physically there is no vortex whose distribution behaves as $1/r$ and which

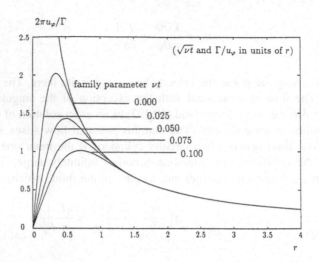

Fig. 10.26 Velocity distribution of a decaying vortex

reaches to infinity. If we have the velocity distribution (10.197) at the time $t = 0$, at a later time it then reads

$$u_\varphi = \frac{\Gamma}{2\pi} \frac{1}{r} \left[1 - \exp\left(-\frac{r^2}{4\nu t} \right) \right].$$ \hfill (10.203)

We obtain this solution from the φ component of the Navier–Stokes equations since no typical length appears in the problem, so r is to be made dimensionless with $(\nu t)^{-1/2}$, and the solution must be a relation between the dimensionless groups

$$\Pi_1 = \frac{u_\varphi r}{\Gamma}, \quad \Pi_2 = \frac{r}{\sqrt{\nu t}}$$

(Fig. 10.26). This flow is no longer irrotational.

The superposition of a potential vortex with a sink (or source) generates a flow whose streamlines are *logarithmic spirals* (*spiral vortex*, Fig. 10.27). The solution of the differential equation for the streamline in polar coordinates

$$\frac{1}{r} \frac{dr}{d\varphi} = \frac{u_r}{u_\varphi} = \frac{m}{\Gamma}$$ \hfill (10.204)

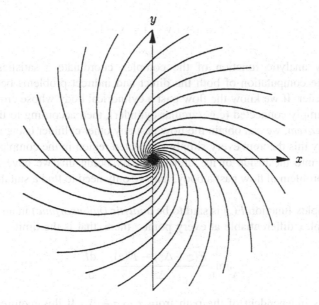

Fig. 10.27 Superposition of sink and potential vortex (logarithmic spiral)

is

$$r = K \exp\left(\frac{m\varphi}{\Gamma}\right). \tag{10.205}$$

This flow is known as "bathtub drainage" and has a technically important application in the potential flow through *radial cascades* (see Fig. 2.9). In the bladeless ring space, far in front of or behind the cascade, the streamlines are logarithmic spirals, but with different values of Γ in front of and behind the cascade. (If the circulation in front of the cascade is Γ_i, then behind the cascade it is

$$\Gamma_o = \Gamma_i + n\,\Gamma_b,$$

where Γ_b is the circulation of a single blade and n is the number of blades in the cascade.)

10.4.2 Complex Potential for Plane Flows

Plane flows differ from other two-dimensional flows (with two independent variables) because the two independent variables x and y can be combined into one complex variable

$$z = x + i\,y, \quad i = \sqrt{-1}. \tag{10.206}$$

Since every analytic function of the complex coordinate z satisfies Laplace's equation, the computation of both the direct and indirect problems becomes considerably easier. If we know the flow past a cylindrical body whose cross-sectional surface is simply connected (e.g. circular cylinder), then according to the *Riemann mapping theorem*, we can obtain the flow past any other cylinder using a *conformal mapping*. By this theorem every simply connected region in the complex plane can by mapped into the inside of the unit circle. By doing this we have in principle solved the problem of flow past a body, and we only need to find a suitable mapping function.

The complex function $F(z)$ is said to be *analytic (holomorphic)* in an open set G, if it is complex differentiable at every point z there, that is the limit

$$\lim_{\Delta z \to 0} \frac{F(z + \Delta z) - F(z)}{\Delta z} = \frac{dF}{dz} \tag{10.207}$$

exists and is independent of the path from z to $z + \Delta z$. If this requirement is not satisfied, the point is a *singular point*.

First we note that along a path parallel to the x-axis

$$\frac{dF}{dz} = \frac{\partial F}{\partial x} \tag{10.208}$$

holds, and along a path parallel to the y-axis

$$\frac{dF}{dz} = \frac{\partial F}{\partial(i\,y)}. \tag{10.209}$$

Since every complex function $F(z)$ is of the form

$$F(z) = \Phi(x, y) + i\psi(x, y), \tag{10.210}$$

we therefore have

$$\frac{\partial F}{\partial x} = \frac{\partial \Phi}{\partial x} + i\frac{\partial \psi}{\partial x} = \frac{1}{i}\frac{\partial \Phi}{\partial y} + \frac{\partial \psi}{\partial y} = \frac{1}{i}\frac{\partial F}{\partial y}. \tag{10.211}$$

Clearly for the derivative to exist it is necessary that

$$\frac{\partial \Phi}{\partial x} = \frac{\partial \psi}{\partial y} \text{ and } \frac{\partial \Phi}{\partial y} = -\frac{\partial \psi}{\partial x} \tag{10.212}$$

hold. The *Cauchy-Riemann differential equations* (10.212) are also sufficient for the existence of the derivative of $F(z)$. We can also show easily that both the real part $\Re(F) = \Phi(x, y)$ and the imaginary part $\Im(F) = \psi(x, y)$ satisfy Laplace's equation. To do this, we differentiate the first differential equation in (10.212) by x and the second by y and add the results. We then see that Φ satisfies Laplace's equation. If we differentiate the first by y and the second by x and subtract the results, we see that the same also holds for ψ. Both functions can therefore serve as the velocity potential of a plane flow. We choose Φ as the velocity potential and shall now consider the physical meaning of ψ. With

$$\vec{u} = \nabla \Phi = \frac{\partial \Phi}{\partial x}\vec{e}_x + \frac{\partial \Phi}{\partial y}\vec{e}_y = u\vec{e}_x + v\vec{e}_y \tag{10.213}$$

because of (10.212) we also have

$$\nabla \psi = \frac{\partial \psi}{\partial x}\vec{e}_x + \frac{\partial \psi}{\partial y}\vec{e}_y = -v\vec{e}_x + u\vec{e}_y. \tag{10.214}$$

From $\nabla \Phi \cdot \nabla \psi = 0$ we conclude that $\nabla \psi$ is perpendicular to the velocity vector $\vec{u}$, and therefore $\psi = $ const are streamlines. We have thus identified ψ as a stream function. (As already mentioned in connection with (10.104) and (10.105), the introduction of a stream function is not restricted to potential flows.) Since an additive constant clearly plays no role in a stream function, we can always adjust it so that

$$\psi = 0 \tag{10.215}$$

is the equation of the body contour. With ψ known, we obtain the velocity vector directly from the formula

$$\vec{u} = \nabla \psi \times \vec{e}_z \quad \text{or} \quad u_i = \epsilon_{ij3}\frac{\partial \psi}{\partial x_j}, \tag{10.216}$$

therefore

$$u = \frac{\partial \psi}{\partial y}, \quad v = -\frac{\partial \psi}{\partial x}, \tag{10.217}$$

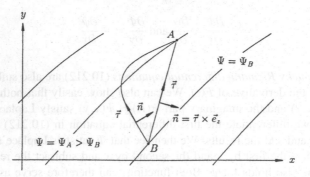

Fig. 10.28 Meaning of the stream function in plane flow

so that the continuity equation

$$\frac{\partial u}{\partial x} + \frac{\partial v}{\partial y} = 0$$

is identically satisfied.

We now calculate the volume flux (per unit depth) between the points A and B (Fig. 10.28)

$$\dot{V} = \int_B^A u_i n_i \, ds \tag{10.218}$$

and to evaluate the integral we write $n_i = \epsilon_{ik3}\tau_k$, where $\tau_k = dx_k/ds$ is the unit vector along the path of integration ds in the direction of increasing ψ (Fig. 10.28).

With (10.216) we then obtain

$$\dot{V} = \int_B^A \left(\epsilon_{ij3} \frac{\partial \psi}{\partial x_j} \epsilon_{ik3} \frac{dx_k}{ds} \right) ds \tag{10.219}$$

or with $\epsilon_{ij3}\epsilon_{ik3} = \delta_{jk}$ also

$$\dot{V} = \int_B^A \frac{\partial \psi}{\partial x_j} dx_j = \int_B^A d\psi = \psi_A - \psi_B. \tag{10.220}$$

This result confirms that the volume flux is independent of the path between A and B and is equal to the difference of the values of the stream function at these points. The velocity components can be most easily calculated using

$$\frac{dF}{dz} = \frac{\partial F}{\partial x} = \frac{\partial \Phi}{\partial x} + i\frac{\partial \psi}{\partial x} = u - iv, \qquad (10.221)$$

where the sign of v is to be noted: dF/dz furnishes the *complex conjugate velocity*

$$\frac{dF}{dz} = \bar{w} = u - iv,$$

that is, the reflection of the *complex velocity* $w = u + iv$ at the real axis.

We shall now look at some examples of complex potentials:

(a) Translational flow:

$$F(z) = (U_\infty - iV_\infty)z, \qquad (10.222)$$

or

$$F(z) = (U_\infty x + V_\infty y) + i\,(U_\infty y - V_\infty x), \qquad (10.223)$$

because of (10.210) therefore

$$\Phi = U_\infty x + V_\infty y, \qquad (10.224)$$

$$\psi = U_\infty y - V_\infty x. \qquad (10.225)$$

The streamlines follow from $\psi = $ const to $y = xV_\infty/U_\infty + C$ and the complex conjugate velocity is

$$\frac{dF}{dz} = U_\infty - iV_\infty. \qquad (10.226)$$

(b) Source flow:

$$F(z) = \frac{m}{2\pi}\ln z \qquad (10.227)$$

or because of $z = re^{i\varphi}$ also

$$F = \frac{m}{2\pi}(\ln r + i\varphi).\qquad(10.228)$$

With (10.210) we obtain the velocity potential and the stream function as

$$\Phi = \frac{m}{2\pi}\ln r,\qquad(10.229)$$

$$\psi = \frac{m}{2\pi}\varphi.\qquad(10.230)$$

The streamlines ψ = const are straight lines through the origin.

(c) Potential vortex:

$$F(z) = -i\frac{\Gamma}{2\pi}\ln z,\qquad(10.231)$$

where the negative sign is needed because we count Γ anticlockwise positive. In polar coordinates we obtain

$$F = -i\frac{\Gamma}{2\pi}(\ln r + i\varphi),\qquad(10.232)$$

therefore

$$\Phi = +\frac{\Gamma}{2\pi}\varphi,\quad\text{and}\qquad(10.233)$$

$$\psi = -\frac{\Gamma}{2\pi}\ln r.\qquad(10.234)$$

The streamlines ψ = const are circles (r = const).

(d) Dipole:

$$F(z) = \frac{M}{2\pi}\frac{1}{z},\qquad(10.235)$$

or

$$F = \frac{M}{2\pi}\frac{1}{r}(\cos\varphi - i\sin\varphi) = \frac{M}{2\pi}\frac{1}{r^2}(x - iy),\qquad(10.236)$$

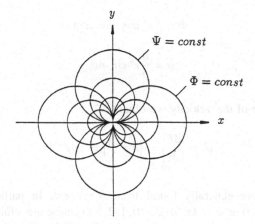

Fig. 10.29 Streamlines and equipotential lines of the plane dipole

from which we read off directly

$$\Phi = +\frac{M}{2\pi}\frac{\cos\varphi}{r} = \frac{M}{2\pi}\frac{1}{r^2}x, \quad \text{and} \tag{10.237}$$

$$\psi = -\frac{M}{2\pi}\frac{\sin\varphi}{r} = -\frac{M}{2\pi}\frac{1}{r^2}y. \tag{10.238}$$

For $\psi = \text{const}$ we obtain with $\sin\varphi = y/r$

$$r^2 = x^2 + y^2 = -\frac{M}{C}y, \tag{10.239}$$

that is, the streamlines are circles which are tangent to the x-axis in the origin
(Fig. 10.29).

(e) Corner flow:

$$F(z) = \frac{a}{n}z^n, \tag{10.240}$$

with $z = re^{i\varphi}$ it follows that

$$F(z) = \frac{a}{n}r^n(\cos n\varphi + i\sin n\varphi), \tag{10.241}$$

and therefore

$$\Phi = \frac{a}{n} r^n \cos n\varphi, \text{ and} \qquad (10.242)$$

$$\psi = \frac{a}{n} r^n \sin n\varphi. \qquad (10.243)$$

For the magnitude of the velocity we obtain

$$|\vec{u}| = \left| \frac{dF}{dz} \right| = \left| a\, z^{n-1} \right| = |a| r^{n-1}. \qquad (10.244)$$

The streamlines are generally found from $\psi = $ const. In particular for $\psi = 0$, therefore $\sin n\varphi = 0$ or $\varphi = k\pi/n\, (k = 0, 1, 2, \ldots)$, these are straight lines through the origin which can represent walls in the flow field. Figure 10.30 shows the streamline plot for different values of the exponent n.

(f) Flow past a circular cylinder (Fig. 10.31):

$$F(z) = U_\infty \left(z + \frac{r_0^2}{z} \right) \qquad (10.245)$$

or again with $z = re^{i\varphi}$

$$F = U_\infty \left(r + \frac{r_0^2}{r} \right) \cos \varphi + iU_\infty \left(r - \frac{r_0^2}{r} \right) \sin \varphi, \qquad (10.246)$$

and therefore

$$\Phi = U_\infty \left(r + \frac{r_0^2}{r} \right) \cos \varphi, \quad \text{and} \qquad (10.247)$$

$$\psi = U_\infty \left(r - \frac{r_0^2}{r} \right) \sin \varphi. \qquad (10.248)$$

We obtain $\psi = 0$ for $r = r_0$ and $\varphi = 0, \pi, \ldots.$ From the complex conjugate velocity

$$\frac{dF}{dz} = U_\infty \left(1 - \frac{r_0^2}{z^2} \right) \qquad (10.249)$$

by $dF/dz = 0$ we find the location of the stagnation points at $z = \pm\, r_0$ and deduce the velocity components

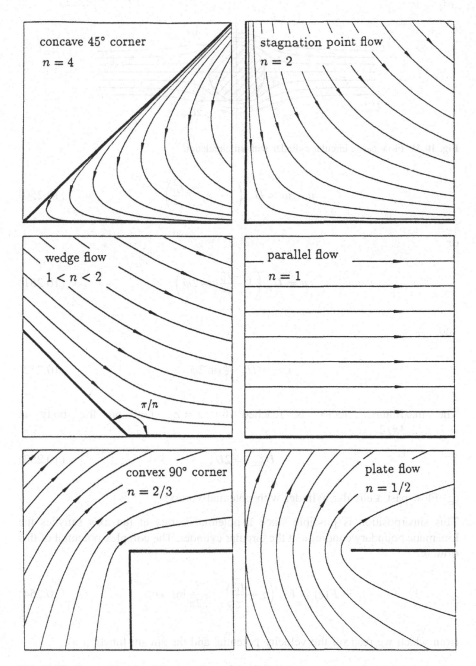

Fig. 10.30 Corner flow for different values of the exponent n

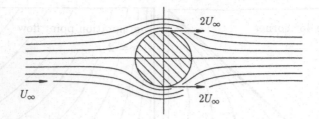

Fig. 10.31 Flow past a circular cylinder without circulation

$$u - iv = U_\infty \left(1 - e^{-i2\varphi} \frac{r_0^2}{r^2} \right) \tag{10.250}$$

or

$$u = U_\infty \left(1 - \frac{r_0^2}{r^2} \cos 2\varphi \right), \tag{10.251}$$

and

$$v = -U_\infty \frac{r_0^2}{r^2} \sin 2\varphi. \tag{10.252}$$

The maximum velocity is reached for $r = r_0$, i.e., on the body at $\varphi = \pi/2, 3\pi/2, \ldots$:

$$U_{max} = 2U_\infty. \tag{10.253}$$

(g) Flow past a circular cylinder with potential vortex:

This superposition is possible since a potential vortex at the axis satisfies the kinematic boundary condition at the circular cylinder. The complex potential of this flow is

$$F(z) = U_\infty \left(z + \frac{r_0^2}{z} \right) - i \frac{\Gamma}{2\pi} \ln(z/r_0), \tag{10.254}$$

from which we read off the velocity potential and the stream function as

$$\Phi = U_\infty \left(r + \frac{r_0^2}{r} \right) \cos \varphi + \frac{\Gamma}{2\pi} \varphi, \quad \text{and} \tag{10.255}$$

$$\psi = U_\infty \left(r - \frac{r_0^2}{z} \right) \sin \varphi - \frac{\Gamma}{2\pi} \ln(r/r_0). \tag{10.256}$$

Since $F(z)$ represents the flow past a circular cylinder for all values of Γ, it is not unique. We obtain a survey of the different flows, by computing the stagnation points on the body contour. From

$$u_\varphi = \frac{1}{r} \frac{\partial \Phi}{\partial \varphi} \bigg|_{r=r_0} = -2U_\infty \sin \varphi + \frac{\Gamma}{2\pi} \frac{1}{r_0} \tag{10.257}$$

the equation for the stagnation points follows

$$\sin \varphi = \frac{\Gamma}{4\pi} \frac{1}{U_\infty} \frac{1}{r_0}. \tag{10.258}$$

Figure 10.32 shows the flow forms for different values of the circulation Γ. The force (per unit depth) on the cylinder in the x-direction vanishes for symmetry reasons, and that in the y-direction is

$$F_y = -\varrho U_\infty \Gamma, \quad (\Gamma < 0). \tag{10.259}$$

The flow field in Fig. 10.32d can be experimentally realized if a rotating cylinder is exposed to a cross-flow with undisturbed U_∞ sufficiently small compared to the circumferential velocity Ωr_0, corresponding to the condition $|\Gamma| > 4\pi r_0 U_\infty$. As we know a rotating cylinder without an external stream in viscous fluid produces a potential vortex, and it is clear that a small enough cross-flow will not lead to separation at the cylinder. As experiments show, the lift calculated from potential theory is already reached at $\Omega r_0/U_\infty > 4$. We call the phenomenon where rotating cylinder in a cross-flow experiences a lift, the *Magnus effect*. It can generally be seen with rotating bodies, as for example, a sliced tennis ball. However this effect is very important in ballistics (spinning missiles). There have been attempts to use rotating cylinders instead of sails on ships (*Flettner rotor*).

10.4.3 Blasius' Theorem

We shall restrict ourselves to steady flows and shall consider a simply connected domain, say the cross-section of a cylinder in a flow (Fig. 10.33). From

$$F_i = - \oint_{(C)} p n_i \, \mathrm{d}s \tag{10.260}$$

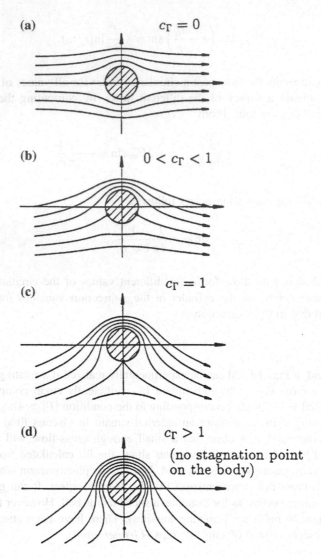

(a)　　　$c_\Gamma = 0$

(b)　　　$0 < c_\Gamma < 1$

(c)　　　$c_\Gamma = 1$

(d)　　　$c_\Gamma > 1$
(no stagnation point
on the body)

Fig. 10.32 Flow past a circular cylinder with clockwise circulation $\Gamma = -4\pi r_0 U_\infty c_\Gamma$

we compute the components of the force per unit depth with $n_i = \epsilon_{ik3} \mathrm{d}x_k / \mathrm{d}s$ (cf. (10.219)) as

$$F_1 = F_x = - \oint_{(C)} p \, \mathrm{d}y \qquad (10.261)$$

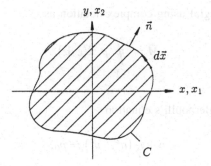

Fig. 10.33 Blasius' theorem

and

$$F_2 = F_y = \oint_{(C)} p \, dx. \tag{10.262}$$

In complex notation

$$z = x + iy, \quad \bar{z} = x - iy$$

we combine the force components as

$$F_x - iF_y = \oint_{(C)} (-i\,p)d\bar{z}. \tag{10.263}$$

The moment on the cylinder about the origin only has a component in the z-direction

$$\vec{M} \cdot \vec{e}_z = M = -\oint_{(C)} \epsilon_{ij3}x_in_jp \, ds = -\oint_{(C)} \epsilon_{ij3}x_i\epsilon_{jk3}p \, dx_k = \oint_{(C)} x_i\delta_{ik}pdx_k \tag{10.264}$$

or

$$M = \oint_{(C)} (x\,p\,dx + y\,p\,dy) \tag{10.265}$$

We write the line integral using complex notation as

$$M = \oint_{(C)} p \, \Re(z \, d\bar{z}).$$

(10.266)

In steady flow, from Bernoulli's equation

$$p + \frac{\varrho}{2}\left(u^2 + v^2\right) = p_0,$$

(10.267)

and from the square of the magnitude of the complex conjugate velocity

$$\left|\frac{dF}{dz}\right|^2 = \frac{dF}{dz}\frac{d\bar{F}}{d\bar{z}} = u^2 + v^2$$

(10.268)

it follows that the pressure is

$$p = p_0 - \frac{\varrho}{2}\frac{dF}{dz}\frac{d\bar{F}}{d\bar{z}}.$$

(10.269)

Therefore we write for the force

$$F_x - i\,F_y = i\frac{\varrho}{2}\oint_{(C)} \frac{dF}{dz}\,d\bar{F},$$

(10.270)

because the closed integral over the constant pressure p_0 vanishes. Since the contour of the body is a curve $\psi = $ const, we have

$$d\bar{F} = d\Phi = dF,$$

(10.271)

and from (10.270) emerges the *first Blasius' theorem*

$$F_x - iF_y = i\frac{\varrho}{2}\oint_{(C)} \left(\frac{dF}{dz}\right)^2 dz.$$

(10.272)

In an analogous manner we obtain from (10.266) the *second Blasius' theorem*

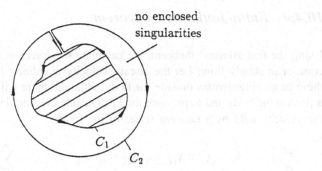

Fig. 10.34 Application of Cauchy's theorem

$$M = -\frac{\varrho}{2}\Re\left(\oint\limits_{(C)} \left(\frac{\mathrm{d}F}{\mathrm{d}z}\right)^2 z\mathrm{d}z\right). \tag{10.273}$$

According to the derivation the integration is to be carried out along the contour of the body. As a consequence of *Cauchy's theorem*

$$\oint\limits_{(C)} f(z)\,\mathrm{d}z = 0 \ \{\text{if } f(z) \text{ is holomorphic on } C \text{ and in the domain enclosed by } C$$

$$\tag{10.274}$$

the integration can also be carried out along any arbitrary closed curve enclosing the body, as long as there are no singularities between the contour of the body and the integration curve. Using the sense of integration in Fig. 10.34 it follows from (10.274) that

$$\oint\limits_{(C_1)} f(z)\mathrm{d}z + \oint\limits_{(C_2)} f(z)\mathrm{d}z = 0 \tag{10.275}$$

or, if C_1 and C_2 are followed in the same sense,

$$\oint\limits_{(C_1)} f(z)\mathrm{d}z = \oint\limits_{(C_2)} f(z)\mathrm{d}z. \tag{10.276}$$

10.4.4   Kutta-Joukowski Theorem

Using the first Blasius' theorem, we calculate the force on a cylinder of arbitrary contour in steady flow. Let the stream velocity at infinity be $U_\infty + iV_\infty$, and let there be no singularities outside the body, although there will be inside, in order to represent the body and to produce the lift. Outside the singularities we can represent the velocity field by a *Laurent series* of the form

$$\frac{dF}{dz} = u - iv = A_0 + A_1 z^{-1} + A_2 z^{-2} + A_3 z^{-3} + \ldots = \sum_{n=0}^{\infty} A_n z^{-n}, \qquad (10.277)$$

which yields the complex potential

$$F(z) = A_0 z + A_1 \ln z - \sum_{n=2}^{\infty} \frac{1}{n-1} A_n z^{-(n-1)} + \text{const.} \qquad (10.278)$$

From the condition at infinity

$$\left. \frac{dF}{dz} \right|_\infty = U_\infty - iV_\infty \qquad (10.279)$$

it follows that

$$A_0 = U_\infty - iV_\infty. \qquad (10.280)$$

In order to calculate the coefficient A_1 we form the integral of $(u - iv)$ around the contour of the body

$$\oint_{(C)} (u - iv)dz = \oint_{(C)} (u - iv)(dx + idy) \qquad (10.281)$$

or

$$\oint_{(C)} (u - iv)dz = \oint_{(C)} \vec{u} \cdot d\vec{x} + i \oint_{(C)} d\psi, \qquad (10.282)$$

where the second integral vanishes since $d\psi$ is zero along the contour of the body. With the definition of the circulation (1.105) we therefore write

$$\oint_{(C)} (u-iv)\mathrm{d}z = \Gamma. \tag{10.283}$$

Since the Laurent series (10.277) has only one essential singularity ($z = 0$), then from the *residue theorem* we have

$$\oint_{(C)} (u-iv)\mathrm{d}z = 2\pi\, i A_1 = \Gamma. \tag{10.284}$$

From this we obtain the complex conjugate velocity in the form

$$u - iv = U_\infty - iV_\infty - i\frac{\Gamma}{2\pi}z^{-1} + \sum_{n=2}^{\infty} A_n\, z^{-n}. \tag{10.285}$$

We now calculate the force on the cylinder using Blasius' theorem (10.272). Because of

$$\left(\frac{\mathrm{d}F}{\mathrm{d}z}\right)^2 = (U_\infty - iV_\infty)^2 - i\frac{\Gamma}{\pi z}(U_\infty - iV_\infty) - \frac{\Gamma^2}{4\pi^2 z^2} + \frac{2A_2}{z^2}(U_\infty - iV_\infty) + \dots \tag{10.286}$$

and by applying the residue theorem we first obtain

$$\oint_{(C)} \left(\frac{\mathrm{d}F}{\mathrm{d}z}\right)^2 \mathrm{d}z = -(2\pi i)i\Gamma\frac{U_\infty - iV_\infty}{\pi} \tag{10.287}$$

and then from (10.272) the *Kutta-Joukowski theorem*

$$F_x - iF_y = i\varrho\Gamma(U_\infty - iV_\infty). \tag{10.288}$$

From this equation we firstly conclude that the lift is perpendicular to the undisturbed stream at infinity, that is, the body experiences no drag, and secondly for a given circulation Γ the lift is independent of the contour of the body.

In a similar manner we obtain for the moment

$$M = -2\pi \, \varrho U_\infty \Re \left[iA_2 \left(1 - i\frac{V_\infty}{U_\infty} \right) \right]; \qquad (10.289)$$

the moment then depends on the complex coefficient A_2 and therefore on the contour of the body.

10.4.5 Conformal Mapping

We know that it is possible to transform the flow past a circular cylinder to the flow past a cylinder of arbitrary contour with the help of conformal mapping. As long as no separation of the boundary layer occurs in the real flow, potential theory will describe the actual flow behavior very well. For this reason the potential flow past a circular cylinder has great technical importance.

The complex analytic mapping function

$$\zeta = f(z), \qquad (10.290)$$

defined at all points z at which $f'(z)$ has a finite nonzero value, maps the z-plane onto the ζ-plane such that the mapping is "similar in the smallest parts". In other words, infinitesimal configurations remain *conformal*, that is, they remain the same. The transformation has the following properties which are easy to prove:

(a) The angle between any two curve elements and its sense of rotation remains the same.
(b) The ratio of two small lengths remains the same, therefore

$$\frac{|\Delta z|}{|\Delta z'|} = \frac{|\Delta \zeta|}{|\Delta \zeta'|}$$

(c) A small element Δz is transformed into the element $\Delta \zeta$ according to

$$\Delta \zeta = \Delta z \frac{d\zeta}{dz}.$$

As an example we shall consider the mapping function (Fig. 10.35).

$$\zeta = z^2 = (x + iy)^2 \qquad (10.291)$$

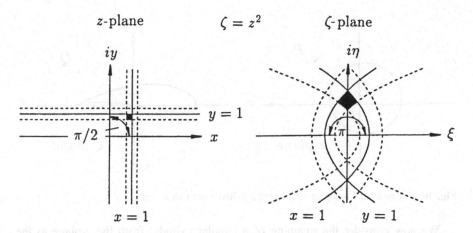

Fig. 10.35 Conformal mapping

It follows that

$$\zeta = \xi + i\eta = (x^2 - y^2) + 2i\,xy \qquad (10.292)$$

therefore

$$\xi = x^2 - y^2, \quad \eta = 2xy. \qquad (10.293)$$

Lines $x = C$ in the z-plane are mapped onto parabolae open to the left, as we see if we eliminate y from the two last equations

$$\xi = C^2 - \frac{\eta^2}{4C^2}. \qquad (10.294)$$

For $C = 0$ (y-axis) the parabola coincides with the negative ξ-axis. Lines $y = C$ are mapped onto parabolae open to the right

$$\xi = \frac{\eta^2}{4C^2} - C^2, \qquad (10.295)$$

where for $C = 0$ (x-axis) the parabola lies along the positive ξ-axis. The origin is a singular point of this mapping. There $f' = d\zeta/dz$ has a simple zero, and the mapping is no longer conformal at this point. At a simple zero the angle between two line elements, such as the x- and y-axes ($\pi / 2$), is doubled in the ζ-plane (π). In general we have: at a zero of order n of $f'(z)$, the angle is altered by a factor $(n + 1)$ (branch point of order n).

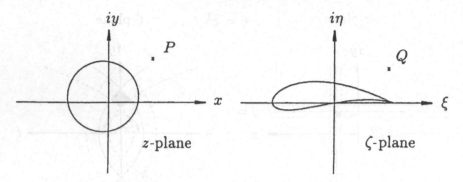

Fig. 10.36 Conformal mapping of a circular cylinder onto an airfoil

We now consider the mapping of a circular cylinder from the z-plane to the ζ-plane. By the mapping function the region outside the cylinder in the z-plane is mapped onto the region outside another cylinder in the ζ-plane (Fig. 10.36).

Let P and Q be corresponding points in the z- and ζ-planes respectively. The potential at the point P is

$$F(z) = \Phi + i\psi. \tag{10.296}$$

The point Q has the same potential, and we obtain it by insertion of the mapping function

$$F(z) = F[z(\zeta)] = F(\zeta). \tag{10.297}$$

We then compute the complex conjugate velocity $\bar{w}_\zeta$ in the ζ-plane from

$$\bar{w}_\zeta(\zeta) = \frac{dF}{d\zeta}. \tag{10.298}$$

The following procedure is often more useful: we consider z to be a parameter and calculate the value of the potential at the point z. With the help of the mapping function $\zeta = f(z)$ we determine the complex coordinate of z which corresponds to z. At this point ζ, the potential then has the same value as at the point z. In order to determine the velocity in the ζ-plane, we form

$$\frac{dF}{d\zeta} = \frac{dF}{dz}\frac{dz}{d\zeta} = \frac{dF}{dz}\left(\frac{d\zeta}{dz}\right)^{-1}, \tag{10.299}$$

or

$$\bar{w}_\zeta(\zeta) = \bar{w}_z(z) \left(\frac{d\zeta}{dz}\right)^{-1}. \tag{10.300}$$

Therefore to compute the velocity at a point in the ζ-plane we divide the velocity at the corresponding point in the z-plane by $d\zeta / dz$. The derivative $dF / d\zeta$ exists at all points with $d\zeta/dz \neq 0$. At singular points with $d\zeta/dz = 0$, the complex conjugate velocity in the ζ-plane $\bar{w}_\zeta(\zeta) = dF/d\zeta$ becomes infinite, if it is not equal to zero at the corresponding point in the z-plane.

10.4.6   Schwarz-Christoffel Transformation

The properties of conformal mappings at singular points mentioned in connection with the mapping function $\zeta = z^2$ can also be used to map the x-axis onto a polygon. We shall consider the mapping given by

$$\frac{d\zeta}{dz} = f'(z) = K(z - x_1)^{\alpha_1/\pi - 1}(z - x_2)^{\alpha_2/\pi - 1}\ldots(z - x_n)^{\alpha_n/\pi - 1}, \tag{10.301}$$

which is known as the *Schwarz-Christoffel transformation*. If we denote the polar angle of a complex number $z = r \exp(i\,\varphi)$ with $\arg(z)$, because of

$$\ln z = \ln r + i \arg(z), \tag{10.302}$$

we read off the logarithm of (10.301)

$$\begin{aligned}\arg(d\zeta) = \arg(dz) + \arg(K) + \left(\frac{\alpha_1}{\pi} - 1\right)\arg(z - x_1) + \\ + \left(\frac{\alpha_2}{\pi} - 1\right)\arg(z - x_2) + \ldots + \left(\frac{\alpha_n}{\pi} - 1\right)\arg(z - x_n).\end{aligned} \tag{10.303}$$

If we move from a point on the x-axis to the left of x_1 (Fig. 10.37) in the direction of increasing x, then the polar angle is $\arg(dz) = 0$. For $x < x_1$ all $(z - x_i)$ in (10.303) are less than zero and real, i.e., $\arg(z - x_i) = \pi$. Therefore $\arg(d\zeta)$ is constant, until the first singularity x_1 is reached. As we move past x_1 the sign of the term $(z - x_1)$ changes, and therefore $\arg(z - x_1)$ decreases abruptly from the value π to 0. Since all the other terms in (10.303) remain unchanged, $\arg(d\zeta)$ changes by the amount $(\alpha_1/\pi - 1) \cdot (-\pi) = \pi - \alpha_1$ and then again remains constant until x_2 is reached. Therefore at the position $\zeta_1 = f(x_1)$ in the transformed plane, the line corresponding to $A - x_1 - x\,(x < x_2)$ is turned by $\pi - \alpha_1$. At $z = x_2$, $\arg(z - x_2)$ jumps by $-\pi$, $\arg(d\zeta)$ therefore by the amount $\pi - \alpha_2$, etc. Between the singular points x_i the corresponding images of the x-axis are straight lines ($\arg(d\zeta) = \text{const}$), and the angle between each

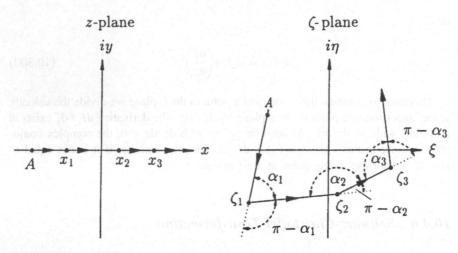

Fig. 10.37 Schwarz-Christoffel transformation

of the straight lines is α_i. The upper half of the z-plane is mapped onto the inside of the polygon in the ζ-plane, where the constant K in (10.301) allows a constant stretching and rotation of the polygon.

As an example we shall treat the transformation

$$\frac{d\zeta}{dz} = K(z+1)^{(1/2-1)}(z-1)^{(3/2-1)} = K\sqrt{\frac{z-1}{z+1}}. \qquad (10.304)$$

The singular points are $x_1 = -1$, $x_2 = 1$, and the associated angles are $\alpha_1 = \pi/2$, $\alpha_2 = 3\pi/2$. The line element on the x-axis to the left of x_1 has, from (10.303), in the ζ-plane a polar angle of

$$\arg(d\zeta) = \arg(dz) + \arg(K) + \left(-\frac{1}{2}\right)\pi + \frac{1}{2}\pi = \arg(K). \qquad (10.305)$$

If we choose K to be a real number, the mapping of the x-axis begins with a straight line parallel to the ξ-axis. For $x_1 < x < x_2$ the polar angle is $\arg(d\zeta) = \pi - \alpha_1 = \pi/2$, i.e., the second straight piece is parallel to the $i\eta$-axis. For $x > x_2$, $\arg(d\zeta) = \pi/2 + (\pi - \alpha_2) = 0$, i.e., the line is again parallel to the ξ-axis. The mapping function for this example can be stated in closed form. From the integration of (10.304) it follows that

$$\zeta = f(z) = K \int \sqrt{\frac{z-1}{z+1}}dz = K\left(\sqrt{z^2-1} - \ln\left(z + \sqrt{z^2-1}\right)\right) + C, \qquad (10.306)$$

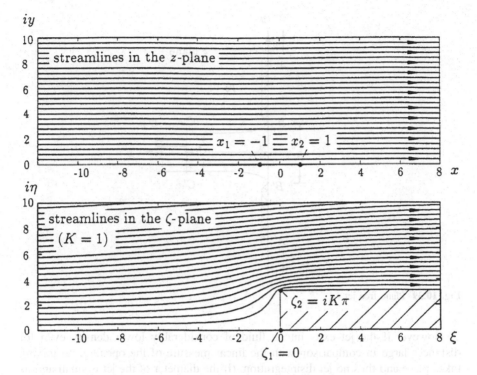

Fig. 10.38 Step in parallel flow

where C occurs as a constant of integration. The image point of the singular point $x_1 = -1$ is $\zeta_1 = -K \ln(-1) + C = -iK\pi + C$, and that of the point $x_2 = 1$ is $\zeta_2 = C$. For $C = iK\pi$, $\zeta_1 = 0$, $\zeta_2 = iK\pi$, and we obtain the configuration shown in Fig. 10.38. Every flow field in the z-plane for which the x-axis is a streamline gives a flow field in the ζ-plane over a step of height $K\pi$. In particular for parallel flow in the plane $F(z) = U_\infty z$, we have $F(\zeta) = U_\infty z(\zeta)$ the complex potential of the flow over a step represented in Fig. 10.38.

10.4.7 Free Jets

In the discussion of the abrupt contraction of a cross-section (Fig. 9.8) we inferred that the fluid separates at the sharp edge, and then no longer follows the wall, but forms a free jet which contracts. The free surface of the jet is unstable and if the surrounding fluid has the same density as the jet (as discussed in the case of cross-section contraction), this instability causes rapid mixing of the jet with the surrounding fluid, as is indicated in Fig. 9.8.

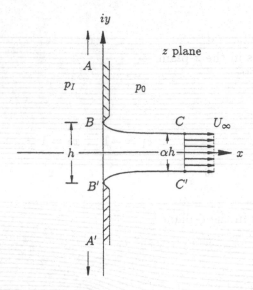

Fig. 10.39 Plane free jet

However if the jet exits into a fluid of considerably lower density even for distances large in comparison with the linear measure of the opening, no mixing takes place and thus no jet disintegration. (If the diameter of the jet is small and its velocity is high, disintegration of the jet can occur directly behind the exit. We shall not discuss this process where the surface tension and the viscosity play a role.)

In *free jets* the shape of the jet is of technical importance, since, for example, the contraction coefficient can be deduced from it. However the computation of the jet flow is in general a difficult problem, since the dynamic boundary condition (4.171) has to be satisfied on the as yet unknown jet boundaries. Only in plane potential flow can problems with free jet boundaries be solved rather simply by conformal mapping.

As the first example we shall compute the jet contraction coefficient of a plane free jet, and to this end shall consider the discharge from a large vessel (Fig. 10.39). The emerging jet contracts from the cross-section $B - B'$ to the cross-section $C - C'$. There the pressure inside the jet is equal to the ambient pressure since the curvature of the streamlines vanishes. The pressure is constant on the free surface of the jet, and the constancy of the velocity then follows from Bernoulli's equation

$$U_\infty = \sqrt{\frac{2}{\varrho}(p_1 - p_0)},\tag{10.307}$$

from which the volume flux (per unit depth) follows as

$$\dot{V} = \alpha h U_\infty. \tag{10.308}$$

We conclude from the curvature of the free surface that the pressure increases as we move towards the center of the jet, and therefore the velocity decreases from its value U_∞ on the edge of the jet towards the middle.

In order to determine the shape of the free jet we use the mapping which results from the definition of the complex conjugate velocity

$$\zeta = f(z) = \frac{dF}{dz} = \bar{w} = u - iv. \tag{10.309}$$

Therefore this function maps the z-plane onto the velocity plane, which is also called the *hodograph plane*.

We shall first examine the course of the streamline from the point A ($x = 0$, $y \to \infty$) to the point B (edge of the container outlet) and then to the point C (Fig. 10.39). From the equality of the potentials at corresponding points in the z- and ζ-planes, it follows directly that streamlines remain streamlines under conformal mapping ($\Psi = \Psi(z) = \Psi[z(\zeta)] = \text{const}$). Therefore the line under consideration is also a streamline in the hodograph plane. On the section of the line $A - B$ we have $u \equiv 0$, and $-v$ increases from zero to the value U_∞; thus its image coincides with the η-axis from $\eta = 0$ to $\eta = -v = U_\infty$. On the contour of the free jet from $B(\bar{w} = iU_\infty)$ to C ($\bar{w} = U_\infty$), $|\bar{w}|$ is, from (10.307), constant equal to U_∞, and so the image of this section of the streamline is the quarter circle sketched in Fig. 10.40. The image of the lower streamline $A' - B' - C'$, on which the velocities are everywhere the complex conjugates of those above, corresponds to a reflection

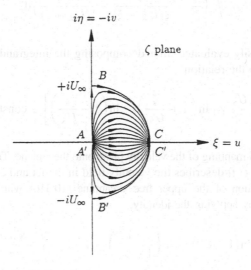

Fig. 10.40 Free jet in the hodograph plane

of the image of $A - B - C$ through the ζ-axis. The upper and lower free surfaces of the jet then form the semicircle

$$\zeta = \bar{w} = U_\infty e^{i\vartheta}, \tag{10.310}$$

with $-\pi/2 \leq \vartheta \leq \pi/2$. The streamlines inside the container and the jet fall into the inside of this semicircle in the right half plane.

The flow field in the hodograph plane can be generated from the superposition of a source at the origin and two sinks at $\zeta = \pm U_\infty$, but then of course it also occupies the ξ-η-plane outside the region of interest. Since only half of the strengths of the source ($m > 0$) and of the right-hand sink ($m < 0$) flow into or out of the semicircle, the strengths are chosen to be $|m| = 2\dot{V} = 2\alpha h U_\infty$, respectively. From (10.227) the complex potential now reads

$$F(\zeta) = \frac{\alpha}{\pi} U_\infty h[\ln \zeta - \ln(\zeta - U_\infty) - \ln(\zeta + U_\infty)]. \tag{10.311}$$

We can easily convince ourselves that the free surface of the jet (10.310) really is a streamline.

Our next step is to determine the mapping function $z = z(\zeta)$, and thus to obtain the free surface in the z-plane. From (10.309) it follows that

$$z = \int \frac{dF}{\zeta} = \int \frac{dF}{d\zeta} \frac{d\zeta}{\zeta}, \tag{10.312}$$

and with (10.311) then

$$z = \frac{\alpha}{\pi} U_\infty h \int \left(\frac{1}{\zeta^2} + \frac{1}{\zeta(U_\infty - \zeta)} - \frac{1}{\zeta(U_\infty + \zeta)} \right) d\zeta. \tag{10.313}$$

The integral is easily evaluated after decomposing the integrand into partial fractions, and leads to the relation

$$z = \frac{\alpha}{\pi} h \left[-\frac{U_\infty}{\zeta} + \ln\left(1 + \frac{\zeta}{U_\infty}\right) - \ln\left(1 - \frac{\zeta}{U_\infty}\right) \right] + \text{const}, \tag{10.314}$$

which is the desired mapping of the velocity plane onto the z-plane. The inverse function $\zeta = \bar{w} = u - iv = f(z)$ describes the velocity field in the jet and container. We now introduce the equation of the upper free streamline (10.310) with $0 \leq \vartheta \leq \pi/2$ into Eq. (10.314), and by applying the identity

$$\ln\left(1 + e^{i\vartheta}\right) - \ln\left(1 - e^{i\vartheta}\right) = \ln\left(\frac{1 + e^{i\vartheta}}{1 - e^{i\vartheta}}\right) = i\frac{\pi}{2} + \ln\left(\frac{\sin\vartheta}{1 - \cos\vartheta}\right) \tag{10.315}$$

obtain the shape of the jet in the z-plane as

$$z(\vartheta) = \frac{\alpha}{\pi} h \left[-e^{-i\vartheta} + \ln\left(\frac{\sin \vartheta}{1 - \cos \vartheta} \right) \right] + K, \quad 0 \leq \vartheta \leq \frac{\pi}{2}. \tag{10.316}$$

We determine the constant of integration K from the condition that at the point B

$$z(\vartheta = \pi/2) = i\frac{h}{2} \tag{10.317}$$

to

$$K = i\frac{h}{2}\left(1 - \frac{2\alpha}{\pi}\right). \tag{10.318}$$

Taking the limit $\vartheta \to 0$ in (10.316) furnishes an equation for the contraction coefficient α. For $\vartheta \to 0$ the real part of z tends to infinity, i.e., the point C in the z-plane lies at infinity. The imaginary part $\Im[z(\vartheta)]$ must, from Fig. 10.39, satisfy the condition

$$\lim_{\vartheta \to 0} \Im[z(\vartheta)] = i\alpha\frac{h}{2}, \tag{10.319}$$

so that from (10.316), $i\alpha h/2 = K$, and thus with (10.318) the contraction coefficient follows

$$\alpha = \frac{\pi}{\pi + 2} \approx 0.61. \tag{10.320}$$

The impact of a plane free jet on an infinite plane wall can also be computed with the method discussed, where here an explicit equation for the free streamline can be given. The flow in the z-plane (Fig. 10.41) has the hodograph of Fig. 10.42. Analogous to the previous example, this field can be represented by two sources at the positions $\zeta = \pm U_\infty$ and two sinks at the positions $\zeta = \pm i U_\infty$. The strength in each case is

$$|m| = 2\dot{V} = 4U_\infty h. \tag{10.321}$$

The complex potential in the ζ-plane therefore, according to (10.227) is

$$F(\zeta) = \frac{2}{\pi} U_\infty h[\ln(\zeta - U_\infty) + \ln(\zeta + U_\infty) - \ln(\zeta - iU_\infty) - \ln(\zeta + iU_\infty)]. \tag{10.322}$$

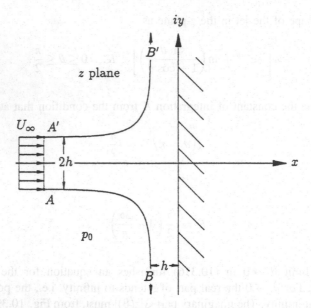

Fig. 10.41 Free jet impact perpendicular to a wall

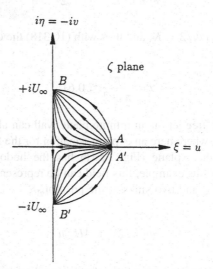

Fig. 10.42 Free jet impact in the hodograph plane

As before we obtain from (10.312)

$$z = \frac{2h}{\pi}\left[\ln\left(\frac{1 - \zeta/U_\infty}{1 + \zeta/U_\infty}\right)\right] - i\ln\left(\frac{1 - i\zeta/U_\infty}{1 + i\zeta/U_\infty}\right), \qquad (10.323)$$

where the integration constant is found, from the condition at the stagnation point

$$z(\zeta = 0) = 0, \qquad (10.324)$$

to be zero. On the lower free streamline we again have

$$\zeta = U_\infty e^{i\vartheta}, \quad 0 \leq \vartheta \leq \frac{\pi}{2} \qquad (10.325)$$

so that by applying the identities

$$\ln\left(\frac{1 - e^{i\vartheta}}{1 + e^{i\vartheta}}\right) = -i\frac{\pi}{2} + \ln\left(\frac{1 - \cos\vartheta}{\sin\vartheta}\right) = -i\frac{\pi}{2} + \ln\left(\tan\frac{\vartheta}{2}\right), \qquad (10.326)$$

and

$$\ln\left(\frac{1 - i\,e^{i\vartheta}}{1 + i\,e^{i\vartheta}}\right) = \ln\left(\frac{1 - e^{i(\vartheta + \pi/2)}}{1 + e^{i(\vartheta + \pi/2)}}\right) = -i\frac{\pi}{2} + \ln\left[\tan\left(\frac{\vartheta}{2} + \frac{\pi}{4}\right)\right] \qquad (10.327)$$

the equation of the free streamline appears in the form

$$z(\vartheta) = x + iy = -h\left\{1 - \frac{2}{\pi}\ln\left[\tan\frac{\vartheta}{2}\right]\right\} - i\,h\left\{1 + \frac{2}{\pi}\ln\left[\tan\left(\frac{\vartheta}{2} + \frac{\pi}{4}\right)\right]\right\} \qquad (10.328)$$

From the real part the relation

$$\tan\frac{\vartheta}{2} = \exp\left[\frac{\pi}{2}\left(1 + \frac{x}{h}\right)\right] \qquad (10.329)$$

follows, and with

$$\ln\left[\tan\left(\frac{\vartheta}{2} + \frac{\pi}{4}\right)\right] = \ln\left(\frac{1 + \tan(\vartheta/2)}{1 - \tan(\vartheta/2)}\right) = 2\,\text{arctanh}\left(\tan\frac{\vartheta}{2}\right) \qquad (10.330)$$

we obtain, from the imaginary part, the explicit equation of the lower free streamline

$$-\frac{y}{h} = 1 + \frac{4}{\pi} \text{arctanh}\left\{ \exp\left[\frac{\pi}{2}\left(1+\frac{x}{h}\right)\right] \right\}, \quad x < -h. \tag{10.331}$$

The upper free streamline is symmetric to this.

10.4.8 Flow Around Airfoils

The main purpose of the conformal mapping lies in the possibility to map the unknown flow past an airfoil to the known flow past a circular cylinder. In this manner we can obtain the direct solution of the flow past a cylinder of arbitrary contour. Although numerical methods of solution of the direct problem have now superseded the method of conformal mapping, it has still retained its fundamental importance. We shall discuss these methods using as an example the *Joukowski mapping*

$$\zeta = f(z) = z + \frac{a^2}{z}. \tag{10.332}$$

The function $f(z)$ maps a circle with radius a in the z-plane onto a "slit" in the ζ-plane. With the complex coordinate of the circle

$$z = a\, e^{i\varphi} \tag{10.333}$$

we obtain

$$\zeta = 2a\cos\varphi \tag{10.334}$$

purely real, i.e., the circle is mapped onto a section of the ξ-axis reaching from $-2a$ to $2a$ (Fig. 10.43). With the complex potential (10.245) of the cylinder flow ($r_0 = a$)

$$F(z) = U_\infty\left(z + \frac{a^2}{z}\right) \tag{10.335}$$

the Joukowski mapping function directly furnishes the potential in the ζ-plane as

$$F(\zeta) = U_\infty\zeta, \tag{10.336}$$

as was indeed expected. Now if we map a circle with radius b which is smaller or larger than the mapping constant a, we obtain an ellipse (Fig. 10.43). If we map a

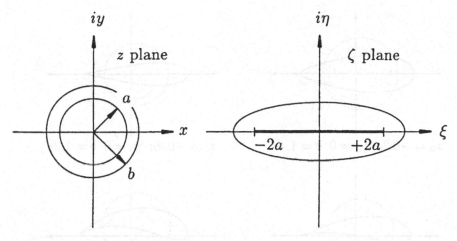

Fig. 10.43 Mapping of a circle onto a slit and ellipse

circle whose midpoint coordinates (x_0, y_0) are not zero, we obtain typical airfoils (Fig. 10.44).

The Joukowski mapping has a singular point at each of the positions $z = \pm a$, as can be seen

$$\frac{d\zeta}{dz} = 1 - \frac{a^2}{z^2}. \tag{10.337}$$

The point $z = -a$ is generally mapped into the interior of the airfoil and is thus of no interest to us. The angle between the two line elements emerging from $z = a$, dz_1 and dz_2 in Fig. 10.45 is π. Since we are dealing with a simple zero, the angle between the corresponding line elements $d\zeta_1$ and $d\zeta_2$ is doubled, and thus is 2π. Therefore the *trailing edge angle* is zero, a typical property of the Joukowski mapping, which is already evident from the mapping of the circle of radius a to a slit.

At the singular point B, the velocity in the ζ-plane becomes infinite if we do not ensure that it is zero at the point B in the z-plane. We accomplish this by choosing the circulation of the cylinder flow such that B lies at a stagnation point. This requirement determines the value of the circulation and prevents a flow past the trailing edge in the ζ-plane, which we already excluded in our discussion of the generation of circulation (Fig. 4.6). If the angle of attack is not too large, the real circulation adjusts itself according to this condition known as *Joukowski's hypothesis* or *Kutta condition*. It enables us to fix the value of the circulation about the cylinder. The circulation about the airfoil is then exactly the same size, because

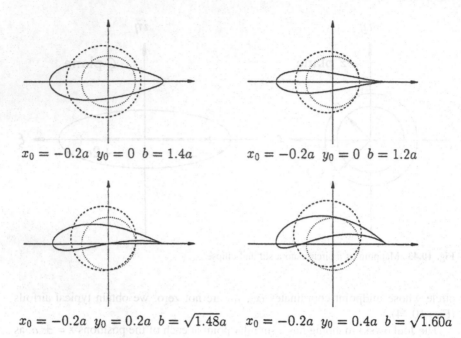

$$x_0 = -0.2a \quad y_0 = 0 \quad b = 1.4a \qquad x_0 = -0.2a \quad y_0 = 0 \quad b = 1.2a$$

$$x_0 = -0.2a \quad y_0 = 0.2a \quad b = \sqrt{1.48}a \qquad x_0 = -0.2a \quad y_0 = 0.4a \quad b = \sqrt{1.60}a$$

Fig. 10.44 Joukowski mapping

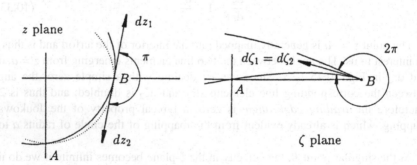

Fig. 10.45 Trailing edge angle of a Joukowski airfoil

$$\Gamma = \oint_{C_\zeta} \bar{w}_\zeta(\zeta)\, \mathrm{d}\zeta = \oint_{C_\zeta} \bar{w}_z(z) \frac{\mathrm{d}z}{\mathrm{d}\zeta} \mathrm{d}\zeta = \oint_{C_z} \bar{w}_z(z)\mathrm{d}z. \qquad (10.338)$$

For a coordinate system $z' = x' + iy'$, whose origin lies in the center of the circle and whose x'-axis denotes the direction of the undisturbed velocity, the complex potential, from (10.254), reads

$$F(z') = U_\infty \left(z' + \frac{r_0^2}{z'} \right) - i \frac{\Gamma}{2\pi} \ln \frac{z'}{r_0}. \tag{10.339}$$

In order to obtain the potential of the flow past a circular cylinder with center at z_0 in a uniform stream at an angle of attack α to the x-axis, we use the transformation

$$z = z_0 + |z'| e^{i(\varphi' + \alpha)} = z_0 + z' e^{i\alpha}, \tag{10.340}$$

which can be read off Fig. 10.46. We insert the transformation

$$z' = (z - z_0) e^{-i\alpha}, \tag{10.341}$$

into (10.339) to get

$$F(z) = U_\infty (z - z_0) e^{-i\alpha} + U_\infty \frac{r_0^2}{z - z_0} e^{i\alpha} - i \frac{\Gamma}{2\pi} \ln \left(\frac{z - z_0}{r_0} e^{-i\alpha} \right). \tag{10.342}$$

The complex conjugate velocity is then

$$\bar{w} = u - iv = U_\infty e^{-i\alpha} - U_\infty e^{i\alpha} \frac{r_0^2}{(z - z_0)^2} - i \frac{\Gamma}{2\pi} \frac{1}{z - z_0}. \tag{10.343}$$

At the point B, i.e., for $z - z_0 = r_0 e^{-i\beta}$, Joukowski's hypothesis requires that $u - iv = 0$ holds, so that (10.343) becomes an equation for the circulation Γ with the solution

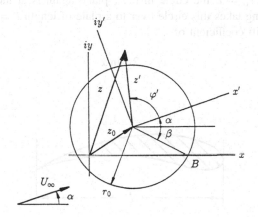

Fig. 10.46 Coordinate transformation

$$\Gamma = -4\pi r_0 U_\infty \sin(\alpha + \beta). \tag{10.344}$$

The value of Γ depends on the airfoil parameters r_0 and β, on the angle of attack α and on the undisturbed velocity U_∞. The mapping function itself need not be known to determine Γ, because as we have already shown, the circulation in the ζ-plane is exactly the same as in the z-plane.

The force per unit depth on the airfoil is calculated from the Kutta-Joukowski theorem (10.288), where we note that the complex conjugate velocity $U_\infty - iV_\infty$ is now to be replaced by $U_\infty \exp(-i\alpha)$. We obtain the complex conjugate force as

$$F_x - iF_y = -i\, 4\pi r_0 \varrho U_\infty^2 e^{-i\alpha} \sin(\alpha + \beta). \tag{10.345}$$

The magnitude of the force is

$$|F| = \sqrt{F_x^2 + F_y^2} = \sqrt{(F_x - iF_y)(F_x + iF_y)}, \tag{10.346}$$

and therefore

$$|F| = 4\pi\, r_0 \varrho U_\infty^2 \sin(\alpha + \beta). \tag{10.347}$$

We denote the dimensionless quantity

$$c_l = \frac{|F|}{(\varrho/2)U_\infty^2 l} = 8\pi \frac{r_0}{l} \sin(\alpha + \beta) \tag{10.348}$$

as the *lift coefficient*, where l is the length, or the *chord*, of the airfoil (Fig. 10.47) which can be calculated from the mapping function.

For $\beta = 0$ and $r_0 = a$ the circle in the z-plane again is at the origin, and the Joukowski mapping takes this circle over to a plate of length $l = 4a$ (Fig. 10.48). We then have a lift coefficient of

$$c_l = 2\pi \sin \alpha. \tag{10.349}$$

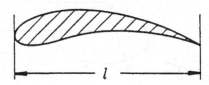

Fig. 10.47 Airfoil length

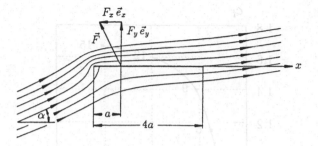

Fig. 10.48 Flow past an infinitesimally thin plate

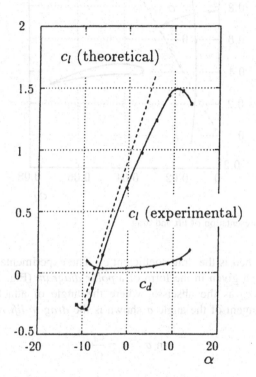

Fig. 10.49 Lift and drag coefficients

A suction force in the negative x-direction arises from the flow past the leading edge, and together with the pressure force perpendicular to the plate gives rise to a *lift force*, which (in agreement with the Kutta-Joukowski theorem) is perpendicular to the undisturbed flow direction, so that the drag force vanishes (d'Alembert's paradox).

The angle $\alpha = -\beta$ is called the *no lift direction* ($c_l = 0$) of the airfoil. In Fig. 10.49 a typical comparison between theoretical and experimental results is

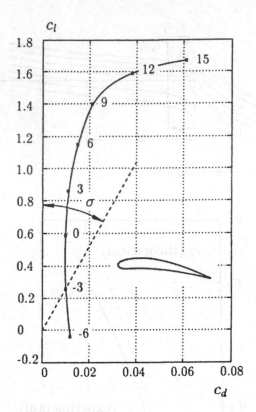

Fig. 10.50 Polar representation of lift and drag

shown. Also sketched is the drag coefficient c_d. The experimental results found in Fig. 10.49 are often given in the form of a *polar diagram* (Fig. 10.50), with c_l as the ordinate and c_d as the abscissa, where the angle of attack a is the curve parameter. The tangent of the angle σ shown is the *drag-to-lift ratio* ϵ

$$\tan \sigma = \epsilon = \frac{c_d}{c_l}. \tag{10.350}$$

The smallest drag-to-lift ratio is given by the tangent to the polar curve at the origin. Beyond a certain angle of attack, the lift decreases and the drag rises. This is due to boundary layer separation on the suction (upper) side of the airfoil. The airfoil is then said to be stalled.

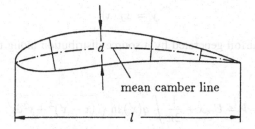

Fig. 10.51 Construction of an airfoil from the mean camber line and symmetric thickness distribution

10.4.9 Approximate Solution for Slender Airfoils in Incompressible Flow

In aerodynamics, airfoils where the length is much larger than the thickness are invariably used, in order to avoid separation. We can generate these airfoils by a symmetric thickness distribution about a mean camber line (Fig. 10.51). For a slender airfoil, i.e.,

$$\frac{d}{l} = \epsilon \ll 1,$$

the flow past the airfoil can be determined by first computing the solution for the symmetric airfoil of the same thickness distribution, then the solution for an infinitesimally thin mean camber line, and finally superimposing both solutions to obtain the flow past the real airfoil. In doing this, of course, an error is involved, but this is only of the order of magnitude $O(\epsilon^2)$, which is negligible for a very slender airfoil. This method leads to an explicit solution of the direct problem, although it has now been superseded by numerical methods. In spite of this we shall discuss the method, since it can serve as an introduction to *perturbation theory*, and some numerical methods are only generalizations of this method.

We first consider the symmetric airfoil (Fig. 10.52) with a contour given by

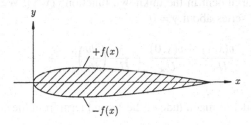

Fig. 10.52 Symmetric airfoil

$$y = \pm f(x) \tag{10.351}$$

and envisage an airfoil generated by a source distribution along the x-axis, so that the potential is

$$\Phi = U_\infty x + \frac{1}{2\pi} \int_0^l q(x') \ln \sqrt{(x-x')^2 + y^2} dx' \tag{10.352}$$

The velocity components generated by the source distribution are denoted by u and v. Since the body is very slender the *perturbation velocities* u and v are small compared to U_∞ and we have

$$\frac{u}{U_\infty} \sim \frac{v}{U_\infty} \sim \epsilon. \tag{10.353}$$

With $F(x, y) = -y \pm f(x) = 0$ the kinematic boundary condition (4.170) reads

$$\pm (U_\infty + u) \frac{df}{dx} - v = 0 \quad \text{(at the wall)}, \tag{10.354}$$

or

$$v = \pm (U_\infty + u) \frac{df}{dx} \quad \text{at} \quad y = \pm f(x). \tag{10.355}$$

Now $f = O(d)$ holds and thus $df/dx = O(\epsilon)$; because of (10.353) we therefore also write

$$\frac{v}{U_\infty} = \pm \frac{df}{dx} + O(\epsilon^2) \quad \text{at} \quad y = \pm f(x). \tag{10.356}$$

In what follows we shall ignore terms of the order $O(\epsilon^2)$. There are still difficulties involved in satisfying the boundary condition (10.356) on the body, since then $f(x)$ occurs as an argument in the unknown function $v(x, y)$. We therefore expand $v(x, y)$ in a Taylor series about $y = 0$

$$\frac{v(x, y)}{U_\infty} = \frac{v(x, 0)}{U_\infty} + \frac{y}{U_\infty} \left(\frac{\partial v}{\partial y} \right)_{y=0} + \cdots, \tag{10.357}$$

and estimate the order of magnitude of the second term from the continuity equation as

$$\frac{\partial u}{\partial x} = -\frac{\partial v}{\partial y} \sim \frac{u}{l}, \quad \text{and} \tag{10.358}$$

$$\frac{y}{U_\infty} \frac{\partial v}{\partial y} \sim \frac{d}{U_\infty} \frac{u}{l} = \frac{u}{U_\infty} \frac{d}{l} \sim \epsilon^2. \tag{10.359}$$

Since we are neglecting terms of the order $O(\epsilon^2)$, it follows that

$$\frac{v(x,y)}{U_\infty} = \frac{v(x,0)}{U_\infty} = \pm\frac{df}{dx} \quad \text{at} \quad y = 0; \tag{10.360}$$

therefore we can satisfy the boundary condition on the x-axis instead of on the body $y = \pm f(x)$. If we denote the upper side of the airfoil with O^+ and the lower side with O^-, we then write (10.356) in the form

$$\frac{v(x,0^+)}{U_\infty} = \frac{df}{dx}, \quad \frac{v(x,0^-)}{U_\infty} = -\frac{df}{dx}. \tag{10.361}$$

The velocity $v(x, y)$ is found from the potential (10.352)

$$v(x,y) = \frac{\partial \Phi}{\partial y} = \frac{1}{2\pi} \int_0^l \frac{q(x')y}{(x - x')^2 + y^2} dx'. \tag{10.362}$$

If we insert (10.362) into (10.361), an *integral equation* is obtained for the unknown source distribution $q(x)$, which can, in fact, be easily solved; we obtain the velocity $v(x, 0)$ by taking the limit $y \to 0$. The integral has a singularity at $x = x'$ and only there is the integrand nonzero for $y \to 0$. Using the transformation

$$\eta = \frac{x - x'}{y}; \quad x' = x - \eta y; \quad \frac{dx'}{y} = -d\eta \tag{10.363}$$

we get a regular integral

$$v(x,y) = -\frac{1}{2\pi} \int_{x/y}^{-(l-x)/y} \frac{q(x - \eta y)}{1 + \eta^2} d\eta, \tag{10.364}$$

thus for $0 < x < l$

$$v(x,0^+) = \lim_{y \to 0^+} [v(x,y)] = \frac{q(x)}{2\pi} \int_{-\infty}^{+\infty} \frac{d\eta}{1 + \eta^2} = \frac{q(x)}{2}. \tag{10.365}$$

The desired source distribution then follows from the boundary condition

$$q(x) = 2\frac{\mathrm{d}f}{\mathrm{d}x} U_\infty.$$ (10.366)

For $v(x, 0^-)$ we correspondingly obtain

$$v(x,0^-) = -\frac{q(x)}{2}$$ (10.367)

or again

$$q(x) = 2\frac{\mathrm{d}f}{\mathrm{d}x} U_\infty.$$ (10.368)

It is easy to show that the closure condition (10.119) is satisfied, and thus the problem is solved. With the known source distribution $q(x)$ the potential is now known, and the velocity and pressure fields follow in the manner already discussed. We note that the solution is in general not uniformly valid in the entire flow field. For airfoils with blunt noses, $\mathrm{d}f / \mathrm{d}x$ is infinite at $x = 0$. From Eq. (10.361) it then follows that $v/U_\infty \to \infty$, so that the assumptions of perturbation theory break down locally. The solution there is no longer valid, and we have to deal with a *singular perturbation problem*.

In order to calculate flow past the mean camber line, we cover it with a continuous vortex distribution, thus replacing the mean camber line with a bound *vortex sheet*. This represents a surface of discontinuity in the tangential velocity. Since the sheet is fixed in space, the jump in the tangential velocity leads to a pressure difference between the upper and lower sides, which gives rise to a force on the mean camber line. (A free vortex sheet, such as appears in unsteady motion of an airfoil, will deform such that the dynamic boundary condition of pressure equality (4.173) is satisfied.) Let the mean camber line be given by

$$y = f(x).$$ (10.369)

With $f_{\max}/l = \epsilon$ the order of magnitude equation

$$\frac{\mathrm{d}f}{\mathrm{d}x} = O(\epsilon)$$ (10.370)

is valid, as is (10.353). Let the angle of attack α be of the order $O(\epsilon)$. Within our approximation we can place the vortex distribution along the x-axis instead of along the mean camber line. For a *vortex intensity* γ opposite to the mathematically positive sense, the infinitesimal vortex strength is

$$d\Gamma = -\gamma(x)dx, \tag{10.371}$$

so that with (10.195) and in analogy to the source distribution we obtain the potential

$$\Phi = U_\infty x + V_\infty y - \frac{1}{2\pi} \int\limits_0^l \gamma(x') \arctan\frac{y}{x - x'} dx', \tag{10.372}$$

with the still unknown vortex intensity $\gamma(x')$. We then obtain the velocity components from (10.372) as before, by taking the differentiation with respect to x and y into the integral. For the perturbation velocities we thus obtain the expressions

$$u(x, y) = +\frac{1}{2\pi} \int\limits_0^l \gamma(x') \frac{y}{(x - x')^2 + y^2} dx' \tag{10.373}$$

and

$$v(x, y) = -\frac{1}{2\pi} \int\limits_0^l \gamma(x') \frac{x - x'}{(x - x')^2 + y^2} dx'. \tag{10.374}$$

Because of the formal equality of the expression for u with that for v in the source distribution (10.362), we can directly obtain the velocity on the x-axis

$$u(x, 0^\pm) = \pm\frac{1}{2}\gamma(x), \tag{10.375}$$

which is equal to the velocity on the mean camber line, up to terms of the order $O(\epsilon^2)$. Therefore the jump in velocity through the vortex sheet is

$$\Delta u = u^+ - u^- = \gamma(x). \tag{10.376}$$

From this, the pressure jump could be computed directly from Bernoulli's equation and then integrated to find the force (per unit depth). However we shall prefer to determine the lift from the Kutta-Joukowski theorem (10.288)

$$F_a = -\varrho\Gamma U_\infty\sqrt{1 + \left(\frac{V_\infty}{U_\infty}\right)^2}, \tag{10.377}$$

with Γ (positive in the anticlockwise sense) from (10.371)

$$\Gamma = \int\limits_0^l -\gamma(x')dx'. \tag{10.378}$$

Since $V_\infty / U_\infty \approx \alpha \sim \epsilon$, we have

$$F_a = \varrho U_\infty \int\limits_0^l \gamma(x')dx' + O(\epsilon^2). \tag{10.379}$$

The implicit form of the mean camber line is $F(x, y) = -y + f(x) = 0$, and we obtain the kinematic boundary condition from (4.170) as

$$(U_\infty + u)\frac{df}{dx} - (V_\infty + v) = 0 \quad \text{at } y = f(x), \tag{10.380}$$

or

$$\alpha + \frac{v}{U_\infty} = \frac{df}{dx}\left(1 + \frac{u}{U_\infty}\right) \quad \text{at } y = f(x). \tag{10.381}$$

By neglecting terms of the order $O(\epsilon^2)$, we can again satisfy the boundary condition on the x-axis instead of on $y = f(x)$, and with (10.374) we extract the equation

$$U_\infty \frac{df}{dx} - \alpha U_\infty = -\frac{1}{2\pi}\int\limits_0^l \frac{\gamma(x')}{x - x'}dx', \tag{10.382}$$

which is a singular *integral equation* of the first kind for the unknown distribution $\gamma(x)$. The integral equation has no unique solution. Here we do not wish to go into the mathematical aspects, but shall only note that the flow past a circular cylinder with circulation (cf. (10.254)) is not unique either. Since this solution can be mapped onto the flow past airfoils, these are also not unique. It is necessary in addition to specify the value of the circulation, i.e., invoke Joukowski's hypothesis. In steady flow this is equivalent to requiring that the velocities on the upper and lower sides be equal at the trailing edge $x = l$

$$\Delta u(x = l) = \gamma(l) = 0. \tag{10.383}$$

In general then there is a flow around the leading edge of the mean camber line. This leads to infinitely large velocities, and to an infinitely large $\gamma(0)$. Only if the local flow direction towards the airfoil (not the undisturbed stream at infinity) is there tangential to the mean camber line, does no flow around this edge occur. We call this the *shock-free incidence*. For the flow past an infinitely thin edge, we find the potential from (10.242) with $n = 1/2$

$$\Phi = 2a\sqrt{r}\cos\frac{\varphi}{2}. \tag{10.384}$$

From this we extract the velocity on the upper side ($\varphi = 0$) of the edge as

$$u^+ = \left.\frac{\mathrm{d}\Phi}{\mathrm{d}r}\right|_{\varphi=0} = \frac{a}{\sqrt{r}} = \frac{a}{\sqrt{x}}, \tag{10.385}$$

and on the lower side ($\varphi = 2\pi$) as

$$u^- = -\frac{a}{\sqrt{x}}. \tag{10.386}$$

Therefore the jump in the tangential velocity is

$$\lim_{x\to 0}\Delta u(x) = \lim_{x\to 0} 2\frac{a}{\sqrt{x}}. \tag{10.387}$$

The function

$$\gamma_0(x) = 2a\sqrt{\frac{l-x}{x}} \tag{10.388}$$

satisfies the required conditions at the leading and trailing edges, because of (10.376), but is not yet the desired function in the domain $0 < x < l$. We subtract the distribution $\gamma_0(x)$ from the desired distribution $\gamma(x)$. The remaining part of the distribution can be expanded into a Fourier series in the coordinate φ, given by

$$x = \frac{l}{2}(1 + \cos\varphi). \tag{10.389}$$

Because $x = 0$ for $\varphi = \pi$ and $x = l$ for $\varphi = 0$, the cosine terms in the series expansion must vanish, since these are not zero for $x = 0$ and $x = l$. We set the constant $a = U_\infty A_0$ and expand $(\gamma - \gamma_0)$ in a Fourier series

$$\gamma(\varphi) - 2U_\infty A_0 \tan\frac{\varphi}{2} = 2U_\infty \sum_{n=1}^{\infty} A_n \sin n\varphi. \tag{10.390}$$

We insert this into the integral Eq. (10.382), use the transformation (10.389) for x' and obtain the integral equation in the form

$$\alpha - \frac{df}{dx} = \frac{1}{\pi} A_0 \int_0^\pi \frac{1 - \cos\varphi'}{\cos\varphi - \cos\varphi'} d\varphi' + \frac{1}{\pi} \sum_{n=1}^{\infty} A_n \int_0^\pi \frac{\sin n\varphi' \sin\varphi'}{\cos\varphi - \cos\varphi'} d\varphi'. \tag{10.391}$$

The integrals can be evaluated with

$$\sin n\varphi' \sin\varphi' = \frac{1}{2}[\cos(n-1)\varphi' - \cos(n+1)\varphi']$$

from the formula

$$\frac{1}{\pi} \int_0^\pi \frac{\cos n\varphi'}{\cos\varphi - \cos\varphi'} d\varphi' = -\frac{\sin n\varphi}{\sin\varphi}, \tag{10.392}$$

and in this manner we see that the left-hand side should be expanded in a cosine series

$$\alpha - \frac{df}{dx} = A_0 + \sum_{n=1}^{\infty} A_n \cos n\varphi. \tag{10.393}$$

As is known, the coefficients are

$$A_0 = \alpha - \frac{1}{\pi} \int_0^\pi \frac{df}{dx}(\varphi) d\varphi, \tag{10.394}$$

and

$$A_n = -\frac{2}{\pi} \int_0^\pi \frac{df}{dx}(\varphi) \cos n\varphi \, d\varphi. \tag{10.395}$$

From (10.379) we determine the lift coefficient c_l to be

$$c_l = \pi(2A_0 + A_1).\tag{10.396}$$

With (10.266) the moment about the leading edge can be determined. It is taken as positive if it tends to increase the angle of attack. Without performing the calculation we shall simply state the moment coefficient

$$c_m = \frac{M}{(\varrho/2)U_\infty^2 l^2} = -\frac{\pi}{4}(2A_0 + 2A_1 + A_2).\tag{10.397}$$

In shock-free incidence, we have $A_0 = 0$, and since $\gamma(\pi)$ remains finite, we have for this case

$$c_l = \pi A_1.\tag{10.398}$$

As an example we shall compute the coefficients for a flat plate, for which $df/dx = 0$, and therefore $A_0 = \alpha$, $A_n = 0$ hold. If follows immediately that

$$c_l = 2\pi\alpha\tag{10.399}$$

(in agreement with (10.349) for small α) and

$$c_m = -\frac{\pi}{2}\alpha = -\frac{1}{4}c_l,\tag{10.400}$$

from which we conclude that the point at which the lift force acts is $x = l\,/\,4$ (cf. Fig. 10.48).

10.4.10 Slender Airfoils in Compressible Flow

As in Sect. 10.4.9 we shall consider slender airfoils $(d/l = \epsilon \ll 1)$. The perturbation velocities u and v are then of the order $O(\epsilon U_\infty)$, and for the potential we assume the form

$$\Phi = U_\infty x + \varphi,\tag{10.401}$$

where φ is the *perturbation potential*, and $u = d\varphi/dx$ and $v = d\varphi/dy$ are the perturbation velocities. We shall start out with the potential Eq. (10.50), in which we replace a^2 from the energy equation

$$a^2 = a_\infty^2 + \frac{\gamma - 1}{2}\left(U_\infty^2 - \frac{\partial \Phi}{\partial x_i}\frac{\partial \Phi}{\partial x_i}\right). \tag{10.402}$$

If we insert (10.401) into the resulting equation and neglect all terms of the order $O(\epsilon^2)$, then after some manipulation we obtain a differential equation for the perturbation potential

$$\left(1 - M_\infty^2\right)\frac{\partial^2 \varphi}{\partial x^2} + \frac{\partial^2 \varphi}{\partial y^2} = (\gamma + 1)M_\infty^2 \frac{u}{U_\infty}\frac{\partial^2 \varphi}{\partial x^2} + (\gamma - 1)M_\infty^2 \frac{u}{U_\infty}\frac{\partial^2 \varphi}{\partial y^2}$$
$$+ 2M_\infty^2 \frac{v}{U_\infty}\frac{\partial^2 \varphi}{\partial x \partial y}, \tag{10.403}$$

in which $M_\infty = U_\infty / a_\infty$. In many practical cases this equation or the original Eq. (10.50) is solved numerically. However here we shall discuss the simplifications arising in the limit $\epsilon \to 0$, since in this case the solution can be found using the methods already known. In the limit $\epsilon \to 0$ the right-hand side vanishes, since each term contains a factor of order $O(\epsilon)$. We obtain the equation

$$\left(1 - M_\infty^2\right)\frac{\partial^2 \varphi}{\partial x^2} + \frac{\partial^2 \varphi}{\partial y^2} = 0, \tag{10.404}$$

which is valid in both the subsonic and supersonic flows. The sign of $\left(1 - M_\infty^2\right)$ determines the type of this partial differential equation. For $M_\infty < 1$ the equation is *elliptic*; for $M_\infty > 1$ it is *hyperbolic*. For $M_\infty \approx 1$ the sign of $\partial^2 \varphi/\partial x^2$ is also affected by the first term on the right-hand side of (10.403), which can then no longer be neglected; for this case we obtain the *transonic perturbation equation*

$$\left(1 - M_\infty^2\right)\frac{\partial^2 \varphi}{\partial x^2} + \frac{\partial^2 \varphi}{\partial y^2} = (\gamma + 1)\frac{M_\infty^2}{U_\infty}\frac{\partial \varphi}{\partial x}\frac{\partial^2 \varphi}{\partial x^2}. \tag{10.405}$$

This equation is nonlinear, and apart from some specific solutions, numerical methods are used to integrate it.

We shall first consider the subsonic flow past a slender airfoil given by $y = f(x)$. Then (10.404) is to be solved subject to the boundary condition (10.356), that is

$$\frac{1}{U_\infty}\frac{\partial \varphi}{\partial y} = \frac{df}{dx} \quad \text{for} \quad y = 0. \tag{10.406}$$

It is clear that (10.404) can be brought to the form of Laplace's equation by a suitable coordinate transformation. We could, for example, transform x (i.e. change the length of the airfoil) and leave y unchanged, or else retain x and transform y (i.e. change the thickness of the airfoil). We choose

$$\bar{y} = y\sqrt{1 - M_\infty^2}; \quad \bar{x} = x \tag{10.407}$$

and for

$$\bar{\varphi} = \varphi\left(1 - M_\infty^2\right) \tag{10.408}$$

obtain from (10.404) Laplace's equation

$$\frac{\partial^2 \bar{\varphi}}{\partial \bar{x}^2} + \frac{\partial^2 \bar{\varphi}}{\partial \bar{y}^2} = 0. \tag{10.409}$$

The equation of the upper surface in the transformed coordinates reads

$$\bar{y} = \sqrt{1 - M_\infty^2}\, f(\bar{x}) = \bar{f}(\bar{x}), \tag{10.410}$$

and with this the boundary condition

$$\frac{1}{U_\infty} \frac{\partial \bar{\varphi}}{\partial \bar{y}} = \frac{d\bar{f}}{d\bar{x}}. \tag{10.411}$$

With (10.409) and (10.411) the solution of compressible flow past an airfoil $y = f(x)$ in the x-y-plane in a stream with undisturbed velocity U_∞ and Mach number M_∞ is reduced to the incompressible flow past a (thinner) airfoil $\bar{y} = \bar{f}(\bar{x})$ in the $\bar{x} - \bar{y}$-plane with the undisturbed velocity U_∞. At corresponding points the perturbation velocities u and v are to be calculated from the perturbation velocities $\bar{u}$ and $\bar{v}$ of the incompressible flow, according to

$$u = \frac{\partial \varphi}{\partial x} = \frac{1}{1 - M_\infty^2} \frac{\partial \bar{\varphi}}{\partial \bar{x}} = \frac{\bar{u}}{1 - M_\infty^2}, \tag{10.412}$$

and

$$v = \frac{\partial \varphi}{\partial y} = \frac{1}{\sqrt{1 - M_\infty^2}} \frac{\partial \bar{\varphi}}{\partial \bar{y}} = \frac{\bar{v}}{\sqrt{1 - M_\infty^2}}. \tag{10.413}$$

Within our approximation we can neglect the change in density in the field, and Bernoulli's equation holds in the form valid for incompressible flow. Neglecting the quadratic terms in the perturbation velocities, the pressure coefficient (10.123) then reads

$$c_p = -\frac{2u}{U_\infty},$$ (10.414)

where with (10.412) the transformation

$$c_p = -\frac{1}{1 - M_\infty^2}\frac{2\bar{u}}{U_\infty} = \frac{1}{1 - M_\infty^2}\bar{c}_p$$ (10.415)

follows, called *Goethert's rule*. Now in practice we often want to know the change in the pressure coefficient as a function of the Mach number for a given airfoil, which is approximately described by the *Prandtl-Glauert rule*

$$c_p(M_\infty) = c_p(0)\frac{1}{\sqrt{1 - M_\infty^2}}.$$ (10.416)

Here $c_p(M_\infty)$ is the pressure coefficient at the Mach number M_∞, for an airfoil, which in incompressible flow has the coefficient $c_p(0)$.

For supersonic flow $(M_\infty^2 - 1) > 0$, (10.404) corresponds to the wave equation

$$\frac{\partial^2 \varphi}{\partial y^2} = (M_\infty^2 - 1)\frac{\partial^2 \varphi}{\partial x^2}.$$ (10.417)

The solution can therefore proceed in analogy to the one-dimensional sound propagation of Sect. 10.1. There is, however, a difference, in so far as there the perturbation is also felt upstream, while this is not possible in supersonic flows. The reason for this is that a perturbation can only propagate with the speed of sound. We shall explain this fact in Chap. 11, and in Sect. 11.4 we shall return to the Eq. (10.417).

Chapter 11
Supersonic Flow

In a supersonic flow, the disturbance caused by a body is perceived only within a bounded range of influence. This is completely analogous to unsteady compressible flow, which also is described by *hyperbolic* differential equations, but there the resulting state of affairs is independent of whether the Mach number is greater than or less than one.

For example, consider a steady flow with a stationary sound source, which sends out a signal at $t = 0$. This signal imparts a small pressure disturbance to the fluid. In a reference frame moving with the flow velocity u, the disturbance spreads out spherically with the velocity of sound a. With respect to a reference frame fixed in space, the sound wave has the position shown in Fig. 11.1 after time t and for $u < a$ (subsonic).

As $t \to \infty$ the sound wave will fill the entire space. If $u > a$ (supersonic) the sound wave has the positions shown in Fig. 11.2 in the fixed frame at successive points in time. We can see from this figure that the sound wave will not reach the entire space as $t \to \infty$. We call the envelope of the waves the *Mach cone* whose angle μ is calculated from

$$\sin \mu = \frac{a}{u} = \frac{1}{M} \tag{11.1}$$

and is called the *Mach angle*. We can also imagine, for example, a very slender body as a source of the disturbance. A thick body, however, will cause a disturbance which is no longer small and then the Mach cone becomes a shock front. The disturbance which originates from the body remains restricted to the region behind the shock surface even in this case.

© Springer Nature Switzerland AG 2020
J. H. Spurk and N. Aksel, *Fluid Mechanics*,
https://doi.org/10.1007/978-3-030-30259-7_11

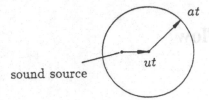

Fig. 11.1 Propagation of a disturbance in subsonic flow

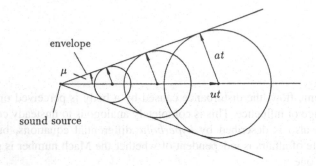

Fig. 11.2 Propagation of a disturbance in supersonic flow

11.1 Oblique Shock Wave

As the first step in the treatment of supersonic flow we want to derive the relations of
an oblique shock wave in two-dimensional flow from those of a one-dimensional,
normal shock wave. To do this we decompose the velocity $\vec{u}_1$ in front of the shock
into its components u_{1n} normal to and u_{1t} tangential to the shock front (Fig. 11.3)

$$u_{1n} = u_1 \sin \Theta, \tag{11.2}$$

$$u_{1t} = u_1 \cos \Theta. \tag{11.3}$$

For an observer who moves with velocity u_{1t} along the shock, the flow velocity in
front of the shock is now normal to the shock. Therefore in his reference frame the

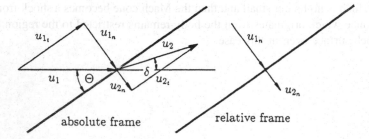

Fig. 11.3 Oblique shock wave

relations of the normal shock wave are valid, where the Mach number in front of the shock is then

$$M_{1n} = \frac{u_{1n}}{a_1} M_1 \sin \Theta. \tag{11.4}$$

The shock relations (9.149), (9.150), and (9.151) can then be carried over to the oblique shock wave, if M_1 there is replaced by M_{1n} from (11.4)

$$\frac{p_2}{p_1} = 1 + 2\frac{\gamma}{\gamma+1} \left(M_1^2 \sin^2 \Theta - 1 \right), \tag{11.5}$$

$$\frac{\varrho_2}{\varrho_1} = \frac{(\gamma+1)M_1^2 \sin^2 \Theta}{2 + (\gamma-1)M_1^2 \sin^2 \Theta}, \tag{11.6}$$

$$\frac{T_2}{T_1} = \frac{\left[2\gamma M_1^2 \sin^2 \Theta - (\gamma-1) \right] \left[2 + (\gamma-1)M_1^2 \sin^2 \Theta \right]}{(\gamma+1)^2 M_1^2 \sin^2 \Theta}. \tag{11.7}$$

Behind the oblique shock, the Mach number is formed with u_2, therefore $M_2 = u_2/a_2$. Since $u_{2n} = u_2 \sin (\Theta - \delta)$ holds, it follows that

$$M_{2n} = \frac{u_{2n}}{a_2} = M_2 \sin(\Theta - \delta). \tag{11.8}$$

Although M_{2n} is smaller than 1, M_2 can be larger than 1. If we again replace M_1 and M_2 by M_{1n} and M_{2n} (using (11.4) and (11.8), respectively), in the relation (9.153) valid for a normal shock, we extract the equation

$$M_2^2 \sin^2(\Theta - \delta) = \frac{\gamma + 1 + (\gamma - 1)\left[M_1^2 \sin^2 \Theta - 1 \right]}{\gamma + 1 + 2\gamma \left[M_1^2 \sin^2 \Theta - 1 \right]}. \tag{11.9}$$

Using the continuity equation, we can transform this into a relation between the wave angle Θ and the deflection angle δ (Fig. 11.4)

$$\tan \delta = \frac{2 \cot \Theta \left[M_1^2 \sin^2 \Theta - 1 \right]}{2 + M_1^2 \left[\gamma + 1 - 2 \sin^2 \Theta \right]}. \tag{11.10}$$

The lower of the two dividing lines sketched in Fig. 11.4 separates the regions where the Mach number M_2 is larger and smaller than one, and the upper line connects the points of maximum deflections. (A diagram of the relation between wave angle Θ and deflection angle δ with the family parameter M_1 is also found in Appendix C.) A shock is called a *strong shock* if the wave angle Θ for a given Mach number M_1 is larger than the angle Θ_{max} associated with the maximum deflection δ_{max}; otherwise we talk of a *weak shock*.

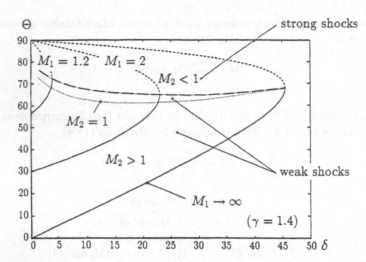

Fig. 11.4 Relation between wave angle and deflection angle

In a weak shock the fluid velocity behind the shock can lie in either the subsonic or the supersonic range, but behind a strong shock the flow is always subsonic. If the deflection angle δ is smaller than δ_{max}, there are then two possible solutions for the shock angle Θ. Which solution actually arises depends on the boundary conditions far behind the shock.

Knowing about the oblique shock wave, we can immediately consider supersonic flow in a corner and around a wedge, providing $\delta < \delta_{max}$ (Fig. 11.5). It is observed that in flows past "slender wedges" $\delta < \delta_{max}$, weak shocks are always attached to the nose. On the one side, the wave angle Θ is limited by the value $\pi/2$ (normal shock) and on the other by the condition $M_1 \sin \Theta \geq 1$ (velocity normal to the shock supersonic). Using (11.1) we then have

$$\sin \Theta \geq \frac{1}{M_1} = \sin \mu_1. \tag{11.11}$$

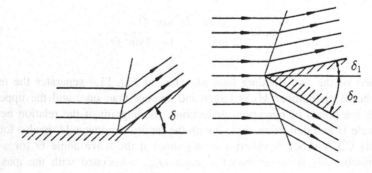

Fig. 11.5 Supersonic flow in a corner and past a wedge

Therefore Θ must be equal or larger than μ, and thus move in the range

$$\mu \leq \Theta \leq \frac{\pi}{2}. \qquad (11.12)$$

For $\Theta = \mu$ the shock deteriorates into a Mach wave. From $M_2 = M_2 (\Theta, \delta, M_1)$ and $\delta = \delta(M_1, \Theta)$ we can eliminate the wave angle Θ to obtain $M_2 (\delta, M_1)$. This relation is given in the form of a diagram in Appendix C.

11.2 Detached Shock Wave

We now consider deflection angles $\delta > \delta_{max}$; these lead to flows past "blunt wedges". If, for a given Mach number M_1 a deflection angle δ larger than δ_{max} arises, a *detached shock* is the only possibility. Both strong and weak shocks are then realized in the shock configuration (Fig. 11.6a). Close to the stagnation streamline the wave angle is around 90° (strong shock, subsonic flow behind the shock), while at greater distances from the body the shock deteriorates into a Mach wave ($\Theta = \mu$, Fig. 11.6b). It is difficult to calculate the resulting flow behind the shock, since subsonic flow, supersonic flow and flow close to the velocity of sound all appear together (*transonic flow*). Behind a curved shock, the flow is no longer homentropic, and from Crocco's theorem (4.157) is no longer irrotational.

The shock behavior derived up to now also holds locally for curved shocks; we recognize this from the fact that no derivatives appear in the shock relations. Then Θ is the local inclination of the shock front.

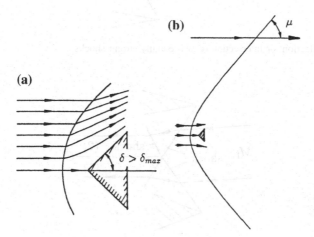

Fig. 11.6 Detached shock wave

11.3   Reflection of Oblique Shock Waves

If a shock meets a wall it is reflected. The strength of the *reflected shock* adjusts itself so that the flow velocity after the shock is again directed parallel to the wall. The reflected shock can be either a weak or strong shock. If the original shock is a weak shock it is generally observed that the reflected shock is also a weak shock. We can interpret this flow as the same as the intersection of two oblique shocks of the same strength where the line of symmetry is replaced by a plane wall (Fig. 11.7).

Downstream from the reflected shocks, the state of the gas is everywhere the same and the flow direction is the same as the flow direction in front of the shock.

If two shocks of different strengths cross (Fig. 11.8), then the reflected shocks must be such that the same pressure and the same flow direction are produced behind each. However all other fluid mechanical and thermodynamical quantities, particularly the magnitude of the flow velocity, can be different from each other in the regions of constant gas state 1 and 2. These are separated from each other by the dot-dashed *contact discontinuity* C which is a streamline.

The contact discontinuity has the property of a vortex layer, i.e., its tangential velocity on this surface changes discontinuously. We conclude from Crocco's theorem that the entropy in regions 1 and 2 is different. We come to the same

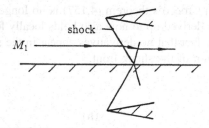

Fig. 11.7 Reflection, or intersection of two equally strong shocks

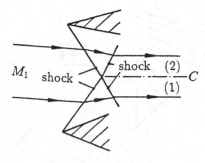

Fig. 11.8 Intersection of two shocks of different strengths

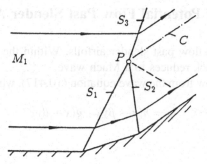

Fig. 11.9 Confluence of two shocks

conclusion if we consider that gas particles on both sides of the contact discontinuity have moved through shocks of different strengths where the change in entropy
is different for each particle.

Similar behavior is observed in a flow past a wall having two successive corners
(Fig. 11.9). At the point P two shocks merge into one. From the above arguments, it
is clear that a contact discontinuity (dot-dashed line) must emanate from the point
P. However another wave (here dashed) must also emerge from P, and it can be
either a weak shock wave or an expansion wave (see Sect. 11.5) for the following
reason: the shock strengths of S_1, S_2 and S_3 are all determined by the slope of the
wall. Since the pressure must be the same on either side of the contact discontinuity
C (the dynamic boundary condition), this can in general only be accompanied by
another wave.

If the wedge angle of Fig. 11.7 is increased, the Mach number behind the shock
becomes smaller. If the wedge angle is large enough, the maximum deflection
associated with the Mach number behind the shock becomes smaller than is necessary to satisfy the boundary condition (parallel flow to the wall) behind the
reflected shock. Then the so-called *Mach reflection* takes place (Fig. 11.10).

The theory of Mach reflection is difficult because the shocks S_1, S_2 and the
contact discontinuity are curved and the state of flow downstream from S_1 and S_2 is
no longer constant. In addition the flow behind the partly normal shock S_2 must be
subsonic, and so the shock configuration also depends on the conditions far behind
the shock.

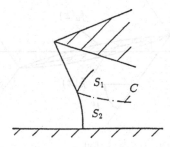

Fig. 11.10 Mach reflection

11.4   Supersonic Potential Flow Past Slender Airfoils

We now return to the flow past slender airfoils. Within the framework of *perturbation theory*, the shock reduces to a Mach wave.

We compute the flow from the wave equation (10.417), whose general solution is

$$\varphi = h(x - \beta y) + g(x + \beta y) \tag{11.13}$$

with

$$\beta = \sqrt{M_\infty^2 - 1}. \tag{11.14}$$

In supersonic flow from the left, disturbances can only spread out to the right and we must have $g \equiv 0$ above the airfoil and $h \equiv 0$ below. We shall first consider only the flow above the upper side of the airfoil (Fig. 11.11), where

$$f(x) = f_u(x).$$

Above the airfoil the *perturbation potential* is

$$\varphi = h(x - \beta y), \tag{11.15}$$

and the component of the perturbation velocity in the y-direction is therefore,

$$v = \frac{\partial \varphi}{\partial y} = -\beta h'(x, y). \tag{11.16}$$

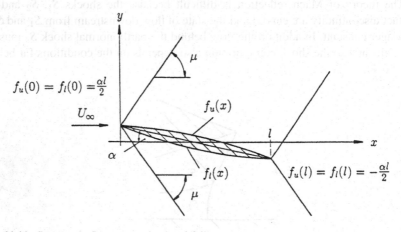

Fig. 11.11 Supersonic flow past a slender airfoil

We introduce (11.16) into the boundary condition (10.406), resulting in

$$v(x,0) = U_\infty \frac{df_u}{dx} = -\beta h'(x,0).$$

(11.17)

From this it follows immediately that

$$h(x) = -U_\infty \frac{f_u(x)}{\beta}$$

(11.18)

and therefore, for the potential at the position $y = 0$, we have

$$\varphi(x) = -U_\infty \frac{f_u(x)}{\beta},$$

(11.19)

or more generally in the whole upper half plane

$$\varphi(x,y) = -U_\infty \frac{f_u(x - \beta y)}{\beta}.$$

(11.20)

In the same manner we find the solution in the lower half plane to be

$$\varphi(x,y) = U_\infty \frac{f_l(x + \beta y)}{\beta},$$

(11.21)

where we now know the solution everywhere.

The fundamental assumption of linear supersonic flow for slender airfoils ($M_\infty \epsilon \ll 1$) allows us to use (10.414) here as well, and we obtain the pressure coefficient on the upper side (in the approximation $y = 0^+$) of the airfoil as

$$c_{p_u} = -\frac{2}{U_\infty} \frac{\partial \varphi}{\partial x} = \frac{2}{\beta} \frac{df_u}{dx},$$

(11.22)

and on the lower side ($y = 0^-$) as

$$c_{p_l} = -\frac{2}{\beta} \frac{df_l}{dx}.$$

(11.23)

Using (10.262) we can also write the force in the y-direction per unit depth as

$$F_y = \oint (p - p_\infty)dx,$$

(11.24)

since p_∞ provides no contribution. From the definition of the lift coefficient, it then follows that

$$c_l = \frac{2F_y}{\varrho_\infty U_\infty^2 l} = \frac{1}{l} \oint c_p dx = \frac{1}{l} \int\limits_0^l \left(c_{p_l} - c_{p_u} \right) dx. \tag{11.25}$$

If we insert the expressions for c_{p_u} and c_{p_l}, the integration yields

$$c_l = \frac{2}{l\beta} [-f_l(l) + f_l(0) - f_u(l) + f_u(0)]. \tag{11.26}$$

Since $f_u(l) = f_l(l) = -\alpha l/2$ and $f_u(0) = f_l(0) = \alpha l/2$ hold, we find a lift coefficient which is independent of the shape of the airfoil

$$c_l = \frac{4\alpha}{\sqrt{M_\infty^2 - 1}}. \tag{11.27}$$

The analogous calculation furnishes the force F_x, or rather the drag coefficient c_d as

$$c_d = \frac{2}{\beta l} \int\limits_0^l \left[\left(\frac{df_l}{dx}\right)^2 + \left(\frac{df_u}{dx}\right)^2 \right] dx, \tag{11.28}$$

which depends on the shape of the airfoil. For the flat plate we find

$$c_l = \frac{4\alpha}{\sqrt{M_\infty^2 - 1}}, \quad \text{and} \quad c_d = \alpha c_l, \tag{11.29}$$

as can be seen from Fig. 11.12.

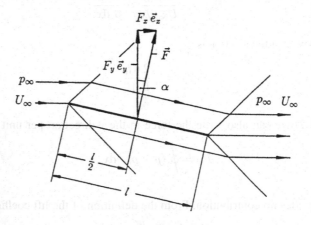

Fig. 11.12 Supersonic flow past an infinitesimally thin plate

11.5   Prandtl-Meyer Flow

We have seen that supersonic flows at concave corners are turned by oblique shock waves, and we now ask what the conditions at a convex corner are. To do this we consider the supersonic flow in Fig. 11.13. Suppose that all flow quantities in the flow to the corner are homogeneous in space, and so no characteristic length can be formed from the data of the flow. Even using the independent and dependent variables, no new dimensionless independent variable can be combined, aside from the angle φ. Since the flow boundary has no typical length either, the solution may also not depend on a length, which means it cannot depend on r. From the continuity equation in polar coordinates (see Appendix B) we then have

$$\frac{u_\varphi}{r}\frac{d\varrho}{d\varphi} + \frac{\varrho}{r}\frac{du_\varphi}{d\varphi} + \varrho\frac{u_r}{r} = 0. \tag{11.30}$$

Euler's equations in polar coordinates now simplify to

$$\frac{u_\varphi}{r}\frac{du_r}{d\varphi} - \frac{u_\varphi^2}{r} = 0 \quad \text{or} \quad \frac{du_r}{d\varphi} = u_\varphi, \tag{11.31}$$

and

$$\frac{u_\varphi}{r}\frac{du_\varphi}{d\varphi} + \frac{u_r u_\varphi}{r} + \frac{1}{\varrho r}\frac{dp}{d\varphi} = 0. \tag{11.32}$$

Finally the entropy equation becomes

$$\frac{u_\varphi}{r}\frac{ds}{d\varphi} = 0. \tag{11.33}$$

Since $u_\varphi \neq 0$ it follows that $ds/d\varphi = 0$. Therefore the flow is homentropic. By Crocco's relation it is then also irrotational. We could now introduce a velocity potential but shall here refrain from doing this. Since the flow is homentropic, $dp/d\varrho = a^2$ holds. The continuity equation becomes

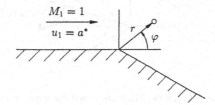

Fig. 11.13 Geometry of the Prandtl-Meyer flow

$$\frac{1}{r}\frac{u_\varphi^2}{a^2}\frac{dp}{d\varphi} + \frac{\varrho}{r}\left(u_\varphi\frac{du_\varphi}{d\varphi} + u_r u_\varphi\right) = 0. \tag{11.34}$$

Equation (11.32) yields

$$\frac{\varrho}{r}\left(u_\varphi\frac{du_\varphi}{d\varphi} + u_r u_\varphi\right) + \frac{1}{r}\frac{dp}{d\varphi} = 0. \tag{11.35}$$

The difference of the last two equations gives us

$$\frac{1}{r}\left(\frac{u_\varphi^2}{a^2} - 1\right)\frac{dp}{d\varphi} = 0. \tag{11.36}$$

Clearly $dp/d\varphi$ cannot vanish in the entire field since then there would be no turning of the flow. In the region where $dp/d\varphi \neq 0$ it follows that $u_\varphi^2 = a^2$, and since φ is measured anticlockwise,

$$u_\varphi = -a. \tag{11.37}$$

With $u = |\vec{u}|$, we see from Fig. 11.14

$$\frac{-u_\varphi}{u} = \frac{a}{u} = \frac{1}{M} = \sin\mu, \tag{11.38}$$

i.e., $\vec{u}$ just forms the Mach angle μ with the r-direction. Therefore, the straight lines $\varphi = $ const are Mach lines, or characteristics. Such a flow where the flow velocity and thermodynamic state along a Mach line are constant is called a *simple wave*. The velocity vector $\vec{u}(\varphi)$ on such a characteristic is turned about the angle

$$v = \mu - \varphi \tag{11.39}$$

from the direction of the incident flow ($M_1 = 1$).

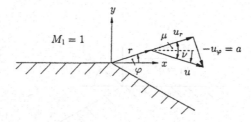

Fig. 11.14 The connection between the Mach angle and the angle of inclination

We shall now restrict ourselves to the calorically perfect gas and form, using $a^2 = \gamma p / \varrho$ and (11.37), the expression

$$d(a^2) = (\gamma - 1)\frac{dp}{\varrho} = 2u_\varphi du_\varphi, \tag{11.40}$$

which we insert into (11.32)

$$\frac{\gamma + 1}{\gamma - 1}\frac{du_\varphi}{d\varphi} = -u_r. \tag{11.41}$$

From (11.31) we further replace u_φ by $du_r/d\varphi$

$$\frac{d^2u_r}{d\varphi^2} + \frac{\gamma - 1}{\gamma + 1}u_r = 0. \tag{11.42}$$

This equation is the equation of the simple harmonic oscillator, whose general solution is

$$u_r = C\sin\left(\sqrt{\frac{\gamma - 1}{\gamma + 1}}\varphi + \varphi_0\right), \tag{11.43}$$

which is here subject to the boundary conditions

$$u_r(\varphi = \pi/2) = 0 \tag{11.44}$$

and

$$u_\varphi(\varphi = \pi/2) = \frac{du_r}{d\varphi}\bigg|_{\frac{\pi}{2}} = -a^*. \tag{11.45}$$

With

$$a^* = \sqrt{\frac{2}{\gamma + 1}}a_t$$

we obtain the solution

$$u_r = \sqrt{\frac{2}{\gamma - 1}}a_t\sin\left(\sqrt{\frac{\gamma - 1}{\gamma + 1}}(\pi/2 - \varphi)\right) \tag{11.46}$$

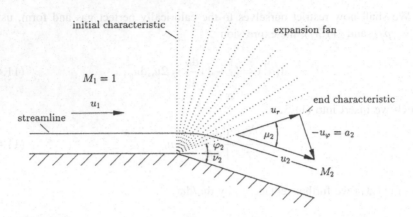

Fig. 11.15 Expansion fan of Prandtl-Meyer flow

for u_r and from (11.31) we find that u_φ is

$$u_\varphi = -\sqrt{\frac{2}{\gamma+1}}\, a_t \cos\left(\sqrt{\frac{\gamma-1}{\gamma+1}}(\pi/2 - \varphi)\right). \tag{11.47}$$

With these the velocity field is known. The domain in which Eqs. (11.46) and (11.47) are valid is limited in after the end characteristic, whose slope $\varphi_2 = \mu_2 - \nu_2$ depends on the turning angle ν_2, the flow is homogeneous again (Fig. 11.15). Equation (11.36) is satisfied here because $dp/d\varphi = 0$.

The characteristics between the start and end characteristics form an "expansion fan", similar to the one we have already met in the case of the suddenly accelerated piston in Sect. 9.3. Since the flow is homentropic

$$\frac{p}{p_t} = \left(\frac{a}{a_t}\right)^{2\gamma/(\gamma-1)} = \left[\sqrt{\frac{2}{\gamma+1}}\cos\left(\sqrt{\frac{\gamma-1}{\gamma+1}}(\pi/2 - \varphi)\right)\right]^{2\gamma/(\gamma-1)} \tag{11.48}$$

holds everywhere and we recognize that for a value of φ

$$\varphi_V = -\frac{\pi}{2}\left(\sqrt{\frac{\gamma+1}{\gamma-1}} - 1\right) \tag{11.49}$$

($\approx -130°$ for $\gamma = 1.4$) vacuum is reached (Fig. 11.16).

For $\varphi = \varphi_V$ the Mach number becomes infinite, i.e., $\mu = 0$, and because of (11.39) the associated turning angle is $\nu_2 = \nu_V = -\varphi_V$. Then a further increase of the turning angle does not change the flow any more. A vacuum forms between the wall and the line $\varphi = \varphi_V$.

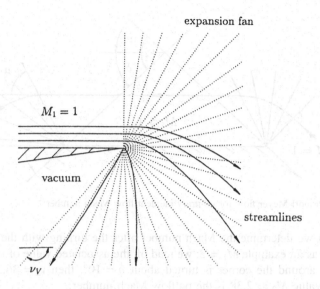

Fig. 11.16 Expansion in a vacuum

In order to calculate the flow for a given flow deflection, the relation between v and the Mach number will first be found. With

$$M^2 = \frac{u_r^2 + u_\varphi^2}{a^2}, \qquad (11.50)$$

the function $M(\varphi)$ is given. Together with $\sin \mu = \sin(v + \varphi) = M^{-1}$, after some easy computation, the relation known as the *Prandtl-Meyer function* appears

$$v = \sqrt{\frac{\gamma+1}{\gamma-1}} \arctan \sqrt{\frac{\gamma-1}{\gamma+1}(M^2-1)} - \arctan \sqrt{M^2-1}, \qquad (11.51)$$

which is tabulated in Appendix C.

We have derived the Prandtl-Meyer function for an incident flow with Mach number $M_1 = 1$, to which the value $v_1 = 0$ belongs. If we wish to know the downstream Mach number M_2 ($M_2 \geq M_1$) for any incident flow Mach number $M_1 > 1$, we first of all determine the angle v_1 associated with M_1 from the table in Appendix C (Fig. 11.17).

If the flow is then turned about δ it holds that

$$v_2 = v_1 + \delta, \qquad (11.52)$$

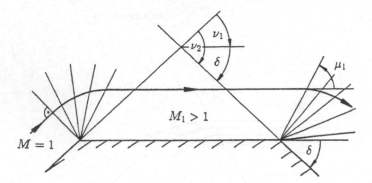

Fig. 11.17 Prandtl-Meyer flow for arbitrary incident flow Mach number

from which we determine the Mach number after the turning with the help of the table. Take as an example $M_1 = 2$: we read off the associated value of v_1 as 26.38°. If the flow around the corner is turned about $\delta = 10°$, then $v_2 = 36.38°$ and the associated value $M_2 \approx 2.38$ is the outflow Mach number.

Of course the same formulae hold if the turning is continuous and also for compression waves (Fig. 11.18). If the Mach lines form an envelope in the case of the concave wall, a shock wave forms at some distance from the wall (Fig. 11.19). This is analogous to unsteady flow where a piston with finite acceleration forms compression waves (see Fig. 9.33).

If $y_w''(0) \geq y_w''(x)$ for all $x > 0$ (Fig. 11.19), the cusp of the envelope lies on the first characteristic which emanates from the point where the turning of the wall begins (the origin in Fig. 11.19), and its coordinates can be explicitly calculated in the same way as before which led to Eqs. (9.223) and (9.225)

$$y_P = \frac{\sin^2(2\mu_1)}{2(\gamma + 1)y_w''(0)}, \tag{11.53}$$

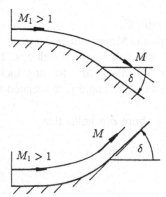

Fig. 11.18 Gradual deviation

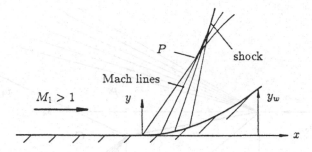

Fig. 11.19 Formation of a shock wave

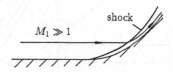

Fig. 11.20 For large incident flow Mach numbers the shock leans against the wall

$$x_P = y_P \cot \mu_1. \tag{11.54}$$

We conclude from this result that for $\mu_1 = \pi/2$, i.e., $M_1 = 1$, the coordinates y_P and x_P tend to zero. In this case a normal shock wave degenerates to a Mach wave which forms at the origin.

For $M_1 \to \infty$ the shock wave moves into the origin, and μ_1 tends to zero so that the shock wave leans against the wall (Fig. 11.20). Between the shock and the wall, the strongly compressed gas moves in a very thin layer along the surface of the body. If $y_w''(0)$ tends to infinity (wall corner) we obtain a shock wave which starts at the corner, that is the case shown in Fig. 11.5.

11.6 Shock Expansion Theory

With the help of the relations for the oblique shock wave and the Prandtl-Meyer function, the supersonic flow past an airfoil may be calculated in a simple manner.

The flow past an inclined plate (Fig. 11.21) is on the upper side at the *leading edge* first turned by a centered Prandtl-Meyer expansion wave and then at the trailing edge by an oblique shock wave, whereas on the lower side the flow is first turned by an oblique shock and then by an expansion wave. A contact discontinuity starts out from the trailing edge which, at small angle of attack, is almost parallel to the undisturbed flow. The wave system behind the plate does not affect the force on the plate. Since the expansion waves reflected by the shocks never reach the plate again, flow quantities along the surface, like Mach number and pressure are

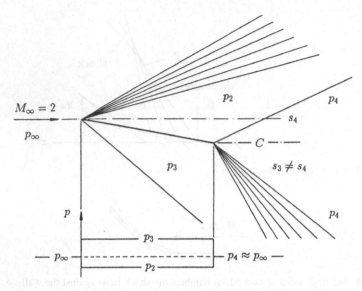

Fig. 11.21 Exact solution for the supersonic flow past a plate

computed exactly within the framework of inviscid flow theory. In contrast to the subsonic case (see Fig. 10.48) the lift force acts in the middle of the airfoil in accordance with Sect. 11.4.

In the same way we find the solution for the supersonic flow past a diamond airfoil (Fig. 11.22). Depending on the geometry and the conditions of the incident flow, the reflected waves can reach the airfoil again. However in shock expansion theory, these reflections are ignored in determining the flow quantities along the

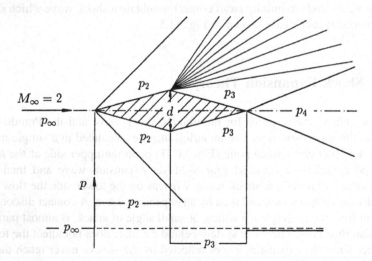

Fig. 11.22 Supersonic flow past a diamond airfoil

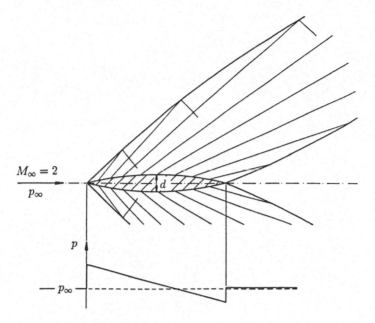

Fig. 11.23 Supersonic flow past an airfoil

surface of the airfoil. If the reflected waves do not meet the airfoil, then this solution is again exact.

As we already know from perturbation theory, an airfoil in supersonic flow has a drag, in spite of the assumption of no viscosity. The value of this per unit depth for the symmetric diamond airfoil in Fig. 11.22 is

$$F_w = (p_2 - p_3)d. \tag{11.55}$$

With an airfoil having continuous surface curvature (Fig. 11.23), the reflected Mach waves will certainly meet the airfoil. Therefore the flow field between the front and rear shocks is not a simple combination of shocks and Prandtl-Meyer flow. On the upper side of the airfoil, besides the waves leaning to the right $(y = x \tan \mu + \text{const})$ of the Prandtl-Meyer flow, waves leaning to the left also occur. Exact calculation of the flow field can be done by the method of characteristics, however if only the data along the airfoil contour are of interest, the following simplified view is sufficient: to determine the flow at the leading edge, we approximate this edge by a wedge, so that the shock and the conditions behind the shock are known. Using the now known initial data, we calculate the flow along the curved surface as a simple Prandtl-Meyer expansion. The trailing edge is then also approximated as a wedge again.

Chapter 12
Boundary Layer Theory

We have shown earlier that the flow past a body found under the assumption of zero viscosity can serve as an approximate solution to viscous flow for large Reynolds' numbers. However this solution is not uniformly valid in the entire field because it breaks down completely near a solid wall to which a real fluid adheres, while the theory of inviscid flow in general yields a nonzero tangential velocity.

The thickness of the boundary layer, i.e., the layer where friction effects cannot be ignored, is proportional to $Re^{-1/2}$. As previously stated, this is so in the laminar case to which we restrict ourselves for the time being. But even in turbulent flow the boundary layer thickness goes to zero in the limit $Re \to \infty$, so that the body "seen" by the flow corresponds to the actual body. The inviscid solution then represents an approximate solution of the Navier–Stokes equations for large Reynolds' numbers, with an error of order $O(Re^{-1/2})$ in the laminar case. The breakdown of the solution directly at the wall nevertheless remains, no matter how large the Reynolds' number is.

The complete approximate solution to the Navier–Stokes equations must be built up from two part solutions valid in different regions. One of these is the solution of the inviscid flow problem, the so-called *outer solution*, and the other is the *inner solution* close to the wall. The inner solution describes the boundary layer flow and must be such that the flow velocity from its value zero at the wall passes asymptotically into the velocity predicted by the outer (inviscid) solution directly at the wall. Because of this nonuniformity, the approximate solution of the Navier–Stokes equation represents an example of a *singular perturbation problem*, as they often appear in applications. An example already mentioned is the approximate solution for the potential flow past a slender airfoil (Sect. 10.4) which only breaks down at the blunt nose of the airfoil and outside this region describes the flow quite accurately. The outer, inviscid solution for large Reynolds' numbers gives important information about, for example, the pressure and velocity distributions, but is not able to predict the drag and makes no statements about where the boundary layer separates, or even if it does so at all. The answer to these questions is obviously

© Springer Nature Switzerland AG 2020
J. H. Spurk and N. Aksel, *Fluid Mechanics*,
https://doi.org/10.1007/978-3-030-30259-7_12

important, and requires the solution of the inner problem, which is the subject of *boundary layer theory*.

The differential equations required for the inner solution can be found systematically from the Navier–Stokes equations within the framework of singular perturbation theory. However, here we proceed along a more intuitive path. In what follows we shall assume that the outer solution is known and so the pressure and velocity distributions are at hand from this solution.

To begin with, we restrict ourselves to incompressible and plane two-dimensional flow and introduce the so-called boundary layer coordinate system, in which x is measured along the surface of the body and y perpendicular to it. If the boundary layer thickness is very small compared to the radius of curvature R of the wall contour $(\delta/R \ll 1)$, the Navier–Stokes equations hold in the same form as in Cartesian coordinates. In the calculation of the inner solution, i.e., of the boundary layer flow, the curvature of the wall then plays no role. The boundary layer developes as if the wall were flat. The wall curvature only manifests itself indirectly through the pressure distribution given by the outer solution.

Since the boundary layer is very thin for large Reynolds' numbers, the following inequalities hold

$$\frac{\partial u}{\partial x} \ll \frac{\partial u}{\partial y} \quad \text{and} \quad \frac{\partial^2 u}{\partial x^2} \ll \frac{\partial^2 u}{\partial y^2}. \tag{12.1}$$

A consequence of the last condition is that the x-component of the Navier–Stokes equations reduces to

$$\frac{\partial u}{\partial t} + u\frac{\partial u}{\partial x} + v\frac{\partial u}{\partial y} = -\frac{1}{\varrho}\frac{\partial p}{\partial x} + \nu\frac{\partial^2 u}{\partial y^2}. \tag{12.2}$$

In order to determine the order of magnitude of the term $u\partial u/\partial x$ in comparison to $v\partial u/\partial y$, we begin with the continuity equation for plane two-dimensional and incompressible flow

$$\frac{\partial u}{\partial x} + \frac{\partial v}{\partial y} = 0, \tag{12.3}$$

and together with (12.1) conclude that $\partial v/\partial y \ll \partial u/\partial y$, so that $v \ll u$ holds. Therefore the second and third terms on the left-hand side in (12.2) are of the same order of magnitude.

While the viscous forces are completely ignored in the outer flow, they do play a role in the boundary layer. The order of magnitude of the boundary layer thickness can be determined by considering the thickness of the layer where the viscous forces are of the same order of magnitude as the inertial forces, e.g., where

$$\frac{u}{\nu}\frac{\partial u/\partial x}{\partial^2 u/\partial y^2} \sim 1. \tag{12.4}$$

In the x-direction, let L be the typical length scale (cf. Fig. 12.1), and if U_∞ is the incident flow velocity, we have the order of magnitude equation

$$u\frac{\partial u}{\partial x} \sim \frac{U_\infty^2}{L}. \tag{12.5}$$

The typical length scale in the y-direction is the average boundary layer thickness δ_0, so that

$$\nu\frac{\partial^2 u}{\partial y^2} \sim \nu\frac{U_\infty}{\delta_0^2}. \tag{12.6}$$

Using (12.5) we then have the estimate

$$\frac{U_\infty^2/L}{\nu U_\infty/\delta_0^2} \sim 1, \tag{12.7}$$

from which we obtain the result (4.38) again

$$\frac{\delta_0}{L} \sim Re^{-\frac{1}{2}}. \tag{12.8}$$

With this result, the individual terms in the equations of motion are reviewed in order to systematically simplify the equations themselves. It follows from the continuity equation that

$$v \sim \frac{\delta_0}{L}U_\infty \text{ and therefore } \quad v \sim U_\infty Re^{-\frac{1}{2}}. \tag{12.9}$$

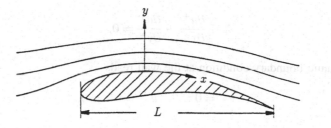

Fig. 12.1 Boundary layer coordinates

To discuss this further we introduce dimensionless quantities, chosen so that they are all of the same order of magnitude

$$u^+ = \frac{u}{U_\infty}, \quad v^+ = \frac{v}{U_\infty}\frac{L}{\delta_0} = \frac{v}{U_\infty}Re^{\frac{1}{2}}, \quad p^+ = \frac{p}{U_\infty^2 \varrho} \tag{12.10}$$

and

$$x^+ = \frac{x}{L}, \quad y^+ = \frac{y}{\delta_0} = \frac{y}{L}Re^{\frac{1}{2}}, \quad t^+ = t\frac{U_\infty}{L}. \tag{12.11}$$

Using these variables the Navier–Stokes equations take on the form

$$\frac{\partial u^+}{\partial t^+} + u^+\frac{\partial u^+}{\partial x^+} + v^+\frac{\partial u^+}{\partial y^+} = -\frac{\partial p^+}{\partial x^+} + \frac{1}{Re}\frac{\partial^2 u^+}{\partial x^{+2}} + \frac{\partial^2 u^+}{\partial y^{+2}}, \tag{12.12}$$

and

$$\frac{1}{Re}\left(\frac{\partial v^+}{\partial t^+} + u^+\frac{\partial v^+}{\partial x^+} + v^+\frac{\partial v^+}{\partial y^+}\right) = -\frac{\partial p^+}{\partial y^+} + \frac{1}{Re^2}\frac{\partial^2 v^+}{\partial x^{+2}} + \frac{1}{Re}\frac{\partial^2 v^+}{\partial y^{+2}}, \tag{12.13}$$

in which all differential expressions have the same order of magnitude, and the order of magnitude of the whole term is controlled by the prefactor.

Since we are looking for an approximate solution for large Reynolds' numbers, we take the limit $Re \to \infty$ and obtain the *boundary layer equations* in dimensionless form

$$\frac{\partial u^+}{\partial t^+} + u^+\frac{\partial u^+}{\partial x^+} + v^+\frac{\partial u^+}{\partial y^+} = -\frac{\partial p^+}{\partial x^+} + \frac{\partial^2 u^+}{\partial y^{+2}}, \tag{12.14}$$

and

$$0 = -\frac{\partial p^+}{\partial y^+}. \tag{12.15}$$

In addition we have the continuity equation which remained unaffected by taking the limit

$$\frac{\partial u^+}{\partial x^+} + \frac{\partial v^+}{\partial y^+} = 0. \tag{12.16}$$

The dynamic boundary condition at the wall reads

$$y^+ = 0: \quad u^+ = v^+ = 0, \tag{12.17}$$

and, since at the outer edge of the boundary layer the velocity u of the inner solution should pass asymptotically into the velocity $U(x, t) = U(x, y = 0, t)$ of the outer solution,

$$y^+ \to \infty: \qquad u^+ \to \frac{U}{U_\infty}. \tag{12.18}$$

We shall address the initial conditions later on, but shall first show that Eqs. (12.14) and (12.15) are much simpler than the Navier–Stokes equations. In the dimensionless boundary layer equations and in the boundary conditions the viscosity does not appear, and therefore the solution is valid for all Reynolds' numbers, as long as they are large enough (always assuming laminar flow) so that the simplifications are justified. Of course in terms of dimensional quantities the solution does change with the Reynolds' number. We read off from (12.10) and (12.11) that u and x do not change if u^+ and x^+ respectively do not change, and that for fixed v^+ and y^+ respectively, v and y are proportional to $Re^{-1/2}$. In the "physical" plane the quantities change with the Reynolds' number as follows: distances and velocities in the y-direction vary proportionally to $Re^{-1/2}$, while in the x-direction they remain constant.

We shall now rewrite the boundary layer equations in dimensional form and shall restrict ourselves to steady flow. These were first stated in this form in 1904 by Prandtl

$$u \frac{\partial u}{\partial x} + v \frac{\partial u}{\partial y} = -\frac{1}{\varrho} \frac{\partial p}{\partial x} + \nu \frac{\partial^2 u}{\partial y^2}, \tag{12.19}$$

$$0 = \frac{\partial p}{\partial y}, \quad \text{and} \tag{12.20}$$

$$\frac{\partial u}{\partial x} + \frac{\partial v}{\partial y} = 0. \tag{12.21}$$

From the second equation of this system of partial differential equations of the *parabolic type* we see that $p = p(x)$. In the remaining equations u and v are the independent variables, while p is no longer to be counted as an unknown. Because of (12.20) the pressure in the boundary layer (inner solution) has the same value as outside it, where it is known from the outer solution. We evaluate the Euler's equation (outer solution) (4.40) at the wall $y = 0$ (region of the inner solution). With vanishing perpendicular velocity at the wall we obtain

$$-\frac{1}{\varrho} \frac{\partial p}{\partial x} = U \frac{\partial U}{\partial x}. \tag{12.22}$$

We note that for $y \to \infty$ only one condition is placed on the component u (cf. 12.18). Because of the parabolic character of the system of equations, an initial distribution must be given

$$x = x_0 : \qquad u = u_0(y), \tag{12.23}$$

as well as the boundary condition (12.17) at the wall. The initial-boundary value problem (12.19–12.22) is defined in a half-open domain $(x = x_0, y = 0, y \to \infty)$ and the solution can be found by continuation of a given velocity profile in growing x-direction.

The system of equations is nonlinear and must in general be solved numerically. The methods of solution can be arranged into field and integral methods. Numerical field methods arise from replacing the differential Eqs. (12.19) and (12.21) by their finite difference forms. We shall go into the integral methods in Sect. 12.4.

Finally, we note that the flow over an oscillating wall (Sect. 6.2.1) and the flow over a wall which is suddenly set in motion (Sect. 6.2.2) are exact solutions of the Navier–Stokes equations of the boundary layer type.

12.1 Solutions of the Boundary Layer Equations

For certain pressure and velocity distributions, the partial differential Eqs. (12.19) to (12.21) can be reduced to ordinary differential equations. The most important cases are the *power law distributions*

$$U(x) = C x^m. \tag{12.24}$$

These correspond to the corner flows (10.240) with $C = |a|$ and $m = n - 1$. Stagnation point flow ($m = 1$, $n = 2$) and parallel flow ($m = 0$, $n = 1$) are of particular interest. However in this connection, flows with exponents in the range $1 < n < 2$ are interesting, since they describe flow past wedges. First we shall consider the particularly simple case $m = 0$, which describes the flow past a semi-infinite plate.

12.1.1 Flat Plate

The outer flow is the unperturbed parallel flow $U = U_\infty$ (Fig. 12.2), and therefore $\partial p / \partial x = 0$ holds. From (12.19) and (12.21) it then follows that

$$u \frac{\partial u}{\partial x} + v \frac{\partial u}{\partial y} = \nu \frac{\partial^2 u}{\partial y^2} \tag{12.25}$$

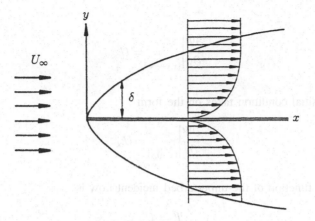

Fig. 12.2 Boundary layer on the flat plate

and

$$\frac{\partial u}{\partial x} + \frac{\partial v}{\partial y} = 0, \tag{12.26}$$

which is to be solved subject to the boundary conditions

$$y = 0, \qquad x > 0: \qquad u = v = 0, \tag{12.27}$$

$$y \to \infty: \qquad u = U_\infty, \tag{12.28}$$

and the initial condition

$$x = 0: \qquad u = U_\infty. \tag{12.29}$$

We note that the inequalities (12.1) do not hold at the leading edge of the plate $(x = 0)$. Hence, it is a *singular point* within the frame of the boundary layer theory.

By introducing the stream function we satisfy the continuity equation identically and from (12.25) obtain the differential equation

$$\frac{\partial \Psi}{\partial y}\frac{\partial^2 \Psi}{\partial x \partial y} - \frac{\partial \Psi}{\partial x}\frac{\partial^2 \Psi}{\partial y^2} = \nu\frac{\partial^3 \Psi}{\partial y^3}. \tag{12.30}$$

The boundary conditions now become

$$\Psi(x, y = 0) = \left.\frac{\partial \Psi}{\partial y}\right|_{(x,y=0)} = 0 \tag{12.31}$$

and

$$\frac{\partial \Psi}{\partial y}\bigg|_{(x,y\to\infty)} = U_\infty, \qquad (12.32)$$

while the initial condition takes on the form

$$\frac{\partial \Psi}{\partial y}\bigg|_{(x=0,y)} = U_\infty. \qquad (12.33)$$

The stream function of the unperturbed incident flow is

$$\Psi = U_\infty y,$$

and we expect that in the boundary layer $\Psi = O(U_\infty \delta_0)$. Because of (12.8) we therefore have

$$\Psi \sim \frac{U_\infty L}{\sqrt{Re}}. \qquad (12.34)$$

We use this result to make the stream function dimensionless with (12.11). It must hold then that

$$\frac{\Psi}{\sqrt{L\nu U_\infty}} = f(x^+, y^+) = f\left(\frac{x}{L}, y\sqrt{\frac{U_\infty}{L\nu}}\right). \qquad (12.35)$$

Since no geometrical length is introduced into this problem of the infinitely long plate, L plays the role here of an *artificial length*. The requirement that this artificial length vanish from the problem leads us to the similarity variables

$$\eta = y\frac{\sqrt{U_\infty/(L\nu)}}{\sqrt{x/L}} = y\sqrt{\frac{U_\infty}{\nu x}}, \qquad (12.36)$$

and

$$\Psi\frac{\sqrt{L/x}}{\sqrt{LU_\infty \nu}} = \frac{\Psi}{\sqrt{\nu U_\infty x}}. \qquad (12.37)$$

Therefore the solution must be of the form

$$\Psi = \sqrt{\nu U_\infty x}\, f(\eta). \qquad (12.38)$$

If we insert this form of solution into (12.30), the differential equation

$$2f''' + ff'' = 0,$$ (12.39)

appears; this is known as the *Blasius' equation*. The boundary conditions on f follow from (12.31) and (12.32) as

$$f(0) = f'(0) = 0$$ (12.40)

and

$$f'(\infty) = 1.$$ (12.41)

Since

$$\eta(y \to \infty, x) = \eta(y, x = 0) = \infty,$$ (12.42)

the initial condition (12.33) also leads to (12.41). The solution of the Blasius' equation with these boundary conditions is a boundary value problem, since conditions are given on both boundaries $\eta = 0$ and $\eta = \infty$. The problem can also be solved numerically as an initial value problem: besides the initial values (12.40) we then lay down a further initial value for f'', say $f''(0) = \alpha$, and try out different values of α until the boundary condition for $\eta = \infty$ is satisfied (*shooting method*). In this manner we find that

$$f''(0) = 0.33206.$$ (12.43)

As well as showing the velocity $f'(\eta) = u/U_\infty$, Fig. 12.3 also shows the functions $f(\eta)$ and $f''(\eta)$. Using (12.43) the shear stress at the wall

$$\tau_w = \eta \frac{\partial u}{\partial y}\bigg|_{y=0} = \eta \sqrt{\frac{U_\infty^3}{\nu x}} f''(0)$$ (12.44)

can be calculated, where η in Eq. (12.44) is the shear viscosity and not the similarity variable of (12.36).

Theoretically the boundary layer reaches to infinity because the transition from the boundary layer to outer flow is asymptotic, and so the geometric boundary layer thickness can be arbitrarily defined. Often the boundary layer thickness is taken as the distance from the wall where $u/U_\infty = 0.99$. As the numerical calculation shows, this value is reached for $\eta \approx 5$. The boundary layer thickness defined in this way is therefore

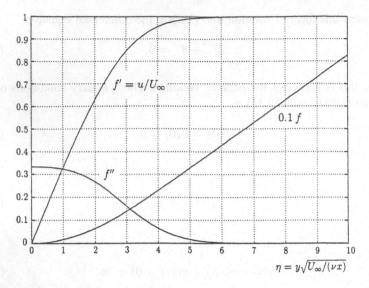

Fig. 12.3 Solution of the Blasius' equation

$$\delta = 5\sqrt{\frac{\nu x}{U_\infty}}.$$ (12.45)

Instead of the geometric boundary layer thickness δ, the uniquely defined *displacement thickness* δ_1 is often preferred

$$\delta_1 = \int\limits_0^\infty \left(1 - \frac{u}{U}\right)\mathrm{d}y,$$ (12.46)

which is a measure of the displacing action of the boundary layer. From the velocity profile u/U_∞, because of $U \equiv U_\infty$, we obtain

$$\delta_1 = 1.7208\sqrt{\frac{\nu x}{U_\infty}}.$$ (12.47)

The outer, inviscid flow does not "see" the infinitesimally thin plate, but instead a half body with the contour (12.47). A measure for the loss of momentum in the boundary layer is the *momentum thickness* δ_2

$$\delta_2 = \int\limits_0^\infty \left(1 - \frac{u}{U}\right)\frac{u}{U}\mathrm{d}y,$$ (12.48)

for which we here obtain the value

$$\delta_2 = 0.664 \sqrt{\frac{\nu x}{U_\infty}}. \tag{12.49}$$

Of course the plate experiences a drag (per unit depth), which for the plate wetted on one side and extending to $x = L$ is found as

$$F_w = \int_0^L \tau_w \mathrm{d}x = 0.664 \varrho U_\infty^2 L \left(\frac{U_\infty L}{\nu}\right)^{-1/2}. \tag{12.50}$$

The formula for the coefficient of friction c_f may be obtained from (12.50)

$$c_f = \frac{F_w}{(\varrho/2)U_\infty^2 L} = \frac{1,33}{\sqrt{Re}}, \tag{12.51}$$

a result which is called *Blasius' friction law*.

12.1.2   Wedge Flows

We consider symmetric wedges as in Fig. 12.4 and shall first deal with the outer inviscid potential flow, whose velocity distribution leads to asymptotic boundary conditions for the inner flow calculation. The outer flow has already been given by the corner flow in Fig. 10.30 in the exponent range $1 \le n \le 2$. We turn the coordinate system in Fig. 12.4 around $\pi - \pi/n$ in the positive direction, to make it comparable to Fig. 10.30. This means that the complex coordinate z is replaced by $z \exp\{-i\pi \, [(n - 1)/n]\}$. The stream function corresponding to (10.243) now reads

$$\Psi = \frac{a}{n} r^n \sin[n\varphi - \pi(n - 1)], \tag{12.52}$$

and $\Psi = 0$ is obtained for the angle

$$\beta = \pi \frac{n-1}{n} = \pi \frac{m}{m+1}$$

as well as for the negative x-axis.

As we reflect the corner flow through the x-axis, it becomes a wedge flow whose velocity distribution is given by (10.244). In the boundary layer coordinates where we measure x along the upper surface of the body and y perpendicular to it, we therefore obtain exactly the power law distribution (12.24).

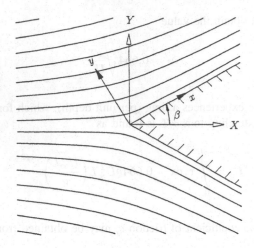

Fig. 12.4 Wedge flow

As is shown by Euler's equation

$$-\frac{1}{\varrho}\frac{\partial p}{\partial x} = mC^2 x^{2m-1}, \qquad (12.53)$$

the pressure gradient here does not in general vanish; in spite of this no typical length appears in wedge flows, and we should not be surprised that using the variables (12.36) and (12.37) leads to a similarity solution here too. The similarity form

$$\Psi = \sqrt{\nu U(x) x}\, f(\eta) \qquad (12.54)$$

with

$$\eta = y\sqrt{\frac{U(x)}{\nu x}} \qquad (12.55)$$

transforms the system (12.19) to (12.21) with (12.53) into the ordinary differential equation

$$f''' + \frac{m+1}{2} f f'' + m\left(1 - f'^2\right) = 0. \qquad (12.56)$$

The solutions of this so-called *Falkner-Skan equation*, which must satisfy the boundary conditions (12.40) and (12.41), are sketched in Fig. 12.5 for different wedge angles which correspond to the exponent range $m = 0$ (i.e. $\beta = 0°$) to $m = 1$ (i.e. $\beta = 90°$). From the figure we take the boundary layer thickness corresponding to $f' = 0.99$ of the two-dimensional stagnation point flow as

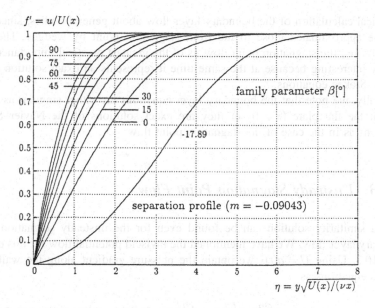

Fig. 12.5 Velocity distribution in the boundary layer of a wedge flow

$$\delta = 2.4\sqrt{\frac{\nu x}{U(x)}} = 2.4\sqrt{\frac{\nu}{a}}, \tag{12.57}$$

where for the stagnation point flow we follow the convention of setting C in (12.24) to a.

The differential Eq. (12.56) with the boundary conditions (12.40) and (12.41) also allows solutions for negative values of m which correspond to flows past convex corners. As was mentioned in connection with Fig. 10.30 and as is directly clear from (12.24), the velocity at $x = 0$ becomes infinite and the solution can only have any physical meaning beyond a certain distance downstream from this position. Since the solutions with negative m are not unique their physical meaning is called into question. In fact there is an infinite number of solutions of the differential Eq. (12.56) which satisfy the boundary conditions and have different values of $f''(0)$, and therefore different values of the shear stress at the wall.

A "plausible" solution with $m = -0.09043$ is included in Fig. 12.5, because this profile represents a *separation profile* as we shall show in Sect. 12.1.4. This small negative value of m makes it clear that the boundary layer separates even for very small positive pressure gradients. Turbulent boundary layers can suffer a considerably higher increase in pressure; a fact which is very important for flow past an airfoil and to which we have already referred in the discussion of the drag on a sphere.

The importance of the solutions of the Falkner-Skan equation also rests in the fact that they also provide the necessary initial distributions (cf. (12.23)) for the numerical calculation of the boundary layer flow about general bodies, since they may be approximated close to the front stagnation point by wedges. The case $\beta = 90°$, i.e., the stagnation point flow, is of the greatest practical importance; it is already interesting because at the same time it represents an exact solution of the Navier–Stokes equations.

Finally, we note that the boundary layer equations are either singular as in the case of the flat plate $(x = 0)$ or they are exact solutions of the Navier–Stokes equations as in the case of the stagnation point flow.

12.1.3 Unsteady Stagnation Point Flow

Now a similarity solution can be found even for the unsteady stagnation point boundary layer flow. We have mentioned the inviscid potential flow for this case in Sect. 10.3. Using $U = a(t)x$ we obtain the pressure gradient along the wall from Euler's equations as

$$-\frac{1}{\varrho}\frac{\partial p}{\partial x} = \frac{\partial U}{\partial t} + U\frac{\partial U}{\partial x} = a^2 x\left(\frac{\dot{a}}{a^2} + 1\right),\qquad(12.58)$$

where we have set $da/dt = \dot{a}$. Here the form of solution (12.54) with (12.55) is

$$\Psi = \sqrt{\nu\, a(t)}\, x\, f(\eta),\qquad(12.59)$$

with

$$\eta = y\sqrt{\frac{a(t)}{\nu}}\qquad(12.60)$$

and transforms the boundary layer Eqs. (12.2), (12.20) and (12.21) using (12.58) into the equation

$$\frac{\dot{a}}{a^2}\left(f' + \frac{\eta}{2}f''\right) + f'^2 - ff'' = \left(\frac{\dot{a}}{a^2} + 1\right) + f'''.\qquad(12.61)$$

This becomes an ordinary differential equations if $\dot{a}/a^2$ is a constant

$$\frac{1}{a^2}\frac{da(t)}{dt} = \text{const},$$

where in particular const $= 0$ gives the steady stagnation point flow. Choosing const $= 1/2$, and integrating $\dot{a}/a^2 = 1/2$ leads to the relation $a(t) = -2/t$, if the

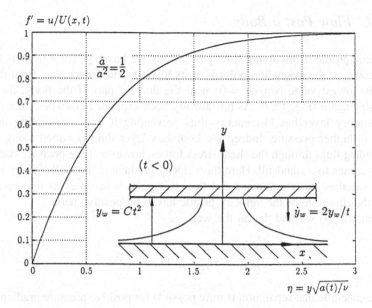

Fig. 12.6 Velocity distribution in the boundary layer of the unsteady stagnation point flow

constant of integration is set to zero. Therefore the velocity of the inviscid potential flow at the edge of the boundary layer is

$$U = -\frac{2x}{t}, \qquad (12.62)$$

which is positive for $t < 0$ and which tends to infinity as $t \to 0$. We can imagine this velocity in inviscid fluid as being produced when the upper wall (see Fig. 12.6) is moved towards the lower wall with the velocity

$$\dot{y}_w = \frac{2y_w}{t},$$

thus carrying out the motion

$$y_w = C t^2.$$

The velocity distribution in the boundary layer for this case is shown in Fig. 12.6.

We also note here that the similarity solution mentioned can be extended to compressible flow.

12.1.4 Flow Past a Body

In general problems involving flow past a body we have $\partial p/\partial x \neq 0$; as is known
the pressure at the stagnation point is at its highest, and it decreases from there to
reach its lowest value ($\partial p/\partial x = 0$) near the thickest part of the body, thereafter
increasing again (Fig. 12.7). As has already been explained elsewhere, the fluid in
the boundary layer has lost energy thus preventing it from penetrating into the
region of higher pressure. Indeed the boundary layer fluid is pulled along by the
surrounding fluid through the shear stress forces, however if the pressure rise is too
large it comes to a standstill. Here the velocity gradient at the wall ($\partial u/\partial y = 0$ for
$y = 0$) vanishes. In two-dimensional flow this point is accepted as the *separation
point*; the curvature of the velocity profile has to be positive here.

From (12.19) we find that at the wall

$$\frac{1}{\varrho}\frac{\partial p}{\partial x} = \nu \frac{\partial^2 u}{\partial y^2} \quad \text{for} \quad y = 0, \tag{12.63}$$

and we conclude that separation is only possible for positive pressure gradients, i.e.,
the separation point (Fig. 12.8) lies in the region where the pressure is rising, as we
have already shown heuristically. From the mathematical point of view the topol-
ogy of the separation point region corresponds to an unstable *saddle point*.

As we have previously stated, only numerical methods can be used for the
general problem of flow past a body. For a given pressure distribution, the boundary
layer calculation cannot in general be carried further than the separation point. The
reason for this is to be found in the parabolic character of the boundary layer
equation. We can only count on the convergence of a numerical algorithm if the
velocity profile stays positive. However there remains a need to develop a computational method which predicts the flow past the separation point. This can

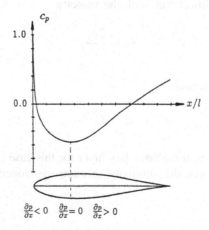

Fig. 12.7 Pressure distribution at an airfoil

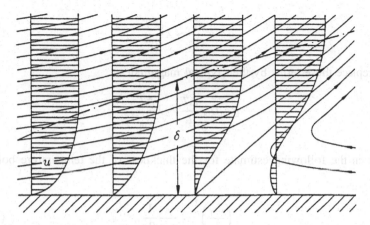

Fig. 12.8 Sketch of the separation region

successfully be done in the so-called *inverse problem*, where instead of the pressure distribution the shear stress distribution is given.

12.2 Temperature Boundary Layer in Forced Convection

In calculating the temperature distribution within the boundary layer we start out with the energy Eq. (4.4), where we first introduce the simplifications possible within boundary layer theory. Because of (12.1), we find the same relation for the dissipation function as for unidirectional flows

$$\Phi = \eta \left(\frac{\partial u}{\partial y} \right)^2,$$ (12.64)

and for the calorically perfect gas, (4.4) assumes the form

$$\varrho \, c_p \frac{\mathrm{D}T}{\mathrm{D}t} - \frac{\mathrm{D}p}{\mathrm{D}t} = \eta \left(\frac{\partial u}{\partial y} \right)^2 + \lambda \frac{\partial^2 T}{\partial y^2},$$ (12.65)

where again $\partial^2 / \partial x^2$ has been neglected compared to $\partial^2 / \partial y^2$.

The *heat transfer* between a body and its surrounding fluid takes place in a layer along the contour of the body, in which, besides convection, i.e., transport of heat by fluid motion, heat conduction plays a role. While heat conduction can usually be ignored in the outer flow, it is of the same order of magnitude as the convection term in (12.65) in the inner layer, called the *temperature boundary layer*; that is

$$\varrho\, c_p u \frac{\partial T}{\partial x} \sim \lambda \frac{\partial^2 T}{\partial y^2}. \tag{12.66}$$

If we replace these terms by their typical magnitudes

$$\varrho\, c_p U_\infty \frac{\Delta T}{L} \sim \lambda \frac{\Delta T}{\delta_{0t}^2}, \tag{12.67}$$

we obtain the following estimate for the thickness of the temperature boundary layer

$$\left(\frac{\delta_{0t}}{L}\right)^2 \sim \frac{\lambda}{c_p \eta} \frac{1}{Re}. \tag{12.68}$$

The dimensionless combination of the material properties λ, c_p, η in the first brackets on the right-hand side is the inverse of the *Prandtl's number*

$$Pr = \frac{c_p \eta}{\lambda}, \tag{12.69}$$

which we met in (4.178), and which, as is clear from (12.8) and (12.68), determines the ratio of the thicknesses of the velocity boundary layer and the temperature boundary layer

$$\frac{\delta_0}{\delta_{0t}} \sim \sqrt{Pr} \tag{12.70}$$

For monatomic gases the kinetic gas theory furnishes the relation between λ and η

$$\lambda = \frac{5}{2} c_v \eta, \tag{12.71}$$

so that for $\gamma = c_p/c_v = 5/3$ the Prandtl's number assumes the value $Pr = 2/3$. For diatomic gases, the Prandtl's number can be calculated from the *formula of Eucken*

$$Pr = \frac{c_p}{c_p + 1.25R}. \tag{12.72}$$

From this we find for the ideal gas that $Pr = 0.74$. This formula does not produce good results for polyatomic gases, and it is recommended to calculate the Prandtl's number from the measured values of η, λ and c_p.

The Prandtl's number for gases is of the order of 1, so that the temperature and velocity boundary layers are of about the same thickness. The Prandtl's number for liquids is considerable larger than 1 (water: $Pr \approx 7$ at a temperature of 20 °C and

1 bar pressure), so the temperature boundary layer is usually smaller than the velocity boundary layer. Compared to this, Pr for liquid metals is much smaller than 1 (mercury: $Pr \approx 0.026$ at a temperature of 20 °C), so that the temperature boundary layer is correspondingly large.

If instead of the thermal conductivity λ we use the *thermal diffusivity*

$$a = \frac{\lambda}{c_p \varrho},$$

the Prandtl's number can be written in the easily remembered form

$$Pr = \frac{\nu}{a}.$$

We can simplify the energy equation even further within the framework of the boundary layer theory. For the dissipation function Φ from (12.64) and the work (per unit volume) of the pressure forces Dp/Dt, we obtain the order of magnitude equations

$$\Phi \sim \eta \left(\frac{U_\infty}{\delta_0} \right)^2 \sim \varrho_\infty \frac{U_\infty^3}{L} \tag{12.73}$$

and

$$\frac{Dp}{Dt} \sim U_\infty \frac{\partial p}{\partial x} \sim \varrho_\infty \frac{U_\infty^3}{L}. \tag{12.74}$$

The estimate shows that both terms are of the same order of magnitude. The ratio of these terms to the convective heat transport

$$\varrho \, c_p u \frac{\partial T}{\partial x} \sim \varrho_\infty c_p U_\infty \frac{T_w - T_\infty}{L} \tag{12.75}$$

is known as Eckert's number Ec

$$Ec = \frac{U_\infty^2}{c_p (T_w - T_\infty)}. \tag{12.76}$$

Eckert's number is the ratio of (twice) the kinetic energy of the unperturbed flow to the enthalpy difference between the wall and the fluid. The largest possible self-heating of the fluid is found from the energy Eq. (4.150) for the calorically perfect gas as

$$c_p(T_t - T_\infty) = \frac{U_\infty^2}{2},$$ (12.77)

or with $a_\infty^2 = \gamma RT_\infty$

$$\frac{T_t - T_\infty}{T_\infty} = \frac{\gamma - 1}{2} M_\infty^2.$$ (12.78)

As we have stated previously, the self-heating of incompressible fluids ($M_\infty \to 0$) is negligible. In heat transfer problems with small Mach numbers, Eckert's number is generally very small and the dissipation Φ as well as the work per unit volume Dp/Dt are negligible, so that we obtain the energy Eq. (12.65) in the form

$$\varrho\, c_p \left(u \frac{\partial T}{\partial x} + v \frac{\partial T}{\partial y} \right) = \lambda \frac{\partial^2 T}{\partial y^2}.$$ (12.79)

In order to solve (12.79) we clearly require the velocity field in the boundary layer. The assumption of incompressibility has the consequence that the equations of motion are decoupled from the energy equation. Therefore we can first solve the equations for the flow boundary layer and then with the velocity distribution resulting from this solve the temperature boundary layer.

However in the case of strong external heating the change in density as a result of the change in temperature must be taken into account. Then the flow is to be treated as a compressible flow even for vanishing Mach numbers, and the decoupling mentioned above in general does not occur. In these circumstances the temperature dependence of the material properties usually has to be taken into account too. In what follows we shall start out from the idea that the temperature differences in the boundary layer are so small that the above effects can be ignored.

We shall consider the heat transfer problem of a flat plate; the system of equations and boundary conditions are summarised below

$$u \frac{\partial u}{\partial x} + v \frac{\partial u}{\partial y} = \nu \frac{\partial^2 u}{\partial y^2},$$ (12.80)

$$\frac{\partial u}{\partial x} + \frac{\partial v}{\partial y} = 0,$$ (12.81)

$$u \frac{\partial T}{\partial x} + v \frac{\partial T}{\partial y} = \frac{\nu}{Pr} \frac{\partial^2 T}{\partial y^2};$$ (12.82)

$$y = 0, x > 0: \quad u = v = 0, T = T_w,$$ (12.83)

$$y \rightarrow \infty : u = U_\infty, T = T_\infty. \tag{12.84}$$

The velocity components u and v follow from (12.38) as

$$u = U_\infty f', \quad \text{and} \tag{12.85}$$

follows, with the boundary conditions

$$v = -\frac{1}{2}\sqrt{\frac{\nu U_\infty}{x}}(f - \eta f'). \tag{12.86}$$

We conclude from (12.80) to (12.82) that the dimensionless temperature too can only be a function of the similarity variables (12.36). Therefore

$$\frac{T_w - T}{T_w - T_\infty} = \Theta(\eta), \tag{12.87}$$

and from (12.82) the equation

$$\Theta'' + \frac{1}{2}Prf\Theta' = 0 \tag{12.88}$$

follows, with the boundary conditions

$$\eta = 0 : \quad \Theta = 0, \tag{12.89a}$$

$$\eta \rightarrow \infty : \quad \Theta = 1. \tag{12.89b}$$

If we first set $\Theta' = F$ as the solution to (12.88) then

$$F = C_1 \exp\left(-\frac{1}{2}Pr\int_0^\eta f \, d\eta\right), \tag{12.90}$$

and further because of (12.89a)

$$\Theta = \int_0^\eta F \, d\eta = C_1 \int_0^\eta \exp\left(-\frac{1}{2}Pr\int_0^\eta f \, d\eta\right) d\eta. \tag{12.91}$$

Taking account of the boundary condition (12.89b) this finally becomes

$$\Theta = \left[\int_0^\eta \exp\left(-\frac{1}{2}Pr\int_0^n f\,d\eta\right)d\eta\right]\left[\int_0^\infty \exp\left(-\frac{1}{2}Pr\int_0^n f\,d\eta\right)d\eta\right]^{-1} \qquad (12.92)$$

Because of (12.39) we also have $f = -2f'''/f''$, so that we can write

$$-\frac{1}{2}Pr\int_0^\eta f\,d\eta = Pr\int_0^\eta \frac{f'''}{f''}\,d\eta = Pr\ln\left(\frac{f''(\eta)}{f''(0)}\right), \qquad (12.93)$$

and (12.92) then becomes

$$\Theta = \left[\int_0^\eta f''^{Pr}d\eta\right]\left[\int_0^\infty f''^{Pr}d\eta\right]^{-1}. \qquad (12.94)$$

The dimensionless temperature Θ is thus known, since $f''(\eta)$ is given from the solution of the Blasius' equation. The solution in the above form was first given by *Pohlhausen*. $\Theta = \Theta\,(\eta, Pr)$ for various values of Pr is shown in Fig. 12.9.

We shall now calculate the only nonzero component of the heat flux vector q_y at the wall

$$q_y(x) = q(x) = -\lambda\frac{\partial T}{\partial y}\bigg|_w, \qquad (12.95)$$

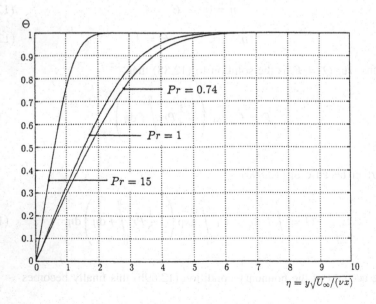

Fig. 12.9 Temperature profiles of the flat plate boundary layer

or

$$q(x) = \lambda(T_w - T_\infty)\frac{d\Theta}{d\eta}\bigg|_w \sqrt{\frac{U_\infty}{\nu x}}. \tag{12.96}$$

From (12.94) is follows that

$$\frac{d\Theta}{d\eta}\bigg|_{\eta=0} = [f''(0)]^{Pr}\left(\int_0^\infty f''^{Pr} d\eta\right)^{-1} = g(Pr), \tag{12.97}$$

so that the heat flux is obtained as

$$q(x) = \lambda(T_w - T_\infty)g(Pr)\sqrt{\frac{U_\infty}{\nu x}}. \tag{12.98}$$

From

$$\dot{Q} = -\iint_{(S)} q_i n_i \, dS = \int_0^L q(x) dx \tag{12.99}$$

we finally find the total heat per unit time and unit width transferred from the plate of length L (on one side of the plate) as

$$\dot{Q} = 2\lambda(T_w - T_\infty)g(Pr)\sqrt{\frac{U_\infty L}{\nu}}, \tag{12.100}$$

or

$$\dot{Q} = 2\lambda(T_w - T_\infty)g(Pr)\sqrt{Re}. \tag{12.101}$$

The function $g(Pr)$ is well approximated by

$$g(Pr) = 0.332 Pr^{1/3}. \tag{12.102}$$

Using this we obtain

$$\dot{Q} = 0.664\lambda \, Pr^{1/3} Re^{1/2}(T_w - T_\infty), \tag{12.103}$$

or else, using the defining equation for *Nusselt's number*

$$\dot{Q} = Nu \, \lambda A \frac{T_w - T_\infty}{L}, \quad A \triangleq L \tag{12.104}$$

we find *Nusselt's relation*

$$Nu = 0.664 Pr^{1/3} Re^{1/2}. \tag{12.105}$$

This is a specific form of the general law valid for forced convection

$$Nu = Nu(Pr, Re). \tag{12.106}$$

12.3   Temperature Boundary Layer in Natural Convection

In the section Hydrostatic Pressure Distribution 5.1, we showed that static equilibrium is only possible if the density gradient is parallel to the vector of the mass body force. If we again choose the coordinate system so that the z-direction is antiparallel to the direction in which gravity acts, then in equilibrium the density can only be a function of z. Close to a heated wall where the density is affected by heating, this condition of static equilibrium in general breaks down, and the fluid is set in motion so that a flow originates close to the wall. Under certain conditions which will be defined more precisely later, this flow has the character of a boundary layer.

In order to derive the equations of motion we start out from the Navier–Stokes equations and split the pressure and density up into their static and dynamic parts

$$p = p_{st} + p_{dyn}, \quad \varrho = \varrho_{st} + \varrho_{dyn}.$$

It follows that

$$\left(\varrho_{st} + \varrho_{dyn}\right) \frac{Du_i}{Dt} = \varrho_{st} k_i - \frac{\partial p_{st}}{\partial x_i} + \varrho_{dyn} k_i - \frac{\partial p_{dyn}}{\partial x_i} + \eta \frac{\partial^2 u_i}{\partial x_j \partial x_j}, \tag{12.107}$$

or, since the hydrostatic equation is given by

$$\frac{\partial p_{st}}{\partial x_i} = \varrho_{st} k_i,$$

we have

$$\left(\varrho_{st} + \varrho_{dyn}\right) \frac{Du_i}{Dt} = \varrho_{dyn} k_i - \frac{\partial p_{dyn}}{\partial x_i} + \eta \frac{\partial^2 u_i}{\partial x_j \partial x_j}. \tag{12.108}$$

We assume that the change in density ϱ_{dyn} is very small $\left(\varrho_{dyn} \ll \varrho_{st}\right)$, so that, using the mass body force of gravity $k_i = g_i$ it follows first that

$$\frac{Du_i}{Dt} = \frac{\varrho_{dyn}}{\varrho_{st}} g_i - \frac{1}{\varrho_{st}} \frac{\partial p_{dyn}}{\partial x_i} + \nu \frac{\partial^2 u_i}{\partial x_j \partial x_j}. \tag{12.109}$$

In the literature this approach is named *Boussinesq approximation*.

Now for the change in density we set

$$\varrho_{dyn} = -\varrho_{st} \beta (T - T_\infty), \tag{12.110}$$

where the thermal expansion coefficient is given by

$$\beta = \left[-\frac{1}{\varrho} \left(\frac{\partial \varrho}{\partial T} \right)_p \right]_\infty, \tag{12.111}$$

and therefore for the ideal gas

$$\beta = \frac{1}{T_\infty}. \tag{12.112}$$

Again we abbreviate p_{dyn} with p and ϱ_{st} with ϱ. In the boundary layer the convective terms are again of the same order of magnitude as the viscosity terms, so that we have

$$\frac{U^2/L}{\nu U/\delta_0^2} \sim 1. \tag{12.113}$$

In this case where there is no velocity U_∞, the typical velocity can only be given indirectly by the data of the problem. The driving force of the flow is the term $\varrho_{dyn} g = \varrho_{st} \beta \Delta T g$, where $\Delta T = |T_w - T_\infty|$. Using the characteristic length L, the typical velocity

$$U = \sqrt{\beta \Delta T g L} \tag{12.114}$$

can be formed. Then from (12.113) we extract

$$\frac{\delta_0}{L} \sim \left(\frac{\nu^2}{g \beta \Delta T L^3} \right)^{1/4}. \tag{12.115}$$

For

$$Gr = \frac{g\,\beta\,\Delta T\,L^3}{\nu^2} \gg 1 \tag{12.116}$$

$\delta_0/L \ll 1$, i.e., the flow has the character of a boundary layer if the dimensionless number Gr (*Grashof's number*) is large. Under this condition, the boundary layer simplifications are valid. Instead of Grashof's number, *Rayleigh's number* $Ra = Gr\,Pr$ is often used.

Let us consider the flow at a vertical semi-infinite, heated plate as an example. The origin lies at the lower edge, x is measured along the plate and y normal to it. Then the vector of the mass body force of gravity has the components $g_x = -g$ and $g_y = 0$. Introducing the dimensionless temperature

$$\Theta = \frac{T - T_\infty}{T_w - T_\infty}, \tag{12.117}$$

the now coupled system of equations for the ideal gas reads

$$\frac{\partial u}{\partial x} + \frac{\partial v}{\partial y} = 0, \tag{12.118}$$

$$u\frac{\partial u}{\partial x} + v\frac{\partial u}{\partial y} = \nu\frac{\partial^2 u}{\partial y^2} + g\Theta\frac{T_w - T_\infty}{T_\infty}, \quad \text{and} \tag{12.119}$$

$$u\frac{\partial \Theta}{\partial x} + v\frac{\partial \Theta}{\partial y} = \frac{\lambda}{\varrho c_p}\frac{\partial^2 \Theta}{\partial y^2}. \tag{12.120}$$

This is to be solved subject to the boundary conditions

$$y = 0: \qquad u = v = 0; \quad \Theta = 1, \tag{12.121}$$

$$y \to \infty: \qquad u = 0; \qquad \Theta = 0. \tag{12.122}$$

Introducing a dimensionless stream function analogous to (12.38), and using

$$U = \sqrt{\frac{g\,\Delta T\,L}{4T_\infty}} \quad \text{and} \quad \Delta T = |T_w - T_\infty| \tag{12.123}$$

we obtain

$$\Psi = 4\left(\frac{\nu^2\,\Delta T\,g\,L^3}{4T_\infty}\right)^{1/4} f\left[\frac{x}{L}, y\left(\frac{\Delta T\,g}{4T_\infty\,\nu^2 L}\right)^{1/4}\right]. \tag{12.124}$$

Since L may not appear in the solution, Ψ must have the following form

$$\Psi = 4\left(\frac{\nu^2 \Delta T\, g\, x^3}{4 T_\infty}\right)^{1/4} \zeta(\eta),\tag{12.125}$$

where

$$\eta = y\left(\frac{\Delta T\, g}{4 T_\infty \nu^2 x}\right)^{1/4}\tag{12.126}$$

is the dimensionless similarity variable of the problem. Writing

$$C = \left(\frac{\Delta T\, g}{4 T_\infty \nu^2}\right)^{1/4},\tag{12.127}$$

we find the following form for the stream function

$$\Psi = 4 C \nu\, x^{3/4} \zeta(\eta).\tag{12.128}$$

The dimensionless temperature can also only be a function of the dimensionless variable η; therefore

$$\Theta(x, y) = \Theta(\eta).\tag{12.129}$$

Setting these into Eqs. (12.119) and (12.120) we obtain the coupled ordinary differential equations

$$\zeta''' + 3\zeta\zeta'' - 2\zeta' + \Theta = 0,\tag{12.130}$$

$$\Theta'' + 3 Pr \zeta \Theta' = 0,\tag{12.131}$$

with the boundary conditions

$$\eta = 0: \quad \zeta = \zeta' = 0;\ \Theta = 1,\tag{12.132}$$

$$\eta \to \infty: \quad \zeta' = 0;\ \Theta = 0.\tag{12.133}$$

This system of equations must be solved numerically. For $Pr = 0.733$ this gives us Nusselt's number as

$$Nu = 0.48 Gr^{1/4}.\tag{12.134}$$

The following formula, which explicitly states the dependency on Prandtl's number, is considered a good approximation

$$Nu = \left(\frac{Ra}{2.43478 + 4.884 Pr^{1/2} + 4.95283 Pr} \right)^{1/4}. \tag{12.135}$$

This is a special form of the general law for the natural convection $Nu = Nu(Pr, Ra)$. In the general case of a mixed convection, namely the superposition of forced and natural convection, the relation holds $Nu = Nu(Re, Pr, Ra)$.

12.4 Integral Methods of Boundary Layer Theory

In order to calculate boundary layers approximately, we often use methods where the equations of motion are not satisfied everywhere in the field but only in integral means across the thickness of the boundary layer. The starting point for these *integral methods* is usually the momentum equation which can be derived by applying the continuity Eq. (2.7) and the balance of momentum (2.43) in its integral form to a section of the boundary layer of length dx (Fig. 12.10).

The infinitesimal mass flux $d\dot{m}$ per unit depth which flows between (1) and (2) into the control volume is

$$d\dot{m} = \dot{m}(x + dx) - \dot{m}(x) = dx \frac{d\dot{m}}{dx} = dx \frac{d}{dx} \int\limits_{0}^{\delta(x)} \varrho\, u\, dy. \tag{12.136}$$

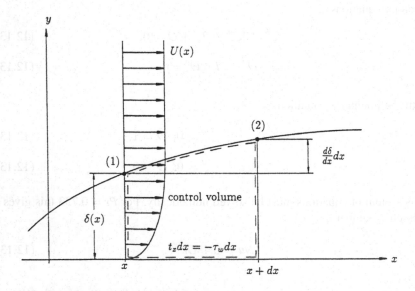

Fig. 12.10 Control volume in the boundary layer

With this mass flux the momentum flux

$$U \, d\dot{m} = U \, dx \frac{d}{dx} \int_0^{\delta(x)} \varrho \, u \, dy \qquad (12.137)$$

in the x-direction is associated so that the component of the balance of momentum in this direction reads

$$-U \frac{d}{dx} \int_0^{\delta(x)} \varrho \, u \, dy + \frac{d}{dx} \int_0^{\delta(x)} \varrho \, u^2 dy = -\frac{dp}{dx} \delta(x) - \tau_w. \qquad (12.138)$$

Again we shall restrict ourselves to incompressible flows for which the integrals appearing in (12.138) can be expressed using the displacement thickness (12.46) and the momentum thickness (12.48)

$$\int_0^{\delta(x)} u \, dy = U(\delta - \delta_1), \quad \text{and} \qquad (12.139)$$

$$\int_0^{\delta(x)} u^2 dy = U^2(\delta - \delta_1 - \delta_2). \qquad (12.140)$$

With $dp/dx = -\varrho \, U \, dU/dx$ the momentum equation can be brought to the form

$$\frac{d\delta_2}{dx} + \frac{1}{U} \frac{dU}{dx} (2\delta_2 + \delta_1) = \frac{\tau_w}{\varrho \, U^2}. \qquad (12.141)$$

We can also obtain this directly by integrating the equation of motion (12.19) over y (from 0 to ∞) and using the continuity equation. This holds for steady incompressible laminar and turbulent boundary layers, but can also be extended to unsteady and compressible flows. Equation (12.141) is an ordinary differential equation for the unknowns δ_1, δ_2 and τ_w. In the laminar case the equations necessary to complete the mathematical description are found by specifying the velocity profile inside the boundary layer. In the turbulent case empirical relations are necessary in addition; in this respect reference is made to Sect. 12.5.

As a simple example of the application of Eq. (12.141) we shall calculate the boundary layer on a flat plate, whose exact solution has already been found in Sect. 12.1.1. For the velocity distribution inside the boundary layer we assume the profile

$$u\left(\frac{y}{\delta(x)}\right) = U\sin\left(\frac{\pi}{2}\frac{y}{\delta(x)}\right), \tag{12.142}$$

from which the ratios for the boundary layer thicknesses take on the values

$$\frac{\delta_1}{\delta} = \int\limits_0^1 \left(1 - \frac{u}{U}\right)\mathrm{d}(y/\delta) = \frac{\pi - 2}{\pi} \tag{12.143}$$

and

$$\frac{\delta_2}{\delta} = \int\limits_0^1 \frac{u}{U}\left(1 - \frac{u}{U}\right)\mathrm{d}(y/\delta) = \frac{4 - \pi}{2\pi}. \tag{12.144}$$

Using (12.142) we find the wall shear stress to be

$$\tau_w = \eta\frac{\partial u}{\partial y}\bigg|_{y=0} = \eta\frac{\pi}{2}\frac{U}{\delta} = \eta\frac{4 - \pi}{4}\frac{U}{\delta_2}, \tag{12.145}$$

where we have made use of (12.144) to eliminate the boundary layer thickness δ. Inserting τ_w from (12.145) into the momentum Eq. (12.141), and recalling that $U \equiv U_\infty$, furnishes the ordinary differential equation

$$\frac{\mathrm{d}\delta_2}{\mathrm{d}x} = \frac{\nu}{U_\infty}\frac{4 - \pi}{4}\frac{1}{\delta_2}, \tag{12.146}$$

in which δ_2 is the only unknown appearing and whose general solution reads

$$\frac{\delta_2^2}{2} = \frac{4 - \pi}{4}\frac{\nu x}{U_\infty} + C. \tag{12.147}$$

We obtain the constant of integration from the momentum thickness at the position $x = 0$. This is zero for the flat plate, so that the solution is

$$\delta_2 = 0.655\sqrt{\frac{\nu x}{U_\infty}}. \tag{12.148}$$

Using (12.143) and (12.144) we obtain the displacement thickness as

$$\delta_1 = \frac{\delta_1/\delta}{\delta_2/\delta}\delta_2 = \frac{2\pi - 4}{4 - \pi}\delta_2 = 1.743\sqrt{\frac{\nu x}{U_\infty}}. \tag{12.149}$$

Comparing with the exact values in (12.49) and (12.47) we see that these results agree very well; the relative error for δ_1 and δ_2 is about 1.3%.

We use the same profile (12.142) to calculate the boundary layer of the two-dimensional stagnation point flow (10.65) along the x-axis, where according to (10.66), $U = ax$. While (12.143) to (12.145) are still valid, from (12.141) we obtain the differential equation linear in δ_2^2

$$\frac{x\,\mathrm{d}\delta_2^2}{2\,\mathrm{d}x} + \frac{4}{4-\pi}\delta_2^2 = \frac{4-\pi}{4}\frac{\nu}{a}. \tag{12.150}$$

The homogeneous solution of this reads

$$\delta_{2H} = C\,x^{-\frac{4}{4-\pi}}. \tag{12.151}$$

Since the boundary layer thickness must remain finite for $x \to 0$, we conclude that the homogeneous solution vanishes ($C = 0$). Therefore the solution of (12.150) only consists of the particular solution

$$\delta_2 = \frac{4-\pi}{4}\sqrt{\frac{\nu}{a}} = 0.215\sqrt{\frac{\nu}{a}}, \tag{12.152}$$

i.e., the momentum thickness and therefore also the boundary layer thickness of the two-dimensional stagnation point flow are constant. Using (12.48) and f' from Fig. 12.5 ($\beta = 90°$) we obtain the exact solution for the momentum thickness

$$\delta_2 = 0.292\sqrt{\frac{\nu}{a}}, \tag{12.153}$$

and comparison shows that the simple velocity profile in (12.142) indeed predicts the constancy of the boundary layer thickness correctly, although it does not lead to good quantitative results.

In flows with a pressure gradient, as in the previous case, fourth order polynomials in y/δ are better since they satisfy the condition (12.63) which has previously not been satisfied. As long as the flow does not separate, this method leads to quite good results, but the separation point is generally not predicted precisely enough by it. Finally, we note that the idea of *integral boundary layer* method was also successfully applied to problems with free surfaces.

12.5   Turbulent Boundary Layers

Restricting ourselves to steady, plane two-dimensional, incompressible flows, the relations for turbulent boundary layers can be obtained from the laminar boundary layer equations by replacing the quantities there with the corresponding mean quantities and adding onto the right-hand side of (12.19) the only important term from the divergence of the Reynolds' stresses, namely $-\varrho\partial\left(\overline{u'v'}\right)/\partial y$. If we further introduce the exchange coefficient A defined in Eq. (7.56), or the eddy viscosity $A/\varrho = \nu_t$, the boundary layer equations read

$$\bar{u}\frac{\partial\bar{u}}{\partial x} + \bar{v}\frac{\partial\bar{u}}{\partial y} = -\frac{1}{\varrho}\frac{\partial\bar{p}}{\partial x} + \frac{\partial}{\partial y}\left[(\nu+\nu_t)\frac{\partial\bar{u}}{\partial y}\right], \quad \text{and} \tag{12.154}$$

$$\frac{\partial\bar{u}}{\partial x} + \frac{\partial\bar{v}}{\partial y} = 0. \tag{12.155}$$

In (12.154) we have ignored the term $\partial\left(\overline{u'^2} - \overline{v'^2}\right)/\partial x$ so that the pressure gradient inside the boundary layer is the same as outside.

These equations are formally the same as the boundary layer equations for laminar flow and are subject to the same boundary conditions. If we use a turbulence model, the numerical field method can also be applied here. If the eddy viscosity according to (7.59) is used, then for example the distribution of the mixing length is required. In the region where the law of the wall is valid (i.e. approximately in the region $y \leq 0.22\,\delta$) the formula (7.60) is often used, but from $y/\delta \approx 0.22$ onwards the ratio l/δ is set constant, approximately equal to $0.22\ \kappa = 0.09$. Within the intermediate layer, (7.60) is no longer valid and must be modified for very small values, for example by multiplying by the factor $[1 - \exp(-y_*/A)]$, where $A \approx 26$.

As well as this there are further modifications of the mixing length formula. These algebraic semi-empirical methods still have the disadvantage that the eddy viscosity vanishes, even for nonzero mixing lengths whenever $\partial\bar{u}/\partial y$ is zero, therefore at places where $\bar{u}$ is at its maximum. The eddy viscosity model (7.59) loses its meaning in such turbulent fields where the mean velocity is homogeneous. In attempting to avoid this problem (and others), higher order methods are used. If we set the typical fluctuation velocity u' not proportional to $(l\,d\bar{u}/dy)$, but to the root of the kinetic energy (per unit mass) of the fluctuating motion

$$k = \frac{1}{2}\left(\overline{u'^2} + \overline{v'^2} + \overline{w'^2}\right), \tag{12.156}$$

we obtain the following expression for the eddy viscosity

$$\nu_t = C\,k^{1/2}L, \tag{12.157}$$

where now L is an integral length scale which essentially represents the mixing length, while C is a dimensionless constant. For the turbulent kinetic energy, a differential equation is then formed which (semi-empirically) accounts for the processes that contribute to the material change of the turbulent energy. Through the solution of the equation, the eddy viscosity at some position now depends on the history of the turbulent kinetic energy of the particle passing this position, and the direct coupling of ν_t to the local field of the mean velocity is avoided. A distribution must still be stated for the length L. Since one differential equation appears in this turbulence model, it is called a one-equation model. If a differential equation is also used for the length L, we are then dealing with models in which two differential equations appear, hence they are called two-equation models. Models which retain the concept of the eddy viscosity cannot be used if $\overline{u'v'}$ vanishes at some position other than $\partial \bar{u}/\partial y$. We can get around this difficulty if instead of Boussinesq's formula, differential equations for the Reynolds' stresses themselves are introduced, sometimes in addition to the equations already mentioned. With increasing number of differential equations in the turbulence model, the number of assumptions required to close the system of equations increases. In addition, the solution of the differential equations demands boundary conditions for the unknown functions, which in certain circumstances may themselves be unknown. However here we shall not discuss the use of turbulence models in the field methods any further.

Apart from the field methods the integral methods mentioned in Sect. 12.4 are also widely used in the description of turbulent boundary layers. As we already know, the velocity distribution in laminar flows can be represented by polynomials in y/δ, something that clearly does not make sense in the turbulent case since the flat profile can only be approximated very badly by polynomials. Instead the power law is more useful in the form

$$\frac{\bar{u}}{U} = \left(\frac{y}{\delta}\right)^{1/n},\tag{12.158}$$

where the exponent $n \approx 7$ but increases slowly with the Reynolds' number. Using this distribution, we calculate the displacement thickness defined in (12.46) and the momentum thickness (12.48) as

$$\delta_1 = \frac{\delta}{n+1}, \quad \text{and}\tag{12.159}$$

$$\delta_2 = \frac{n\delta}{(n+1)(n+2)};\tag{12.160}$$

thus for $n = 7$

$$\delta_1 = \frac{1}{8}\delta \quad \text{and} \quad \delta_2 = \frac{7}{72}\delta.\tag{12.161}$$

From (12.141) we then obtain the differential equation for the boundary layer thickness on the flat plate

$$\frac{\tau_w}{\varrho\, U_\infty^2} = \frac{7}{72}\frac{\mathrm{d}\delta}{\mathrm{d}x},\tag{12.162}$$

which indeed cannot be solved since the wall shear stress is not known, and it is necessary to refer back to empirical data. In the Reynolds' number range in which the *1/7–power law* is valid, the following empirical relation (*Blasius' law*) also holds

$$\frac{\tau_w}{\varrho\, U_\infty^2} = 0.0225 \left(\frac{\nu}{U_\infty \delta}\right)^{1/4}, \tag{12.163}$$

with which the boundary layer thicknesses become

$$\frac{\delta}{x - x_0} = 0.37 Re_x^{-1/5}, \tag{12.164}$$

$$\frac{\delta_1}{x - x_0} = 0.046 Re_x^{-1/5}, \quad \text{and} \tag{12.165}$$

$$\frac{\delta_2}{x - x_0} = 0.036 Re_x^{-1/5}, \tag{12.166}$$

where Re_x is the Reynolds' number formed with the length $x - x_0$

$$Re_x = U_\infty \frac{x - x_0}{\nu} \tag{12.167}$$

and x_0 is the fictitious distance from the leading edge of the plate at which the thickness of the turbulent boundary layer would be zero; this position does not coincide with the leading edge of the plate. First a laminar boundary layer forms from the leading edge of the plate. At a certain displacement thickness δ_1, more precisely at the Reynolds' number formed with this displacement thickness, the boundary layer becomes unstable for the first time (indifference point $x = x_I$, $U_\infty \delta_1/\nu \approx 520$). The fully turbulent boundary layer is established through a *transition region* between the indifference point x_I and the transition point ($x = x_{tr}$). The length of this transition region depends on the disturbances of the incident flow. If we extrapolate the turbulent boundary layer forwards using the boundary layer thickness found at x_{tr}, we find the *fictitious starting point* x_0 of the boundary layer (see Fig. 12.11).

For very large plate lengths L, x_0 can be ignored compared to L. In this case, using (12.141), we find for the drag per unit depth for a plate wetted on one side

$$F_w = \int_0^L \tau_w dx = \varrho\, U_\infty^2 \delta_2(L). \tag{12.168}$$

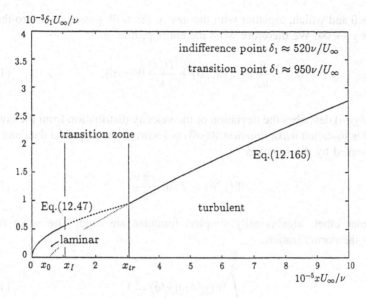

Fig. 12.11 Definition of the fictitious starting point

For the *friction coefficient* c_f, the formula

$$c_f = \frac{F_w}{(\varrho/2)U_\infty^2 L} = 0.072 Re_L^{-1/5} \tag{12.169}$$

follows, where Re_L is the Reynolds' number formed with the plate length L ($Re = U_\infty L/\nu$). The local coefficient of friction c_f' is then defined as

$$c_f' = \frac{\tau_w}{(\varrho/2)U_\infty^2}, \tag{12.170}$$

which, with (12.141) furnishes the expression

$$c_f' = 2\frac{d\delta_2}{dx} = 0.0576 Re_x^{-1/5} \tag{12.171}$$

directly. The formulae stated are restricted to the range where Blasius' law is valid, which, expressed in terms of Re_L lies approximately in the interval

$$5 \cdot 10^5 < Re_L < 10^7. \tag{12.172}$$

In order to make more precise statements, we use the universal law of the wall (7.46), which however is only valid close to the wall. For the whole boundary layer, the law of the wall is to be supplemented by a distribution which is so chosen that it vanishes for $y \to 0$ and which, together with the law of the wall, passes over into the outer flow for $y \to \infty$. We therefore write the entire profile as

$$\frac{\bar{u}}{u_*} = f(yu_*/\nu) + \frac{\Pi(x)}{\kappa}W(y/\delta), \qquad (12.173)$$

where $W(y/\delta)$ describes the deviation of the velocity distribution from the law of the wall. This so-called *wake function* $W(y/\delta)$ is known from empirical data and is well approximated by the function

$$W(y/\delta) = 2\sin^2\left(\frac{\pi}{2}\frac{y}{\delta}\right). \qquad (12.174)$$

Sometimes other, algebraically simpler, formulae are used. The wake function satisfies the normalization

$$\int\limits_0^1 W(y/\delta)\mathrm{d}(y/\delta) = 1 \qquad (12.175)$$

and the boundary conditions

$$W(0) = 0, \qquad W(1) = 2. \qquad (12.176)$$

The change of the distribution $\bar{u}/u_*$ with x is now handled by the *profile parameter* $\Pi(x)$, which is dependent on the pressure gradient. If we only use the logarithmic wall law (7.70), from (12.173) we extract the equation for $y = \delta$

$$\frac{U}{u_*} = \frac{1}{\kappa}\ln(\delta u_*/\nu) + B + 2\frac{\Pi}{\kappa} \qquad (12.177)$$

or

$$\frac{U - \bar{u}}{u_*} = -\frac{1}{\kappa}\ln(y/\delta) + \frac{\Pi}{\kappa}[2 - W(y/\delta)]. \qquad (12.178)$$

The last equation is called the *velocity defect law*. For constant Π this is the same as the defect law (7.79) of pipe flow. Equation (12.177) directly represents a relation between the shear stress at the wall and the profile parameter Π. With the definition of the local friction coefficient and $\tau_w = \rho u_*^2$ we write this equation in the form

$$\sqrt{\frac{2}{c_f'}} = \frac{U}{u_*} = \frac{1}{\kappa}\ln\left(\frac{\delta U}{\nu}\sqrt{\frac{c_f'}{2}}\right) + B + 2\frac{\Pi}{\kappa}. \qquad (12.179)$$

If we ignore the effect of the viscous sublayer in integrating and use the definition of the displacement thickness δ_1 from (12.173) we obtain the relation

$$\frac{\delta_1}{\delta} = (1+\Pi)\frac{u_*}{U\kappa} = \sqrt{\frac{c_f'}{2}}\frac{1+\Pi}{\kappa},$$ (12.180)

and correspondingly for the momentum thickness

$$\frac{\delta_2}{\delta} = \sqrt{\frac{c_f'}{2}}\frac{1+\Pi}{\kappa} - \frac{2+3.18\Pi+1.5\Pi^2}{\kappa^2}\frac{c_f'}{2}.$$ (12.181)

In the last equations, the unknowns c_f', δ, δ_1, δ_2 and Π appear. Including the balance of momentum (12.141) we then have four equations available for the five unknowns, so that a further empirical relation is needed

$$\Pi \approx 0.8(\beta+0.5)^{3/4},$$ (12.182)

in which β is the *equilibrium parameter*

$$\beta = \frac{\delta_1}{\tau_w}\frac{\partial p}{\partial x} = -\frac{\delta_1}{\delta_2}\frac{2}{c_f'}\frac{\delta_2}{U}\frac{dU}{dx}.$$ (12.183)

With this we now have five equations for the five unknowns and for a given velocity profile the turbulent boundary layer can by calculated by numerical methods where the initial values of the quantities to be calculated must be given.

The integral methods of which the above exposition is a simple example are often equivalent to the field methods for turbulent boundary layers (although this is not so in the laminar case). This is probably due to the large amount of empirical data that enters into the calculation. In the application to the turbulent boundary layer on a flat plate ($U \equiv U_\infty$), we set $\Pi \approx 0.55$ (instead of $\Pi = 0.476$ from (12.182)) and rewrite the momentum Eq. (12.141) with $Re_{\delta_2} = U_\infty\delta_2/\nu$ and $Re_x = U_\infty x/\nu$

$$\frac{d\delta_2}{dx} = \frac{dRe_{\delta_2}}{dRe_x} = \frac{c_f'}{2}.$$ (12.184)

We now represent c_f' as a function of the Reynolds' number Re_{δ_2}, where we replace δ in (12.179) by δ_2 using the relation (12.181). We can describe the result of the numerical integration of (12.184) using the formula

$$Re_{\delta_2} = 0.0142Re_x^{6/7}.$$ (12.185)

If we insert this result into (12.184) the local coefficient of friction is found to be

$$c_f' = 0.024 Re_x^{-1/7}. \tag{12.186}$$

This formula is valid in the domain

$$10^5 < Re_x < 10^9.$$

As is clear, the calculation of the coefficient of friction and the boundary layer thicknesses is, even in the case of the flat plate, rather complicated. Therefore we wish to derive simpler formulae for this case, based on dimensional considerations. We assume that the logarithmic law of the wall is valid in the entire boundary layer. Then we must insert $\Pi = 0$ into (12.173) and instead of (12.177) we obtain

$$\frac{U_\infty}{u_*} = \frac{1}{\kappa} \ln(\delta u_*/\nu) + B. \tag{12.187}$$

The boundary layer thickness δ cannot yet be represented as a function of x from this equation, since the shear stress τ_w and therefore u_* depend on x, so that δ must satisfy a relation of the form

$$\delta = \delta(x, u_*, U_\infty). \tag{12.188}$$

For dimensional reasons the relation takes on the form

$$\frac{\delta}{x} = f(u_*/U_\infty). \tag{12.189}$$

The slope of the boundary layer is of the order v'/U_∞, and since v' is of the order u_*, it follows that

$$\frac{d\delta}{dx} \sim \frac{u_*}{U_\infty}. \tag{12.190}$$

If u_* only weakly depends on x then

$$\delta \sim x u_*/U_\infty \tag{12.191}$$

is valid, in accordance with (12.189), and where we assumed that the turbulent boundary layer begins at the position $x = 0$. Therefore the boundary layer grows

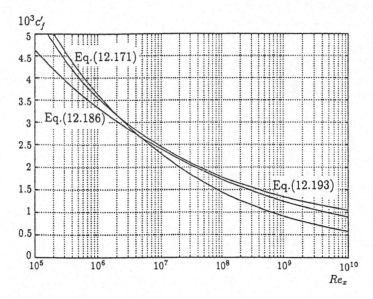

Fig. 12.12 Drag formulae

proportionally to x while the more precise formula (12.185) gives a slightly weaker growth proportional to $x^{6/7}$. We insert the result (12.191) into (12.187) and obtain

$$\frac{U_\infty}{u_*} = \frac{1}{\kappa}\ln\left[(U_\infty x/\nu)(u_*/U_\infty)^2\right] + B, \tag{12.192}$$

from which, with the universal constant $B \approx 5$, we extract the friction law

$$c_f'^{-1/2} = 1.77\ln\left(Re_x c_f'\right) + 2.3. \tag{12.193}$$

The three different friction formulae (12.171), (12.186) and (12.193) are compared in Fig. 12.12.

Chapter 13
Creeping Flows

In this chapter, we investigate steady flows at low Reynolds' number, restricting ourselves to incompressible flows, which of course include gas flows at low Mach number. The equations of motion are already known (see (4.35)), and may be expressed in the form

$$\nabla p = \eta \Delta \vec{u}. \tag{13.1}$$

To this we add the continuity equation (2.5)

$$\nabla \cdot \vec{u} = 0. \tag{13.2}$$

The pressure dependence may be eliminated by taking the curl of Eq. (13.1), that is to say operating on (13.1) by $\nabla \times$. Then if we use the vector identity (4.10) in which $\vec{u}$ is replaced by $2\vec{\omega}$, we obtain, using (4.14), the relation

$$\Delta(\nabla \times \vec{u}) = 2\Delta \vec{\omega} = -2\nabla \times (\nabla \times \vec{\omega}) = 0. \tag{13.3}$$

Taking the divergence of Eq. (13.1) reduces it to the Laplace equation for the pressure

$$\Delta p = 0. \tag{13.4}$$

13.1 Plane and Axially-Symmetric Flows

It is well known that stream functions may be defined for plane and axially-symmetric flows; the continuity equation (13.2) is then eliminated because it is identically satisfied by the stream function. In the case of plane flows, taking

© Springer Nature Switzerland AG 2020
J. H. Spurk and N. Aksel, *Fluid Mechanics*,
https://doi.org/10.1007/978-3-030-30259-7_13

the velocity components from (10.217) shows that the only non-vanishing component of curl $\vec{u}$, using Appendix B.1, is

$$2\vec{\omega} = \operatorname{curl}\vec{u} = \left(\frac{\partial v}{\partial x} - \frac{\partial u}{\partial y}\right)\vec{e}_z = -\left(\frac{\partial^2\Psi}{\partial x^2} + \frac{\partial^2\Psi}{\partial y^2}\right)\vec{e}_z = -\Delta\Psi\,\vec{e}_z. \tag{13.5}$$

Using this, we obtain the stream-function equation

$$\Delta(\Delta\psi) = 0, \tag{13.6}$$

or alternatively

$$\nabla^4\,\Psi = 0. \tag{13.7}$$

This equation is known as the *biharmonic (bipotential) equation*, the name being indicated by the form of (13.6).

To obtain the corresponding equations for the axially-symmetric stream function, we start with (13.3) in the form

$$\nabla \times (\nabla \times \vec{\omega}) = 0; \tag{13.8}$$

the equation is then derived by repeated application of the operation $\nabla\times$, given in Appendix B.3.

In Sect. 13.1.3 below, we seek to calculate a creeping flow past a sphere. In order to satisfy the no-slip condition on the surface of the sphere, we express the equation in spherical polar coordinates, for which the surface of the sphere is then a coordinate surface. Next it follows from Appendix B.3 that

$$2\vec{\omega} = \operatorname{curl}\vec{u} = \frac{1}{r}\left\{\frac{\partial(ru_\vartheta)}{\partial r} - \frac{\partial u_r}{\partial\vartheta}\right\}\vec{e}_\varphi \tag{13.9}$$

and using (10.104) and (10.105), that

$$\vec{\omega} = -\frac{1}{2r\sin\vartheta}\left\{\frac{\partial^2}{\partial r^2} + \frac{\sin\vartheta}{r^2}\frac{\partial}{\partial\vartheta}\left(\frac{1}{\sin\vartheta}\frac{\partial}{\partial\vartheta}\right)\right\}\Psi\vec{e}_\varphi = -\frac{1}{2r\,\sin\vartheta}E^2\Psi\vec{e}_\varphi,$$

$$\tag{13.10}$$

where the expression behind the last equality sign defines the operator E^2. We note that the unit vector $\vec{e}_\varphi$ in Eq. (13.9) is not constant with consequences for the differential operator. We now calculate $\nabla \times \vec{\omega}$ by replacing $\vec{u}$ by $\vec{\omega}$ in the expression for curl $\vec{u}$ given in Appendix B.3. Observing that only $w_\varphi \neq 0$ and that the derivative with respect to φ of every component vanishes yields the equation

$$\nabla \times \vec{\omega} = -\frac{1}{2r^2 \sin \vartheta} \frac{\partial}{\partial \vartheta} \left(E^2 \Psi \right) \vec{e}_r + \frac{1}{2r \sin \vartheta} \frac{\partial}{\partial r} \left(E^2 \Psi \right) \vec{e}_\vartheta. \tag{13.11}$$

Proceeding with $\nabla \times \vec{\omega}$ in a similar way to that used above for $\vec{\omega}$ gives

$$\nabla \times (\nabla \times \vec{\omega}) = \frac{1}{2r \sin \vartheta} \left\{ \frac{\partial^2}{\partial r^2} \left(E^2 \Psi \right) + \frac{\sin \vartheta}{r^2} \frac{\partial}{\partial \vartheta} \left(\frac{1}{\sin \vartheta} \frac{\partial}{\partial \vartheta} \left(E^2 \Psi \right) \right) \right\} \vec{e}_\varphi = 0,$$

$$\tag{13.12}$$

and

$$E^2 \left(E^2 \Psi \right) = 0. \tag{13.13}$$

For the sake of completeness we state also the operator E^2 in cylindrical coordinates (Appendix B.2)

$$E^2 = \frac{\partial^2}{\partial z^2} + \frac{\partial^2}{\partial r^2} - \frac{1}{r} \frac{\partial}{\partial r}. \tag{13.14}$$

For axially-symmetric flows in the plane $z = 0$, we have

$$E^2 = \frac{\partial^2}{\partial x^2} + \frac{\partial^2}{\partial y^2} - \frac{1}{y} \frac{\partial}{\partial y}, \tag{13.15}$$

where x, y are Cartesian coordinates as in Sect. 10.3.1.

13.1.1 Examples of Plane Flows

We consider here creeping corner flows and include the corresponding potential flows in and around corners already discussed in Chap. 10. As in the above cases, we therefore consider those flows whose boundaries are given by the lines $\varphi =$ constant and for which polar coordinates r, φ can be used.

Each flow is described, as has been seen above, in terms of its stream function $\Psi(r, \varphi)$, in which dimensional considerations require a dimensional constant to be present. We start with

$$\Psi = A r^n f(\varphi), \tag{13.16}$$

for which the biharmonic equation is separable. The constant A has the dimension $(length^{2-n}/time)$. Just as in the corner flows of Chap. 10, these flows are only valid in the locality of the corner. The constant A depends on the "driving force" outside the region of validity of the local solution; it can only be determined when the flow in the entire region is known. We now restrict ourselves to those cases for which the constant n is a whole-number. This is for example the case when the driving force is produced by a boundary which moves with a given velocity. The interpretation of the constant A is obvious when one or both boundaries of the flow are in motion.

We consider now the flow generated by a plane, inclined at an angle φ_0 to the x-axis, moving at velocity U parallel to the x-axis and scraping fluid off a stationary wall coincident with the x-axis. In a moving frame in which the scraper is at rest the flow is stationary, with the lower wall moving under the scraper with velocity: $-U$. When the polar form of the Laplace operator is twice repeated on the expression

$$\Psi = -U\,r f(\varphi), \tag{13.17}$$

we obtain the ordinary differential equation

$$\frac{U}{r^3}\left(f + 2(f'' + f'''')\right) = 0, \tag{13.18}$$

whose general solution is

$$f = (C_1 + \varphi\, C_2)\cos(\varphi) + (C_3 + \varphi\, C_4)\sin(\varphi). \tag{13.19}$$

Now the polar form of Eq. (13.2), that is to say

$$\frac{\partial(u_r\, r)}{\partial r} + \frac{\partial u_\varphi}{\partial \varphi} = 0 \tag{13.20}$$

yields the necessary and sufficient condition for the total differential

$$d\Psi = -u_\varphi dr + u_r\, r d\varphi, \tag{13.21}$$

from which follow the velocity components

$$u_r = \frac{1}{r}\frac{\partial \Psi}{\partial \varphi} = -U f'(\varphi), \quad u_\varphi = -\frac{\partial \Psi}{\partial r} = U f(\varphi). \tag{13.22}$$

The no-slip conditions on the wall and scraper lead to the boundary conditions

$$f(0) = 0, \quad f'(0) = 1 \quad \text{and} \quad f(\varphi_0) = 0, \quad f'(\varphi_0) = 0. \tag{13.23}$$

The particular solution which satisfies these boundary conditions is found from (13.19) to be

$$f(\varphi) = \frac{2\varphi \sin \varphi_0 \, \sin(\varphi_0 - \varphi) + 2\varphi_0(\varphi - \varphi_0) \sin \varphi}{2\varphi_0^2 - 1 + \cos(2\varphi_0)}. \tag{13.24}$$

When $\varphi_0 = \pi/2$ the stream function becomes

$$\Psi = U \, r \frac{4\varphi \cos \varphi - \pi^2 \sin \varphi + 2\pi\varphi \sin \varphi}{\pi^2 - 4}; \tag{13.25}$$

the streamlines of the flow are shown in Fig. 13.1. The shear stress on the wall may be evaluated by using Appendix B.2; it is

$$\tau_{r\varphi}(0) = \eta \frac{1}{r} \frac{\partial u_r(0)}{\partial \varphi} = \eta \frac{1}{r^2} \frac{\partial^2 \Psi(0)}{\partial \varphi^2} = \eta \frac{U}{r} \frac{4\pi}{\pi^2 - 4}. \tag{13.26}$$

This shows that the force necessary to move the scraper with velocity U, which is obtained by integrating the shear force, is logarithmically infinite.

Of course this result arises from the infinitesimally small gap between scraper and wall; in reality this gap must naturally be finite; however, it is clear that the force driving the scraper increases with decreasing gap size.

A flow closely related to the above arises in the case of a heavy fluid with a free surface that is bounded by a plane which is inclined at an angle $-\varphi_0$ to the horizontal and moves with speed U as in Fig. 13.2. The general solution (13.19) for $f(\varphi)$, then holds. The continuity of stress on the free surface requires that the shear stress vanish there, that is $f''(0) = 0$. Since the free surface is a streamline it follows that $f(0) = 0$; also, because the fluid adheres to the moving wall, $f(-\varphi_0) = 0, f'(-\varphi_0) = 1$. Under these conditions the particular solution takes the form

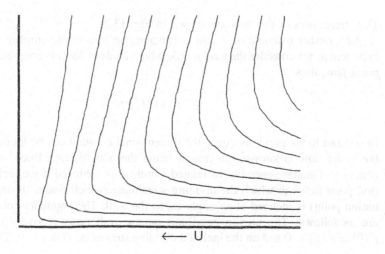

Fig. 13.1 Streamlines in the neighbourhood of the point of intersection of the scraper and wall, $\varphi_0 = \pi/2$

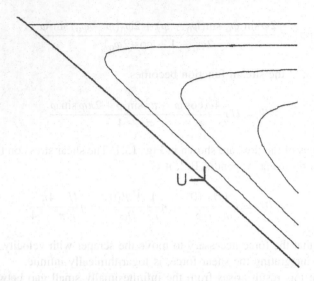

Fig. 13.2 Streamlines in the neighbourhood of the point of intersection of a free surface and wall
$\varphi_0 = \pi/4$

$$f(\varphi) = \frac{2(\varphi_0 \cos \varphi_0 \sin \varphi - \varphi \cos \varphi \sin \varphi_0)}{2\varphi_0 - \sin(2\varphi_0)}.$$ (13.27)

When $\varphi_0 = \pi/4$, the stream function is

$$\Psi = \frac{U r(\pi \sin \varphi - 4\varphi \cos \varphi)}{2^{1/2}(\pi - 2)}.$$ (13.28)

The streamlines of the flow are shown in Fig. 13.2.

As a further example of a flow corresponding to a whole-number value of the exponent n, we consider the case $n = 3$, which leads to *Stokes' creeping stagnation point flow*, thus

$$\Psi = A r^3 f(\varphi).$$ (13.29)

In contrast to the previous flows, the dimensional constant can be found only when the entire flow is known, the reason being that the "driving force" acts at large distances. Greater generality of related solutions is achieved if we include stagnation point flows in which the dividing streamline (which passes through the stagnation point) is inclined at an angle φ_0 to the wall. The boundary conditions then are as follows: The no-slip condition on the wall (x-axis) give $f(0) = f(\pi) = f'(0) = f'(\pi) = 0$ and on the inclined dividing streamline $f(\varphi_0) = 0$. The boundary conditions at $\varphi = \pi$ being obviously redundant, only three conditions stand. When these are satisfied, we find from (13.19) that the stream function is

$$\Psi = \frac{-A\,r^3 \sin(\varphi - \varphi_0) \sin^2 \varphi}{\sin \varphi_0}, \qquad (13.30)$$

in which the undetermined constant has been absorbed into A. Figures 13.3 and 13.4 show the streamlines for $\varphi_0 = \pi/2$, and $\varphi_0 = \pi/4$, respectively, in both cases A is positive. To compare the Stokes' creeping stagnation point flow with that of the corresponding incompressible potential flow (given in Sect. 10.3.1), and with that

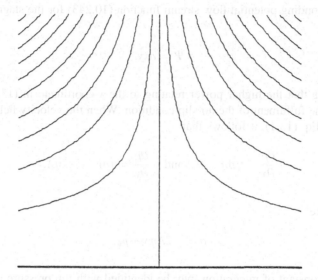

Fig. 13.3 Creeping flow streamlines in the neighbourhood of a stagnation point, $\varphi_0 = \pi/2$

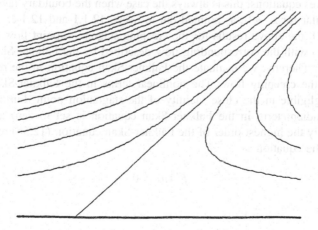

Fig. 13.4 Creeping flow streamlines in the neighbourhood of a stagnation point, $\varphi_0 = \pi/4$

of the boundary layer, which is an exact solution of the Navier-Stokes equations, we introduce Cartesian coordinates with $x = r\cos\varphi$, $y = r\sin\varphi$. By restricting ourselves to the case $\varphi_0 = \pi/2$, we obtain from (13.30)

$$\Psi = Axy^2, \tag{13.31}$$

with Cartesian velocity components

$$u = \frac{\partial\Psi}{\partial y} = 2Axy \quad\text{and}\quad v = -\frac{\partial\Psi}{\partial x} = -Ay^2. \tag{13.32}$$

The corresponding potential flow stream function (10.243) for the stagnation point flow is

$$\Psi = axy.$$

It is obvious that the higher power relating to the y-coordinate in (13.31) is necessary for the fulfilment of the no-slip condition. When the velocity field (13.32) is inserted in Eq. (13.1), it follows that

$$\frac{\partial p}{\partial x} = \eta\Delta u = 0 \quad\text{and}\quad \frac{\partial p}{\partial y} = \eta\Delta v = -2\eta A, \tag{13.33}$$

and therefore

$$p = -2A\eta\,y + p_w, \tag{13.34}$$

where the constant of integration may be identified with the pressure on the wall.

It was noted that the stagnation point boundary layer is an exact solution of the Navier-Stokes equations; this is always the case when the boundary layer equations remain regular for $x \to 0$, as pointed out in Sects. 12.1.1 and 12.1.2.

We match the above results for the creeping stagnation point flow (inner solution) (13.31) with the corresponding flow described by the Falkner-Skan equation with $m = 1$. Thereby, we evaluate the Falkner-Skan equation in the region of the validity of the creeping flow. The nonlinear terms in the Falkner-Skan equation become negligible in the close vicinity of the stagnation point. Furthermore, the pressure gradient term in the Falkner-Skan equation is set to zero according to (13.33). Only the highest order of the Falkner-Skan equation (12.56) remains, thus we obtain the equation

$$f'''(\eta) = 0, \tag{13.35}$$

in which η is now the similarity variable defined by (12.55). Integration of (13.35) subject to the boundary conditions (12.40) leads to

$$f = c\eta^2 = c\frac{a}{\nu}y^2, \tag{13.36}$$

in which $c = f''(0)/2$ is a dimensionless constant whose value is known only when the solution of the complete Falkner-Skan equation has been calculated. Using (12.54) the stream function of (13.36) then becomes

$$\Psi = ca\sqrt{\frac{a}{\nu}}xy^2. \tag{13.37}$$

The combination of constants (13.37) corresponds and is dimensionally equivalent to the constant A in (13.31). The component of pressure gradient $\partial p/\partial y$ does not vanish in the Navier-Stokes equation; when the solution of (12.56) has been found it may be determined by using the y-component of the Navier-Stokes equation; with the non-linear terms neglected, Eq. (13.1) is again appropriate for the calculation of pressure and leads of course to the result (13.34).

The dimensional constant a of a local potential flow is fixed when the potential flow around the body is known. For example, the potential flow around a circular cylinder as given by (10.257) without circulation ($\Gamma = 0$), thus

$$u_\varphi = -2U_\infty \sin \varphi = -2U_\infty \sin\left(\pi - \frac{x}{r_0}\right). \tag{13.38}$$

Here x is the circumferential distance along the cylinder measured clockwise from the front stagnation point. The velocity $-u_\varphi$ then corresponds to the velocity u in the positive x-direction in the body coordinates, thus

$$u = 2U_\infty \sin\frac{x}{r_0} = 2U_\infty \frac{x}{r_0} + O\left((x/r_0)^2\right). \tag{13.39}$$

For small x/r_0, the coefficient a is determined by comparison with the x-component of the plane potential stagnation point flow (10.66), thus

$$a = 2U_\infty/r_0. \tag{13.40}$$

Of course the theoretical value of the velocity is not attained experimentally. At higher Reynolds' numbers the flow separates and goes through a series of different forms which have been described already in Sect. 10.3. Separation especially, which is responsible for the form drag, influences the entire flow around the cylinder so that in the region around the forward stagnation point Reynolds'-number-dependent departures occur; measurements of these indicate a deviation of roughly 10% at a Reynolds' number of about 20,000. There is, however, better agreement between the theoretical predictions and the results of experiments when the body is streamlined and separation therefore prevented; a small shape drag nevertheless arises from the displacement thickness of the boundary layer. The flow "sees" a body which is enlarged by the displacement thickness, and this gives rise to an additional pressure term which may be determined using potential theory. The altered pressure component gives rise to another force which no longer satisfies the D'Alembert paradox and leads therefore to a shape drag; this is considerably smaller than the frictional drag, from which it is difficult to distinguish, since both resistances are proportional to $Re^{-1/2}$ in the laminar flow, see (12.47) and (12.51) respectively.

We now consider the flow in the neighbourhood of the intersection of fixed walls; for example, around wedges or flows in corners. The no-slip condition then holds on both walls, and there are therefore four homogeneous boundary conditions to be satisfied. In general the exponent n in (13.16) is no longer a whole-number. Substituting the relation (13.16) into the biharmonic equation gives

$$\Psi = A\, r^{n-4} \left(n^2(n-2)^2 f(\varphi) + 2(n(n-2)+2)f''(\varphi) + f''''(\varphi) \right). \tag{13.41}$$

The general solution for $f(\varphi)$ is then

$$f(\varphi) = B_1 e^{i(n-2)\varphi} + B_2 e^{-i(n-2)\varphi} + C_3 \cos\, n\varphi + C_4 \sin\, n\varphi, \tag{13.42}$$

or

$$f(\varphi) = C_1 \cos(n-2)\varphi + C_2 \sin(n-2)\varphi + C_3 \cos\, n\varphi + C_4 \sin\, n\varphi, \tag{13.43}$$

where the constants are complex. When the four boundary conditions are imposed on this solution, a system of four homogeneous equations with the unknown coefficients C_i is obtained. A unique but trivial solution $C_i = 0$ is obtained when the determinant D of the matrix of coefficients is non-zero. Non-trivial solutions arise when D vanishes. Because of this additional condition only three of the four equations are independent; thus only the ratios of the coefficients are determined. The equation $D = 0$ is a transcendental equation for n which has more than one solution. On physical grounds we are interested only in those solutions for which the velocity at the point of intersection vanishes; this occurs when $n > 1$ or $\Re(n) > 1$, when n is complex. It is to be expected that the solution which corresponds to the smallest real part is dominant in the corner. In determining the roots of D an iterative process such as Newton's method is recommended; for this process a good starting value, if necessary complex, must be used. If simultaneously the streamlines of a known flow are plotted, then the correct exponent can be found after some trials.

Since the biharmonic equation is linear, it is convenient to discuss the symmetric and antisymmetric parts of Eq. (13.41) separately. The general solution can then be formed by superposition. The symmetric part of the equation, namely

$$f(\varphi) = C_1 \cos(n - 2)\varphi + C_3 \cos n\varphi \qquad (13.44)$$

leads to an antisymmetric velocity field. In this case applying the boundary condition on the walls ($\varphi = \pm\varphi_0$) yields $f(\pm\varphi_0) = 0$, $f'(\pm\varphi_0) = 0$, and thus

$$\begin{aligned}
C_1 \cos(n - 2)\varphi_0 + C_3 \cos n\varphi_0 &= 0, \\
C_1(n - 2) \sin(n - 2)\varphi_0 + C_3 n \sin n\varphi_0 &= 0,
\end{aligned} \qquad (13.45)$$

where n satisfies the equation

$$D = -(\sin 2\varphi_0 + (n - 1) \sin 2(n - 1)\varphi_0) = 0. \qquad (13.46)$$

From (13.45) it follows that

$$C_3 = -C_1 \cos(n - 2)\varphi_0 / \cos n\varphi_0 \qquad (13.47)$$

and hence

$$\Psi = Ar^n(\cos(n - 2)\varphi - \cos(n - 2)\varphi_0 \cos n\varphi/(\cos n\varphi_0)), \qquad (13.48)$$

where C_1 has been absorbed into A.

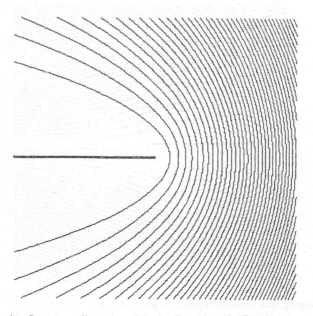

Fig. 13.5 Creeping flow streamlines around the leading edge of a flat plate, $\varphi_0 = \pi$

Equation (13.46) indicates that there are infinitely many solutions in the case $\varphi_0 = \pi$, the smallest non-trivial value being, $n = 1.5$. The streamlines corresponding to this value are shown in Fig. 13.5; this flow is in fact that around the leading edge of an infinitely thin flat plate.

The antisymmetric part of the general solution (13.43) may be processed in the same way. When $\varphi_0 = \pi$ the exponent is again $n = 3/2$. The streamlines in Fig. 13.6 correspond to a symmetric flow. The velocity at the point $r = 0$ is zero, which is true for all these flows; the shear stress at this point is finite.

Flows around a wedge are found when $\pi/2 < \varphi_0 < \pi$. The symmetric flow is the creeping flow around a wedge; this is related to the wedge flows at large Reynolds' numbers which have already been discussed in Sect. 12.1.2. On the other hand, the antisymmetric flow occurs in the flow around a sharp leading edge.

It is surprising that there are no more real solutions in the case of a sharp corner for which $\varphi_0 < \approx 73°$. The streamlines for this limiting value are shown in Fig. 13.7. Apart from the trivial solution $n = 1$, there is only one solution, which is found to be $n \approx 2.76$. One can think of the flow in the sharp corner being produced by a rotating cylinder which is far from the intersection point of both walls and drives the flow into the corner. The fluid velocity falls as the corner is approached, but is only zero on the walls. Obviously, it is more difficult to push the flow into the corner as the angle becomes smaller.

The volume flux simply cannot vanish in the flows, and streamlines must either end on the wall bounding the flow or must form closed curves. Each closed streamline encloses a region of circulating flow, similar to that of a rotating cylinder, which tries to drive the underlying flow into the corner. There is in fact an

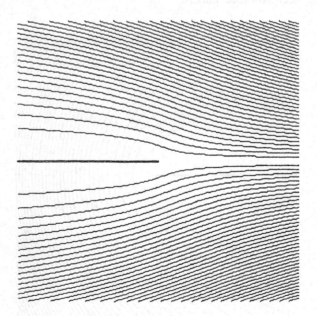

Fig. 13.6 Creeping flow streamlines around a plate, $\varphi_0 = n$

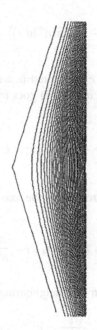

Fig. 13.7 Creeping flow streamlines in a sharp corner, $\varphi_0 = 73°$

infinite series of circulating flow cells (*eddies*), and by reason of the continuity of the velocity field these must rotate in alternating senses. They are separated from one another by null streamlines which end on the walls; moving through a null streamline changes the sign of the stream function. Figure 10.7 has already suggested the series of alternating vortices for the flow in a right angled corner.

The zeros of the stream function on the line of symmetry for one of the complex-exponent solutions of (13.48), where $n = n' + in''$, may be found from the relation

$$\Psi = Ar^{n' + in''}(a'(n, \varphi_0) + ia''(n, \varphi_0)) = 0. \tag{13.49}$$

The meaning of the complex number $a' + ia''$ follows by comparing (13.48) with (13.49). Using the relation

$$r^{n'} r^{in''} = r^{n'} e^{\ln r\,(in'')} = r^{n'}[\cos(n'' \ln r) + i \sin(n'' \ln r)], \tag{13.50}$$

we obtain

$$\Psi = A(a' + ia'')r^{n'}(\cos(n'' \ln r) + i \sin(n'' \ln r)) = 0. \tag{13.51}$$

The real part of Ψ which has physical significance, is

$$\Re[\Psi] = A\,r^{n'}[a'\cos(n''\ln r) - a''\sin(n''\ln r)] = C\,r^{n'}\cos(n''\ln r + \delta) = 0,$$
$$(13.52)$$

where δ is a phase angle. When r (measured in arbitrary units) runs from $r = 1$ to $r = 0$, the argument of the cosine in (13.52) goes from δ to minus infinity, and thus the stream function has zeros at

$$n''\ln r + \delta = -\left(\frac{1}{2} + k\right)\pi; \quad k = 0, 1, 2, \ldots \qquad (13.53)$$

Let r_k denote the distance measured along the axis of symmetry from the origin to the kth zero, then, using (13.53)

$$\ln r_k - \ln r_{k+1} = \frac{\pi}{n''} \qquad (13.54)$$

and so the distance ratio between two neighboring zeros is

$$\frac{r_k}{r_{k+1}} = e^{\pi/n''}. \qquad (13.55)$$

The difference $r_k - r_{k+1}$ may be used as a measure for the cell size. The stream function has extrema at the points

$$n''\ln r + \delta = -\pi l; \quad l = 0, 1, 2, \ldots \qquad (13.56)$$

and the ratio of neighboring r-coordinates is

$$\frac{r_l}{r_{l+1}} = e^{\pi/n''}. \qquad (13.57)$$

The value of the extremum of the stream function in a cell serves as a measure of the "strength" of the cell. The ratio of two neighboring cell (eddy) strengths is therefore

$$\left|\frac{\Re[\Psi_l]}{\Re[\Psi_{l+1}]}\right| = \frac{r_l^{n'}}{r_{l+1}^{n'}} = e^{\pi n'/n''}. \qquad (13.58)$$

As is evident from (13.46), the exponent n depends on the angle φ_0. The imaginary part n'' is zero for the critical value $\varphi_0 \approx 73°$. One can interpret this as the distance between cells becoming infinite.

When $\varphi_0 < 73°$, the imaginary part increases and the distance between adjacent cells becomes smaller as does the ratio of their strengths. Let the angle φ_0 tend to zero so that a plane channel comes into being; a channel in which a point other than the apex is fixed on both walls and the angle φ_0 is then allowed to tend to zero. Then both cell sizes and strengths tend to constant values.

We now consider the *corner flow* sketched in Fig. 10.7 as a concrete example of cell sizes and relative distances between cells. For the right-angled corner $\varphi_0 = \pi/4$ the solution of (13.46) is found to be $n = 3.7396 + 1.1191i$. The streamlines of this flow are displayed in Fig. 13.8 (see also Fig. 10.7).

In Fig. 13.8, two complete cells (eddies) and the boundary of a third cell may be clearly identified. The ratio of the distances between successive zeros (or the ratio of the distances between two extrema) is given by $\exp(\pi/n'') = 16.56$ which one can also gather from Fig. 13.8. The streamlines are shown for the following values of the stream function: $\left(0, 10^{-10}, 10^{-9}, -10^{-6}, 8 \times 10^{-6}, -3.6 \times 10^{-5}, 10^{-5}, 10^{-4}\right)$, the units being arbitrary. The ratio of cell strengths is $\exp(\pi n'/n'') \approx 3.6 \times 10^5$; these correspond roughly to the respective values of the stream function at the centre of the cells.

We refrain from discussing further streamlines corresponding to small angles; these can be readily solved using the results considered above. We confine ourselves to the observation that cell (eddy) formation is nature's quickest way of

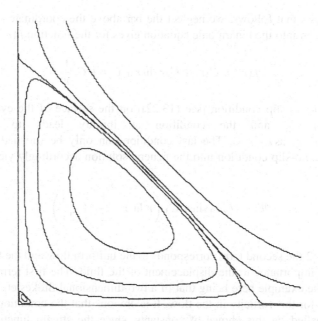

Fig. 13.8 Creeping corner flow streamlines in a rectangular corner

reducing the fluid flow in a sharp corner. If quantitative estimates of such flows are lacking because the constant A is unknown, we can for most practical purposes assume that the flow in such corners is essentially stagnant.

13.1.2   Plane Creeping Flow Around a Body (Stokes' Paradox)

The flow around a body plays a very significant role in fluid dynamics. By plane flow we mean here the flow about cylindrical bodies at very small Reynolds' numbers. We confine ourselves to a flow around a circular cylinder, where clearly the fundamental problem of creeping flow is evident.

The stream function for a corresponding potential flow around a cylinder is made up of two parts, namely (1) the uniform flow, i.e., $U_\infty r \sin \varphi$, and (2) the dipole flow $-U_\infty r_0^2/r \sin \varphi$; part (2) is responsible for the displacement of fluid by the cylinder. We expect these two parts to be present in the creeping flow as well as the potential flow. We introduce dimensionless coordinates with $\bar{r} = r/r_0$, and write the general form of the solution as

$$\Psi = U_\infty r_0 f(\bar{r}) \sin \varphi. \tag{13.59}$$

In the analysis that follows, we neglect the bar above the coordinate r.

Substitution into the biharmonic equation gives for the function $f(r)$ the solution

$$f(r) = C_4 r^3 + C_3 r \ln r + C_2 r + C_1 \frac{1}{r}. \tag{13.60}$$

Applying the no-slip condition (see (13.22)) on the surface of the cylinder gives $f(1) = f'(1) = 0$, and the condition at infinity leads to the result $\Psi \propto U_\infty r_0 \sin \varphi$ as $r \to \infty$. The last condition can only be satisfied if $C_4 = 0$. Inserting the no-slip condition into the general solution accordingly yields the new form

$$\Psi = U_\infty r_0 \sin \varphi \, C_3 \left(r \ln r - \frac{1}{2} r + \frac{1}{2} \frac{1}{r} \right), \tag{13.61}$$

when $C_3 = -2$ the second term corresponds to the uniform flow and the third term to a dipole of importance to the displacement of the fluid. The first term is called a Stokeslet, the example here being that of a two-dimensional Stokeslet. This term is responsible for the vorticity in the flow. It is obvious that the conditions at infinity are not satisfied by this choice of constants, since the stream function diverges logarithmically there. In fact, the condition at infinity cannot be satisfied for any choice of constants; thus no creeping-flow solution past a cylinder exists which satisfies the condition at infinity. This fact is called *Stokes' Paradox*.

The divergence of the solution originates from the circumstance that the disturbance due to the Stokeslet does not die away. In the case of the flow around a sphere, there exists a solution which is made possible because the three-dimensional disturbances involved die away faster. We now proceed to discuss this latter case.

13.1.3 Creeping Flow Around a Sphere

It is obviously helpful to express the boundary conditions of a sphere in terms of spherical polar coordinates, since then $r = r_0$ constitutes a coordinate surface. Previously, we introduced the dimensionless coordinate $\bar{r} = r/r_0$; in the following analysis we will omit the bar.

In spherical polar coordinates, the stream function takes the form

$$\Psi = U_\infty \frac{r_0^2}{2} \sin^2 \vartheta f(r), \tag{13.62}$$

which, when substituted in (13.13), yields the equation for $f(r)$

$$r^4 f''''(r) - 4r^2 f''(r) + 8r f'(r) - 8f(r) = 0. \tag{13.63}$$

Substituting $f(r) = r^m$ in this equation and determining m, we get the general solution in the form

$$f(r) = \frac{C_1}{r} + C_2 r + C_3 r^2 + C_4 r^4. \tag{13.64}$$

The stream function at infinity entails the conditions $C_4 = 0$ and $C_3 = 1$. The boundary conditions on the surface of the sphere, i.e., $f(1) = 0$ and $f'(1) = 0$, determine the constants $C_1 = 1/2$ and $C_2 = -3/2$ the solution is thus

$$\Psi = U_\infty \frac{r_0^2}{2} \sin^2 \vartheta \left(\frac{1}{2} \frac{1}{r} - \frac{3}{2} r + r^2 \right). \tag{13.65}$$

The streamlines are shown in Fig. 13.9. It may be of interest to compare these with the corresponding streamlines of a potential flow around a sphere. We can obtain the relevant stream function from the general solution (13.64) by eliminating the Stokeslet, i.e., setting $C_2 = 0$ and setting the coefficients of the dipole singularity and the uniform flow to $C_1 = -1$, $C_3 = 1$ respectively. Alternatively, one can obtain the velocity components from the potential function (10.149) of the flow past a sphere and with (10.105) obtain a differential equation which together with (10.104) leads to the stream function

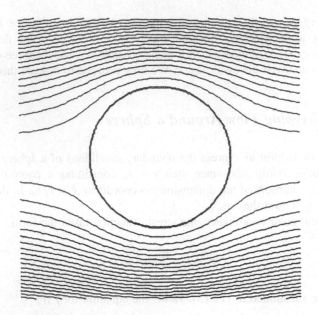

Fig. 13.9 Creeping flow streamlines past a sphere

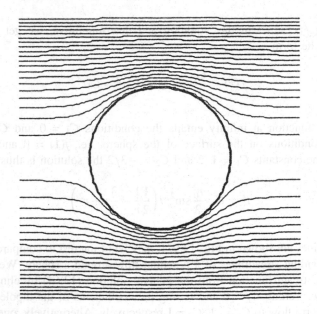

Fig. 13.10 Potential flow streamlines past a sphere

$$\Psi = U_\infty \frac{r_0^2}{2}\sin^2 \vartheta\left(r^2 - \frac{1}{r}\right). \tag{13.66}$$

The streamlines of this flow are shown in Fig. 13.10. As we have already remarked (Sect. 10.3.1), the streamlines in both cases are symmetric but differ in detail. From the list of examples of creeping flows in Sect. 13.1.1 it becomes evident that the drag on a sphere in creeping flows has a special significance. (It is well known that with the help of the resistance formula the charge of an electron was determined with great accuracy.) To determine the force, we use the stress vector (2.29a) in the form

$$t_i = \left(-p\delta_{ij} + 2\eta\, e_{ij}\right)n_j = \left(-p\delta_{ij} + \eta\left(\frac{\partial u_i}{\partial x_j} + \frac{\partial u_j}{\partial x_i}\right)\right)n_j. \tag{13.67}$$

One can now find the components of the rate-of-deformation tensor and those of the normal to the surface (Appendix B.3), which will enable the stress vector on the surface of the sphere to be integrated. We prefer, however, to proceed with a familiar result which we now explain. Consider a surface element to which fluid adheres, an element which coincides with the coordinate surface of an orthogonal coordinate system (not necessarily Cartesian, but an orthogonal system to which (13.67) is applicable). The normal to the surface then has only one component, for example $n_3 = 1$, and accordingly the terms $(\partial u_j/\partial x_i)n_j$ reduce to $\partial u_3/\partial x_i$. Because of the no-slip condition u_j is zero over the whole element, as is the variation of the velocity components with x_1 and x_2.

The only remaining term is $\partial u_3/\partial x_3$ which vanishes as a consequence of the continuity equation. We can therefore replace (13.67) by

$$t_i = \left(-p\delta_{ij} + \eta\left(\frac{\partial u_i}{\partial x_j} - \frac{\partial u_j}{\partial x_i}\right)\right)n_j = -pn_i - \eta 2\omega_k \varepsilon_{ijk} n_j \tag{13.68}$$

or

$$\vec{t} = -p\vec{n} + \eta\, 2\vec{\omega} \times \vec{n}. \tag{13.69}$$

Using the stream function (13.65) we now obtain form (13.10)

$$2\vec{\omega} = -\frac{3U_\infty}{2}\frac{\sin\vartheta}{r_0 r^2}\vec{e}_\varphi, \tag{13.70}$$

in which r is still dimensionless. Still Eq. (13.1) may now be used to calculate the pressure. We first calculate $\nabla \times \vec{\omega}$ using Eq. (13.11), and then substitute into (4.11), getting from (13.1)

$$\nabla p = \frac{3\eta \, U_\infty \cos \vartheta}{r_0 \, r^3} \vec{e}_r + \frac{3\eta \, U_\infty \sin \vartheta}{2r_0 \, r^3} \vec{e}_\vartheta \tag{13.71}$$

which on integration gives

$$p = -\frac{3\eta \, U_\infty \cos \vartheta}{2r_0 \, r^2} + p_0, \tag{13.72}$$

where $p(r \to \infty) = p_0$.

The stress vector on the surface of the sphere $(\vec{n} = \vec{e}_r, \ r = 1)$ is thus

$$\vec{t} = \left(\eta \frac{3U_\infty \cos \vartheta}{2r_0} - p_0 \right) \vec{e}_r - \eta \frac{3U_\infty \sin \vartheta}{2r_0} \vec{e}_\varphi \times \vec{e}_r. \tag{13.73}$$

The component of the stress vector in the direction of flow, $\vec{U}_\infty = U_\infty \vec{e}_x$, is

$$\vec{t} \cdot \vec{e}_x = \eta \left(\frac{3U_\infty \cos^2 \vartheta}{2r_0} + \frac{3U_\infty \sin^2 \vartheta}{2r_0} \right) - p_0 \cos \vartheta, \tag{13.74}$$

as follows from the identities $\vec{e}_\varphi \times \vec{e}_r \cdot \vec{e}_x = \vec{e}_\vartheta \cdot \vec{e}_x = -\sin \vartheta$ and $\vec{e}_r \cdot \vec{e}_x = \cos \vartheta$.

The viscous term in (13.74) is obviously constant over the surface of the sphere. The pressure term does not contribute to the force on the sphere, as can be verified by direct integration, and it follows that the expression for the force, namely F, on the sphere in the direction of flow is

$$F = \eta \frac{3U_\infty}{2 \, r_0} \cdot 4\pi r_0^2 = 6\pi \eta r_0 U_\infty. \tag{13.75}$$

It is immediately evident that the force vanishes for the potential flow $(\vec{\omega} = 0)$, which is in agreement with the D'Alembert's paradox. The force in (13.75) is proportional to the first power of the velocity, and also to the body's linear dimensions. These properties hold equally when the body has a general form. Of course the force is in general no longer in the direction of flow, so that (13.75) takes the form

$$\vec{F} = \mathbf{A} \cdot \vec{U}_\infty \eta \, d, \tag{13.76}$$

in which $\mathbf{A}$ is a tensor which depends on the body's form and for complicated shapes can be determined only numerically.

When the sphere is itself a fluid, other boundary conditions come into play. If η' is the viscosity of the fluid drop, then we state without derivation that

$$F = 6\pi\eta\, r_0 U_\infty \frac{2\eta + \eta'}{3\eta + \eta'}. \tag{13.77}$$

For drops of very large viscosity ($\eta' \gg \eta$), such as droplets of water in air, (13.77) reduces to (13.75).

It is astonishing that the streamlines around the sphere are symmetric with respect to the plane $\vartheta = \pi/2$ even though the sphere experiences a force. From the balance of momentum we should expect a momentum deficit in the wake behind the sphere. That this is not evident is a consequence of the fact that the creeping solution is not valid for large r. The flow (13.65) is produced by the diffusion of vorticity (see (13.3)). The vorticity diffuses throughout the fluid, and at large distances from the source (namely, the sphere) it is plausible that convection plays a role and with it inertia effects. This is the case when the typical convective term $\vec{u} \cdot \nabla\omega$ in the vorticity transport equation (4.15) becomes comparable with the diffusion term $\nu\Delta\omega$: the former has order of magnitude $O(U_\infty\omega/r)$ and the latter $O(\nu\omega/r^2)$. It follows that the ratio of these terms is

$$\frac{U_\infty r}{\nu} = \frac{U_\infty r_0}{\nu} \frac{r}{r_0} = Re \frac{r}{r_0}. \tag{13.78}$$

The convective terms accordingly become arbitrarily large at large distances from the sphere, no matter how small the Reynolds' number is. The creeping-flow solution is thus no longer valid at large distances from the sphere, though it is an approximate solution for small Reynolds' numbers which is valid in the neighbourhood of the sphere. This situation is closely related to the singular nature of potential flow, which for large Reynolds' numbers is a consistent approximation but breaks down close to the wall, where it must be replaced by a boundary layer flow. Creeping flow is in fact a singular perturbation problem and can be treated systematically by *singular perturbation methods*.

Long before these methods were known, Oseen, in 1910, discovered the outer flow using a heuristic approach which provided a uniformly valid solution at small Reynolds' numbers. In the *Oseen approximation*, the convection terms were replaced by $U_\infty \vec{e}_x \cdot \nabla\vec{u}$, thus linearising the equation, whose solution is still so complicated that we will not discuss it here: suffice it to say that the method leads also to the uniformly valid solution for the low-Reynolds'-number flow around a circular cylinder.

In the breakdown of the Stokes solution at large distances we see the reason why the solution's region of validity is restricted at relatively small Reynolds' numbers.

Using the drag coefficient (3.11), we obtain with $W = F$ and $L = r_0\sqrt{\pi}$ the expression

$$c_w \frac{W}{\frac{\varrho}{2} U_\infty^2 \pi r_0^2} = \frac{24}{Re},$$ (13.79)

where the convention is used that the Reynolds' number is defined in terms of the diameter of the sphere. Figure 10.21, which is in log-log form, indicates that the experimentally measured values deviate from the straight line corresponding to (13.79) when $Re > 1$. Actually, this is already the case when $Re \approx 0.6$.

Appendix A
Introduction to Cartesian Tensors

In this text book a certain knowledge of tensors has been assumed. We restrict ourselves to *Cartesian tensors*, since all equations in fluid mechanics can in principle be developed in Cartesian coordinate systems. The most important elements of Cartesian tensors are summarized in this chapter; otherwise the literature should be consulted.

A.1 Summation Convention

When dealing with quantities in index notation we make use of *Einstein's summation convention*, which states that all indices which appear twice within an expression are to be summed. In $\mathcal{R}^3$ the summation indices run from 1 to 3

$$P = F_i u_i = \sum_{i=1}^{3} F_i u_i,$$

$$t_i = \tau_{ji} n_j = \sum_{j=1}^{3} \tau_{ji} n_j,$$

$$\vec{x} = x_i \vec{e}_i = \sum_{i=1}^{3} x_i \vec{e}_i.$$

Indices which appear twice are called *dummy indices*. Since they vanish after carrying out the summation, they may be arbitrarily named

$$F_i u_i = F_k u_k = F_j u_j,$$
$$x_i \vec{e}_i = x_l \vec{e}_l = x_m \vec{e}_m.$$

© Springer Nature Switzerland AG 2020
J. H. Spurk and N. Aksel, *Fluid Mechanics*,
https://doi.org/10.1007/978-3-030-30259-7

As well as the dummy indices, single indices can also appear in equations. These *free indices* must be identical in all terms of the same equation

$$t_i = \tau_{ji} n_j,$$
$$\vec{e}_i = a_{ij} \vec{g}_j,$$
$$a_{ij} = b_{ik} c_{kj} + d_{ijl} n_l.$$

Otherwise they may be arbitrarily named

$$t_m = \tau_{jm} n_j,$$
$$t_k = \tau_{mk} n_m.$$

In order to by unambiguous, the summation convention requires that an index appears no more than twice within an expression. A forbidden expression would be

$$t_i = a_{ij} b_{ij} n_j \text{ (wrong!)},$$

but the following would be allowed

$$t_i = -p \delta_{ij} n_j + 2 \eta e_{ij} n_j.$$

A.2 Cartesian Tensors

A tensor consists of *tensor components* and *basis vectors*. The number of linearly independent basis vectors gives the *dimension of the tensor space*. In three-dimensional space $\mathcal{R}^3$, from which, in what follows, we shall always start from, there are three linearly independent vectors, which along with three linear factors are in the position to determine a point in space uniquely. Such a set of three vectors which span a (not necessarily orthogonal) *coordinate system* can be used as a set of basis vectors. If these basis vectors are functions of position, the coordinate system which they span is called a *curvilinear coordinate system*. (Think for example of polar coordinates where the direction of the basis vectors is a function of the polar angle.) As basis vectors we choose fixed, orthogonal *unit vectors*, which we denote by $\vec{e}_i$ ($i = 1, 2, 3$). The coordinate system spanned by these is the *Cartesian coordinate system* with the coordinate axes x_i ($i = 1, 2, 3$).

We differentiate between tensors of different orders. Tensors of order zero are *scalars*. Since a scalar is completely independent of the choice of coordinate system, no basis vector is needed to describe it. Tensors of order one are *vectors*. The example of the position vector,

$$\vec{x} = \sum_{i=1}^{3} x_i \vec{e}_i = x_i \vec{e}_i, \tag{A.1}$$

shows that each component of a tensor of order one appears along with one basis vector.

Tensors of order two (*dyadics*) can be thought of as being formed from two vectors $\vec{a}$ and $\vec{b}$ multiplied together, so that each term $a_i \vec{e}_i$ of the vector $\vec{a}$ is multiplied with each term $b_j \vec{e}_j$ of the vector $\vec{b}$

$$T = \vec{a}\vec{b} = \sum_{i=1}^{3} \sum_{j=1}^{3} a_i b_j \vec{e}_i \vec{e}_j = a_i b_j \vec{e}_i \vec{e}_j. \tag{A.2}$$

This product is called the *dyadic product*, and is not to be confused with the inner product $\vec{a} \cdot \vec{b}$ (whose result is a scalar), or the outer product $\vec{a} \times \vec{b}$ (whose result is a vector). Since the dyadic product is not commutative, the basis vectors $\vec{e}_i \vec{e}_j$ in (A.2) may not be interchanged, since $a_i b_j \vec{e}_j \vec{e}_i$ would correspond to the tensor $\vec{b}\vec{a}$. If we denote the components of the tensor T with t_{ij} in (A.2) we obtain

$$\mathbf{T} = t_{ij} \vec{e}_i \vec{e}_j. \tag{A.3}$$

Therefore to every component of a second order tensor there belong two basis vectors $\vec{e}_i$ and $\vec{e}_j$. In $\mathcal{R}^3$ nine of these basis vector pairs form the so called *basis* of the tensor.

Completely analogously tensors of any order may be formed: the dyadic product of a tensor of order n and one of order m forms a tensor of order $(m + n)$. The basis of an nth order tensor in $\mathcal{R}^3$ consists of 3^n products each of n basis vectors.

Since the basis vectors for Cartesian tensors (unit vectors $\vec{e}_i$) are constant, it suffices to give the components of a tensor if a Cartesian coordinate system has already been layed down. Therefore, for a vector $\vec{x}$ it is enough to state the components

$$x_i \, (i = 1, 2, 3),$$

and a second order tensor $\mathbf{T}$ is fully described by its components

$$t_{ij} \, (i, j = 1, 2, 3).$$

Therefore, if we talk about the tensor t_{ij}, we shall tacitly mean the tensor given in (A.3).

The notation in which the mathematical relations between tensors are expressed solely by their components is the *Cartesian index notation*. Because we assume fixed and orthonormal basis vectors $\vec{e}_i$, Cartesian index notation is only valid for

Cartesian coordinate systems. It is possible to develop this to general curvilinear coordinate systems, but we refer for this to the more advanced literature.

The components of tensors up to the second order may be written in the form of *matrices*, so for example

$$\mathbf{T} \stackrel{\triangle}{=} \begin{bmatrix} t_{11} & t_{12} & t_{13} \\ t_{21} & t_{22} & t_{23} \\ t_{31} & t_{32} & t_{33} \end{bmatrix}. \tag{A.4}$$

Note however that not every matrix is a tensor.

In order to derive some rules we shall digress from the pure index notation and carry the basis vectors along, using a *mixed notation*. First we shall deal with the *inner product (scalar product)*

$$\vec{a} \cdot \vec{b} = (a_i \vec{e}_i) \cdot (b_j \vec{e}_j) = a_i b_j (\vec{e}_i \cdot \vec{e}_j). \tag{A.5}$$

Because of the orthogonality of the unit vectors, the product $\vec{e}_i \cdot \vec{e}_j$ is different from zero only if $i = j$. If we expand (A.5) we can easily convince ourselves that it is enough to carry out the summation

$$\vec{a} \cdot \vec{b} = a_i b_i = a_j b_j. \tag{A.6}$$

Clearly within a summation, the product $\vec{e}_i \cdot \vec{e}_j$ will cause the index on one of the two vector components to be exchanged. We can summarize all possible products $\vec{e}_i \cdot \vec{e}_j$ into a second order tensor

$$\delta_{ij} = \vec{e}_i \cdot \vec{e}_j = \begin{cases} 1 \text{ for } i = j \\ 0 \text{ for } i \neq j \end{cases} \tag{A.7}$$

This tensor is called the *Kronecker delta*, or because of its properties stated above, the *exchange symbol*. Multiplying a tensor with the Kronecker delta brings about an exchange of index in this tensor

$$a_{ij}\delta_{jk} = a_{ik}, \tag{A.8}$$

$$a_i b_j \delta_{ij} = a_i b_i = a_j b_j. \tag{A.9}$$

Applying the Kronecker delta in (A.5) therefore furnishes the inner product in Cartesian index notation

$$\vec{a} \cdot \vec{b} = a_i b_j \delta_{ij} = a_i b_i. \tag{A.10}$$

We now consider the *outer product* (*vector product*) of two vectors

$$\vec{c} = \vec{a} \times \vec{b} = (a_i\vec{e}_i) \times (b_j\vec{e}_j) = a_ib_j(\vec{e}_i \times \vec{e}_j). \tag{A.11}$$

Now the outer product of two orthogonal unit vectors is zero if $i = j$, since this is outer product of parallel vectors. If $i \neq j$, the outer product of the two unit vectors is the third unit vector, possibly with negative sign. It easily follows that the relation

$$\vec{e}_i \times \vec{e}_j = \epsilon_{ijk}\vec{e}_k \tag{A.12}$$

holds if we define ϵ_{ijk} as a third order tensor having the following properties

$$\epsilon_{ijk} = \begin{cases} +1 \text{ if } ijk \text{ is an even permutation (i.e 123, 231, 312)} \\ -1 \text{ if } ijk \text{ is an odd permutation (i.e 321, 213, 132).} \\ 0 \quad \text{ if atleast two indices are equal} \end{cases} \tag{A.13}$$

We call ϵ_{ijk} the *epsilon tensor* or the *permutation symbol*. Inserting (A.12) into (A.11) leads to

$$\vec{c} = a_ib_j\epsilon_{ijk}\vec{e}_k. \tag{A.14}$$

We read off the components of $\vec{c}$ from this equation as

$$c_k = \epsilon_{ijk}a_ib_j, \tag{A.15}$$

where we have used the fact that the order of the factors is arbitrary; we are dealing with components, that is, just numbers.

We shall now examine the behavior of a tensor if we move from a Cartesian coordinate system with basis vectors $\vec{e}_i$ to another with basis vectors $\vec{e}'_i$. The "dashed" coordinate system arises from rotating (and possibly also from translating) the original coordinate system. If we are dealing with a zeroth order tensor, that is a scalar, it is clear that the value of this scalar (e.g. the density of a fluid particle) cannot depend of the coordinate system. The same holds for tensors of all orders. A tensor can only have a physical meaning if it is independent of the choice of coordinate system. This is clear in the example of the position vector of a point. If $\vec{x}$ and $\vec{x}'$ denote the same arrow (Fig. A.1) in the "dashed" and the "undashed" coordinate systems, then

$$\vec{x}' = \vec{x}, \tag{A.16}$$

that is,

$$x'_i\vec{e}'_i = x_i\vec{e}_i. \tag{A.17}$$

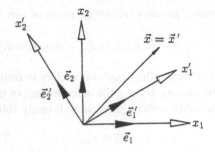

Fig. A.1 Rotation of the coordinate system

To decompose the vector $\vec{x}$ into its components relative to the dashed coordinate system, we form the scalar product with $\vec{e}_j'$ and obtain

$$x_i'\vec{e}_i' \cdot \vec{e}_j' = x_i\vec{e}_i \cdot \vec{e}_j'. \tag{A.18}$$

The scalar product of the unit vectors in the same (dashed) coordinate system $\vec{e}_i' \cdot \vec{e}_j'$, using (A.7), furnishes just δ_{ij}. The scalar product of the unit vectors of the dashed and undashed coordinate systems forms the matrix

$$a_{ij} = \vec{e}_i \cdot \vec{e}_j' \tag{A.19a}$$

or

$$a_{ij} = \cos\left(\angle x_i, x_j'\right). \tag{A.19b}$$

We call the matrix a_{ij} the *rotation matrix*. It is not associated with a basis and therefore is not a tensor. Inserting (A.19a) into (A.18) leads to the desired transformation law for the components of a vector

$$x_j' = a_{ij}x_i. \tag{A.20}$$

If we take the scalar product of (A.17) with $\vec{e}_j$ we decompose the vector $\vec{x}$ into its components relative to the undashed system and thus we obtain the inverse

$$x_j = a_{ji}x_i'. \tag{A.21}$$

The transformation and its inverse may look formally the same, but we note that in (A.20) we sum over the first index and in (A.21) over the second.

Knowing the transformation law for the components we can easily derive that for the basis vectors. To do this we relabel the dummy indices on the right-hand side of (A.17) as j so that we can insert (A.21). We obtain the equation

$$x_i'\vec{e}_i' = x_i'a_{ji}\vec{e}_j, \tag{A.22}$$

from which, using the fact that x_i' is arbitrary (independent variable), we can read off the transformation as $\vec{e}_i' = a_{ji}\vec{e}_j$. In order to be able to compare this with the components (A.20), we relabel the index i as j (and *vice versa*), and therefore write

$$\vec{e}_j' = a_{ij}\vec{e}_i. \tag{A.23}$$

We see that for Cartesian coordinate systems both the components and the basis vectors of a tensor obey the same transformation laws. Thus we take the inverse directly from (A.21) as

$$\vec{e}_j = a_{ji}\vec{e}_i', \tag{A.24}$$

where we could also have obtained this formally be inserting (A.20) into (A.17).

Before we consider the transformation laws for tensors of a higher order we shall take note of one well known property of the rotation matrix. To do this we exchange the indices in the transformation (A.20) (e.g.: $x_i' = a_{ki}x_k$), insert this into (A.21) and thus obtain

$$x_j = a_{ji}a_{ki}x_k. \tag{A.25}$$

Since the vector components are independent variables we can read off the following identity from (A.25)

$$a_{ji}a_{ki} = \delta_{jk}, \tag{A.26a}$$

which reads

$$\mathbf{AA}^{\mathrm{T}} = \mathbf{I} \tag{A.26b}$$

in matrix notation. Since $\mathbf{AA}^{-1} = \mathbf{I}$ is the equation which determines the inverse of $\mathbf{A}$, we conclude from (A.26a) that the transpose of the rotation matrix is equal to its inverse (orthogonal matrix).

The transformation for the components of a tensor of arbitrary order results from the transformations for the unit vectors (A.23) and (A.24). For clarity we shall restrict ourselves to a second order tensor whose basis we express in terms of the basis of the dashed coordinate system using the transformation (A.24) as

$$\mathbf{T} = t_{ij}\vec{e}_i\vec{e}_j = t_{ij}a_{ik}a_{jl}\vec{e}_k'\vec{e}_l'. \tag{A.27}$$

Because of $\mathbf{T} = \mathbf{T}' = t_{kl}'\vec{e}_k'\vec{e}_l'$ we can read off the components in the rotated system directly from (A.27) as

$$t'_{kl} = a_{ik}a_{jl}t_{ij}.$$ (A.28)

If in **T'** we replace the basis vectors using (A.23), we obtain

$$t_{kl} = a_{ki}a_{lj}t'_{ij}.$$ (A.29)

The same procedure is carried out for tensors of any order. The transformation behavior of tensor components is characteristic of them and therefore is used as the definition of a tensor. If we drop the basis vectors and use pure Cartesian index notation, the transformation behavior is the only criterion by which we can decide if a given expression is a tensor. Let us take an example: we shall examine whether the *gradient* of a scalar function is a tensor of order one. The equation $\vec{u} = \nabla \Phi$ reads in index notation

$$u_i = \frac{\partial \Phi}{\partial x_i},$$ (A.30)

or in the rotated coordinate system

$$u'_j = \frac{\partial \Phi}{\partial x'_j}.$$ (A.31)

If $\vec{u}$ is a first order tensor, using the transformation (A.20) should transform (A.30) into (A.31)

$$u'_j = a_{ij}u_i = a_{ij}\frac{\partial \Phi}{\partial x_i},$$ (A.32)

or using the chain rule,

$$u'_j = a_{ij}\frac{\partial \Phi}{\partial x'_k}\frac{\partial x'_k}{\partial x_i}.$$ (A.33)

By $x'_k = a_{jk}x_j$ we have

$$\frac{\partial x'_k}{\partial x_i} = a_{jk}\frac{\partial x_j}{\partial x_i},$$ (A.34)

and since x_j and x_i are independent variables for $i \neq j$, we write

$$\frac{\partial x_j}{\partial x_i} = \delta_{ij},$$ (A.35)

so that we replace (A.34) with

$$\frac{\partial x_k'}{\partial x_i} = a_{ik}. \tag{A.36}$$

We should note that the result of (A.35) is the Kronecker delta and it is therefore a second order tensor, and should not be confused with (A.36), whose result is the rotation matrix and is therefore not a tensor. If we insert (A.36) into (A.33), we obtain

$$u_j' = a_{ij}a_{ik}\frac{\partial \Phi}{\partial x_k'}, \tag{A.37}$$

which, because of (A.26a), is identical to

$$u_j' = \delta_{jk}\frac{\partial \Phi}{\partial x_k'} = \frac{\partial \Phi}{\partial x_j'}. \tag{A.38}$$

This result corresponds to (A.31), and so the gradient of a scalar function is a first order tensor.

The gradient of a tensor of the nth order comes from forming the dyadic product with the Nabla operator and is therefore a tensor of the $(n + 1)$th degree. An important example of this in fluid mechanics is the velocity gradient

$$\nabla \vec{u} = \left(\vec{e}_i\frac{\partial}{\partial x_i}\right)(u_j\vec{e}_j) = \frac{\partial u_j}{\partial x_i}\vec{e}_i\vec{e}_j. \tag{A.39}$$

This is a second order tensor with the components

$$\nabla \vec{u} \hat{=} t_{ij} = \frac{\partial u_j}{\partial x_i}. \tag{A.40}$$

The coordinate with respect to which we differentiate is given by the first index of t_{ij} (the row index in matrix representation) and the component of $\vec{u}$ is determined by the second index (the column index). In index notation we usually write the velocity gradient as $\partial u_i/\partial x_j$, that is in matrix representation as the transpose of (A.40). Although the matrix representation is not needed in index notation, in going from matrix equations to index notation (or *vice versa*), we should be aware of the sequence of indices determined by (A.39).

The *divergence* of the velocity vector (or of another first order tensor) reads $\partial u_i/\partial x_i$ in index notation, and formally corresponds with the scalar product of the Nabla operator with the vector $\vec{u}$. Thus symbolically the divergence reads $\nabla \cdot \vec{u}$ or else div$\vec{u}$. The result is a scalar. In general, the divergence of an nth order tensor is an $(n - 1)$th order tensor. Therefore the divergence of a scalar is not defined. An important quantity in fluid mechanics is the divergence of the stress tensor $\partial \tau_{ji}/\partial x_j$, which is a vector.

Every second order tensor can be decomposed into a symmetric and an antisymmetric part. From the identity

$$t_{ij} = \frac{1}{2}\left(t_{ij} + t_{ji}\right) + \frac{1}{2}\left(t_{ij} - t_{ji}\right) \tag{A.41}$$

we obtain the symmetric tensor

$$c_{ij} = \frac{1}{2}\left(t_{ij} + t_{ji}\right), \tag{A.42}$$

and the antisymmetric tensor

$$b_{ij} = \frac{1}{2}\left(t_{ij} - t_{ji}\right). \tag{A.43}$$

We can see that the symmetric part satisfies $c_{ij} = c_{ji}$ and the antisymmetric part satisfies $b_{ji} = -b_{ji}$. It follows immediately for the antisymmetric tensor that its diagonal elements (where $i = j$) must be zero. While a symmetric tensor has six independent components, an antisymmetric tensor is fully described by three components

$$[b_{ij}] = \begin{bmatrix} 0 & b_{12} & b_{13} \\ -b_{12} & 0 & b_{23} \\ -b_{13} & -b_{23} & 0 \end{bmatrix}. \tag{A.44}$$

In this connection we wish to refer to an important property of the ϵ tensor. To do this we multiply the decomposition of a second order tensor with the ϵ tensor

$$p_k = \epsilon_{ijk} t_{ij} = \epsilon_{ijk} c_{ij} + \epsilon_{ijk} b_{ij}, \tag{A.45}$$

where c_{ij} and b_{ij} are again the symmetric and antisymmetric parts respectively of t_{ij}. We rewrite this equation as follows

$$p_k = \frac{1}{2}\left(\epsilon_{ijk} c_{ij} + \epsilon_{ijk} c_{ji}\right) + \frac{1}{2}\left(\epsilon_{ijk} b_{ij} - \epsilon_{ijk} b_{ji}\right), \tag{A.46}$$

which is allowable because of the properties of c_{ij} and b_{ij}. We now exchange the dummy indices in the second expression in brackets

$$p_k = \frac{1}{2}\left(\epsilon_{ijk} c_{ij} + \epsilon_{jik} c_{ij}\right) + \frac{1}{2}\left(\epsilon_{ijk} b_{ij} - \epsilon_{jik} b_{ij}\right). \tag{A.47}$$

From the definition of the ϵ tensor (A.13) it follows that $\epsilon_{ijk} = -\epsilon_{jik}$, so that the first bracket vanishes. We obtain the equation

$$p_k = \epsilon_{ijk}b_{ij},$$ (A.48a)

which written in matrix form reads

$$\begin{bmatrix} p_1 \\ p_2 \\ p_3 \end{bmatrix} = \begin{bmatrix} b_{23} - b_{32} \\ b_{31} - b_{13} \\ b_{12} - b_{21} \end{bmatrix} = 2 \begin{bmatrix} b_{23} \\ -b_{13} \\ b_{12} \end{bmatrix}.$$ (A.48b)

Applying the ϵ tensor to an arbitrary second order tensor using (A.45) therefore leads to the three independent components of the antisymmetric part of the tensor (compare (A.48b) with (A.44)). From this we conclude that application of the ϵ tensor to a symmetric tensor furnishes the null vector

$$\epsilon_{ijk}c_{ij} = 0, \text{ if } c_{ij} = c_{ji}.$$ (A.49)

Here follow four identities of the ϵ tensor, given without proof

$$\epsilon_{ikm}\epsilon_{jln} = \det \begin{bmatrix} \delta_{ij} & \delta_{il} & \delta_{in} \\ \delta_{kj} & \delta_{kl} & \delta_{kn} \\ \delta_{mj} & \delta_{ml} & \delta_{mn} \end{bmatrix}.$$ (A.50)

Contraction by multiplication with δ_{mn} (setting $m = n$) leads to

$$\epsilon_{ikn}\epsilon_{jln} = \det \begin{bmatrix} \delta_{ij} & \delta_{il} \\ \delta_{kj} & \delta_{kl} \end{bmatrix}.$$ (A.51)

Table A.1 Summary of the most important rules of calculation in vector and index notation

Operation	Symbolic notation	Cartesian index notation
Scalar product	$c = \vec{a} \cdot \vec{b}$	$c = \delta_{ij}a_ib_j = a_ib_i$
	$\vec{c} = \vec{a} \cdot \mathbf{T}$	$c_k = \delta_{ij}a_it_{jk} = a_it_{ik}$
Vector product	$\vec{c} = \vec{a} \times \vec{b}$	$c_i = \epsilon_{ijk}a_jb_k$
Dyadic product	$\mathbf{T} = \vec{a}\vec{b}$	$t_{ij} = a_ib_j$
Gradient of a scalar field	$\vec{c} = \text{grad } a = \nabla a$	$c_i = \frac{\partial a}{\partial x_i}$
Gradient of a vector field	$\mathbf{T} = \text{grad } \vec{a} = \nabla\vec{a}$	$t_{ij} = \frac{\partial a_j}{\partial x_i}$
Divergence of a vector field	$c = \text{div } \vec{a} = \nabla \cdot \vec{a}$	$c = \frac{\partial a_i}{\partial x_i}$
Divergence of a tensor field	$\vec{c} = \text{div } \mathbf{T} = \nabla \cdot \mathbf{T}$	$c_i = \frac{\partial t_{ji}}{\partial x_j}$
Curl of a vector field	$\vec{c} = \text{curl } \vec{a} = \nabla \times \vec{a}$	$c_i = \epsilon_{ijk}\frac{\partial a_k}{\partial x_j}$
Laplace operator on a scalar	$c = \Delta\varphi = \nabla \cdot \nabla\varphi$	$c = \frac{\partial^2 \varphi}{\partial x_i \partial x_i}$

Contracting again by multiplying with δ_{kl} furnishes (Table A.1)

$$\epsilon_{ikn}\epsilon_{jkn} = 2\delta_{ij}, \tag{A.52}$$

and finally for $i = j$

$$\epsilon_{ikn}\epsilon_{ikn} = 2\delta_{ii} = 6. \tag{A.53}$$

Appendix B
Curvilinear Coordinates

In applications it is often useful to use curvilinear coordinates. In order to derive the component equation for curvilinear coordinates we can start from general tensor calculus, which is valid in all coordinate systems. However, if we restrict ourselves to curvilinear but orthogonal coordinates, we can move relatively easily from the corresponding equations in symbolic notation to the desired component equations. Since it is orthogonal coordinate systems which are needed in almost all applications, we shall indeed restrict ourselves to these.

We consider the curvilinear orthogonal coordinates q_1, q_2, q_3, which can be calculated from the Cartesian coordinates x_1, x_2 and x_3

$$q_1 = q_1(x_1, x_2, x_3),$$
$$q_2 = q_2(x_1, x_2, x_3),$$
$$q_3 = q_3(x_1, x_2, x_3),$$

or in short

$$q_i = q_i(x_j) \tag{B.1}$$

We assume that (B.1) has a unique inverse

$$x_i = x_i(q_j) \tag{B.2a}$$

or

$$\vec{x} = \vec{x}(q_j) \tag{B.2b}$$

© Springer Nature Switzerland AG 2020
J. H. Spurk and N. Aksel, *Fluid Mechanics*,
https://doi.org/10.1007/978-3-030-30259-7

If q_2 and q_3 are kept constant, the vector $\vec{x} = \vec{x}(q_1)$ describes a curve in space which is the coordinate curve q_1. $\partial\vec{x}/\partial q_1$ is the tangent vector to this curve. The corresponding unit vector in the direction of increasing q_1 reads

$$\vec{e}_1 = \frac{\partial\vec{x}/\partial q_1}{|\partial\vec{x}/\partial q_1|}. \tag{B.3}$$

If we set $|\partial\vec{x}/\partial q_1| = b_1$, we see that

$$\frac{\partial\vec{x}}{\partial q_1} = \vec{e}_1\, b_1 , \tag{B.4}$$

and in the same way

$$\frac{\partial\vec{x}}{\partial q_2} = \vec{e}_2\, b_2 , \tag{B.5}$$

$$\frac{\partial\vec{x}}{\partial q_3} = \vec{e}_3 b_3, \tag{B.6}$$

with $b_2 = |\partial\vec{x}/\partial q_2|$ and $b_3 = |\partial\vec{x}/\partial q_3|$.

Because of $\vec{x} = \vec{x}(q_j)$ it follows that

$$d\vec{x} = \frac{\partial\vec{x}}{\partial q_1}dq_1 + \frac{\partial\vec{x}}{\partial q_2}dq_2 + \frac{\partial\vec{x}}{\partial q_3}dq_3 = b_1 dq_1\vec{e}_1 + b_2 dq_2\vec{e}_2 + b_3 dq_3\vec{e}_3, \tag{B.7}$$

and, since the basis vectors are orthogonal to each other, the square of the line element is

$$d\vec{x} \cdot d\vec{x} = b_1^2 dq_1^2 + b_2^2 dq_2^2 + b_3^2 dq_3^2. \tag{B.8}$$

For the volume element dV (Fig. B.1) we have

$$dV = b_1 dq_1\vec{e}_1 \cdot (b_2 dq_2\vec{e}_2 \times b_3 dq_3\vec{e}_3) = b_1 b_2 b_3 dq_1 dq_2 dq_3. \tag{B.9}$$

The q_1 surface element of the volume element dV (i.e. the surface element normal to the q_1 direction) is then

$$dS_1 = |b_2 dq_2\vec{e}_2 \times b_3 dq_3\vec{e}_3| = b_2 b_3 dq_2 dq_3. \tag{B.10}$$

In a similar manner we find for the remaining surface elements

$$dS_2 = b_3 b_1 dq_3 dq_1, \tag{B.11}$$

$$dS_3 = b_1 b_2 dq_1 dq_2. \tag{B.12}$$

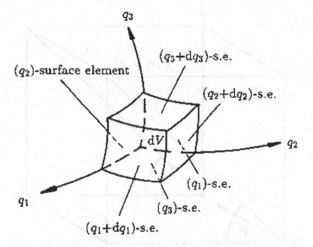

Fig. B1 Volume element in the curvilinear orthogonal coordinate system

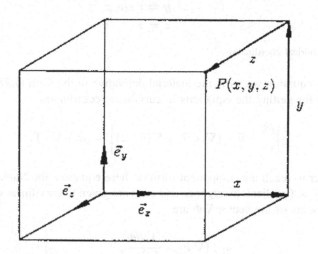

Fig. B.2 Cartesian coordinates

The continuity equation, Cauchy's equation of motion and the entropy equation read symbolically

$$\frac{\partial \varrho}{\partial t} + \vec{u} \cdot \nabla \varrho + \varrho \nabla \cdot \vec{u} = 0,$$

$$\varrho \frac{D\vec{u}}{Dt} = \varrho \vec{k} + \nabla \cdot \mathbf{T}, \quad \text{and}$$

$$\varrho T \left[\frac{\partial s}{\partial t} + \vec{u} \cdot \nabla s \right] = \Phi + \nabla \cdot (\lambda \nabla T).$$

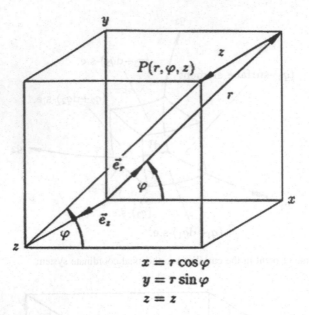

$$x = r \cos \varphi$$
$$y = r \sin \varphi$$
$$z = z$$

Fig. B.3 Cylindrical coordinates

In Cauchy's equation we write the material derivative in the form (1.78), as this is more useful for getting the equations in curvilinear coordinates

$$\varrho \left[\frac{\partial \vec{u}}{\partial t} - \vec{u} \times (\nabla \times \vec{u}) + \nabla \left(\vec{u}^2 / 2 \right) \right] = \varrho \vec{k} + \nabla \cdot \mathbf{T}. \tag{B.13}$$

Now in order to reach the component form of these equation, the Nabla operations $\nabla, \nabla \cdot$, and $\nabla \times$ (gradient, divergence and curl) are given in curvilinear coordinates. The components of the vector $\nabla \Phi$ are

$$q_1 : (\nabla \Phi)_1 = \frac{1}{b_1} \frac{\partial \Phi}{\partial q_1},$$

$$q_2 : (\nabla \Phi)_2 = \frac{1}{b_2} \frac{\partial \Phi}{\partial q_2}, \quad \text{and} \tag{B.14}$$

$$q_3 : (\nabla \Phi)_3 = \frac{1}{b_3} \frac{\partial \Phi}{\partial q_3}.$$

If u_1, u_2 and u_3 are the components of the vector $\vec{u}$ in the direction of increasing q_1, q_2 and q_3, we have

$$\nabla \cdot \vec{u} = \frac{1}{b_1 b_2 b_3} \left[\frac{\partial}{\partial q_1} (b_2 b_3 u_1) + \frac{\partial}{\partial q_2} (b_3 b_1 u_2) + \frac{\partial}{\partial q_3} (b_1 b_2 u_3) \right]. \tag{B.15}$$

Since the basis vectors are orthonormal, the Laplace operator $\Delta = \nabla \cdot \nabla = \nabla^2$ can be easily calculated, if, in (B.15) we identify the components of $\vec{u}$ with the components of ∇

$$\Delta = \frac{1}{b_1 b_2 b_3} \left\{ \frac{\partial}{\partial q_1} \left[\frac{b_2 b_3}{b_1} \frac{\partial}{\partial q_1} \right] + \frac{\partial}{\partial q_2} \left[\frac{b_3 b_1}{b_2} \frac{\partial}{\partial q_2} \right] + \frac{\partial}{\partial q_3} \left[\frac{b_1 b_2}{b_3} \frac{\partial}{\partial q_3} \right] \right\}. \tag{B16}$$

$\nabla \times \vec{u}$ has the components

$$q_1 : (\nabla \times \vec{u})_1 = \frac{1}{b_2 b_3} \left[\frac{\partial}{\partial q_2} (b_3 u_3) - \frac{\partial}{\partial q_3} (b_2 u_2) \right],$$

$$q_2 : (\nabla \times \vec{u})_2 = \frac{1}{b_3 b_1} \left[\frac{\partial}{\partial q_3} (b_1 u_1) - \frac{\partial}{\partial q_1} (b_3 u_3) \right], \tag{B.17}$$

$$q_3 : (\nabla \times \vec{u})_3 = \frac{1}{b_1 b_2} \left[\frac{\partial}{\partial q_1} (b_2 u_2) - \frac{\partial}{\partial q_2} (b_1 u_1) \right].$$

The components of the divergence of the stress tensor are

$$q_1 : \quad (\nabla \cdot \mathbf{T})_1 = \frac{1}{b_1 b_2 b_3} \left[\frac{\partial}{\partial q_1} (b_2 b_3 \tau_{11}) + \frac{\partial}{\partial q_2} (b_3 b_1 \tau_{21}) + \frac{\partial}{\partial q_3} (b_1 b_2 \tau_{31}) \right]$$
$$+ \frac{\tau_{21}}{b_1 b_2} \frac{\partial b_1}{\partial q_2} + \frac{\tau_{31}}{b_1 b_3} \frac{\partial b_1}{\partial q_3} - \frac{\tau_{22}}{b_1 b_2} \frac{\partial b_2}{\partial q_1} - \frac{\tau_{33}}{b_1 b_3} \frac{\partial b_3}{\partial q_1},$$

$$q_2 : \quad (\nabla \cdot \mathbf{T})_2 = \frac{1}{b_1 b_2 b_3} \left[\frac{\partial}{\partial q_1} (b_2 b_3 \tau_{12}) + \frac{\partial}{\partial q_2} (b_3 b_1 \tau_{22}) + \frac{\partial}{\partial q_3} (b_1 b_2 \tau_{32}) \right]$$
$$+ \frac{\tau_{32}}{b_2 b_3} \frac{\partial b_2}{\partial q_3} + \frac{\tau_{12}}{b_2 b_1} \frac{\partial b_2}{\partial q_1} - \frac{\tau_{33}}{b_2 b_3} \frac{\partial b_3}{\partial q_2} - \frac{\tau_{11}}{b_1 b_2} \frac{\partial b_1}{\partial q_2},$$

$$q_3 : \quad (\nabla \cdot \mathbf{T})_3 = \frac{1}{b_1 b_2 b_3} \left[\frac{\partial}{\partial q_1} (b_2 b_3 \tau_{13}) + \frac{\partial}{\partial q_2} (b_3 b_1 \tau_{23}) + \frac{\partial}{\partial q_3} (b_1 b_2 \tau_{33}) \right]$$
$$+ \frac{\tau_{13}}{b_3 b_1} \frac{\partial b_3}{\partial q_1} + \frac{\tau_{23}}{b_3 b_2} \frac{\partial b_3}{\partial q_2} - \frac{\tau_{11}}{b_3 b_1} \frac{\partial b_1}{\partial q_3} - \frac{\tau_{22}}{b_3 b_2} \frac{\partial b_2}{\partial q_3}.$$

$$\tag{B.18}$$

Here for example the stress component τ_{13} is the component in the direction of increasing q_3 which acts on the surface whose normal is in the direction of increasing q_1.

The Cauchy-Poisson law in symbolic form holds for the components of the stress

$$\mathbf{T} = (-p + \lambda^* \nabla \cdot u)\mathbf{I} + 2\eta \mathbf{E}.$$

The components of the rate of deformation tensor are given by

$$e_{11} = \frac{1}{b_1}\frac{\partial u_1}{\partial q_1} + \frac{u_2}{b_1 b_2}\frac{\partial b_1}{\partial q_2} + \frac{u_3}{b_3 b_1}\frac{\partial b_1}{\partial q_3},$$

$$e_{22} = \frac{1}{b_2}\frac{\partial u_2}{\partial q_2} + \frac{u_3}{b_2 b_3}\frac{\partial b_2}{\partial q_3} + \frac{u_1}{b_1 b_2}\frac{\partial b_2}{\partial q_1},$$

$$e_{33} = \frac{1}{b_3}\frac{\partial u_3}{\partial q_3} + \frac{u_1}{b_3 b_1}\frac{\partial b_3}{\partial q_1} + \frac{u_2}{b_2 b_3}\frac{\partial b_3}{\partial q_2},$$

$$2e_{32} = \frac{b_3}{b_2}\frac{\partial(u_3/b_3)}{\partial q_2} + \frac{b_2}{b_3}\frac{\partial(u_2/b_2)}{\partial q_3} = 2e_{23},$$

$$2e_{13} = \frac{b_1}{b_3}\frac{\partial(u_1/b_1)}{\partial q_3} + \frac{b_3}{b_1}\frac{\partial(u_3/b_3)}{\partial q_1} = 2e_{31}, \quad \text{and}$$

$$2e_{21} = \frac{b_2}{b_1}\frac{\partial(u_2/b_2)}{\partial q_1} + \frac{b_1}{b_2}\frac{\partial(u_1/b_1)}{\partial q_2} = 2e_{12}.$$

(B.19)

As an example of how to apply this we consider spherical coordinates r, ϑ, φ with the velocity components $u_r, u_\vartheta, u_\varphi$. The relation between Cartesian and spherical coordinates is given by the transformation (cf. Fig. B.4)

$$x = r\cos\vartheta,$$
$$y = r\sin\vartheta\cos\varphi, \qquad (B.20)$$
$$z = r\sin\vartheta\sin\varphi.$$

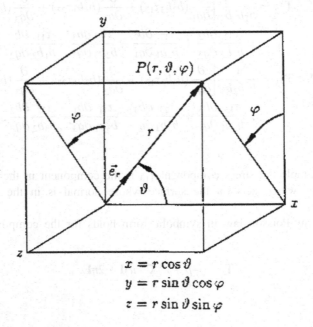

$$x = r\cos\vartheta$$
$$y = r\sin\vartheta\cos\varphi$$
$$z = r\sin\vartheta\sin\varphi$$

Fig. B.4 Spherical coordinates

The x-axis is the polar axis and ϑ is the polar angle. With

$$q_1 = r, \quad q_2 = \vartheta, \quad \text{and} \quad q_3 = \varphi \tag{B.21}$$

it follows that

$$
\begin{aligned}
b_1 &= \left\{ \cos^2 \vartheta + \sin^2 \vartheta \left(\sin^2 \varphi + \cos^2 \varphi \right) \right\}^{1/2} = 1, \\
b_2 &= \left\{ r^2 \sin^2 \vartheta + r^2 \cos^2 \vartheta \left(\cos^2 \varphi + \sin^2 \varphi \right) \right\}^{1/2} = r, \\
b_3 &= \left\{ r^2 \sin^2 \vartheta \left(\sin^2 \varphi + \cos^2 \varphi \right) \right\}^{1/2} = r \sin \vartheta.
\end{aligned}
\tag{B.22}
$$

The line element reads

$$\mathrm{d}\vec{x} = \mathrm{d}r\vec{e}_r + r\mathrm{d}\vartheta\vec{e}_\vartheta + r \sin \vartheta \mathrm{d}\varphi \vec{e}_\varphi, \tag{B.23}$$

and the volume element is

$$\mathrm{d}V = r^2 \sin \vartheta \mathrm{d}r \mathrm{d}\vartheta \mathrm{d}\varphi. \tag{B.24}$$

For the surface elements we obtain

$$
\begin{aligned}
\mathrm{d}S_r &= r^2 \sin \vartheta \mathrm{d}\vartheta \mathrm{d}\varphi, \\
\mathrm{d}S_\vartheta &= r \sin \vartheta \mathrm{d}r \mathrm{d}\varphi, \\
\mathrm{d}S_\varphi &= r\mathrm{d}r\mathrm{d}\vartheta.
\end{aligned}
\tag{B.25}
$$

The components of $\mathrm{grad}\Phi = \nabla \Phi$ are

$$
\begin{aligned}
r &: (\nabla \Phi)_r = \frac{\partial \Phi}{\partial r}, \\
\vartheta &: (\nabla \Phi)_\vartheta = \frac{1}{r} \frac{\partial \Phi}{\partial \vartheta}, \\
\varphi &: (\nabla \Phi)_\varphi = \frac{1}{r \sin \vartheta} \frac{\partial \Phi}{\partial \varphi}.
\end{aligned}
\tag{B.26}
$$

For $\mathrm{div}\, \vec{u} = \nabla \cdot \vec{u}$ it follows that

$$\nabla \cdot \vec{u} = \left(r^2 \sin \vartheta \right)^{-1} \left[\frac{\partial}{\partial r} \left(r^2 \sin \vartheta u_r \right) + \frac{\partial}{\partial \vartheta} \left(r \sin \vartheta u_\vartheta \right) + \frac{\partial}{\partial \varphi} \left(r u_\varphi \right) \right]. \tag{B.27}$$

The components of $\mathrm{curl}\, \vec{u} = \nabla \times \vec{u}$ are

$$r : (\nabla \times \vec{u})_r = \left(r^2 \sin \vartheta\right)^{-1} \left[\frac{\partial}{\partial \vartheta}\left(r \sin \vartheta u_\varphi\right) - \frac{\partial}{\partial \varphi}\left(r u_\vartheta\right)\right],$$

$$\vartheta : (\nabla \times \vec{u})_\vartheta = \left(r \sin \vartheta\right)^{-1} \left[\frac{\partial}{\partial \varphi}\left(u_r\right) - \frac{\partial}{\partial r}\left(r \sin \vartheta u_\varphi\right)\right], \tag{B.28}$$

$$\varphi : (\nabla \times \vec{u})_\varphi = r^{-1} \left[\frac{\partial}{\partial r}\left(r u_\vartheta\right) - \frac{\partial}{\partial \vartheta}\left(u_r\right)\right].$$

We now wish to calculate the r-component of the Navier–Stokes equations. To do this we require the r-component of $\vec{u} \times (\nabla \times \vec{u})$ and of $\nabla \cdot \mathbf{T}$

$$\{\vec{u} \times (\nabla \times \vec{u})\}_r = \frac{1}{r} u_\vartheta \left[\frac{\partial}{\partial r}\left(r u_\vartheta\right) - \frac{\partial}{\partial \vartheta}\left(u_r\right)\right] - \frac{1}{r \sin \vartheta} u_\varphi \left[\frac{\partial}{\partial \varphi}\left(u_r\right) - \frac{\partial}{\partial r}\left(r \sin \vartheta u_\varphi\right)\right],$$

$$\tag{B.29}$$

$$(\nabla \cdot \mathbf{T})_r = \frac{1}{r^2 \sin \vartheta} \left[\frac{\partial}{\partial r}\left(r^2 \sin \vartheta \tau_{rr}\right) + \frac{\partial}{\partial \vartheta}\left(r \sin \vartheta \tau_{\vartheta r}\right) + \frac{\partial}{\partial \varphi}\left(r \tau_{\varphi r}\right)\right]$$

$$- \frac{1}{r}\left(\tau_{\vartheta\vartheta} + \tau_{\varphi\varphi}\right), \tag{B.30}$$

where, from (3.1b) for incompressible flow

$$\begin{aligned}
\tau_{rr} &= -p + 2\eta e_{rr}, \\
\tau_{\vartheta\vartheta} &= -p + 2\eta e_{\vartheta\vartheta}, \\
\tau_{\varphi\varphi} &= -p + 2\eta e_{\varphi\varphi}, \\
\tau_{\vartheta r} &= 2\eta e_{\vartheta r}, \\
\tau_{\varphi r} &= 2\eta e_{\varphi r}, \quad \text{and} \\
\tau_{\varphi\vartheta} &= 2\eta e_{\varphi\vartheta}.
\end{aligned} \tag{B.31}$$

The components of the rate of deformation tensor are

$$e_{rr} = \partial u_r / \partial r,$$

$$e_{\vartheta\vartheta} = \frac{1}{r}\{\partial u_\vartheta / \partial \vartheta + u_r\},$$

$$e_{\varphi\varphi} = \frac{1}{r \sin \vartheta}\left(\partial u_\varphi / \partial \varphi\right) + \frac{1}{r}\left(u_r + u_\vartheta \cot \vartheta\right),$$

$$2e_{\varphi\vartheta} = 2e_{\vartheta\varphi} = \sin \vartheta \frac{\partial}{\partial \vartheta}\left[\frac{1}{r \sin \vartheta} u_\varphi\right] + \frac{1}{\sin \vartheta}\frac{\partial}{\partial \varphi}\left[\frac{1}{r} u_\vartheta\right], \tag{B.32}$$

$$2e_{r\varphi} = 2e_{\varphi r} = \frac{1}{r \sin \vartheta}\partial u_r / \partial \varphi + r \sin \vartheta \frac{\partial}{\partial r}\left[\frac{1}{r \sin \vartheta} u_\varphi\right], \quad \text{and}$$

$$2e_{\vartheta r} = 2e_{r\vartheta} = r \frac{\partial}{\partial r}\left[\frac{1}{r} u_\vartheta\right] + \frac{1}{r}\partial u_r / \partial \vartheta.$$

By inserting these equations into Cauchy's equation, we obtain the r-component of the Navier-Stokes equations for incompressible flow

$$\varrho \left\{ \frac{\partial u_r}{\partial t} - \frac{u_\vartheta}{r} \left[\frac{\partial(ru_\vartheta)}{\partial r} - \frac{\partial u_r}{\partial \vartheta} \right] \right.$$

$$+ \frac{u_\varphi}{r \sin \vartheta} \left[\frac{\partial u_r}{\partial \varphi} - \frac{\partial(r \sin \vartheta u_\varphi)}{\partial r} \right] + \frac{1}{2} \frac{\partial \left(u_r^2 + u_\vartheta^2 + u_\varphi^2 \right)}{\partial r} \right\}$$

$$= \varrho k_r + \frac{1}{r^2 \sin \vartheta} \left\{ \frac{\partial}{\partial r} \left[r^2 \sin \vartheta \left\{ -p + \underline{\underline{\eta \frac{\partial u_r}{\partial r}}} + \underline{\eta \frac{\partial u_r}{\partial r}} \right\} \right] \right.$$

$$+ \frac{\partial}{\partial \vartheta} \left[r^2 \sin \vartheta \eta \underline{\frac{\partial(u_\vartheta/r)}{\partial r}} + \sin \vartheta \eta \underline{\frac{\partial u_r}{\partial \vartheta}} \right] + \frac{\partial}{\partial \varphi} \left[\underline{\frac{\eta}{\sin \vartheta} \frac{\partial u_r}{\partial \varphi}} \right]$$

$$\left. + r^2 \sin \vartheta \eta \underline{\frac{\partial}{\partial r} \left(\frac{1}{r \sin \vartheta} u_\varphi \right)} \right\} + \frac{p}{r} - \frac{2\eta}{r^2} \left[\underline{\underline{\frac{\partial u_\vartheta}{\partial \vartheta}}} + \underline{\underline{u_r}} \right]$$

$$+ \frac{p}{r} - \frac{2\eta}{r^2 \sin \vartheta} \frac{\partial u_\varphi}{\partial \varphi} - \frac{2\eta}{r^2} (u_r + u_\vartheta \cot \vartheta).$$

(B.33)

All terms containing p together result in $-\partial p/\partial r$. In spherical coordinates the Laplace operator reads

$$\Delta = \frac{1}{r^2} \frac{\partial}{\partial r} \left[r^2 \frac{\partial}{\partial r} \right] + \frac{1}{r^2 \sin \vartheta} \left[\frac{\partial}{\partial \vartheta} \left(\sin \vartheta \frac{\partial}{\partial \vartheta} \right) + \frac{1}{\sin \vartheta} \frac{\partial^2}{\partial \varphi^2} \right]. \quad (B.34)$$

We see that the doubly underlined terms can be written together as the differential operator $\eta \Delta u_r$. For the singly underlined terms we can write

$$\eta \frac{\partial}{\partial r} \left\{ \frac{1}{r^2 \sin \vartheta} \left[\frac{\partial}{\partial r} \left(r^2 \sin \vartheta u_r \right) + \frac{\partial}{\partial \vartheta} \left(r \sin \vartheta u_\vartheta \right) + \frac{\partial}{\partial \varphi} \left(ru_\varphi \right) \right] \right\};$$

we can convince ourselves of this by differentiating it out. The expression in curly brackets is, by (B.27) equal to $\nabla \cdot \vec{u}$, and in incompressible flow is zero (Fig. B.2).

If we carry out all the differentiation on the left-hand side we find

$$\varrho \left\{ \frac{\partial u_r}{\partial t} + u_r \frac{\partial u_r}{\partial r} + \frac{1}{r} u_\vartheta \frac{\partial u_r}{\partial \vartheta} + \frac{1}{r \sin \vartheta} u_\varphi \frac{\partial u_r}{\partial \varphi} - \frac{u_\vartheta^2 + u_\varphi^2}{r} \right\}$$

$$= \varrho k_r - \frac{\partial p}{\partial r} + \eta \left\{ \Delta u_r - \frac{2}{r^2} \left[u_r + \frac{\partial u_\vartheta}{\partial \vartheta} + u_\vartheta \cos \vartheta + \frac{1}{\sin \vartheta} \frac{\partial u_\varphi}{\partial \varphi} \right] \right\}$$

(B.35)

as the r-component of the Navier-Stokes equations. The remaining components are obtained in the same manner. We shall now summarize the results for Cartesian, cylindrical and spherical coordinates.

B.1   Cartesian Coordinates

(a) Unit vectors:

$$\vec{e}_x, \vec{e}_y, \vec{e}_z$$

(b) Position vector $\vec{x}$:

$$\vec{x} = x\vec{e}_x + y\vec{e}_y + z\vec{e}_x$$

(c) Velocity vector u:

$$\vec{u} = u\vec{e}_x + v\vec{e}_y + w\vec{e}_z$$

(d) Line element:

$$d\vec{x} = dx\vec{e}_x + dy\vec{e}_y + dz\vec{e}_z$$

(e) Surface elements:

$$dS_x = dydz$$
$$dS_y = dxdz$$
$$dS_x = dxdy$$

(f) Volume element:

$$dV = dxdydz$$

(g) Gradient of the scalar Φ:

$$\text{grad } \Phi = \nabla\Phi = \frac{\partial\Phi}{\partial x}\vec{e}_x + \frac{\partial\Phi}{\partial y}\vec{e}_y + \frac{\partial\Phi}{\partial z}\vec{e}_z$$

(h) Laplace operator on the scalar Φ:

$$\Delta\Phi = \nabla\cdot\nabla\Phi = \frac{\partial^2\Phi}{\partial x^2} + \frac{\partial^2\Phi}{\partial y^2} + \frac{\partial^2\Phi}{\partial z^2}$$

(i) Divergence of the vector $\vec{u}$:

$$\text{div}\,\vec{u} = \nabla \cdot \vec{u} = \frac{\partial u}{\partial x} + \frac{\partial v}{\partial y} + \frac{\partial w}{\partial z}$$

(j) Curl of the vector $\vec{u}$:

$$\text{curl}\,\vec{u} = \nabla \times \vec{u} = \left[\frac{\partial w}{\partial y} - \frac{\partial v}{\partial z}\right]\vec{e}_x + \left[\frac{\partial u}{\partial z} - \frac{\partial w}{\partial x}\right]\vec{e}_y + \left[\frac{\partial v}{\partial x} - \frac{\partial u}{\partial y}\right]\vec{e}_z$$

(k) Laplace operator on the vector $\vec{u}$:

$$\Delta\vec{u} = \nabla \cdot \nabla\vec{u} = \Delta u\vec{e}_x + \Delta v\vec{e}_y + \Delta w\vec{e}_z$$

(l) Divergence of the stress tensor **T**:

$$\begin{aligned}
\text{div }\mathbf{T} = \nabla \cdot \mathbf{T} &= (\partial\tau_{xx}/\partial x + \partial\tau_{yx}/\partial y + \partial\tau_{zx}/\partial z)\vec{e}_x \\
&+ (\partial\tau_{xy}/\partial x + \partial\tau_{yy}/\partial y + \partial\tau_{zy}/\partial z)\vec{e}_y \\
&+ (\partial\tau_{xz}/\partial x + \partial\tau_{yz}/\partial y + \partial\tau_{zz}/\partial z)\vec{e}_z
\end{aligned}$$

(m) Rate of deformation tensor **E**:

$$\begin{aligned}
e_{xx} &= \partial u/\partial x \\
e_{yy} &= \partial v/\partial y \\
e_{zz} &= \partial w/\partial z \\
2e_{xy} = 2e_{yx} &= \partial u/\partial y + \partial v/\partial x \\
2e_{xz} = 2e_{zx} &= \partial u/\partial z + \partial w/\partial x \\
2e_{yz} = 2e_{zy} &= \partial v/\partial z + \partial w/\partial y
\end{aligned}$$

(n) Continuity equation:

$$\frac{\partial\varrho}{\partial t} + \frac{\partial}{\partial x}(\varrho u) + \frac{\partial}{\partial y}(\varrho v) + \frac{\partial}{\partial z}(\varrho w) = 0$$

(o) Navier–Stokes equations (with ϱ, η = const):

$$\begin{aligned}
x &: \varrho(\partial u/\partial t + u\partial u/\partial x + v\partial u/\partial y + w\partial u/\partial z) = \varrho k_x - \partial p/\partial x + \eta\Delta u \\
y &: \varrho(\partial v/\partial t + u\partial v/\partial x + v\partial v/\partial y + w\partial v/\partial z) = \varrho k_y - \partial p/\partial y + \eta\Delta v \\
z &: \varrho(\partial w/\partial t + u\partial w/\partial x + v\partial w/\partial y + w\partial w/\partial z) = \varrho k_z - \partial p/\partial z + \eta\Delta w
\end{aligned}$$

B.2   Cylindrical Coordinates

(a) Unit vectors:

$$\vec{e}_r = + \cos \varphi \vec{e}_x + \sin \varphi \vec{e}_y$$
$$\vec{e}_\varphi = - \sin \varphi \vec{e}_x + \cos \varphi \vec{e}_y$$
$$\vec{e}_z = \vec{e}_z$$

(b) Position vector $\vec{x}$:

$$\vec{x} = r\vec{e}_r + z\vec{e}_z$$

(c) Velocity vector $\vec{u}$:

$$\vec{u} = u_r\vec{e}_r + u_\varphi\vec{e}_\varphi + u_z\vec{e}_z$$

(d) Line element:

$$d\vec{x} = dr\vec{e}_r + rd\varphi\vec{e}_\varphi + dz\vec{e}_z$$

See Fig. B.3.

(e) Surface elements:

$$dS_r = rd\varphi dz$$
$$dS_\varphi = drdz$$
$$dS_z = rdrd\varphi$$

(f) Volume element:

$$dV = rdrd\varphi dz$$

(g) Gradient of the scalar Φ:

$$\text{grad } \Phi = \nabla \Phi = \frac{\partial \Phi}{\partial r}\vec{e}_r + \frac{1}{r}\frac{\partial \Phi}{\partial \varphi}\vec{e}_\varphi + \frac{\partial \Phi}{\partial z}\vec{e}_z$$

(h) Laplace operator on the scalar Φ:

$$\Delta \Phi = \nabla \cdot \nabla \Phi = \frac{\partial^2 \Phi}{\partial r^2} + \frac{1}{r}\frac{\partial \Phi}{\partial r} + \frac{1}{r^2}\frac{\partial^2 \Phi}{\partial \varphi^2} + \frac{\partial^2 \Phi}{\partial z^2}$$

(i) Divergence of the vector $\vec{u}$:

$$\operatorname{div}\vec{u} = \nabla \cdot \vec{u} = \frac{1}{r}\left\{\frac{\partial(u_r r)}{\partial r} + \frac{\partial u_\varphi}{\partial \varphi} + \frac{\partial(u_z r)}{\partial z}\right\}$$

(j) Curl of the vector $\vec{u}$:

$$\operatorname{curl}\vec{u} = \nabla \times \vec{u} = \left\{\frac{1}{r}\frac{\partial u_z}{\partial \varphi} - \frac{\partial u_\varphi}{\partial z}\right\}\vec{e}_r + \left\{\frac{\partial u_r}{\partial z} - \frac{\partial u_z}{\partial r}\right\}\vec{e}_\varphi$$
$$+ \frac{1}{r}\left\{\frac{\partial(u_\varphi r)}{\partial r} - \frac{\partial u_r}{\partial \varphi}\right\}\vec{e}_z$$

(k) Laplace operator on the vector $\vec{u}$:

$$\Delta\vec{u} = \nabla \cdot \nabla \vec{u} = \left\{\Delta u_r - \frac{1}{r^2}\left[u_r + 2\frac{\partial u_\varphi}{\partial \varphi}\right]\right\}\vec{e}_r$$
$$+ \left\{\Delta u_\varphi - \frac{1}{r^2}\left[u_\varphi - 2\frac{\partial u_r}{\partial \varphi}\right]\right\}\vec{e}_\varphi + \Delta u_z\vec{e}_z$$

(l) Divergence of the stress tensor **T**:

$$\operatorname{div}\mathbf{T} = \nabla \cdot \mathbf{T} = \left\{\frac{1}{r}\frac{\partial(\tau_{rr}r)}{\partial r} + \frac{1}{r}\frac{\partial \tau_{\varphi r}}{\partial \varphi} + \frac{\partial \tau_{zr}}{\partial z} - \frac{\tau_{\varphi\varphi}}{r}\right\}\vec{e}_r$$
$$+ \left\{\frac{1}{r}\frac{\partial(\tau_{r\varphi}r)}{\partial r} + \frac{1}{r}\frac{\partial \tau_{\varphi\varphi}}{\partial \varphi} + \frac{\partial \tau_{z\varphi}}{\partial z} + \frac{\tau_{r\varphi}}{r}\right\}\vec{e}_\varphi$$
$$+ \left\{\frac{1}{r}\frac{\partial(\tau_{rz}r)}{\partial r} + \frac{1}{r}\frac{\partial \tau_{\varphi z}}{\partial \varphi} + \frac{\partial \tau_{zz}}{\partial z}\right\}\vec{e}_z$$

(m) Rate of deformation tensor **E**:

$$e_{rr} = \frac{\partial u_r}{\partial r}$$

$$e_{\varphi\varphi} = \frac{1}{r}\frac{\partial u_\varphi}{\partial \varphi} + \frac{1}{r}u_r$$

$$e_{zz} = \frac{\partial u_z}{\partial z}$$

$$2e_{r\varphi} = 2e_{\varphi r} = r\frac{\partial(r^{-1}u_\varphi)}{\partial r} + \frac{1}{r}\frac{\partial u_r}{\partial \varphi}$$

$$2e_{rz} = 2e_{zr} = \frac{\partial u_r}{\partial z} + \frac{\partial u_z}{\partial r}$$

$$2e_{\varphi z} = 2e_{z\varphi} = \frac{1}{r}\frac{\partial u_z}{\partial \varphi} + \frac{\partial u_\varphi}{\partial z}$$

(n) Continuity equation:

$$\frac{\partial \varrho}{\partial t} + \frac{1}{r}\frac{\partial}{\partial r}\left(\varrho u_r r\right) + \frac{1}{r}\frac{\partial}{\partial \varphi}\left(\varrho u_\varphi\right) + \frac{\partial}{\partial z}\left(\varrho u_z\right) = 0$$

(o) Navier-Stokes equations (with ϱ, η = const)

$$r : \varrho\left\{\frac{\partial u_r}{\partial t} + u_r\frac{\partial u_r}{\partial r} + u_z\frac{\partial u_r}{\partial z} + \frac{1}{r}\left[u_\varphi\frac{\partial u_r}{\partial \varphi} - u_\varphi^2\right]\right\}$$

$$= \varrho k_r - \frac{\partial p}{\partial r} + \eta\left\{\Delta u_r - \frac{1}{r^2}\left[u_r + 2\frac{\partial u_\varphi}{\partial \varphi}\right]\right\}$$

$$\varphi : \varrho\left\{\frac{\partial u_\varphi}{\partial t} + u_r\frac{\partial u_\varphi}{\partial r} + u_z\frac{\partial u_\varphi}{\partial z} + \frac{1}{r}\left[u_\varphi\frac{\partial u_\varphi}{\partial \varphi} + u_r u_\varphi\right]\right\}$$

$$= \varrho k_\varphi - \frac{1}{r}\frac{\partial p}{\partial \varphi} + \eta\left\{\Delta u_\varphi - \frac{1}{r^2}\left[u_\varphi - 2\frac{\partial u_r}{\partial \varphi}\right]\right\}$$

$$z : \varrho\left\{\frac{\partial u_z}{\partial t} + u_r\frac{\partial u_z}{\partial r} + u_z\frac{\partial u_z}{\partial z} + \frac{1}{r}u_\varphi\frac{\partial u_z}{\partial \varphi}\right\} = \varrho k_z - \frac{\partial p}{\partial z} + \eta\Delta u_z$$

B.3 Spherical Coordinates

(a) Unit vectors:

$$\vec{e}_r = \cos\vartheta\vec{e}_x + \sin\vartheta\cos\varphi\vec{e}_y + \sin\vartheta\sin\varphi\vec{e}_z$$
$$\vec{e}_\vartheta = -\sin\vartheta\vec{e}_x + \cos\vartheta\cos\varphi\vec{e}_y + \cos\vartheta\sin\varphi\vec{e}_z$$
$$\vec{e}_\varphi = -\sin\varphi\vec{e}_y + \cos\varphi\vec{e}_z$$

(b) Position vector $\vec{x}$:

$$\vec{x} = r\vec{e}_r$$

(c) Velocity vector $\vec{u}$:

$$\vec{u} = u_r\vec{e}_r + u_\vartheta\vec{e}_\vartheta + u_\varphi\vec{e}_\varphi$$

(d) Line element:

$$d\vec{x} = dr\vec{e}_r + rd\vartheta\vec{e}_\vartheta + r\sin\vartheta d\varphi\vec{e}_\varphi$$

(e) Surface elements:

$$dS_r = r^2 \sin \vartheta d\vartheta d\varphi$$
$$dS_\vartheta = r \sin \vartheta dr d\varphi$$
$$dS_\varphi = r dr d\vartheta$$

(f) Volume element:

$$dV = r^2 \sin \vartheta dr d\vartheta d\varphi$$

(g) Gradient of the scalar Φ:

$$\text{grad } \Phi = \nabla \Phi = \frac{\partial \Phi}{\partial r} \vec{e}_r + \frac{1}{r} \frac{\partial \Phi}{\partial \vartheta} \vec{e}_\vartheta + \frac{1}{r \sin \vartheta} \frac{\partial \Phi}{\partial \varphi} \vec{e}_\varphi$$

(h) Laplace operator on the scalar Φ:

$$\Delta \Phi = \nabla \cdot \nabla \Phi = \frac{1}{r^2} \frac{\partial}{\partial r} \left[r^2 \frac{\partial \Phi}{\partial r} \right] + \frac{1}{r^2 \sin \vartheta} \frac{\partial}{\partial \vartheta} \left[\sin \vartheta \frac{\partial \Phi}{\partial \vartheta} \right] + \frac{1}{r^2 \sin^2 \vartheta} \frac{\partial^2 \Phi}{\partial \varphi^2}$$

(i) Divergence of the vector $\vec{u}$:

$$\text{div } \vec{u} = \nabla \cdot \vec{u} = \frac{1}{r^2 \sin \vartheta} \left\{ \frac{\partial (r^2 \sin \vartheta u_r)}{\partial r} + \frac{\partial (r \sin \vartheta u_\vartheta)}{\partial \vartheta} + \frac{\partial (r u_\varphi)}{\partial \varphi} \right\}$$

(j) Curl of the vector $\vec{u}$:

$$\text{curl } \vec{u} = \frac{1}{r^2 \sin \vartheta} \left\{ \frac{\partial (r \sin \vartheta u_\varphi)}{\partial \vartheta} - \frac{\partial (r u_\vartheta)}{\partial \varphi} \right\} \vec{e}_r$$
$$+ \frac{1}{r \sin \vartheta} \left\{ \frac{\partial u_r}{\partial \varphi} - \frac{\partial (r \sin \vartheta u_\varphi)}{\partial r} \right\} \vec{e}_\vartheta$$
$$+ \frac{1}{r} \left\{ \frac{\partial (r u_\vartheta)}{\partial r} - \frac{\partial u_r}{\partial \vartheta} \right\} \vec{e}_\varphi$$

(k) Laplace operator on the vector $\vec{u}$:

$$\Delta \vec{u} = \left\{ \Delta u_r - \frac{2}{r^2} \left[u_r + \frac{\partial u_\vartheta}{\partial \vartheta} + u_\vartheta \cot \vartheta + \frac{1}{\sin \vartheta} \frac{\partial u_\varphi}{\partial \varphi} \right] \right\} \vec{e}_r$$
$$+ \left\{ \Delta u_\vartheta + \frac{2}{r^2} \frac{\partial u_r}{\partial \vartheta} - \frac{1}{r^2 \sin^2 \vartheta} \left[u_\vartheta + 2 \cos \vartheta \frac{\partial u_\varphi}{\partial \varphi} \right] \right\} \vec{e}_\vartheta$$
$$+ \left\{ \Delta u_\varphi - \frac{1}{r^2 \sin^2 \vartheta} \left[u_\varphi - 2 \sin \vartheta \frac{\partial u_r}{\partial \varphi} - 2 \cos \vartheta \frac{\partial u_\vartheta}{\partial \varphi} \right] \right\} \vec{e}_\varphi$$

(l) Divergence of the stress tensor **T**:

$$\nabla \cdot \mathbf{T} = \left\{ \frac{1}{r^2 \sin \vartheta} \left[\frac{\partial(r^2 \sin \vartheta \tau_{rr})}{\partial r} + \frac{\partial(r \sin \vartheta \tau_{\vartheta r})}{\partial \vartheta} + \frac{\partial(r\tau_{\varphi r})}{\partial \varphi} \right] \right.$$
$$\left. - \frac{\tau_{\vartheta\vartheta} + \tau_{\varphi\varphi}}{r} \right\} \vec{e}_r$$
$$+ \left\{ \frac{1}{r^2 \sin \vartheta} \left[\frac{\partial(r^2 \sin \vartheta \tau_{r\vartheta})}{\partial r} + \frac{\partial(r \sin \vartheta \tau_{\vartheta\vartheta})}{\partial \vartheta} + \frac{\partial(r\tau_{\varphi\vartheta})}{\partial \varphi} \right] \right.$$
$$\left. + \frac{\tau_{r\vartheta} - \tau_{\varphi\varphi} \cot \vartheta}{r} \right\} \vec{e}_\vartheta$$
$$+ \left\{ \frac{1}{r^2 \sin \vartheta} \left[\frac{\partial(r^2 \sin \vartheta \tau_{r\varphi})}{\partial r} + \frac{\partial(r \sin \vartheta \tau_{\vartheta\varphi})}{\partial \vartheta} + \frac{\partial(r\tau_{\varphi\varphi})}{\partial \varphi} \right] \right.$$
$$\left. + \frac{\tau_{r\varphi} + \tau_{\vartheta\varphi} \cot \vartheta}{r} \right\} \vec{e}_\varphi$$

(m) Rate of deformation tensor **E**:

$$e_{rr} = \frac{\partial u_r}{\partial r}$$

$$e_{\vartheta\vartheta} = \frac{1}{r} \frac{\partial u_\vartheta}{\partial \vartheta} + \frac{1}{r} u_r$$

$$e_{\varphi\varphi} = \frac{1}{r \sin \vartheta} \frac{\partial u_\varphi}{\partial \varphi} + \frac{1}{r} (u_r + u_\vartheta \cos \vartheta)$$

$$2e_{\varphi\vartheta} = 2e_{\vartheta\varphi} = \sin \vartheta \frac{\partial}{\partial \vartheta} \left[\frac{1}{r \sin \vartheta} u_\varphi \right] + \frac{1}{\sin \vartheta} \frac{\partial}{\partial \varphi} \left[\frac{1}{r} u_\vartheta \right]$$

$$2e_{r\varphi} = 2e_{\varphi r} = \frac{1}{r \sin \vartheta} \frac{\partial u_r}{\partial \varphi} + r \sin \vartheta \frac{\partial}{\partial r} \left[\frac{1}{r \sin \vartheta} u_\varphi \right]$$

$$2e_{\vartheta r} = 2e_{r\vartheta} = r \frac{\partial}{\partial r} \left[\frac{1}{r} u_\vartheta \right] + \frac{1}{r} \frac{\partial u_r}{\partial \vartheta}$$

(n) Continuity equation:

$$\frac{\partial \varrho}{\partial t} + \frac{1}{r^2 \sin \vartheta} \left[\frac{\partial}{\partial r} \left(r^2 \sin \vartheta \varrho u_r \right) + \frac{\partial}{\partial \vartheta} \left(r \sin \vartheta \varrho u_\vartheta \right) + \frac{\partial}{\partial \varphi} \left(r \varrho u_\varphi \right) \right] = 0$$

(o) Navier-Stokes equations (with $\varrho, \eta = $ const):

$$r: \quad \varrho\left\{\frac{\partial u_r}{\partial t} + u_r\frac{\partial u_r}{\partial r} + \frac{1}{r}u_\vartheta\frac{\partial u_r}{\partial \vartheta} + \frac{1}{r\sin\vartheta}u_\varphi\frac{\partial u_r}{\partial \varphi} - \frac{u_\vartheta^2 + u_\varphi^2}{r}\right\}$$

$$= \varrho k_r - \frac{\partial p}{\partial r} + \eta\left\{\Delta u_r - \frac{2}{r^2}\left[u_r + \frac{\partial u_\vartheta}{\partial \vartheta} + u_\vartheta\cot\vartheta + \frac{1}{\sin\vartheta}\frac{\partial u_\varphi}{\partial \varphi}\right]\right\}$$

$$\vartheta: \quad \varrho\left\{\frac{\partial u_\vartheta}{\partial t} + u_r\frac{\partial u_\vartheta}{\partial r} + \frac{1}{r}u_\vartheta\frac{\partial u_\vartheta}{\partial \vartheta} + \frac{1}{r\sin\vartheta}u_\varphi\frac{\partial u_\vartheta}{\partial \varphi} + \frac{u_r u_\vartheta - u_\varphi^2\cot\vartheta}{r}\right\}$$

$$= \varrho k_\vartheta - \frac{1}{r}\frac{\partial p}{\partial \vartheta} + \eta\left\{\Delta u_\vartheta + \frac{2}{r^2}\frac{\partial u_r}{\partial \vartheta} - \frac{1}{r^2\sin^2\vartheta}\left[u_\vartheta + 2\cos\vartheta\frac{\partial u_\varphi}{\partial \varphi}\right]\right\}$$

$$\varphi: \quad \varrho\left\{\frac{\partial u_\varphi}{\partial t} + u_r\frac{\partial u_\varphi}{\partial r} + \frac{1}{r}u_\vartheta\frac{\partial u_\varphi}{\partial \vartheta} + \frac{1}{r\sin\vartheta}u_\varphi\frac{\partial u_\varphi}{\partial \varphi} + \frac{u_\varphi u_r + u_\vartheta u_\varphi\cot\vartheta}{r}\right\}$$

$$= \varrho k_\varphi - \frac{1}{r\sin\vartheta}\frac{\partial p}{\partial \varphi} + \eta\left\{\Delta u_\varphi - \frac{1}{r^2\sin^2\vartheta}\left[u_\varphi - 2\cos\vartheta\frac{\partial u_\vartheta}{\partial \varphi} - 2\sin\vartheta\frac{\partial u_r}{\partial \varphi}\right]\right\}$$

Appendix C
Tables and Diagrams for Compressible Flow

See Table C.1.

Table C.1 Pressure, density, temperature and area ratio as dependent on the Mach number for calorically perfect gas ($\gamma = 1.4$)

M	p/p_t	ϱ/ϱ_t	T/T_t	a/a_t	A^*/A
0.000	1.000000	1.000000	1.000000	1.000000	0.000000
0.010	0.999930	0.999950	0.999980	0.999990	0.017279
0.020	0.999720	0.999800	0.999920	0.999960	0.034552
0.030	0.999370	0.999550	0.999820	0.999910	0.051812
0.040	0.998881	0.999200	0.999680	0.999840	0.069054
0.050	0.998252	0.998751	0.999500	0.999750	0.086271
0.060	0.997484	0.998202	0.999281	0.999640	0.103456
0.070	0.996577	0.997554	0.999021	0.999510	0.120605
0.080	0.995533	0.996807	0.998722	0.999361	0.137711
0.090	0.994351	0.995961	0.998383	0.999191	0.154767
0.100	0.993032	0.995018	0.998004	0.999002	0.171767
0.110	0.991576	0.993976	0.997586	0.998792	0.188707
0.120	0.989985	0.992836	0.997128	0.998563	0.205579
0.130	0.988259	0.991600	0.996631	0.998314	0.222378
0.140	0.986400	0.990267	0.996095	0.998046	0.239097
0.150	0.984408	0.988838	0.995520	0.997758	0.255732
0.160	0.982284	0.987314	0.994906	0.997450	0.272276
0.170	0.980030	0.985695	0.994253	0.997122	0.288725
0.180	0.977647	0.983982	0.993562	0.996776	0.305071
0.190	0.975135	0.982176	0.992832	0.996409	0.321310
0.200	0.972497	0.980277	0.992064	0.996024	0.337437

(continued)

© Springer Nature Switzerland AG 2020
J. H. Spurk and N. Aksel, *Fluid Mechanics*,
https://doi.org/10.1007/978-3-030-30259-7

Table C.1 (continued)

M	p/p_t	ϱ/ϱ_t	T/T_t	a/a_t	A^*/A
0.210	0.969733	0.978286	0.991257	0.995619	0.353445
0.220	0.966845	0.976204	0.990413	0.995195	0.369330
0.230	0.963835	0.974032	0.989531	0.994752	0.385088
0.240	0.960703	0.971771	0.988611	0.994289	0.400711
0.250	0.957453	0.969421	0.987654	0.993808	0.416197
0.260	0.954085	0.966984	0.986660	0.993308	0.431539
0.270	0.950600	0.964460	0.985629	0.992789	0.446734
0.280	0.947002	0.961851	0.984562	0.992251	0.461776
0.290	0.943291	0.959157	0.983458	0.991695	0.476661
0.300	0.939470	0.956380	0.982318	0.991120	0.491385
0.310	0.935540	0.953521	0.981142	0.990526	0.505943
0.320	0.931503	0.950580	0.979931	0.989915	0.520332
0.330	0.927362	0.947559	0.978684	0.989285	0.534546
0.340	0.923117	0.944460	0.977402	0.988637	0.548584
0.350	0.918773	0.941283	0.976086	0.987971	0.562440
0.360	0.914330	0.938029	0.974735	0.987287	0.576110
0.370	0.909790	0.934700	0.973350	0.986585	0.589593
0.380	0.905156	0.931297	0.971931	0.985865	0.602883
0.390	0.900430	0.927821	0.970478	0.985128	0.615979
0.400	0.895614	0.924274	0.968992	0.984374	0.628876
0.410	0.890711	0.920657	0.967474	0.983602	0.641571
0.420	0.885722	0.916971	0.965922	0.982813	0.654063
0.430	0.880651	0.913217	0.964339	0.982008	0.666348
0.440	0.875498	0.909398	0.962723	0.981185	0.678424
0.450	0.870267	0.905513	0.961076	0.980345	0.690287
0.460	0.864960	0.901566	0.959398	0.979489	0.701937
0.470	0.859580	0.897556	0.957689	0.978616	0.713371
0.480	0.854128	0.893486	0.955950	0.977727	0.724587
0.490	0.848607	0.889357	0.954180	0.976821	0.735582
0.500	0.843019	0.885170	0.952381	0.975900	0.746356
0.510	0.837367	0.880927	0.950552	0.974963	0.756906
0.520	0.831654	0.876629	0.948695	0.974010	0.767231
0.530	0.825881	0.872279	0.946808	0.973041	0.777331
0.540	0.820050	0.867876	0.944894	0.972056	0.787202
0.550	0.814165	0.863422	0.942951	0.971057	0.796846
0.560	0.808228	0.858920	0.940982	0.970042	0.806260
0.570	0.802241	0.854371	0.938985	0.969012	0.815444
0.580	0.796206	0.849775	0.936961	0.967968	0.824398
0.590	0.790127	0.845135	0.934911	0.966908	0.833119
0.600	0.784004	0.840452	0.932836	0.965834	0.841609

(continued)

Table C.1 (continued)

M	p/p_t	ϱ/ϱ_t	T/T_t	a/a_t	A^*/A
0.610	0.777841	0.835728	0.930735	0.964746	0.849868
0.620	0.771639	0.830963	0.928609	0.963643	0.857894
0.630	0.765402	0.826160	0.926458	0.962527	0.865688
0.640	0.759131	0.821320	0.924283	0.961396	0.873249
0.650	0.752829	0.816443	0.922084	0.960252	0.880579
0.660	0.746498	0.811533	0.919862	0.959094	0.887678
0.670	0.740140	0.806590	0.917616	0.957923	0.894545
0.680	0.733758	0.801616	0.915349	0.956739	0.901182
0.690	0.727353	0.796612	0.913059	0.955541	0.907588
0.700	0.720928	0.791579	0.910747	0.954331	0.913765
0.710	0.714485	0.786519	0.908414	0.953107	0.919715
0.720	0.708026	0.781434	0.906060	0.951872	0.925437
0.730	0.701552	0.776324	0.903685	0.950624	0.930932
0.740	0.695068	0.771191	0.901291	0.949363	0.936203
0.750	0.688573	0.766037	0.898876	0.948091	0.941250
0.760	0.682071	0.760863	0.896443	0.946807	0.946074
0.770	0.675562	0.755670	0.893991	0.945511	0.950678
0.780	0.669050	0.750460	0.891520	0.944203	0.955062
0.790	0.662536	0.745234	0.889031	0.942885	0.959228
0.800	0.656022	0.739992	0.886525	0.941554	0.963178
0.810	0.649509	0.734738	0.884001	0.940214	0.966913
0.820	0.643000	0.729471	0.881461	0.938862	0.970436
0.830	0.636496	0.724193	0.878905	0.937499	0.973749
0.840	0.630000	0.718905	0.876332	0.936126	0.976853
0.850	0.623512	0.713609	0.873744	0.934743	0.979750
0.860	0.617034	0.708306	0.871141	0.933349	0.982443
0.870	0.610569	0.702997	0.868523	0.931946	0.984934
0.880	0.604117	0.697683	0.865891	0.930533	0.987225
0.890	0.597680	0.692365	0.863245	0.929110	0.989317
0.900	0.591260	0.687044	0.860585	0.927677	0.991215
0.910	0.584858	0.681722	0.857913	0.926236	0.992920
0.920	0.578476	0.676400	0.855227	0.924785	0.994434
0.930	0.572114	0.671079	0.852529	0.923325	0.995761
0.940	0.565775	0.665759	0.849820	0.921857	0.996901
0.950	0.559460	0.660443	0.847099	0.920380	0.997859
0.960	0.553169	0.655130	0.844366	0.918894	0.998637
0.970	0.546905	0.649822	0.841623	0.917400	0.999238
0.980	0.540668	0.644520	0.838870	0.915898	0.999663
0.990	0.534460	0.639225	0.836106	0.914389	0.999916
1.000	0.528282	0.633938	0.833333	0.912871	1.000000

(continued)

Table C.1 (continued)

M	p/p_t	ϱ/ϱ_t	T/T_t	a/a_t	A^*/A
1.000	0.528282	0.633938	0.833333	0.912871	1.000000
1.010	0.522134	0.628660	0.830551	0.911346	0.999917
1.020	0.516018	0.623391	0.827760	0.909813	0.999671
1.030	0.509935	0.618133	0.824960	0.908273	0.999263
1.040	0.503886	0.612887	0.822152	0.906726	0.998697
1.050	0.497872	0.607653	0.819336	0.905172	0.997975
1.060	0.491894	0.602432	0.816513	0.903611	0.997101
1.070	0.485952	0.597225	0.813683	0.902044	0.996077
1.080	0.480047	0.592033	0.810846	0.900470	0.994907
1.090	0.474181	0.586856	0.808002	0.898890	0.993593
1.100	0.468354	0.581696	0.805153	0.897303	0.992137
1.110	0.462567	0.576553	0.802298	0.895711	0.990543
1.120	0.456820	0.571427	0.799437	0.894113	0.988815
1.130	0.451114	0.566320	0.796572	0.892509	0.986953
1.140	0.445451	0.561232	0.793701	0.890899	0.984963
1.150	0.439829	0.556164	0.790826	0.889284	0.982845
1.160	0.434251	0.551116	0.787948	0.887664	0.980604
1.170	0.428716	0.546090	0.785065	0.886039	0.978242
1.180	0.423225	0.541085	0.782179	0.884409	0.975762
1.190	0.417778	0.536102	0.779290	0.882774	0.973167
1.200	0.412377	0.531142	0.776398	0.881134	0.970459
1.210	0.407021	0.526205	0.773503	0.879490	0.967643
1.220	0.401711	0.521292	0.770606	0.877842	0.964719
1.230	0.396446	0.516403	0.767707	0.876189	0.961691
1.240	0.391229	0.511539	0.764807	0.874532	0.958562
1.250	0.386058	0.506701	0.761905	0.872872	0.955335
1.260	0.380934	0.501888	0.759002	0.871207	0.952012
1.270	0.375858	0.497102	0.756098	0.869539	0.948597
1.280	0.370828	0.492342	0.753194	0.867867	0.945091
1.290	0.365847	0.487609	0.750289	0.866192	0.941497
1.300	0.360914	0.482903	0.747384	0.864514	0.937819
1.310	0.356029	0.478225	0.744480	0.862832	0.934057
1.320	0.351192	0.473575	0.741576	0.861148	0.930217
1.330	0.346403	0.468954	0.738672	0.859461	0.926299
1.340	0.341663	0.464361	0.735770	0.857771	0.922306
1.350	0.336971	0.459797	0.732869	0.856078	0.918242
1.360	0.332328	0.455263	0.729970	0.854383	0.914107
1.370	0.327733	0.450758	0.727072	0.852685	0.909905
1.380	0.323187	0.446283	0.724176	0.850985	0.905639
1.390	0.318690	0.441838	0.721282	0.849283	0.901310

(continued)

Table C.1 (continued)

M	p/p_t	ϱ/ϱ_t	T/T_t	a/a_t	A^*/A
1.400	0.314241	0.437423	0.718391	0.847579	0.896921
1.410	0.309840	0.433039	0.715502	0.845874	0.892474
1.420	0.305489	0.428686	0.712616	0.844166	0.887972
1.430	0.301185	0.424363	0.709733	0.842457	0.883416
1.440	0.296929	0.420072	0.706854	0.840746	0.878810
1.450	0.292722	0.415812	0.703978	0.839034	0.874154
1.460	0.288563	0.411583	0.701105	0.837320	0.869452
1.470	0.284452	0.407386	0.698236	0.835605	0.864706
1.480	0.280388	0.403220	0.695372	0.833889	0.859917
1.490	0.276372	0.399086	0.692511	0.832173	0.855087
1.500	0.272403	0.394984	0.689655	0.830455	0.850219
1.510	0.268481	0.390914	0.686804	0.828736	0.845315
1.520	0.264607	0.386876	0.683957	0.827017	0.840377
1.530	0.260779	0.382870	0.681115	0.825297	0.835405
1.540	0.256997	0.378896	0.678279	0.823577	0.830404
1.550	0.253262	0.374955	0.675448	0.821856	0.825373
1.560	0.249573	0.371045	0.672622	0.820135	0.820315
1.570	0.245930	0.367168	0.669801	0.818414	0.815233
1.580	0.242332	0.363323	0.666987	0.816693	0.810126
1.590	0.238779	0.359511	0.664178	0.814971	0.804998
1.600	0.235271	0.355730	0.661376	0.813250	0.799850
1.610	0.231808	0.351982	0.658579	0.811529	0.794683
1.620	0.228389	0.348266	0.655789	0.809808	0.789499
1.630	0.225014	0.344582	0.653006	0.808088	0.784301
1.640	0.221683	0.340930	0.650229	0.806368	0.779088
1.650	0.218395	0.337311	0.647459	0.804648	0.773863
1.660	0.215150	0.333723	0.644695	0.802929	0.768627
1.670	0.211948	0.330168	0.641939	0.801211	0.763382
1.680	0.208788	0.326644	0.639190	0.799494	0.758129
1.690	0.205670	0.323152	0.636448	0.797777	0.752869
1.700	0.202594	0.319693	0.633714	0.796061	0.747604
1.710	0.199558	0.316264	0.630987	0.794347	0.742335
1.720	0.196564	0.312868	0.628267	0.792633	0.737064
1.730	0.193611	0.309502	0.625555	0.790920	0.731790
1.740	0.190698	0.306169	0.622851	0.789209	0.726517
1.750	0.187824	0.302866	0.620155	0.787499	0.721245
1.760	0.184990	0.299595	0.617467	0.785791	0.715974
1.770	0.182195	0.296354	0.614787	0.784083	0.710707
1.780	0.179438	0.293145	0.612115	0.782378	0.705444
1.790	0.176720	0.289966	0.609451	0.780674	0.700187

(continued)

Table C.1 (continued)

M	p/p_t	ϱ/ϱ_t	T/T_t	a/a_t	A^*/A
1.800	0.174040	0.286818	0.606796	0.778971	0.694936
1.810	0.171398	0.283701	0.604149	0.777270	0.689692
1.820	0.168792	0.280614	0.601511	0.775571	0.684457
1.830	0.166224	0.277557	0.598881	0.773874	0.679230
1.840	0.163691	0.274530	0.596260	0.772179	0.674014
1.850	0.161195	0.271533	0.593648	0.770486	0.668810
1.860	0.158734	0.268566	0.591044	0.768794	0.663617
1.870	0.156309	0.265628	0.588450	0.767105	0.658436
1.880	0.153918	0.262720	0.585864	0.765418	0.653270
1.890	0.151562	0.259841	0.583288	0.763733	0.648118
1.900	0.149240	0.256991	0.580720	0.762050	0.642981
1.910	0.146951	0.254169	0.578162	0.760369	0.637859
1.920	0.144696	0.251377	0.575612	0.758691	0.632755
1.930	0.142473	0.248613	0.573072	0.757016	0.627668
1.940	0.140283	0.245877	0.570542	0.755342	0.622598
1.950	0.138126	0.243170	0.568020	0.753671	0.617547
1.960	0.135999	0.240490	0.565509	0.752003	0.612516
1.970	0.133905	0.237839	0.563006	0.750337	0.607504
1.980	0.131841	0.235215	0.560513	0.748674	0.602512
1.990	0.129808	0.232618	0.558030	0.747014	0.597542
2.000	0.127805	0.230048	0.555556	0.745356	0.592593
2.010	0.125831	0.227505	0.553091	0.743701	0.587665
2.020	0.123888	0.224990	0.550637	0.742049	0.582761
2.030	0.121973	0.222500	0.548192	0.740400	0.577879
2.040	0.120087	0.220037	0.545756	0.738753	0.573020
2.050	0.118229	0.217601	0.543331	0.737110	0.568186
2.060	0.116399	0.215190	0.540915	0.735469	0.563375
2.070	0.114597	0.212805	0.538509	0.733832	0.558589
2.080	0.112823	0.210446	0.536113	0.732197	0.553828
2.090	0.111075	0.208112	0.533726	0.730566	0.549093
2.100	0.109353	0.205803	0.531350	0.728937	0.544383
2.110	0.107658	0.203519	0.528983	0.727312	0.539699
2.120	0.105988	0.201259	0.526626	0.725690	0.535041
2.130	0.104345	0.199025	0.524279	0.724071	0.530410
2.140	0.102726	0.196814	0.521942	0.722456	0.525806
2.150	0.101132	0.194628	0.519616	0.720844	0.521229
2.160	0.099562	0.192466	0.517299	0.719235	0.516679
2.170	0.098017	0.190327	0.514991	0.717629	0.512157
2.180	0.096495	0.188212	0.512694	0.716027	0.507663
2.190	0.094997	0.186120	0.510407	0.714428	0.503197

(continued)

Table C.1 (continued)

M	p/p_t	ϱ/ϱ_t	T/T_t	a/a_t	A^*/A
2.200	0.093522	0.184051	0.508130	0.712832	0.498759
2.210	0.092069	0.182004	0.505863	0.711240	0.494350
2.220	0.090640	0.179981	0.503606	0.709652	0.489969
2.230	0.089232	0.177980	0.501359	0.708067	0.485617
2.240	0.087846	0.176001	0.499122	0.706485	0.481294
2.250	0.086482	0.174044	0.496894	0.704907	0.477000
2.260	0.085139	0.172110	0.494677	0.703333	0.472735
2.270	0.083817	0.170196	0.492470	0.701762	0.468500
2.280	0.082515	0.168304	0.490273	0.700195	0.464293
2.290	0.081234	0.166433	0.488086	0.698631	0.460117
2.300	0.079973	0.164584	0.485909	0.697071	0.455969
2.310	0.078731	0.162755	0.483741	0.695515	0.451851
2.320	0.077509	0.160946	0.481584	0.693963	0.447763
2.330	0.076306	0.159158	0.479437	0.692414	0.443705
2.340	0.075122	0.157390	0.477300	0.690869	0.439676
2.350	0.073957	0.155642	0.475172	0.689327	0.435677
2.360	0.072810	0.153914	0.473055	0.687790	0.431708
2.370	0.071681	0.152206	0.470947	0.686256	0.427769
2.380	0.070570	0.150516	0.468850	0.684726	0.423859
2.390	0.069476	0.148846	0.466762	0.683200	0.419979
2.400	0.068399	0.147195	0.464684	0.681677	0.416129
2.410	0.067340	0.145563	0.462616	0.680159	0.412309
2.420	0.066297	0.143950	0.460558	0.678644	0.408518
2.430	0.065271	0.142354	0.458510	0.677133	0.404758
2.440	0.064261	0.140777	0.456471	0.675626	0.401026
2.450	0.063267	0.139218	0.454442	0.674123	0.397325
2.460	0.062288	0.137677	0.452423	0.672624	0.393653
2.470	0.061326	0.136154	0.450414	0.671129	0.390010
2.480	0.060378	0.134648	0.448414	0.669638	0.386397
2.490	0.059445	0.133159	0.446425	0.668150	0.382814
2.500	0.058528	0.131687	0.444444	0.666667	0.379259
2.510	0.057624	0.130232	0.442474	0.665187	0.375734
2.520	0.056736	0.128794	0.440513	0.663712	0.372238
2.530	0.055861	0.127373	0.438562	0.662240	0.368771
2.540	0.055000	0.125968	0.436620	0.660772	0.365333
2.550	0.054153	0.124579	0.434688	0.659309	0.361924
2.560	0.053319	0.123206	0.432766	0.657849	0.358543
2.570	0.052499	0.121849	0.430853	0.656394	0.355192
2.580	0.051692	0.120507	0.428949	0.654942	0.351868
2.590	0.050897	0.119182	0.427055	0.653494	0.348573

(continued)

Table C.1 (continued)

M	p/p_t	ϱ/ϱ_t	T/T_t	a/a_t	A^*/A
2.600	0.050115	0.117871	0.425170	0.652051	0.345307
2.610	0.049346	0.116575	0.423295	0.650611	0.342068
2.620	0.048589	0.115295	0.421429	0.649176	0.338858
2.630	0.047844	0.114029	0.419572	0.647744	0.335675
2.640	0.047110	0.112778	0.417725	0.646316	0.332521
2.650	0.046389	0.111542	0.415887	0.644893	0.329394
2.660	0.045679	0.110320	0.414058	0.643474	0.326294
2.670	0.044980	0.109112	0.412239	0.642058	0.323222
2.680	0.044292	0.107918	0.410428	0.640647	0.320177
2.690	0.043616	0.106738	0.408627	0.639239	0.317159
2.700	0.042950	0.105571	0.406835	0.637836	0.314168
2.710	0.042295	0.104418	0.405052	0.636437	0.311204
2.720	0.041650	0.103279	0.403278	0.635042	0.308266
2.730	0.041016	0.102152	0.401513	0.633650	0.305355
2.740	0.040391	0.101039	0.399757	0.632263	0.302470
2.750	0.039777	0.099939	0.398010	0.630880	0.299611
2.760	0.039172	0.098851	0.396272	0.629501	0.296779
2.770	0.038577	0.097777	0.394543	0.628126	0.293972
2.780	0.037992	0.096714	0.392822	0.626755	0.291190
2.790	0.037415	0.095664	0.391111	0.625389	0.288435
2.800	0.036848	0.094626	0.389408	0.624026	0.285704
2.810	0.036290	0.093601	0.387714	0.622667	0.282999
2.820	0.035741	0.092587	0.386029	0.621312	0.280319
2.830	0.035201	0.091585	0.384352	0.619962	0.277663
2.840	0.034669	0.090594	0.382684	0.618615	0.275033
2.850	0.034146	0.089616	0.381025	0.617272	0.272426
2.860	0.033631	0.088648	0.379374	0.615934	0.269844
2.870	0.033124	0.087692	0.377732	0.614599	0.267286
2.880	0.032625	0.086747	0.376098	0.613268	0.264753
2.890	0.032134	0.085813	0.374473	0.611942	0.262242
2.900	0.031652	0.084889	0.372856	0.610619	0.259756
2.910	0.031176	0.083977	0.371248	0.609301	0.257293
2.920	0.030708	0.083075	0.369648	0.607986	0.254853
2.930	0.030248	0.082183	0.368056	0.606676	0.252436
2.940	0.029795	0.081302	0.366472	0.605370	0.250043
2.950	0.029349	0.080431	0.364897	0.604067	0.247672
2.960	0.028910	0.079571	0.363330	0.602768	0.245323
2.970	0.028479	0.078720	0.361771	0.601474	0.242997
2.980	0.028054	0.077879	0.360220	0.600183	0.240693
2.990	0.027635	0.077048	0.358678	0.598897	0.238412
3.000	0.027224	0.076226	0.357143	0.597614	0.236152

Table C.2 Pressure, density, temperature, total pressure and Mach number M_2 behind a normal shock as dependent on the Mach number M_1 in front of the shock for calorically perfect gas ($\gamma = 1.4$)

M_1	p_2/p_1	ϱ_2/ϱ_1	T_2/T_1	p_{t2}/p_{t1}	M_2
1.000	1.000000	1.000000	1.000000	1.000000	1.000000
1.010	1.023450	1.016694	1.006645	0.999999	0.990132
1.020	1.047133	1.033442	1.013249	0.999990	0.980520
1.030	1.071050	1.050240	1.019814	0.999967	0.971154
1.040	1.095200	1.067088	1.026345	0.999923	0.962026
1.050	1.119583	1.083982	1.032843	0.999853	0.953125
1.060	1.144200	1.100921	1.039312	0.999751	0.944445
1.070	1.169050	1.117903	1.045753	0.999611	0.935977
1.080	1.194133	1.134925	1.052169	0.999431	0.927713
1.090	1.219450	1.151985	1.058564	0.999204	0.919647
1.100	1.245000	1.169082	1.064938	0.998928	0.911770
1.110	1.270783	1.186213	1.071294	0.998599	0.904078
1.120	1.296800	1.203377	1.077634	0.998213	0.896563
1.130	1.323050	1.220571	1.083960	0.997768	0.889219
1.140	1.349533	1.237793	1.090274	0.997261	0.882042
1.150	1.376250	1.255042	1.096577	0.996690	0.875024
1.160	1.403200	1.272315	1.102872	0.996052	0.868162
1.170	1.430383	1.289610	1.109159	0.995345	0.861451
1.180	1.457800	1.306927	1.115441	0.994569	0.854884
1.190	1.485450	1.324262	1.121719	0.993720	0.848459
1.200	1.513333	1.341615	1.127994	0.992798	0.842170
1.210	1.541450	1.358983	1.134267	0.991802	0.836014
1.220	1.569800	1.376364	1.140541	0.990731	0.829987
1.230	1.598383	1.393757	1.146816	0.989583	0.824083
1.240	1.627200	1.411160	1.153094	0.988359	0.818301
1.250	1.656250	1.428571	1.159375	0.987057	0.812636
1.260	1.685533	1.445989	1.165661	0.985677	0.807085
1.270	1.715050	1.463413	1.171952	0.984219	0.801645
1.280	1.744800	1.480839	1.178251	0.982682	0.796312
1.290	1.774783	1.498267	1.184557	0.981067	0.791084
1.300	1.805000	1.515695	1.190873	0.979374	0.785957
1.310	1.835450	1.533122	1.197198	0.977602	0.780929
1.320	1.866133	1.550546	1.203533	0.975752	0.775997
1.330	1.897050	1.567965	1.209880	0.973824	0.771159
1.340	1.928200	1.585379	1.216239	0.971819	0.766412
1.350	1.959583	1.602785	1.222611	0.969737	0.761753
1.360	1.991200	1.620182	1.228997	0.967579	0.757181
1.370	2.023050	1.637569	1.235398	0.965344	0.752692

(continued)

Table C.2 (continued)

M_1	p_2/p_1	ϱ_2/ϱ_1	T_2/T_1	p_{t2}/p_{t1}	M_2
1.380	2.055133	1.654945	1.241814	0.963035	0.748286
1.390	2.087450	1.672307	1.248245	0.960652	0.743959
1.400	2.120000	1.689655	1.254694	0.958194	0.739709
1.410	2.152783	1.706988	1.261159	0.955665	0.735536
1.420	2.185800	1.724303	1.267642	0.953063	0.731436
1.430	2.219050	1.741600	1.274144	0.950390	0.727408
1.440	2.252533	1.758878	1.280665	0.947648	0.723451
1.450	2.286250	1.776135	1.287205	0.944837	0.719562
1.460	2.320200	1.793370	1.293765	0.941958	0.715740
1.470	2.354383	1.810583	1.300346	0.939012	0.711983
1.480	2.388800	1.827770	1.306947	0.936001	0.708290
1.490	2.423450	1.844933	1.313571	0.932925	0.704659
1.500	2.458333	1.862069	1.320216	0.929786	0.701089
1.510	2.493450	1.879178	1.326884	0.926586	0.697578
1.520	2.528800	1.896258	1.333574	0.923324	0.694125
1.530	2.564383	1.913308	1.340288	0.920003	0.690729
1.540	2.600200	1.930327	1.347025	0.916624	0.687388
1.550	2.636250	1.947315	1.353787	0.913188	0.684101
1.560	2.672533	1.964270	1.360573	0.909697	0.680867
1.570	2.709050	1.981192	1.367384	0.906151	0.677685
1.580	2.745800	1.998079	1.374220	0.902552	0.674553
1.590	2.782783	2.014931	1.381081	0.898901	0.671471
1.600	2.820000	2.031746	1.387969	0.895200	0.668437
1.610	2.857450	2.048524	1.394882	0.891450	0.665451
1.620	2.895133	2.065264	1.401822	0.887653	0.662511
1.630	2.933050	2.081965	1.408789	0.883809	0.659616
1.640	2.971200	2.098627	1.415783	0.879920	0.656765
1.650	3.009583	2.115248	1.422804	0.875988	0.653958
1.660	3.048200	2.131827	1.429853	0.872014	0.651194
1.670	3.087050	2.148365	1.436930	0.867999	0.648471
1.680	3.126133	2.164860	1.444035	0.863944	0.645789
1.690	3.165450	2.181311	1.451168	0.859851	0.643147
1.700	3.205000	2.197719	1.458330	0.855721	0.640544
1.710	3.244783	2.214081	1.465521	0.851556	0.637979
1.720	3.284800	2.230398	1.472741	0.847356	0.635452
1.730	3.325050	2.246669	1.479991	0.843124	0.632962
1.740	3.365533	2.262893	1.487270	0.838860	0.630508
1.750	3.406250	2.279070	1.494579	0.834565	0.628089
1.760	3.447200	2.295199	1.501918	0.830242	0.625705
1.770	3.488383	2.311279	1.509287	0.825891	0.623354

(continued)

Table C.2 (continued)

M_1	p_2/p_1	ϱ_2/ϱ_1	T_2/T_1	p_{t2}/p_{t1}	M_2
1.780	3.529800	2.327310	1.516686	0.821513	0.621037
1.790	3.571450	2.343292	1.524117	0.817111	0.618753
1.800	3.613333	2.359223	1.531577	0.812684	0.616501
1.810	3.655450	2.375104	1.539069	0.808234	0.614281
1.820	3.697800	2.390934	1.546592	0.803763	0.612091
1.830	3.740383	2.406712	1.554146	0.799271	0.609931
1.840	3.783200	2.422439	1.561732	0.794761	0.607802
1.850	3.826250	2.438112	1.569349	0.790232	0.605701
1.860	3.869533	2.453733	1.576998	0.785686	0.603629
1.870	3.913050	2.469301	1.584679	0.781125	0.601585
1.880	3.956800	2.484815	1.592392	0.776548	0.599568
1.890	4.000783	2.500274	1.600138	0.771959	0.597579
1.900	4.045000	2.515680	1.607915	0.767357	0.595616
1.910	4.089450	2.531030	1.615725	0.762743	0.593680
1.920	4.134133	2.546325	1.623568	0.758119	0.591769
1.930	4.179049	2.561565	1.631444	0.753486	0.589883
1.940	4.224200	2.576749	1.639352	0.748844	0.588022
1.950	4.269583	2.591877	1.647294	0.744195	0.586185
1.960	4.315200	2.606949	1.655268	0.739540	0.584372
1.970	4.361050	2.621964	1.663276	0.734879	0.582582
1.980	4.407133	2.636922	1.671317	0.730214	0.580816
1.990	4.453450	2.651823	1.679392	0.725545	0.579072
2.000	4.500000	2.666667	1.687500	0.720874	0.577350
2.010	4.546783	2.681453	1.695642	0.716201	0.575650
2.020	4.593800	2.696181	1.703817	0.711527	0.573972
2.030	4.641049	2.710851	1.712027	0.706853	0.572315
2.040	4.688533	2.725463	1.720270	0.702180	0.570679
2.050	4.736249	2.740016	1.728548	0.697508	0.569063
2.060	4.784200	2.754511	1.736860	0.692839	0.567467
2.070	4.832383	2.768948	1.745206	0.688174	0.565890
2.080	4.880799	2.783325	1.753586	0.683512	0.564334
2.090	4.929450	2.797643	1.762001	0.678855	0.562796
2.100	4.978333	2.811902	1.770450	0.674203	0.561277
2.110	5.027450	2.826102	1.778934	0.669558	0.559776
2.120	5.076799	2.840243	1.787453	0.664919	0.558294
2.130	5.126383	2.854324	1.796006	0.660288	0.556830
2.140	5.176199	2.868345	1.804594	0.655666	0.555383
2.150	5.226249	2.882307	1.813217	0.651052	0.553953
2.160	5.276533	2.896209	1.821875	0.646447	0.552541
2.170	5.327050	2.910052	1.830569	0.641853	0.551145

(continued)

Table C.2 (continued)

M_1	p_2/p_1	ϱ_2/ϱ_1	T_2/T_1	p_{t2}/p_{t1}	M_2
2.180	5.377800	2.923834	1.839297	0.637269	0.549766
2.190	5.428783	2.937557	1.848060	0.632697	0.548403
2.200	5.480000	2.951220	1.856859	0.628136	0.547056
2.210	5.531450	2.964823	1.865693	0.623588	0.545725
2.220	5.583133	2.978365	1.874563	0.619053	0.544409
2.230	5.635050	2.991848	1.883468	0.614531	0.543108
2.240	5.687200	3.005271	1.892408	0.610023	0.541822
2.250	5.739583	3.018634	1.901384	0.605530	0.540552
2.260	5.792200	3.031937	1.910396	0.601051	0.539295
2.270	5.845049	3.045179	1.919443	0.596588	0.538053
2.280	5.898133	3.058362	1.928527	0.592140	0.536825
2.290	5.951449	3.071485	1.937645	0.587709	0.535612
2.300	6.005000	3.084548	1.946800	0.583294	0.534411
2.310	6.058783	3.097551	1.955991	0.578897	0.533224
2.320	6.112799	3.110495	1.965218	0.574517	0.532051
2.330	6.167049	3.123379	1.974480	0.570154	0.530890
2.340	6.221533	3.136202	1.983779	0.565810	0.529743
2.350	6.276249	3.148967	1.993114	0.561484	0.528608
2.360	6.331199	3.161671	2.002485	0.557177	0.527486
2.370	6.386383	3.174316	2.011892	0.552889	0.526376
2.380	6.441799	3.186902	2.021336	0.548621	0.525278
2.390	6.497449	3.199429	2.030815	0.544372	0.524192
2.400	6.553332	3.211896	2.040332	0.540144	0.523118
2.410	6.609450	3.224304	2.049884	0.535936	0.522055
2.420	6.665800	3.236653	2.059473	0.531748	0.521004
2.430	6.722383	3.248944	2.069098	0.527581	0.519964
2.440	6.779200	3.261175	2.078760	0.523435	0.518936
2.450	6.836250	3.273347	2.088459	0.519311	0.517918
2.460	6.893533	3.285461	2.098193	0.515208	0.516911
2.470	6.951050	3.297517	2.107965	0.511126	0.515915
2.480	7.008800	3.309514	2.117773	0.507067	0.514929
2.490	7.066783	3.321453	2.127618	0.503030	0.513954
2.500	7.125000	3.333333	2.137500	0.499015	0.512989
2.510	7.183449	3.345156	2.147418	0.495022	0.512034
2.520	7.242133	3.356922	2.157373	0.491052	0.511089
2.530	7.301049	3.368629	2.167365	0.487105	0.510154
2.540	7.360199	3.380279	2.177394	0.483181	0.509228
2.550	7.419583	3.391871	2.187460	0.479280	0.508312
2.560	7.479199	3.403407	2.197562	0.475402	0.507406
2.570	7.539049	3.414885	2.207702	0.471547	0.506509

(continued)

Table C.2 (continued)

M_1	p_2/p_1	ϱ_2/ϱ_1	T_2/T_1	p_{t2}/p_{t1}	M_2
2.580	7.599133	3.426307	2.217879	0.467715	0.505620
2.590	7.659449	3.437671	2.228092	0.463907	0.504741
2.600	7.719999	3.448980	2.238343	0.460123	0.503871
2.610	7.780783	3.460232	2.248631	0.456362	0.503010
2.620	7.841799	3.471427	2.258955	0.452625	0.502157
2.630	7.903049	3.482567	2.269317	0.448912	0.501313
2.640	7.964532	3.493651	2.279716	0.445223	0.500477
2.650	8.026249	3.504679	2.290153	0.441557	0.499649
2.660	8.088199	3.515651	2.300626	0.437916	0.498830
2.670	8.150383	3.526569	2.311137	0.434298	0.498019
2.680	8.212800	3.537431	2.321685	0.430705	0.497216
2.690	8.275450	3.548239	2.332270	0.427135	0.496421
2.700	8.338333	3.558991	2.342892	0.423590	0.495634
2.710	8.401449	3.569690	2.353552	0.420069	0.494854
2.720	8.464800	3.580333	2.364249	0.416572	0.494082
2.730	8.528383	3.590923	2.374984	0.413099	0.493317
2.740	8.592199	3.601459	2.385756	0.409650	0.492560
2.750	8.656249	3.611941	2.396565	0.406226	0.491810
2.760	8.720532	3.622369	2.407412	0.402825	0.491068
2.770	8.785049	3.632744	2.418296	0.399449	0.490332
2.780	8.849799	3.643066	2.429217	0.396096	0.489604
2.790	8.914783	3.653335	2.440176	0.392768	0.488882
2.800	8.980000	3.663552	2.451173	0.389464	0.488167
2.810	9.045449	3.673716	2.462207	0.386184	0.487459
2.820	9.111133	3.683827	2.473279	0.382927	0.486758
2.830	9.177049	3.693887	2.484388	0.379695	0.486064
2.840	9.243199	3.703894	2.495535	0.376486	0.485375
2.850	9.309583	3.713850	2.506720	0.373302	0.484694
2.860	9.376199	3.723755	2.517942	0.370140	0.484019
2.870	9.443048	3.733608	2.529202	0.367003	0.483350
2.880	9.510132	3.743411	2.540499	0.363890	0.482687
2.890	9.577449	3.753163	2.551834	0.360800	0.482030
2.900	9.644999	3.762864	2.563207	0.357733	0.481380
2.910	9.712782	3.772514	2.574618	0.354690	0.480735
2.920	9.780800	3.782115	2.586066	0.351670	0.480096
2.930	9.849050	3.791666	2.597552	0.348674	0.479463
2.940	9.917533	3.801167	2.609076	0.345701	0.478836
2.950	9.986250	3.810619	2.620637	0.342750	0.478215
2.960	10.055200	3.820021	2.632236	0.339823	0.477599
2.970	10.124383	3.829375	2.643874	0.336919	0.476989

(continued)

Table C.2 (continued)

M_1	p_2/p_1	ϱ_2/ϱ_1	T_2/T_1	p_{t2}/p_{t1}	M_2
2.980	10.193799	3.838679	2.655549	0.334038	0.476384
2.990	10.263450	3.847935	2.667261	0.331180	0.475785
3.000	10.333333	3.857143	2.679012	0.328344	0.475191

Table C.3 Prandtl-Meyer function and Mach angle as dependent on the Mach number for calorically perfect gas (stated for v and μ in degrees)

M	v	μ	M	N	μ
1.000	0.0000	90.0000	2.000	26.3798	30.0000
1.010	0.0447	81.9307	2.010	26.6550	29.8356
1.020	0.1257	78.6351	2.020	26.9295	29.6730
1.030	0.2294	76.1376	2.030	27.2033	29.5123
1.040	0.3510	74.0576	2.040	27.4762	29.3535
1.050	0.4874	72.2472	2.050	27.7484	29.1964
1.060	0.6367	70.6300	2.060	28.0197	29.0411
1.070	0.7973	69.1603	2.070	28.2903	28.8875
1.080	0.9680	67.8084	2.080	28.5600	28.7357
1.090	1.1479	66.5534	2.090	28.8290	28.5855
1.100	1.3362	65.3800	2.100	29.0971	28.4369
1.110	1.5321	64.2767	2.110	29.3644	28.2899
1.120	1.7350	63.2345	2.120	29.6309	28.1446
1.130	1.9445	62.2461	2.130	29.8965	28.0008
1.140	2.1600	61.3056	2.140	30.1613	27.8585
1.150	2.3810	60.4082	2.150	30.4253	27.7177
1.160	2.6073	59.5497	2.160	30.6884	27.5785
1.170	2.8385	58.7267	2.170	30.9507	27.4406
1.180	3.0743	57.9362	2.180	31.2121	27.3043
1.190	3.3142	57.1756	2.190	31.4727	27.1693
1.200	3.5582	56.4427	2.200	31.7325	27.0357
1.210	3.8060	55.7354	2.210	31.9914	26.9035
1.220	4.0572	55.0520	2.220	32.2494	26.7726
1.230	4.3117	54.3909	2.230	32.5066	26.6430
1.240	4.5694	53.7507	2.240	32.7629	26.5148
1.250	4.8299	53.1301	2.250	33.0184	26.3878
1.260	5.0931	52.5280	2.260	33.2730	26.2621
1.270	5.3590	51.9433	2.270	33.5268	26.1376
1.280	5.6272	51.3752	2.280	33.7796	26.0144
1.290	5.8977	50.8226	2.290	34.0316	25.8923
1.300	6.1703	50.2849	2.300	34.2828	25.7715
1.310	6.4449	49.7612	2.310	34.5331	25.6518

(continued)

Table C.3 (continued)

M	v	μ	M	N	μ
1.320	6.7213	49.2509	2.320	34.7825	25.5332
1.330	6.9995	48.7535	2.330	35.0310	25.4158
1.340	7.2794	48.2682	2.340	35.2787	25.2995
1.350	7.5607	47.7945	2.350	35.5255	25.1843
1.360	7.8435	47.3321	2.360	35.7715	25.0702
1.370	8.1276	46.8803	2.370	36.0165	24.9572
1.380	8.4130	46.4387	2.380	36.2607	24.8452
1.390	8.6995	46.0070	2.390	36.5041	24.7342
1.400	8.9870	45.5847	2.400	36.7465	24.6243
1.410	9.2756	45.1715	2.410	36.9881	24.5154
1.420	9.5650	44.7670	2.420	37.2289	24.4075
1.430	9.8553	44.3709	2.430	37.4687	24.3005
1.440	10.1464	43.9830	2.440	37.7077	24.1945
1.450	10.4381	43.6028	2.450	37.9458	24.0895
1.460	10.7305	43.2302	2.460	38.1831	23.9854
1.470	11.0235	42.8649	2.470	38.4195	23.8822
1.480	11.3169	42.5066	2.480	38.6551	23.7800
1.490	11.6109	42.1552	2.490	38.8897	23.6786
1.500	11.9052	41.8103	2.500	39.1236	23.5782
1.510	12.1999	41.4718	2.510	39.3565	23.4786
1.520	12.4949	41.1395	2.520	39.5886	23.3799
1.530	12.7901	40.8132	2.530	39.8199	23.2820
1.540	13.0856	40.4927	2.540	40.0503	23.1850
1.550	13.3812	40.1778	2.550	40.2798	23.0888
1.560	13.6770	39.8683	2.560	40.5085	22.9934
1.570	13.9728	39.5642	2.570	40.7363	22.8988
1.580	14.2686	39.2652	2.580	40.9633	22.8051
1.590	14.5645	38.9713	2.590	41.1894	22.7121
1.600	14.8604	38.6822	2.600	41.4147	22.6199
1.610	15.1561	38.3978	2.610	41.6392	22.5284
1.620	15.4518	38.1181	2.620	41.8628	22.4377
1.630	15.7473	37.8428	2.630	42.0855	22.3478
1.640	16.0427	37.5719	2.640	42.3074	22.2586
1.650	16.3379	37.3052	2.650	42.5285	22.1702
1.660	16.6328	37.0427	2.660	42.7488	22.0824
1.670	16.9276	36.7842	2.670	42.9682	21.9954
1.680	17.2220	36.5296	2.680	43.1868	21.9090
1.690	17.5161	36.2789	2.690	43.4045	21.8234
1.700	17.8099	36.0319	2.700	43.6215	21.7385
1.710	18.1034	35.7885	2.710	43.8376	21.6542
1.720	18.3964	35.5487	2.720	44.0529	21.5706

(continued)

Table C.3 (continued)

M	v	μ	M	N	μ
1.730	18.6891	35.3124	2.730	44.2673	21.4876
1.740	18.9814	35.0795	2.740	44.4810	21.4053
1.750	19.2732	34.8499	2.750	44.6938	21.3237
1.760	19.5646	34.6235	2.760	44.9059	21.2427
1.770	19.8554	34.4003	2.770	45.1171	21.1623
1.780	20.1458	34.1802	2.780	45.3275	21.0825
1.790	20.4357	33.9631	2.790	45.5371	21.0034
1.800	20.7251	33.7490	2.800	45.7459	20.9248
1.810	21.0139	33.5377	2.810	45.9539	20.8469
1.820	21.3021	33.3293	2.820	46.1611	20.7695
1.830	21.5898	33.1237	2.830	46.3675	20.6928
1.840	21.8768	32.9207	2.840	46.5731	20.6166
1.850	22.1633	32.7204	2.850	46.7779	20.5410
1.860	22.4492	32.5227	2.860	46.9820	20.4659
1.870	22.7344	32.3276	2.870	47.1852	20.3914
1.880	23.0190	32.1349	2.880	47.3877	20.3175
1.890	23.3029	31.9447	2.890	47.5894	20.2441
1.900	23.5861	31.7569	2.900	47.7903	20.1713
1.910	23.8687	31.5714	2.910	47.9905	20.0990
1.920	24.1506	31.3882	2.920	48.1898	20.0272
1.930	24.4318	31.2072	2.930	48.3884	19.9559
1.940	24.7123	31.0285	2.940	48.5863	19.8852
1.950	24.9920	30.8519	2.950	48.7833	19.8149
1.960	25.2711	30.6774	2.960	48.9796	19.7452
1.970	25.5494	30.5050	2.970	49.1752	19.6760
1.980	25.8269	30.3347	2.980	49.3700	19.6072
1.990	26.1037	30.1664	2.990	49.5640	19.5390
2.000	26.3798	30.0000	3.000	49.7574	19.4712

Diagram C.1 Relation between wave angle Θ and deflection angle δ for an oblique shock, and calorically perfect gas ($\gamma = 1.4$)

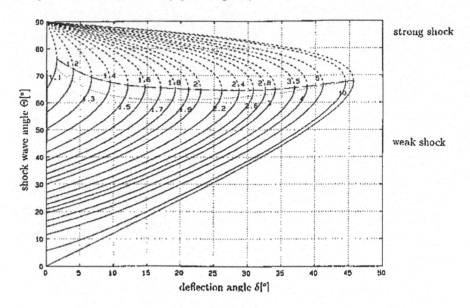

Diagram C.2 Relation between Mach number M_2 behind an oblique shock and deflection angle δ, for calorically perfect gas ($\gamma = 1.4$)

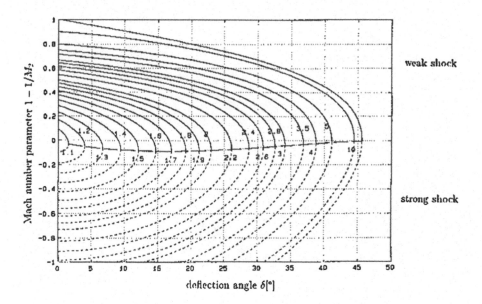

Diagram C.1 Relation between wave angle θ and deflection angle σ for an oblique shock, and calorically perfect gas (κ = 1.4)

Diagram C.2 Relation between Mach number M₂ behind an oblique shock and deflection angle σ, for calorically perfect gas (γ = 1.4)

Appendix D
Physical Properties of Air and Water

See Tables D.1, D.2, D.3, D.4, D.5, D.6.

Table D.1 Dynamic viscosity η [in 10^{-6} kg/(m s)] of dry air

p (bar)	t (°C)								
	-50	0	25	50	100	200	300	400	500
1	14.55	17.10	18.20	19.25	21.60	25.70	29.20	32.55	35.50
5	14.63	17.16	18.26	19.30	21.64	25.73	29.23	32.57	35.52
10	14.74	17.24	18.33	19.37	21.70	25.78	29.27	32.61	35.54
50	16.01	18.08	19.11	20.07	22.26	26.20	29.60	32.86	35.76
100	18.49	19.47	20.29	21.12	23.09	26.77	30.05	33.19	36.04
200	25.19	23.19	23.40	23.76	24.98	28.03	31.10	34.10	36.69
300	32.68	27.77	27.25	27.28	27.51	29.67	32.23	34.93	37.39
400	39.78	32.59	31.41	30.98	30.27	31.39	33.44	35.85	38.15
500	46.91	37.29	35.51	34.06	32.28	33.15	34.64	36.86	38.96

© Springer Nature Switzerland AG 2020
J. H. Spurk and N. Aksel, *Fluid Mechanics*,
https://doi.org/10.1007/978-3-030-30259-7

Table D.2 Kinematic viscosity v [in 10^{-8} m^2/s] of dry air

p (bar)	t (°C)								
	−50	0	25	50	100	200	300	400	500
1	931.1	1341.	1558.	1786.	2315.	3494.	4809.	6295.	7886.
5	186.1	268.5	312.2	358.1	464.2	700.5	964.1	1262.	1580.
10	93.03	134.5	156.5	179.6	232.8	351.4	483.6	632.8	792.1
50	19.11	27.74	32.39	37.19	48.13	72.43	99.35	129.5	161.8
100	10.53	14.82	17.23	19.72	25.34	37.75	51.48	66.77	83.15
200	7.402	9.140	10.33	11.57	14.33	20.68	27.83	35.74	44.00
300	7.274	7.916	8.615	9.455	11.15	15.34	20.11	25.42	31.03
400	7.633	7.687	8.112	8.693	9.825	12.84	16.38	20.38	24.64
500	8.188	7.762	8.005	8.273	8.962	11.44	14.21	17.45	20.87

Table D.3 Thermal conductivity λ [in 10^{-3} W/(m K)] of dry air

p (bar)	t (°C)								
	−50	0	25	50	100	200	300	400	500
1	20.65	24.54	26.39	28.22	31.81	38.91	45.91	52.57	58.48
5	20.86	24.68	26.53	28.32	31.89	38.91	45.92	52.56	58.42
10	21.13	24.88	26.71	28.47	32.00	38.94	45.96	52.57	58.36
50	24.11	27.15	28.78	30.26	33.53	40.34	46.86	53.41	58.98
100	28.81	30.28	31.53	32.75	35.60	42.00	48.30	54.56	60.07
200	41.96	38.00	37.90	38.21	39.91	45.18	50.69	56.62	61.96
300	54.84	46.84	45.38	44.56	44.81	48.54	53.06	58.70	63.74
400	65.15	55.30	52.83	51.29	49.97	52.59	55.91	60.95	65.56
500	73.91	62.92	59.80	57.40	54.70	55.66	58.60	62.86	67.24

Table D.4 Dynamic viscosity η [in 10^{-6} kg/(m s)] of water

p (bar)	t (°C)								
	0	20	50	100	150	200	300	400	500
1	1750.	1000.	544.0	12.11	14.15	16.18	20.25	24.30	28.40
10	1750.	1000.	544.0	279.0	181.0	15.85	20.22	24.40	28.50
50	1750.	1000.	545.0	280.0	182.0	135.0	20.06	25.00	28.90
100	1750.	1000.	545.0	281.0	183.0	136.0	90.50	25.80	29.50
150	1740.	1000.	546.0	282.0	184.0	137.0	91.70	26.90	30.30
200	1740.	999.0	546.0	283.0	185.0	138.0	93.00	28.60	31.10
300	1740.	998.0	547.0	285.0	188.0	141.0	95.50	45.70	32.70
400	1730.	997.0	548.0	287.0	190.0	143.0	98.10	62.80	36.90
500	1720.	996.0	549.0	289.0	192.0	145.0	101.0	69.30	42.20

Table D.5 Kinematic viscosity v [in 10^{-6} m^2/s] of water

p (bar)	t (°C)								
	0	20	50	100	150	200	300	400	500
1	1.750	1.000	0.551	20.50	27.40	35.20	53.40	75.40	101.0
10	1.750	1.000	0.550	0.291	0.197	3.260	5.220	7.480	10.10
50	1.750	1.000	0.550	0.292	0.198	0.156	0.909	1.450	2.020
100	1.740	0.998	0.549	0.292	0.198	0.156	0.126	0.681	0.967
150	1.730	0.995	0.549	0.292	0.199	0.157	0.126	0.421	0.630
200	1.720	0.992	0.548	0.293	0.199	0.157	0.127	0.285	0.459
300	1.720	0.987	0.547	0.293	0.202	0.159	0.127	0.128	0.284
400	1.700	0.981	0.545	0.294	0.203	0.160	0.128	0.120	0.207
500	1.680	0.977	0.544	0.295	0.204	0.162	0.130	0.120	0.164

Table D.6 Thermal conductivity λ [in 10^{-3} W/(m K)] of water

p (bar)	t (°C)								
	0	20	50	100	150	200	300	400	500
1	569.0	604.0	643.0	24.80	28.60	33.10	43.30	54.50	66.60
10	570.0	604.0	644.0	681.0	687.0	35.00	44.20	55.20	67.20
50	573.0	608.0	647.0	684.0	690.0	668.0	52.10	59.30	70.50
100	577.0	612.0	651.0	688.0	693.0	672.0	545.0	67.40	75.70
150	581.0	616.0	655.0	691.0	696.0	676.0	559.0	81.80	82.50
200	585.0	620.0	659.0	695.0	700.0	681.0	571.0	106.0	91.50
300	592.0	627.0	666.0	701.0	706.0	689.0	592.0	263.0	117.0
400	599.0	634.0	672.0	707.0	713.0	697.0	609.0	388.0	153.0
500	606.0	640.0	678.0	713.0	720.0	704.0	622.0	437.0	202.0

References

1. Aris, R.: Vectors, Tensors and the Basic Equations of Fluid Mechanics. Prentice-Hall Inc, Englewood Cliffs, New Jersey (1962)
2. Batchelor, G.K.: An Introduction to Fluid Dynamics. Cambridge University Press, Cambridge (1967)
3. Becker, E.: Gasdynamik. Teubner, Stuttgart (1966)
4. Becker, E.: Technische Strömungslehre. Teubner, Stuttgart (1982)
5. Becker, E., Bürger, W.: Kontinuumsmechanik. Teubner, Stuttgart (1975)
6. Betz, A.: Einführung in die Theorie der Strömungsmaschinen. G. Braun, Karlsruhe (1959)
7. Betz, A.: Konforme Abbildung. Berlin etc.: Springer, 2. Auflage (1964)
8. Bird, R.B.: Transport Phenomena. New York etc.: Wiley (1960)
9. Bird, R.B., Armstrong, R.C., Hassager, O.: Dynamics of Polymeric Liquids. New York etc.: Wiley (1977)
10. Bridgman, P.: Dimensional Analysis. Yale-University Press, Yale (1920)
11. Böhme, G.: Strömungsmechanik nicht-newtonscher Fluide. Teubner, Stuttgart (1981)
12. Cameron, A.: Principles of Lubrication. Longmans Green & Co., London (1966)
13. Chapman, S., Cowling, T.G.: The Mathematical Theory of Non-Uniform Gases. Cambridge University Press, Cambridge (1970)
14. Courant, R., Friedrichs, K.O.: Supersonic Flow and Shock Waves. Intersience Publishers, New York (1948)
15. de Groot, S.R.: Thermodynamik irreversibler Prozesse. BI-Hochschultaschenbücher, Mannheim (1960)
16. Emmons, H.W.: Fundamentals of Gas Dynamics. Princeton University Press, Princeton (1958)
17. Eringen, A.C.: Mechanics of Continua. New York etc.: Wiley (1967)
18. Flügge, S. (Hrsg.): Handbuch der Physik Bd. III/3 1965, Bd. VIII/1 1959, Bd. VIII/2 1963, Bd. IX 1960. Berlin etc.: Springer (1960)
19. Focken, C.M.: Dimensional Methods and their Applications. Edward Arnold & Co., London (1953)
20. Goldstein, S. (Ed.): *Modern Developments in Fluid Dynamics*. New York: Dover Publications, 1965 Vol 2.
21. Jeffreys, H.: Cartesian Tensors. Cambridge University Press, Cambridge (1969)
22. Richter, J.: Höhere Mathematik für den Praktiker. Frankfurt/M: Harri Deutsch, 12. Auflage (1978)
23. Karamcheti, K.: Principles of Ideal-Fluid Aerodynamics. Wiley, New York (1966)
24. Klingbeil, E.: Tensorrechnung für Ingenieure. Bibliographisches Institut, Mannheim (1966)

© Springer Nature Switzerland AG 2020
J. H. Spurk and N. Aksel, *Fluid Mechanics*,
https://doi.org/10.1007/978-3-030-30259-7

25. Lamb, H.: Hydrodynamics. Cambridge University Press, Cambridge (1932)
26. Landau, L.D., Lifschitz, E.M.: Lehrbuch der theoretischen Physik Vol. VI: Hydrodynamik. Berlin: Akademie-Verlag, 3. Auflage (1974)
27. Landolt-Börnstein: Zahlenwerte und Funktionen aus Physik, Chemie... Berlin etc.: Springer (1956)
28. Langhaar, H.L.: Dimensional Analysis and Theory of Models. Wiley, New York etc. (1951)
29. Liepmann, H.W., Roshko, A.: Elements of Gasdynamics. Wiley, New York etc. (1957)
30. Loitsiansky, L.G.: Laminare Grenzschichten. Akademie-Verlag, Berlin (1967)
31. Milne-Thomson, L.M.: Theoretical Hydrodynamics. Mac Millan & Co., London (1949)
32. Monin, A.S., Yaglom, A.M.: Statistical Fluid Mechanics, Mechanics of Turbulence, vol. 2. Cambridge: The MIT Press (1975)
33. Pinkus, O., Sternlicht, B.: Theory of Hydrodynamic Lubrication. McGraw-Hill, New York etc. (1961)
34. Prager, W.: Einführung in die Kontinuumsmechanik. Birkhäuser, Basel (1961)
35. Prandtl, L.: Gesammelte Abhandlungen. Berlin etc.: Springer, 3 Teile (1961)
36. Prandtl, L., Oswatitsch, K., Wieghardt, K.: Führer durch die Strömungslehre. Braunschweig: Vieweg, 8. Auflage (1984)
37. Rotta, J.C.: Turbulente Strömungen. Teubner, Stuttgart (1972)
38. Schlichting, H.: Grenzschicht-Theorie. Karlsruhe: Braun 8. Auflage (1982)
39. Schlichting, H., Truckenbrodt, E.: Aerodynamik des Flugzeuges. Springer, Berlin etc. (1967)
40. Sedov, L.I.: Similarity and Dimensional Methods in Mechanics. Academic Press, New York (1959)
41. Shapiro, A.H.: The Dynamics and Thermodynamics of Compressible Fluid Flow, vol. 2. New York: The Ronald Press Company (1953)
42. Sommerfeld, A: Vorlesungen über Theoretische Physik Bd. 2: Mechanik der deformierbaren Medien. Leipzig: Geest u. Portig (1964)
43. Spurk, J.H.: Dimensionsanalyse in der Strömungslehre. Springer, Berlin etc. (1992)
44. Tietjens, O.: Strömungslehre. Berlin etc.: Springer, 2 Bde (1960)
45. van Dyke, M.: Perturbation Methods in Fluid Mechanics. The Parabolic Press, Stanford (1975)
46. White, F.M.: Viscous Fluid Flow. McGraw-Hill, New York etc. (1974)
47. Wieghardt, K.: Theoretische Strömungslehre. Teubner, Stuttgart (1965)
48. Wylie, C.R., Barett, L.C.: Advanced Engineering Mathematics, 5th edn, New York etc.: McGraw-Hill (1985)
49. Yih, C.: Fluid Mechanics. McGraw-Hill, New York etc. (1969)

Index

© Springer Nature Switzerland AG 2020
J. H. Spurk and N. Aksel, *Fluid Mechanics*,
https://doi.org/10.1007/978-3-030-30259-7

Printed in the United States
By Bookmasters